Analog

Filter

Design

M. E. VAN VALKENBURG

Department of Electrical Engineering
University of Illinois at Urbana-Champaign

Holt, Rinehart and Winston

New York   Chicago   San Francisco   Philadelphia
Montreal   Toronto   London   Sydney   Tokyo
Mexico City   Rio de Janeiro   Madrid

Address correspondence to:
383 Madison Avenue, New York, NY 10017

Library of Congress Cataloging in Publication Data

Van Valkenburg, M.E. (Mac Elwyn)
  Analog filter design.

  (HRW series in electrical and computer engineering)
  Bibliography: p.
  Includes index.
  1. Electric filters, Active. 2. Operational
amplifiers. I. Title. II. Series.
TK7872.F5V38   621.3815'324   81-23774
ISBN 0-03-059246-1   AACR2

Printed in the United States of America
1 2 3   144   9 8 7 6 5 4 3 2 1

CBS COLLEGE PUBLISHING
Holt, Rinehart and Winston
The Dryden Press
Saunders College Publishing

Analog
Filter
Design

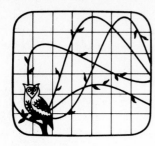

**HRW
Series in
Electrical and
Computer Engineering**

M. E. Van Valkenburg, Series Editor

Preface

Filters are essential in systems electrical engineers are called upon to design. Thus, the topic of filters is appropriate for study by undergraduates preparing to enter their professions and by practicing engineers wishing to extend their skills. This book was written to fill such needs.

The course at the University of Illinois from which this book evolved was offered to juniors and seniors for three hours of credit; an additional hour was offered for the associated laboratory. Students taking the course had, as a minimum background, the first course in circuits and a course that included the study of Laplace transforms. Because the book is intended for undergraduate use, sophisticated mathematics has been avoided in favor of algebraic derivations. In addition, the sequence of topics treated is such that design is stressed beginning with Chapter 2.

In organizing the material for the course, the decision was made to concentrate on inductorless filters in which the active element is the operational amplifier (op amp). Such filters are required for voice and data communications systems, for which the size and weight of inductors would make their use prohibitive. I thus exclude passive *LCR* filters except as prototypes from which an active equivalent is then found. A benefit of this decision is that complicated realizability conditions are avoided and design becomes relatively simple.

Another advantage of this choice is that the usefulness of the op amp is stressed. Two developments that have profoundly affected the practice of electrical engineering are the microprocessor for digital systems and the op amp for analog systems. It seems imperative that our students have experience with both. Most universities now offer courses in microprocessors. However, students ordinarily encounter op amps only briefly in their study of electronic circuits, and this seems inadequate.

Chapter 2 introduces the use of the op amp in analog operations: addition, subtraction, multiplication, and integration. This chapter may be skipped by students who have covered similar material in electronics courses. Chapters 3 and 4 constitute a review of sinusoidal steady-state topics recast to provide an introduction to first-order filters. Chapter 5 features the universal biquad, and through it the standard forms of response of filters: lowpass, bandpass, highpass, bandstop, and allpass.

In the chapters that follow, the functions of a filter are studied in combination with the frequency-response approximations used by filter designers: Butterworth, Chebyshev, Bessel–Thompson, inverse Chebyshev, and

Cauer (or elliptic). The sequence of topics was chosen to maintain student interest and to provide practice in design. Once this interest is established, two important topics are introduced: sensitivity and frequency transformations.

Chapter 14 treats the synthesis of doubly terminated passive ladders. This together with Chapter 11 constitute an introduction to passive filter design. An important conclusion is reached in Chapter 14 relating to sensitivity. This is the result due to Orchard—the passive ladder circuit has low sensitivity. On this basis, the study of simulated ladder circuits is undertaken in Chapters 15 through 17. Simulation of the passive ladder is accomplished in three ways: by introducing new elements which make it possible to exclude the inductor, through the simulation of the block diagram representation of the ladder, and through the simulation of resistors using switched capacitors.

A pleasant discovery has been made by students in classes on analog filters: synthesis or design is much simpler than analysis. In analysis, any but the simplest circuits will require the use of a computer or hours of tedious hand calculations. The design topics of this book require at most a hand-held calculator, preferably one with an inverse hyperbolic function key. For faculty willing to experiment with the curriculum, the book might be used immediately following a first circuits course in which the sinusoidal steady state has been introduced.

The course on analog filter design is most successful when taught with an associated laboratory. There students may observe the amazing agreement of theory based on the ideal op amp and measurement, as long as the frequency range is sufficiently low. At some point in the experiments, students will observe the need for better models and so be prepared for Chapter 20.

I am indebted to the hundreds of students over the years who participated in testing the material of this book, and to my colleagues in the circuits and systems area at the University of Illinois who have provided a stimulating milieu for discussions. The book was written for the most part while I was on leave from the University of Illinois. It is my pleasure to acknowledge the congenial atmosphere and assistance provided by my hosts: Professor Bharat Kinariwala of the University of Hawaii at Manoa, Professor Carl H. Durney of the University of Utah, and Professor Donald E. Kirk of the Naval Postgraduate School, Monterey. I benefited from comments by Professor Ronald A. Rohrer who used a preliminary version of the notes for a class he taught at the University of Maine. At the University of Illinois, the course has been taught jointly with Professor E. I. El-Masry, who has been generous in making helpful suggestions. Philip R. Geffe of Scientific-Atlanta has introduced me to many of the topics of the book through his writings; he also offered specific suggestions which were most helpful. Teaching assistants and former students who helped shape the coverage include Peter L. Chu, J. M. Cioffi, Leon Garza, and Gordon Jacobs. It has been a pleasure to work with Paul Becker in the production of the book. Finally, I express my special thanks to my wife Evelyn for her assistance in proofreading and for patience and encouragement during the writing of the book.

M. E. Van Valkenburg

Contents

Analog
Filter
Design

CHAPTER 1

Introduction

The subject of this book is a special class of electric wave filters which make use of the remarkable properties of the operational amplifier. Such filters are sometimes called *active* filters, but more often *analog* filters, an expression that distinguishes them from *digital* filters. In modern integrated circuit technology, both analog and digital filters may be implemented on the same chip. This introductory chapter provides some background that will be useful in the studies that follow.

1.1 THE CIRCUIT DESIGN PROBLEM

Figure 1.1 shows a circuit with a voltage source v_1 connected to the excitation terminals 1-1'. The response terminals 2-2' are characterized by the voltage v_2. If the circuit is operating in the sinusoidal steady state, then the two voltages may be represented by the phasors

$$|V_1| \angle \theta_1 \quad \text{and} \quad |V_2| \angle \theta_2 \quad (1.1)$$

These two phasor quantities may be used to define a transfer function. Our choice for such a transfer function will always be

$$\text{Transfer function} = \frac{\text{output quantity}}{\text{input quantity}} \quad (1.2)$$

This choice is important since many authors use the reciprocal definition and even construct tables using their definition. Then the magnitude of the transfer function is

$$\frac{|V_2|}{|V_1|} = |T| \quad (1.3)$$

and the phase is

$$\theta = \theta_2 - \theta_1 \quad (1.4)$$

When we make measurements using the configuration of Fig. 1.1, we ordinarily keep $|V_1|$ constant, and we also select the phase reference such that $\theta_1 = 0$. Then the variations of $|V_2|$ and θ_2 with changes in frequency ω, in radians per second, or f, in hertz, constitute the *frequency response* of the circuit. Unless the circuit

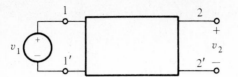

FIGURE 1.1

contains only resistors, both the magnitude and the phase will change in value with frequency.

Now if we know the circuit and the form of the input, we may determine the output v_2. This is known as *circuit analysis*. However, if we know the input and the output or the ratio, such as $|T|$ and θ_2, and we wish to know the circuit, this is known as *circuit design* (or synthesis). Throughout this textbook we will be concerned with design, but will resort to analysis when we find it necessary to characterize some circuit or combination of circuit elements. The circuits that we design, having a prescribed variation of the magnitude, phase, or related quantities as a function of frequency, are known as *filters*.

Filtering has a commonly accepted meaning of separation—something retained, something rejected. In electrical engineering we filter signals, usually voltages. Any signal may be thought of as made up of packets of signal, each at a specified frequency. Thus if a signal is made up of two tones, one at high frequency, such as that produced by a piccolo, and one of low frequency, such as that of a tuba, we can imagine a filtering action by which one tone or the other is suppressed. You will come to understand that any signal may be thought of as made up of components, each at a given frequency. It is as if any signal were generated by a large number of sinusoidal voltage sources connected in series, each characterized by a frequency and an amplitude and phase. This concept will be formalized in your education when you study Fourier analysis.

We reject components of a signal by designing a filter that provides *attenuation* over a band of frequencies, and we retain components of a signal through the absence of attenuation or perhaps even *gain*. It is important that we distinguish between measures of attenuation and gain. We define attenuation as

$$\alpha = -20 \log |T| \text{ dB}, \qquad |T| \leq 1 \tag{1.5}$$

where $|T|$ is defined by Eq. (1.3). The unit of attenuation is the *decibel* (dB), and the minus sign is introduced so that α is positive when the circuit provides loss, as implied by $|T| \leq 1$. We use a different symbol for the case when $|T| > 1$, which implies gain:

$$A = 20 \log |T| \text{ dB}, \qquad |T| > 1 \tag{1.6}$$

We may solve Eq. (1.5) for $|T|$ through the following steps. First we divide by -20 such that

$$-\frac{\alpha}{20} = \log |T| \tag{1.7}$$

Taking the antilogarithm of this equation, gives

$$|T| = 10^{-\alpha/20} \tag{1.8}$$

This equation may be solved by using the y^z key on a calculator when written in the form

$$|T| = (10^{0.05})^{-\alpha} \quad \text{or} \quad |T| = \frac{1}{(10^{0.05})^{\alpha}} \tag{1.9}$$

Similarly, Eq. (1.6) may be solved to give

$$|T| = (10^{0.05})^A \tag{1.10}$$

Using these last two equations, you may verify entries in Table 1.1.

It is sometimes useful to memorize some values from the chart to serve as "rules of thumb." Using Eq. (1.9), it is seen that

1 dB implies $\approx 10\%$ decrease in $|T|$ (from 1.0 to 0.891)
2 dB implies $\approx 20\%$ decrease in $|T|$ (from 1.0 to 0.794)
3 dB implies $\approx 30\%$ decrease in $|T|$ (from 1.0 to 0.708)
6 dB implies $\approx 50\%$ decrease in $|T|$ (from 1.0 to 0.501)

Each additional 6 dB of α reduces $|T|$ by 1/2.

1.2   KINDS OF FILTERS AND DESCRIPTIVE TERMINOLOGY

Filters are classified according to the functions they are to perform, in terms of ranges of frequencies, as *pass bands* and *stop bands*. In the ideal case a pass band is such that $|T| = 1$ and $\alpha = 0$, while in a stop band $|T| = 0$ and $\alpha = \infty$. The patterns of pass band and stop band, which give rise to the names of the four most common filters are shown in Fig. 1.2. In terms of the four parts, the filters illustrated in Fig. 1.2 are defined as follows:

1. A *lowpass* filter characteristic is one in which the pass band extends from $\omega = 0$ to $\omega = \omega_0$, where ω_0 is known as the *cutoff frequency* (Fig. 1.2a).
2. A *highpass* filter is the complement of the lowpass filter in that the frequency range from 0 to ω_0 is a stop band, while from ω_0 to infinity is a pass band (Fig. 1.2b).
3. A *bandpass* filter is one in which frequencies extending from ω_1 to ω_2 are passed, while all other frequencies are stopped (Fig. 1.2c).
4. The *stop-band* filter is the complement of the bandpass filter where the frequencies from ω_1 to ω_2 are stopped and all others are passed (Fig. 1.2d). These filters are sometimes known as *notch* filters.

There will be other kinds of filters introduced as our study progresses, but filtering action can be visualized in terms of these basic four types of filters.

It is now possible to realize the ideal characteristics of Fig. 1.2 with a finite number of elements. Realistic filter characteristics which correspond to the four basic types are shown in Fig. 1.3. We will see later that the sharpness of the transition from stop band to pass band can be controlled to some extent in the design of the filters.

TABLE 1.1 Values of $|T|$

dB	A	α	dB	A	α	dB	A	α
0	1.000	1.0000	4.0	1.585	0.6310	8.0	2.512	0.3981
0.1	1.012	0.9886	4.1	1.603	0.6237	8.1	2.541	0.3936
0.2	1.023	0.9772	4.2	1.622	0.6166	8.2	2.570	0.3890
0.3	1.035	0.9661	4.3	1.641	0.6095	8.3	2.600	0.3846
0.4	1.047	0.9550	4.4	1.660	0.6026	8.4	2.630	0.3802
0.5	1.059	0.9441	4.5	1.679	0.5957	8.5	2.661	0.3758
0.6	1.072	0.9333	4.6	1.698	0.5888	8.6	2.692	0.3715
0.7	1.084	0.9226	4.7	1.718	0.5821	8.7	2.723	0.3673
0.8	1.096	0.9120	4.8	1.738	0.5754	8.8	2.754	0.3631
0.9	1.109	0.9016	4.9	1.758	0.5689	8.9	2.786	0.3589
1.0	1.122	0.8913	5.0	1.778	0.5623	9.0	2.818	0.3548
1.1	1.135	0.8810	5.1	1.799	0.5559	9.1	2.851	0.3508
1.2	1.148	0.8710	5.2	1.820	0.5495	9.2	2.884	0.3467
1.3	1.161	0.8610	5.3	1.841	0.5433	9.3	2.917	0.3428
1.4	1.175	0.8511	5.4	1.862	0.5370	9.4	2.951	0.3388
1.5	1.189	0.8414	5.5	1.884	0.5309	9.5	2.985	0.3350
1.6	1.202	0.8318	5.6	1.905	0.5248	9.6	3.020	0.3311
1.7	1.216	0.8222	5.7	1.928	0.5188	9.7	3.055	0.3273
1.8	1.230	0.8128	5.8	1.950	0.5129	9.8	3.090	0.3236
1.9	1.245	0.8035	5.9	1.972	0.5070	9.9	3.126	0.3199
2.0	1.259	0.7943	6.0	1.995	0.5012	10.0	3.162	0.3162
2.1	1.274	0.7852	6.1	2.018	0.4955	10.1	3.199	0.3126
2.2	1.288	0.7762	6.2	2.042	0.4898	10.2	3.236	0.3090
2.3	1.303	0.7674	6.3	2.065	0.4842	10.3	3.273	0.3055
2.4	1.318	0.7586	6.4	2.089	0.4786	10.4	3.311	0.3020
2.5	1.334	0.7499	6.5	2.113	0.4732	10.5	3.350	0.2985
2.6	1.349	0.7413	6.6	2.138	0.4677	10.6	3.388	0.2951
2.7	1.365	0.7328	6.7	2.163	0.4624	10.7	3.428	0.2917
2.8	1.380	0.7244	6.8	2.188	0.4571	10.8	3.467	0.2884
2.9	1.396	0.7161	6.9	2.213	0.4519	10.9	3.508	0.2851
3.0	1.413	0.7079	7.0	2.239	0.4467	11.0	3.548	0.2818
3.1	1.429	0.6998	7.1	2.265	0.4416	11.1	3.589	0.2786
3.2	1.445	0.6918	7.2	2.291	0.4365	11.2	3.631	0.2754
3.3	1.462	0.6839	7.3	2.317	0.4315	11.3	3.673	0.2723
3.4	1.479	0.6761	7.4	2.344	0.4266	11.4	3.715	0.2692
3.5	1.496	0.6683	7.5	2.371	0.4217	11.5	3.758	0.2661
3.6	1.514	0.6607	7.6	2.399	0.4169	11.6	3.802	0.2630
3.7	1.531	0.6531	7.7	2.427	0.4121	11.7	3.846	0.2600
3.8	1.549	0.6457	7.8	2.455	0.4074	11.8	3.890	0.2570
3.9	1.567	0.6383	7.9	2.483	0.4027	11.9	3.936	0.2541

In the chapters to follow we will switch from $|T|$ characteristics to attenuation characteristics, depending on the characteristic that seems to make the point more clearly. The attenuation characteristics corresponding to those of Fig. 1.3 are shown in Fig. 1.4. The two quantities are, of course, related by Eq. (1.5).

dB	A	α	dB	A	α
12.0	3.981	0.2512	16.0	6.310	0.1585
12.1	4.027	0.2483	16.1	6.383	0.1567
12.2	4.074	0.2455	16.2	6.457	0.1549
12.3	4.121	0.2427	16.3	6.531	0.1531
12.4	4.169	0.2399	16.4	6.607	0.1514
12.5	4.217	0.2371	16.5	6.683	0.1496
12.6	4.266	0.2344	16.6	6.671	0.1479
12.7	4.315	0.2317	16.7	6.839	0.1462
12.8	4.365	0.2291	16.8	6.918	0.1445
12.9	4.416	0.2265	16.9	6.998	0.1429
13.0	4.467	0.2239	17.0	7.079	0.1413
13.1	4.519	0.2213	17.1	7.161	0.1396
13.2	4.571	0.2188	17.2	7.244	0.1380
13.3	4.624	0.2163	17.3	7.328	0.1365
13.4	4.677	0.2138	17.4	7.413	0.1349
13.5	4.732	0.2113	17.5	7.499	0.1334
13.6	4.786	0.2089	17.6	7.586	0.1318
13.7	4.842	0.2065	17.7	7.674	0.1303
13.8	4.898	0.2042	17.8	7.762	0.1288
13.9	4.955	0.2018	17.9	7.852	0.1274
14.0	5.012	0.1995	18.0	7.943	0.1259
14.1	5.070	0.1972	18.1	8.035	0.1245
14.2	5.129	0.1950	18.2	8.128	0.1230
14.3	5.188	0.1928	18.3	8.222	0.1216
14.4	5.248	0.1905	18.4	8.318	0.1202
14.5	5.309	0.1884	18.5	8.414	0.1189
14.6	5.370	0.1862	18.6	8.511	0.1175
14.7	5.433	0.1841	18.7	8.610	0.1161
14.8	5.495	0.1820	18.8	8.710	0.1148
14.9	5.559	0.1799	18.9	8.811	0.1135
15.0	5.623	0.1778	19.0	8.913	0.1122
15.1	5.689	0.1758	19.1	9.016	0.1109
15.2	5.754	0.1738	19.2	9.120	0.1096
15.3	5.821	0.1718	19.3	9.226	0.1084
15.4	5.888	0.1698	19.4	9.333	0.1072
15.5	5.957	0.1679	19.5	9.441	0.1059
15.6	6.026	0.1660	19.6	9.550	0.1047
15.7	6.095	0.1641	19.7	9.661	0.1035
15.8	6.166	0.1622	19.8	9.772	0.1023
15.9	6.237	0.1603	19.9	9.886	0.1012

Since it is impossible to realize filters with characteristics like those shown in Fig. 1.2 with abrupt changes from pass to stop and from stop to pass, we must learn to cope with realistic filter characteristics illustrated in Figs. 1.3 and 1.4. The way this will be accomplished is explained in Fig. 1.5 in comparison with

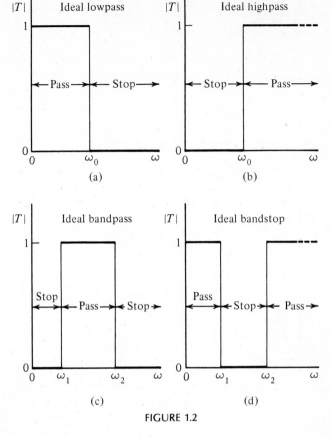

FIGURE 1.2

Fig. 1.4. We will specify the characteristics we require by the different definitions of pass band and stop band:

1. A pass band is one in which the attenuation is always less than a value designated as α_{max}.
2. A stop band is one in which the attenuation is always greater than a value designated as α_{min}.
3. Bands of frequencies between the stop bands and pass bands so defined are known as *transition bands*.

In terms of Fig. 1.5a we see that the pass band extends from $\omega = 0$ to $\omega = \omega_p$, the range of frequencies from ω_p to ω_s is the transition band, and all frequencies greater than ω_s constitute the stop band. In this figure as well as in Fig 1.5b we have used the subscripts p and s to indicate the edges of the pass bands and stop bands. The same concept applies to the bandpass and bandstop cases shown in Fig. 1.5c and d. Here there are two transition bands.

We will use the attenuation characteristics shown in Fig. 1.5 as the filter *specifications* in later chapters. In terms of Fig. 1.5a the design problem will be as follows: Given the four quantities α_{max}, α_{min}, ω_p, and ω_s, find an attenuation speci-

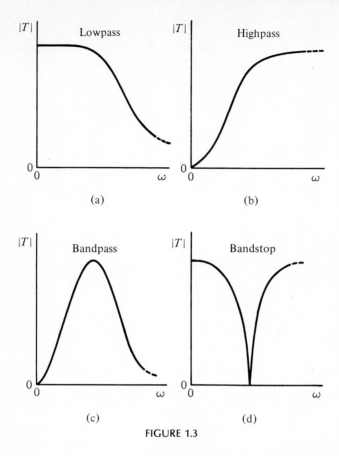

FIGURE 1.3

fication that satisfies the four requirements. The form of the solution is indicated in Fig. 1.6 by the dashed lines. In terms of Fig. 1.6a, we will have found the required attenuation characteristic $\alpha(\omega)$ if at the frequency ω_p, $\alpha \leq \alpha_{max}$, and at the frequency ω_s, $\alpha \geq \alpha_{min}$. This kind of description relates in analogous ways to the characteristics shown in Fig. 1.6b, c, and d.

The response of Fig. 1.6a is repeated in Fig. 1.7. Again we see the frequencies ω_p and ω_s, and the attenuation values α_{max} and α_{min} along with an indication of the pass band, the transition band, and the stop band. Here ω_0 designates the half-power frequency at which $\alpha = 3$ dB, and the corresponding $|T|$ has the value 0.707. The frequency range from 0 to ω_0 is sometimes called the *bandwidth* (BW). The portion of the curve at high frequencies is sometimes called the *skirt*, and the rate at which it increases is known as the *rolloff*. If frequency is plotted on logarithmic coordinates, then it is known as a *Bode plot*, and the rolloff or asymptotic slope is measured in multiples of 6 dB per octave or 20 dB per decade. This property is examined in greater detail in Chapter 4.

Figure 1.8 shows similar quantities for the case of the bandpass filter. We note that the pass band is defined as the band of frequencies that extends from ω_1 to ω_2 where the attenuation is less than α_{max}, while the bandwidth is defined in terms of the half-power frequencies and extends from ω_5 to ω_6. In the case of the

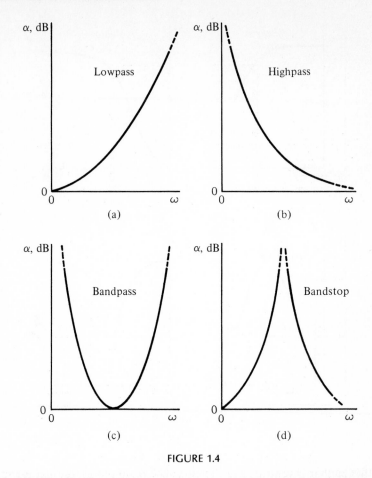

FIGURE 1.4

bandpass filter there is a lower stop band and an upper stop band. We will generally assume that α_{min} is the same for both stop bands, although this is not a necessary condition.

Another factor must be considered in design. The attenuation curves shown thus far have had a minimum value of $\alpha = 0$ dB. Since we are using active elements, this need not be the case, for the active elements may provide gain. If it is necessary to meet the specifications exactly, then it will be necessary to provide a circuit to reduce the gain. We call this unwanted gain the *insertion gain*. If the circuit provides excess attenuation, this is called *insertion loss*. These two conditions are illustrated in Fig. 1.9, in which the characteristic bandpass curve is shifted up or down, but the shape is not changed.

1.3 WHY WE USE ANALOG FILTERS

The basic concepts of the electric wave filter were invented in 1915 independently by K. W. Wagner in Germany and George A. Campbell in the United States. In

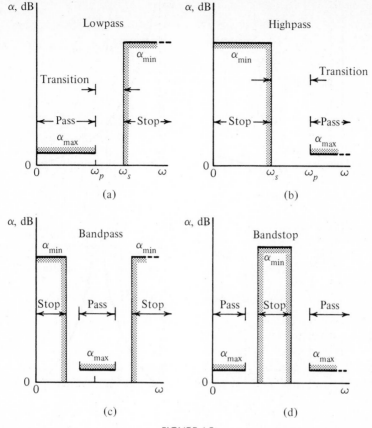

FIGURE 1.5

the years since that invention, the theory and realization techniques have been developed to a high state of perfection. Making an active filter became a possibility with the invention of the vacuum tube, and Black, Bode, and others made use of the vacuum tube and feedback in the early 1930s to provide a theory for active filter amplifiers. The present era of extensive use of active filters is due to the development of an inexpensive monolithic operational amplifier by Widlar in 1967.

The practical active filter makes use of operational amplifiers together with capacitors and resistors, and is generally known as an analog filter. How do we decide whether to use an analog filter in preference to filters made up of passive elements? In answering that question, we must consider such factors as:

1. The range of frequency of operation
2. Sensitivity to parameter changes and stability
3. Weight and size of the realization
4. Availability of voltage sources for the operational amplifiers

There is also the matter of whether to choose an analog or a digital filter for a particular application.

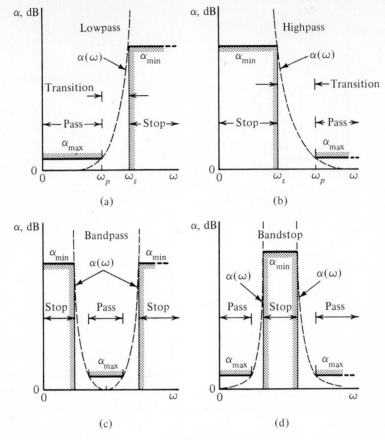

FIGURE 1.6

A comparison of the frequency range of operation of three kinds of filters is shown in Fig. 1.10. Compared to the passive filter, we see that analog filters can be realized for lower frequencies, but not for higher frequencies. This may be a temporary limitation since better operational amplifiers may become available at

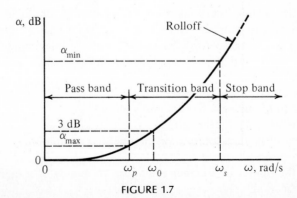

FIGURE 1.7

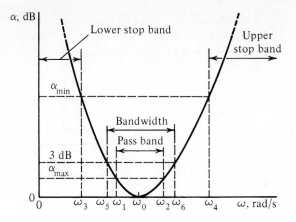

FIGURE 1.8

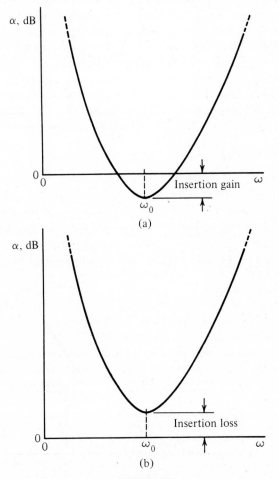

FIGURE 1.9

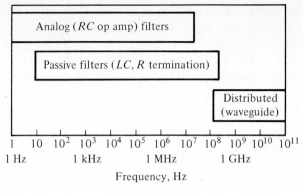

FIGURE 1.10

low prices. But for frequencies near to or in the microwave range, only distributed filters, waveguide or coaxial cable filters, may be used.

If high sensitivity is an important consideration, then passive realizations often have an advantage. This is considered in greater detail in Chapter 9.

Analog filters may be realized on a chip, thus they are superior when considerations of weight and size are important. This is a factor in the design of filters for low-frequency applications where passive filters require large inductors.

Finally, analog filters do require voltage sources ranging from 1 V to about 15 V for the proper operation of the operational amplifier. Whether such voltages are available without maintenance is an important consideration.

1.4   SOME CHOICES

It will become clear as the study progresses that filter design is primarily a frequency-domain matter and that we seldom make reference to time-domain quantities such as rise time or overshoot. Design specifications and physical measurements are made in terms of frequency f in hertz. However, it turns out to be much more convenient to use radian frequency ω rather than f. We will follow the practice of using ω as long as possible and then convert to f only in the last step. Experience in design will show the advantages of this choice.

We will make extensive use of both magnitude and frequency scaling and also of normalized values of frequency as well as element values. Skill in using scaling is important in filter design, as we will show by examples. This means that small values for resistors and capacitors and the uniform choice of $\omega_0 = 1$ rad/s as a characterizing frequency will appear in all design steps, except for the last. In that step the denormalized values will be determined from the equations

$$L_{\text{new}} = \frac{k_m}{k_f} L_{\text{old}} \tag{1.11}$$

$$C_{\text{new}} = \frac{1}{k_m k_f} C_{\text{old}} \tag{1.12}$$

$$R_{\text{new}} = k_m R_{\text{old}} \tag{1.13}$$

These equations are derived and the concepts are explained in Appendix A.

Finally, it will become clear in the chapters to follow that ordinarily there is no unique solution to a design problem. One of the decisions that the designer will have to make is that of element size. Skill in such choices will come with experience. Until then, the following guidelines are offered for the passive elements:

1. *Capacitors*

	Largest	Smallest
Readily realizable	1 μF	5 pF
Practical	10 μF	0.2 pF
Marginally practical	500 μF	0.5 pF

2. *Inductors*

	Largest	Smallest
Readily realizable	1 mH	1 μH
Practical	10 mH	0.1 μH
Marginally practical	1 H	50 nH

3. *Resistors.* Resistor size will depend on the quality of the operational amplifier used and on power dissipation considerations. As a guideline,

Preferred range	1–100 kΩ
Lower limit	0.1–1 kΩ
Upper limit	100–500 kΩ

These choices are tentative and depend on the state of the art.

CHAPTER 2

Resistor

Operational-

Amplifier

Circuits

In this chapter we introduce the operational amplifier (op amp). Op amps are never used alone, but only in combination with other circuit elements that provide feedback, determine gain, and so on. The simplest element used in the combination is the resistor, and this is the combination studied in this chapter. We will make use of a simple model for the op amp, postponing consideration of more adequate models until Chapter 20. The ultimate test of a model is its success in the laboratory. Experience has shown that even the simplest model allows us to predict circuit operation, provided only that the frequency range is kept sufficiently low.

2.1 OPERATIONAL AMPLIFIERS AND SIMPLE MODELS

The differential amplifier is familiar in modern electronics and differs from ordinary amplifiers in that two inputs are provided. Its operation is such that the output voltage is the difference of the two input voltages multiplied by an overall gain. In terms of the quantities defined in Fig. 2.1,

$$v_2(t) = A[v_+(t) - v_-(t)] \qquad (2.1)$$

These voltages are general, and will be so assumed in this chapter. Beginning with the next chapter, we use capital letters implying rms or effective values of sinusoidal waveforms. The results of this chapter apply in this case with V's replacing v's. When v_+ and v_- include noise, then an interesting property of differential amplifiers becomes apparent. If $v_+(t)$ and $v_-(t)$ have the same noise, then these signals cancel in forming $v_2(t)$, and are known as *common-mode* signals. On the other hand, if $v_+(t)$ and $v_-(t)$ are different, then each component is multiplied by A. Such signals are known as *differential-mode* signals.

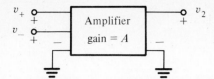

FIGURE 2.1

Differential amplifiers of unusually high gain were developed in the early 1940s, especially by George Philbrick and his associates. These were intended for use in analog simulation and such applications as radar and control systems. To John Ragazzini goes the credit for coining the name operational amplifier in 1947. Early units employed vacuum tubes and were both bulky and expensive. The trend toward extremely small and inexpensive op amps began in the 1960s when Philbrick, Burr-Brown, and other companies developed the modular solid-state units. The modern monolithic op amp, early versions of which were the LM 101 and the μA 709, was designed in 1967 by R. J. Widlar. Quad op amps (four on one chip) were announced in 1972, and there is little doubt that the number of op amps on a single chip will continue to increase with the passage of time.

The construction of an integrated-circuit op amp is shown in Fig. 2.2. A simplified wiring diagram of the integrated-circuit realization of the op amp is shown in Fig. 2.3a. It shows a myriad of components—resistors, capacitors, transistors—in an intricate arangement. Pin connections are shown in Fig. 2.3b and c. In actual use there are external connections and adjustments to be taken into account, as pictured in Fig. 2.3c.

A simplified model* which adequately represents the op amp for many applications is illustrated in Fig. 2.4. It shows v_+ and v_- of Fig. 2.1 and the corresponding currents i_+ and i_-. The terminal marked v_+ is called the *noninverting terminal*, and that marked v_- is the *inverting terminal*. The controlled voltage source produces the voltage v_2 given by Eq. (2.1), R_i is the input resistance, and R_o the output resistance. The quantity A is the amplifier gain. Typical parameter values for many practical op amps are as follows:

$$A = 10^5 \text{ (with a range of } 10^4-10^6)$$
$$R_i = 100 \text{ k}\Omega$$
$$R_o = 100 \ \Omega$$
$$|v_+ - v_-| < 1 \text{ mV (typically a few microvolts)}$$

Idealized input–output characteristics of an op amp are shown in Fig. 2.5. Saturation of the output voltage takes place when it reaches the supply voltage V_{CC}, which is typically less than 15 V. Thus linear operation requires the input voltage to be

$$|v_+ - v_-| < \frac{V_{CC}}{A} \tag{2.2}$$

This means that the op amp operates with input voltages that are typically a few microvolts.

---

* A more adequate model, which is sufficiently accurate for use in computer simulation, used in computer programs such as SPICE, is described by G. R. Boyle, B. M. Cohn, D. O. Pederson, and J. E. Solomon, "Macromodeling of Integrated Circuit Operational Amplifiers," *IEEE J. Solid-State Circuits*, vol. SC-9, pp. 353–363, Dec. 1974.

Chip dimensions and bonding diagram
dimensions shown in inches (mm)

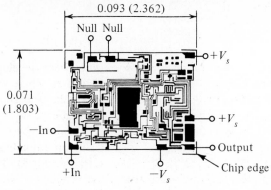

FIGURE 2.2

To go from the model described in Fig. 2.4 to the one we will use for most applications, we assume limiting values of the parameters. We will assume a model in which

$$A = \infty$$
$$R_i = \infty$$
$$R_o = 0$$

These assumptions will imply that

$$v_+ = v_- \quad \text{and} \quad i_+ = i_- = 0 \tag{2.3}$$

The symbol which we will use for this model is shown in Fig. 2.6b. The triangular shape is to suggest the unilateral nature of the device: the difference of input voltages determines the output voltage, but a voltage applied at the output does not influence the input.

2.2 THE OPERATIONAL AMPLIFIER AND RESISTIVE FEEDBACK

Figure 2.7 shows the op amp encased in a harness made up of two resistors. First let us consider the qualitative operation of this circuit in the time domain. Assume that v_1 is applied with a relatively small value. This causes the output to assume the value $v_2 = Av_1$. Assume that there is a short time delay in the op amp, so that v_2 increases, as shown in Fig. 2.8. As soon as v_2 appears, current will flow in the series resistors R_2 and R_1. But these resistors actually form a voltage-dividing circuit causing a voltage v_a to appear at the − terminal of the op amp. If the design is such that v_a is positive, then the op-amp voltage may be made to vanish,

$$v_x = v_a - v_1 = 0 \tag{2.4}$$

and since $v_2 = Av_x$, it will not have a large value. When $v_x = 0$, then we write

$$v_1 = v_a = \frac{R_1}{R_1 + R_2} v_2 \tag{2.5}$$

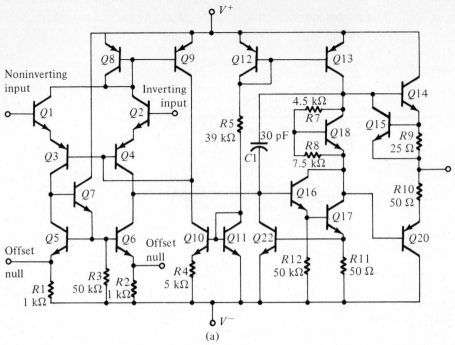

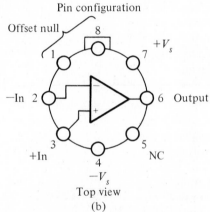

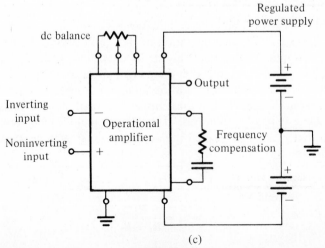

FIGURE 2.3

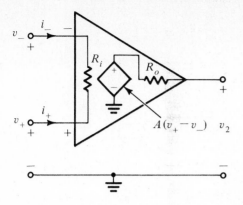

FIGURE 2.4

and the gain of the circuit will be

$$\frac{v_2}{v_1} = 1 + \frac{R_2}{R_1} \tag{2.6}$$

Thus we see that the circuit of Fig. 2.7 is an amplifier with a gain which may be designed by simply selecting the proper values of R_1 and R_2. The choice $R_2 = 9$ kΩ and $R_1 = 1$ kΩ, for example, gives an amplifier with a gain of 10. Through feedback provided by the resistors the amplifier is stabilized and the output is finite, while the voltage at the input to the op amp is made vanishingly small. If we should mix our connections in the circuit of Fig. 2.7 by interchanging the + and − terminals, then the current i is reversed so that

$$i = \frac{-v_2}{R_1 + R_2} \tag{2.7}$$

Then v_a will always be negative, and it will not be possible for $v_a - v_1$ to vanish. In this case the output v_2 will increase without limit and destroy the op amp. Properly connected, this circuit is an important one, and is known as a *noninverting amplifier*. If we have a noninverting amplifier, we must obviously have an inverting one, which we consider next.

The circuit of Fig. 2.9 is different from that of Fig. 2.7 in that the input voltage v_1 is applied at the terminal of the op amp. The action of this circuit is very

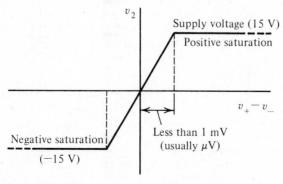

FIGURE 2.5

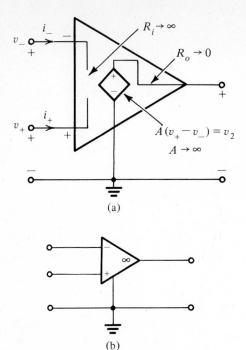

(a)

(b)

FIGURE 2.6

similar to that just considered. Again, we desire to make $v_x = 0$. To find this voltage, we solve the series circuit for the voltage at node a. From Kirchoff's voltage law we have

$$v_1 = v_2 + i(R_1 + R_2) \tag{2.8}$$

or

$$i = \frac{v_1 - v_2}{R_1 + R_2} \tag{2.9}$$

Since the voltage at node a is $v_1 + R_1 i$, we have

$$v_x = v_1 - R_1 \left(\frac{v_1 - v_2}{R_1 + R_2} \right) = 0 \tag{2.10}$$

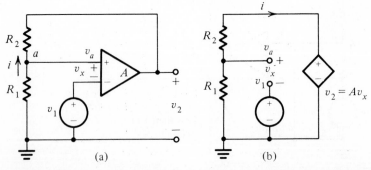

FIGURE 2.7 The noninverting amplifier arranged in a form convenient for analysis.

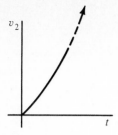

FIGURE 2.8

or

$$\frac{v_2}{v_1} = \frac{-R_2}{R_1} \qquad (2.11)$$

which represents a gain of R_2/R_1 and a sign reversal which we call inverting. Again, the op amp has been harnessed, and design is easily accomplished by the proper choices of R_1 and R_2.

2.3 BLOCK DIAGRAMS AND FEEDBACK

We have a question remaining: why should A have an infinite value? (At least a very large value.) Before answering this, we digress to introduce the representation of circuits by block diagrams as an aid to visualizing feedback. The equation

$$Tv_1 = v_2 \qquad (2.12)$$

may be represented by the block diagram shown in Fig. 2.10. Both the equation and the block diagram tell us that when v_1 is multiplied by T, the result is v_2. The arrows associated with the block indicate the unilateral nature of the operation. For example, Ohm's law

$$Ri = v \qquad (2.13)$$

may be represented by the block diagram of Fig. 2.11.

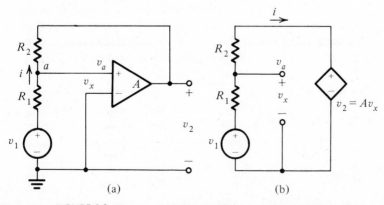

(a)   (b)

FIGURE 2.9   Two representations of the inverting amplifier.

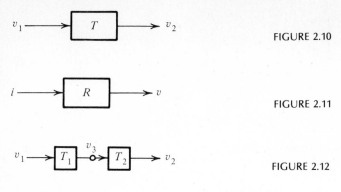

$v_1 \longrightarrow \boxed{T} \longrightarrow v_2$

FIGURE 2.10

$i \longrightarrow \boxed{R} \longrightarrow v$

FIGURE 2.11

$v_1 \longrightarrow \boxed{T_1} \overset{v_3}{\multimap} \boxed{T_2} \longrightarrow v_2$

FIGURE 2.12

The equation

$$Tv_1 = T_1T_2v_1 = v_2 \tag{2.14}$$

implies the cascade connection of blocks shown in Fig. 2.12. The identification of a new variable v_3 permits us to write

$$T_1v_1 = v_3 \quad \text{and} \quad T_2v_3 = v_2 \tag{2.15}$$

which combine to give Eq. (2.14).

The next block diagram operation, which is addition and subtraction, is represented in Fig. 2.13a. We assume that v_1 is positive, but that v_3 may be either positive or negative, as indicated by the ± sign. This diagram is equivalent to the equation

$$v_2 = v_1 \pm v_3 \tag{2.16}$$

indicating that v_2 is either the sum or the difference of v_1 and v_3. Figure 2.13b illustrates a *pick-off point*. A line drawn from another line, frequently with a heavy dot for emphasis, indicates that v_2 follows two paths. Here v_2 is an output, but it is picked off and connected to another part of the system.

The connection of blocks in the form of Fig. 2.14 is known as an *elementary feedback system*. Here v_1 is the input and v_2 the output, and feedback is accomplished by feeding back v_2 through block H. Then Hv_2 is either added or subtracted from v_1 to form e. A negative sign implies negative feedback, a positive sign positive feedback. If $H = 1$ and negative feedback is used, then e is the difference between input and output, or the *error*. Many feedback systems are designed to minimize this error so that the output will follow the input with small error. Two equations describe the block diagram system of Fig. 2.14:

$$v_2 = Ae \tag{2.17}$$

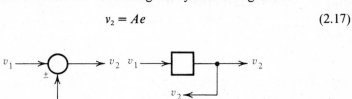

(a)   (b)

FIGURE 2.13

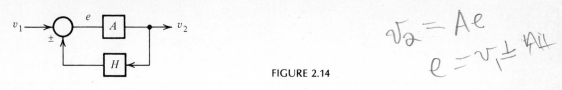

FIGURE 2.14

and

$$e = v_1 \pm H v_2 \tag{2.18}$$

Substituting the second of these equations into the first and rearranging the result, we have

$$\frac{v_2}{v_1} = \frac{A}{1 \mp AH} \tag{2.19}$$

or

$$\frac{v_2}{v_1} = \frac{1}{1/A \mp H} \tag{2.20}$$

Here the + sign indicates negative feedback, the − sign positive feedback.

It has long been recognized (since the late 1920s) that if A becomes very large, corresponding to a high gain, then $1/A$ is negligible compared to H, and Eq. (2.20) becomes

$$\frac{v_2}{v_1} = \frac{1}{\mp H} \tag{2.21}$$

This will indeed be the case when A represents an op amp with very high gain. We begin to see how we might use an amplifier with infinite gain.

2.4   BLOCK DIAGRAM REPRESENTATION OF THE NONINVERTING AMPLIFIER

In this section and the next we examine further the last equation, Eq. (2.21). To begin, study the four circuits of Fig. 2.15 and convince yourself that they are all really identical to that given in Fig. 2.7. The topology of circuits sometimes fools you, but with a little experience you will recognize this circuit in its various disguises routinely. The circuit is known as a noninverting amplifier, "noninverting" being necessary because there exists an inverting amplifier. To understand its operation, let the op amp gain be A for the time being. The output of the op amp was given in Eq. (2.1) as

$$v_2 = A(v_+ - v_-) \tag{2.22}$$

where these voltages are identified in Fig. 2.15a. From the same figure, note that $v_- = v_x$ and $v_+ = v_1$, so v_2 in Eq. (2.22) becomes

$$v_2 = A(v_1 - v_x) \tag{2.23}$$

This equation is represented in the circuit of Fig. 2.16 in controlled-source form.

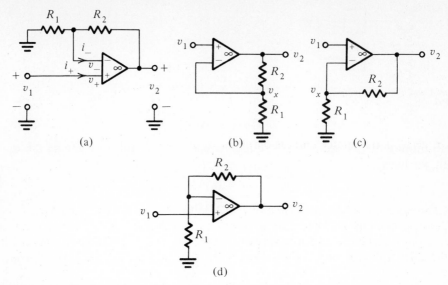

FIGURE 2.15 The figure shows four equivalent ways in which the circuit diagram of the non-inverting amplifier may be drawn.

To find v_x in Eq. (2.23) we recognize that it is determined by v_2 and a voltage-divider circuit, redrawn in Fig. 2.17 for clarity. The voltage-divider equation applied to this simple circuit gives

$$v_x = \frac{R_1}{R_1 + R_2} v_2 \qquad (2.24)$$

Substituting this equation into Eq. (2.23), we have

$$v_2 = A v_1 - \frac{A R_1}{R_1 + R_2} v_2 \qquad (2.25)$$

This equation may be rearranged in the form

$$\frac{v_2}{v_1} = \frac{A}{1 + A\, R_1/(R_1 + R_2)} \qquad (2.26)$$

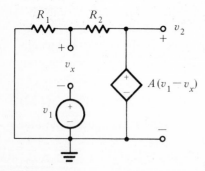

FIGURE 2.16

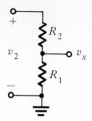

FIGURE 2.17

or, dividing numerator and denominator by A,

$$\frac{v_2}{v_1} = \frac{1}{1/A + R_1/(R_1 + R_2)} \tag{2.27}$$

Comparing these last two equations with those found for the elementary feedback system, Eqs. 2.19 and 2.20, we see that the circuit of Fig. 2.15 is operating as a negative feedback system having the form given in Fig. 2.18. Here A represents the gain of the op amp. As $A \rightarrow \infty$, the limiting form of Eq. (2.27) is

$$\frac{v_2}{v_1} = 1 + \frac{R_2}{R_1} \tag{2.28}$$

which is a positive constant. Any value of gain larger than 1 is easily realized by the appropriate choices of R_1 and R_2.

Another representation of Eq. (2.28) is a circuit shown in Fig. 2.19, which is equivalent to that of Fig. 2.15. There we see that the output voltage is controlled by the input, and hence the controlled-source representation is required. The input terminals 1-1' appear as an open circuit since the input current $i_+ = 0$ and

$$R_{in} = \frac{v_1}{i_+} = \infty \tag{2.29}$$

The resistance measured at terminals 2-2' is the internal resistance of the controlled source which is zero,

$$R_{out} = 0 \tag{2.30}$$

These properties apply only to the model we have used, of course, with infinite gain provided by the op amp. But they are approximated by more realistic models.

The properties we have found are summarized in Table 2.1.

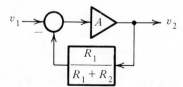

FIGURE 2.18

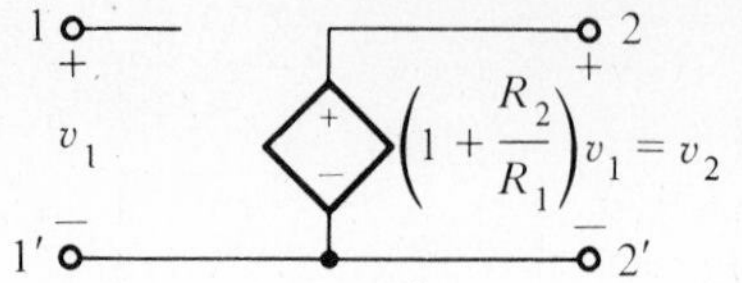

FIGURE 2.19

2.5   THE VOLTAGE FOLLOWER

A voltage-follower circuit is a special case of the noninverting amplifier circuit of Fig. 2.15. In Eq. (2.28), with $R_1 = \infty$, we have

$$\frac{v_2}{v_1} = 1 \qquad \text{or} \qquad v_2 = v_1 \tag{2.31}$$

so that output and input voltages are identical. How does the voltage follower differ from simply two parallel connecting wires? The answer is that the non-inverting amplifier properties $R_{\text{in}} = \infty$ and especially $R_{\text{out}} = 0$ also apply here, and in addition the controlled-source nature of the circuit, shown by Fig. 2.20, is such that it provides *isolation* between input and output. For this reason the circuit is sometimes called a unity-gain *buffer*.

The steps in the evolution of a noninverting amplifier to a voltage follower are shown in Fig. 2.21. Starting with Fig. 2.21a we let $R_1 = \infty$ and obtain the circuit of Fig. 2.21b. Since $i_- = 0$, there is no voltage drop across R_2, and so it may be replaced by a short circuit, as shown in Fig. 2.21c. Thus we have the voltage follower.

To better understand the operation of the voltage-follower circuits, we modify Fig. 2.19 to the form shown in Fig. 2.22. As before, with finite gain A for the op amp,

$$v_2 = A(v_1 - v_x) = A(v_1 - v_2) \tag{2.32}$$

since $v_x = v_2$. Rearranging this equation gives the voltage ratio or gain

$$\frac{v_2}{v_1} = \frac{A}{1 + A} = \frac{1}{1/A + 1} \tag{2.33}$$

The block diagram representation of this equation is shown in Fig. 2.23. As the gain A of the op amp becomes infinite, the voltage ratio of Eq. (2.33) becomes 1, as found previously by Eq. (2.31). Further, the input voltage to the op amp is

$$v_+ - v_- = v_1 - v_2 = v_1 - v_1 = 0 \tag{2.34}$$

and the voltage is zero as before. Properties of the voltage follower are summarized in Table 2.1.

The voltage follower's main role is to provide isolation of parts of a circuit when it is required that the two parts should not interact. This buffer action is possible because of low output resistance and high input resistance. To show the meaning of these statements, consider a signal generator which produces a voltage of 0.1 V. The internal resistance of the generator is $R_S = 1000 \ \Omega$. A load resistance of 100 Ω is to be connected to this source, as shown in Fig. 2.24. With the connection made as illustrated in Fig. 2.24a, voltage-divider action gives a load

TABLE 2.1   Catalog of op-amp analog operations

Name—Function	Block diagram	Circuit	Controlled source	Equation	Conditions
Voltage follower "buffer"				$v_2 = v_1$	
Inverting amplifier				$v_2 = -Kv_1$	$K = \dfrac{R_2}{R_1}$
Noninverting amplifier				$v_2 = Kv_1$	$K = 1 + \dfrac{R_2}{R_1}$
Voltage adder–multiplier				$v_2 = -(K_1 v_1 + K_3 v_3)$	$K_1 = \dfrac{R_2}{R_3}$ $K_3 = \dfrac{R_2}{R_1}$
Voltage subtracter–multiplier				$v_2 = K_3 v_3 - K_1 v_1$	$K_3 = \dfrac{1 + \dfrac{R_2}{R_1}}{1 + \dfrac{R_a}{R_b}}$ $K_1 = \dfrac{R_2}{R_1}$
Voltage subtracter–multiplier				$v_2 = K(v_3 - v_1)$	$K = \dfrac{R_2}{R_1}$ $R_1 = R_2$

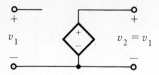

FIGURE 2.20

voltage of

$$v_2 = \frac{100}{1000 + 100} 0.1 = 0.009 \text{ V} \tag{2.35}$$

However, if a voltage follower is employed, as illustrated by Fig. 2.24b, there is no voltage drop across R_s since there is no input current, and the full voltage of 0.1 V appears across the load, assuming that the maximum output current of the op amp is not exceeded. This illustrates *buffer* action for which the voltage follower is useful as a building block in circuit design.

2.6 BLOCK DIAGRAM REPRESENTATION OF THE INVERTING AMPLIFIER

The second fundamental op-amp circuit we will study is the *inverting amplifier* shown in Fig. 2.25. The name of the amplifier comes from the negative sign of v_2/v_1. If the input voltage v_1 is a sine wave as shown in Fig. 2.26a, then the non-inverting amplifier has an amplified output as shown in Fig. 2.26b. However, the output of the inverting amplifier is indeed inverted, as shown in Fig. 2.26c.

The circuit is analyzed making use of Kirchhoff's voltage law. In the circuit of Fig. 2.25b we apply Kirchhoff's law to the path *abcde* and equate voltage rises and drops. As we traverse the path chosen, we sense the polarity assigned to each element, with − to + considered a rise and + to − a drop. Then

$$v_1 = iR_1 + iR_2 + v_2 \tag{2.36}$$

Solving for i,

$$i = \frac{v_1 - v_2}{R_1 + R_2} \tag{2.37}$$

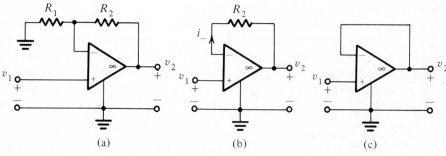

FIGURE 2.21

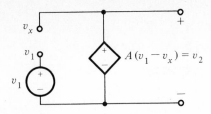

FIGURE 2.22

Similarly applying Kirchhoff's voltage law around the path *abca* gives

$$v_1 = iR_1 + v_a \tag{2.38}$$

or

$$v_a = v_1 - iR_1 \tag{2.39}$$

Substituting Eq. (2.37) for i into Eq. (2.39) and rearranging, we have

$$v_a = \frac{R_2}{R_1 + R_2} v_1 + \frac{R_1}{R_1 + R_2} v_2 \tag{2.40}$$

But from Eq. (2.1) we know that with this connection to the op amp,

$$v_2 = -Av_a \tag{2.41}$$

Combining these two equations, and solving for v_2/v_1, we have

$$\frac{v_2}{v_1} = \frac{-A[R_2/(R_1 + R_2)]}{1 + A[R_1/(R_1 + R_2)]} \tag{2.42}$$

or

$$\frac{v_2}{v_1} = \frac{-R_2/R_1}{\{1/[AR_1/(R_1 + R_2)]\} + 1} \tag{2.43}$$

Comparing this pair of equations with Eqs. (2.19) and (2.20) and with the block diagram structure of Fig. 2.14, we see that the inverting amplifier may be represented as in Fig. 2.27. As $A \to \infty$, then from the block diagram, or from Eq. (2.43),

$$\frac{v_2}{v_1} = \frac{-R_2}{R_1} \tag{2.44}$$

As was the case for the noninverting amplifier, the finite gain of the amplifier is controlled by the two resistors R_1 and R_2.

The gain functions v_2/v_1 for the two amplifiers studied differ by both the sign of R_2/R_1 and the constant 1. However, the difference that gives rise to the names of these basic amplifiers, inverting and noninverting, is the sign of Eq. (2.44). We will most often interpret this in terms of a sine-wave input to the amplifier, shown in Fig. 2.26a, normalized to have unit amplitude. In terms of this input, v_2 for the

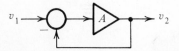

FIGURE 2.23

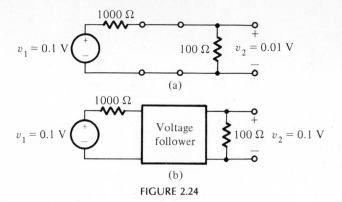

(a)

(b)

FIGURE 2.24

noninverting amplifier is that of Fig. 2.26b, while the inverting amplifier has an output as shown in Fig. 2.26c.

2.7 THE VIRTUAL SHORT

Our studies thus far have shown the wondrous consequences of resistor feed-back. This resistor feedback creates a stable circuit with an output-to-input ratio fixed by the resistors. The very large gain makes operation of the circuit relatively independent of changes in the gain of the op amp. Resistor feedback causes special circumstances at the input to the op amp—no current and no voltage. Now zero current normally implies an open circuit, while zero voltage normally implies a short circuit, as depicted in Fig. 2.29. But here we have *both*, a very special condition. We next show that resistor feedback creates another special condition in the inverting amplifier, causing a difference in the inverting and noninverting amplifiers with respect to input resistance.

Consider once more the inverting amplifier as shown in Fig. 2.30a. There two additional connections to the op amp are shown, which provide the bias voltages. This detail is normally left to laboratory implementation, and the equivalent op-amp circuit is shown in Fig. 2.30b. The controlled-source equivalent of Fig. 2.30b is shown in Fig. 2.30c. We note that the op-amp connection to ground, marked *a* in Fig. 2.30, is normally omitted or left implied. Observe the three wires in Fig. 2.30b and c. Connection *a* carries current, while connections *b* and *c* to the

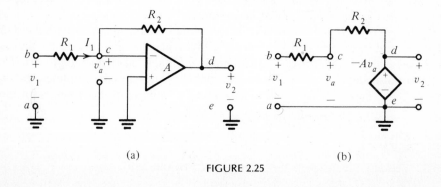

(a)   (b)

FIGURE 2.25

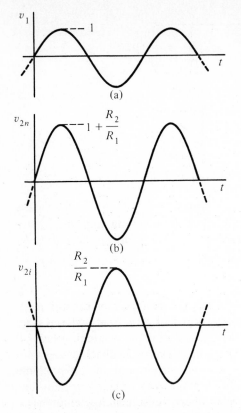

(a)

(b)

(c)   FIGURE 2.26

input of the op amp do not, at least in the ideal case. In the actual case the current is very small. In Fig. 2.30c we wish to calculate the input resistance, marked
R_i', excluding R_1.

Part of Fig. 2.30c is shown in greater detail in Fig. 2.31, with the addition of

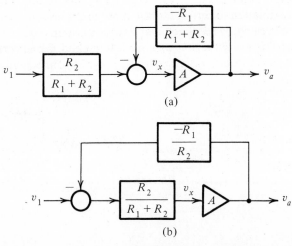

(a)

(b)

FIGURE 2.27

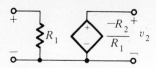

FIGURE 2.28

R_i and R_o from the more complete model of the op amp. At node 2 we employ Kirchhoff's current law to write

$$i_1 + i_2 + i_3 = i_1 + \frac{(0 - v_x)}{R_i} + \frac{-Av_x - v_x}{R_o + R_2} = 0 \tag{2.45}$$

From this equation we solve for the ratio of v_x and i_1,

$$R_i' = \frac{v_x}{i_1} = \frac{1}{1/[(R_o + R_2)/(A + 1)] + 1/R_i} \tag{2.46}$$

Various approximations may be made starting from Eq. (2.46). If R_i is so large that the term containing R_i can be ignored, then

$$R_i' \cong \frac{R_o + R_2}{A + 1} \tag{2.47}$$

Since R_2 is ordinarily in the kilohm range of values, and R_o is typically 10 Ω, and also 1 can be neglected compared to A, then

$$R_i' \cong \frac{R_2}{A} \tag{2.48}$$

Finally if A is very large indeed, then we may use the approximate value

$$R_i' \cong 0 \tag{2.49}$$

If we calculate R_i' using $R_2 = 10$ kΩ and the values given as typical for some op amps, then

$$R_i' = 0.1 \tag{2.50}$$

which is an approximation to zero. Thus the input to the op amp is a short circuit. Since this short is created by feedback and does not actually appear in the circuit diagram, it is known as a *virtual short*. This virtual short is depicted in Fig. 2.32

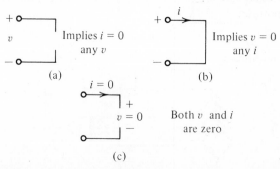

FIGURE 2.29

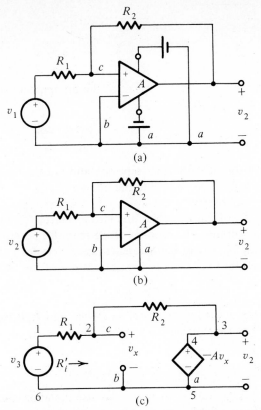

(a)

(b)

(c)

FIGURE 2.30   Circuit representations relating to the explanation of the virtual short.

together with the special condition we found for the input to the op amp. From this figure it is clear that

$$R_{in} = \frac{v_1}{i_1} = R_1 + R_i' \cong R_1 \qquad (2.51)$$

The fact that R_{in} is finite for the inverting amplifier and infinite for the non-inverting amplifier is a significant difference which will be important in design. We will favor the noninverting amplifier in applications for which a high input resistance is important.

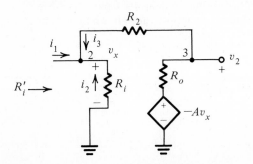

FIGURE 2.31

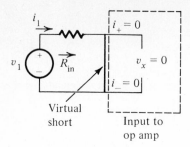

Virtual
short

Input to
op amp

FIGURE 2.32

The controlled-source equivalent representation of the inverting amplifier is shown in Fig. 2.33 with R_1 connected to the input and $R_{out} = 0$, the internal resistance of the controlled source.

We will make use of the concept of the virtual short to ground to rederive v_2/v_1 for the inverting amplifier. Here we apply Kirchhoff's current law at the node marked A in Fig. 2.34. Using the reference directions for current shown and equating the current into the node to the current out of the node, we have,

$$\frac{v_1 - v_x}{R_1} = \frac{v_x - v_2}{R_2} \tag{2.52}$$

But $v_x = 0$ due to the virtual short to ground, and so we modify the equation to the form

$$\frac{v_2}{v_1} = -\frac{R_2}{R_1} \tag{2.53}$$

which is identical to Eq. (2.44). The derivation is shorter, however, since it makes use of the virtual short and $v_x = 0$ rather than deriving that fact. The principle of the virtual short will be used frequently in studies that follow.

2.8 VOLTAGE FOLLOWER ANALYSIS WITH A BETTER OPERATIONAL-AMPLIFIER MODEL

Thus far we have dealt with the simple op-amp model we will use throughout most of the remaining chapters—one with infinite gain, infinite R_{in} and zero R_{out}. Before proceeding we digress to ask the question: How do the results we have obtained compare with those we might expect in circuit design when we use op

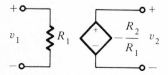

FIGURE 2.33

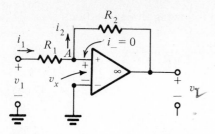

FIGURE 2.34

amps which we might purchase for that purpose? We will obtain an answer to this question for the specific case of the voltage follower and accept as fact that the results apply to other op-amp configurations.

The voltage follower shown in Fig. 2.35a differs from that previously considered in that it includes R_S, the resistance of the source, and a load resistor R_L. Suppose that we use the model in Fig. 2.35b for the op amp shown in Fig. 2.35a with finite R_i and nonzero R_o. The result of superimposing Fig. 2.35a and b is shown in Fig. 2.35c, which is the equivalent circuit of the voltage follower.

Analysis of this circuit is accomplished by the Kirchhoff laws and is straight-

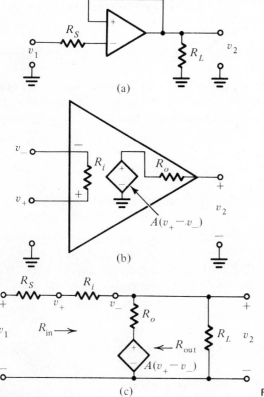

(a)

(b)

(c)   FIGURE 2.35

forward although lengthy. The result are the following three equations*:

$$\frac{v_2}{v_1} = \frac{R_L(R_o + AR_i)}{R_L(R_o + AR_i) + (R_i + R_S)(R_o + R_L)} \tag{2.54}$$

$$R_{\text{in}} = R_i \frac{(A + 1)R_L + R_o}{R_o + R_L} + \frac{R_o R_L}{R_o + R_L} \tag{2.55}$$

$$R_{\text{out}} = \frac{R_o(R_i + R_S)}{R_o + (1 + A)R_i + R_S} \tag{2.56}$$

Typical values for these parameters are as follows:

$$\begin{aligned} A &= 10^5 \\ R_S &= R_L = 1 \text{ k}\Omega \\ R_i &= 100 \text{ k}\Omega \\ R_o &= 10 \text{ }\Omega \end{aligned} \tag{2.57}$$

Substituting these values into Eq. (2.54), we find that

$$\frac{v_2}{v_1} = 0.999989 \cdots \approx 1 \tag{2.58}$$

Under the condition that $R_L \gg R_o$, Eq. (2.55) assumes the simple form

$$R_{\text{in}} \cong AR_i \tag{2.59}$$

and R_{out} similarly has a simple form when $AR_i \gg R_o$ and $R_i \gg R_S$,

$$R_{\text{out}} \cong \frac{R_o}{A} \tag{2.60}$$

Then using the numerical values of Eqs. (2.57), we find that for the op amp modeled in Fig. 2.35c,

$$\frac{v_2}{v_1} = 0.999989, \qquad R_{\text{in}} = 10^{10}, \qquad R_{\text{out}} = 10^{-3} \tag{2.61}$$

which compare well with the ideal values of 1, ∞, and 0. Hence we can use the simple model of the op amp with some confidence that calculated values will agree closely with those measured in the laboratory.

We can also estimate the input voltage and current to the op amp. The output voltage will not exceed that of the supply voltage for the op amp which is typically 15 V. Thus

$$|v_+ - v_-| < \frac{V_{\text{supply}}}{A} \tag{2.62}$$

For $A = 10^5$, the input voltage will thus be less than 1.5 μV. Similarly, the input current will be

$$i_+ = i_- = \frac{v_+ - v_-}{R_i} \tag{2.63}$$

* W. G. Oldham and S. E. Schwarz, *An Introduction to Electronics,* Holt, New York, 1972, chap. 12.

which for the values being used will be less than 0.15 pA (picoamperes), indeed small and so appropriately considered to be zero.

The approximate relationships we have derived and the corresponding ones for the inverting and noninverting amplifiers are shown in Table 2.2.

2.9 ANALOG ADDITION AND SUBTRACTION

Op-amp circuits to add and subtract voltages were highly developed in connection with analog computers in the early 1940s. The ideas involved are simple but useful in the work to follow. The circuit of Fig. 2.36 has two inputs v_1 and v_3, and one output v_2. At node A Kirchhoff's current law permits us to equate the currents in to the currents out, giving

$$\frac{v_1 - v_x}{R_1} + \frac{v_3 - v_x}{R_3} = \frac{v_x - v_2}{R_2} \tag{2.64}$$

TABLE 2.2

Circuits	Approximate equations
Voltage follower	$T = 1$ $R_{in} = AR_i$ $R_{out} = \dfrac{R_o}{A}$
Noninverting	$T = 1 + \dfrac{R_2}{R_1}$ $R_{in} = AR_i \dfrac{1}{1 + R_2/R_1}$ $R_{out} = \dfrac{R_o}{A}\left(1 + \dfrac{R_2}{R_1}\right)$
Inverting	$T = \dfrac{-R_2}{R_1}$ $R_{in} = R_1$ $R_{out} = \dfrac{R_o}{A}\left(1 + \dfrac{R_2}{R_1}\right)$

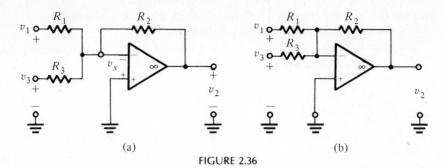

(a)   (b)

FIGURE 2.36

But $v_x = 0$ by the principle of the virtual short circuit, and so this equation may be simplified to

$$v_2 = -\left(\frac{R_2}{R_1} v_1 + \frac{R_2}{R_3} v_3 \right) \tag{2.65}$$

which indicates that v_1 and v_3 are multiplied by constants and then added to give $-v_2$. If $R_1 = R_3$ and if we let $R_2/R_1 = K$, then

$$v_2 = -K(v_1 + v_3) \tag{2.66}$$

If $K = 1$, we do have simple summation and inversion of the voltages. Thus a network equivalent of Fig. 2.36 is that of Fig. 2.37, which involves a controlled-voltage source. We note that the input resistances are R_1 and R_3, which are not infinite as is the case with the noninverting amplifier.

The result found for two inputs may be extended to any number of inputs. If there are three, for example, then Eq. (2.64) with $v_x = 0$ becomes

$$\frac{v_1}{R_1} + \frac{v_3}{R_3} + \frac{v_4}{R_4} = \frac{-v_2}{R_2} \tag{2.67}$$

or

$$-v_2 = \frac{R_2}{R_1} v_1 + \frac{R_2}{R_3} v_3 + \frac{R_2}{R_4} v_4 \tag{2.68}$$

As an example, if it is required to design a circuit to realize

$$v_2 = -3v_1 - 2v_3 \tag{2.69}$$

then this may be accomplished by making $R_2/R_1 = 3$ and $R_2/R_3 = 2$. If we let $R_2 = 60 \ \Omega$, then the design requires that $R_1 = 20 \ k\Omega$ and $R_3 = 30 \ k\Omega$.

The differential amplifier circuit shown in Fig. 2.38 is a basic circuit for

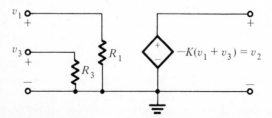

FIGURE 2.37

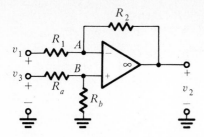

FIGURE 2.38

forming the difference of voltages. We will show that it will permit us to form an analog circuit to realize the equation

$$v_2 = K_3 v_3 - K_1 v_1 \qquad (2.70)$$

where the constants K_1 and K_3 are determined by resistor values. Before analyzing the circuit, consider the situation at the input terminals to the op amp, shown in enlarged form in Fig. 2.39. The voltage between nodes A and B is zero, as we have discussed earlier, which means that

$$v_+ = v_- \qquad (2.71)$$

Thus the voltage at node A is identical to the voltage at node B. Without the op amp you might be tempted to connect the two nodes together. This cannot be done, of course, and it is necessary to apply Kirchoff's current law separately at each node. The voltage at both nodes is indicated as v_a. Since $i_- = 0$, as shown in Fig. 2.39, we write at node A

$$\frac{v_1 - v_a}{R_1} = \frac{v_a - v_2}{R_2} \qquad (2.72)$$

At node B the voltage is determined by the voltage-divider equation as

$$v_a = \frac{R_b}{R_a + R_b} v_3 \qquad (2.73)$$

If we eliminate v_a from these two equations, then we obtain the following expression for v_2 in terms of voltages v_1 and v_3:

$$v_2 = \frac{R_2}{R_1} \frac{R_1/R_2 + 1}{R_a/R_b + 1} (v_3 - v_1) \qquad (2.74)$$

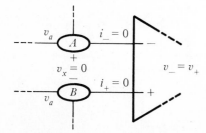

FIGURE 2.39

From this result we see that if $R_a = R_1$ and $R_2 = R_b$, then

$$v_2 = \frac{R_2}{R_1}(v_3 - v_1) \tag{2.75}$$

This simplified circuit is shown in Fig. 2.40. Further simplification results when $R_2 = R_1$ and

$$v_2 = v_3 - v_1 \tag{2.76}$$

This information is summarized in Table 2.1.

These equations permit the routine design of difference amplifiers. For example, if we require that

$$v_2 = 2v_3 - 3v_1 \tag{2.77}$$

then Eq. (2.74) may be used to determine $R_2/R_1 = 3$ and $R_a/R_b = 1$, so that a suitable design would result with $R_2 = 30$ kΩ and $R_1 = R_a = R_b = 10$ kΩ. For many applications we will wish to realize Eq. (2.76), and in this case R_1, R_2, R_b, and R_a will have identical values—say 10 kΩ.

Finally, with respect to the circuit of Fig. 2.40, we see that the input resistance at terminal pair 1-0 is the ratio of v_1 and i_1 with $v_3 = 0$, and that is R_1. Similarly, the input resistance at terminal pair 3-0 is found with $v_1 = 0$ and is $R_1 + R_2$. The voltage v_2 is a controlled voltage given by Eq. (2.75). These facts are summarized in the circuit of Fig. 2.40b, and they are also contained in Table 2.1.

2.10 APPLICATIONS OF OP-AMP RESISTOR CIRCUITS

Thus far in this chapter, we have described circuits which provided amplification, isolation, and analog addition and subtraction. In this section we give additional examples of applications of resistor–op-amp circuits.

Figure 2.41 is similar to Fig. 2.40a shown earlier. In this circuit R_2' represents a pressure transducer. In an application, changes in pressure cause R_2 to change to $R_2' = R_2 + \Delta R_2$. We will show that this circuit will produce an output voltage which is proportional to this measure of pressure change ΔR_2 and to the battery voltage, V_0.

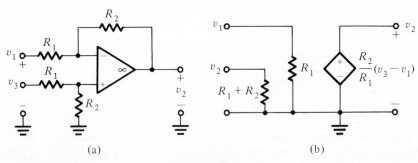

(a)   (b)

FIGURE 2.40

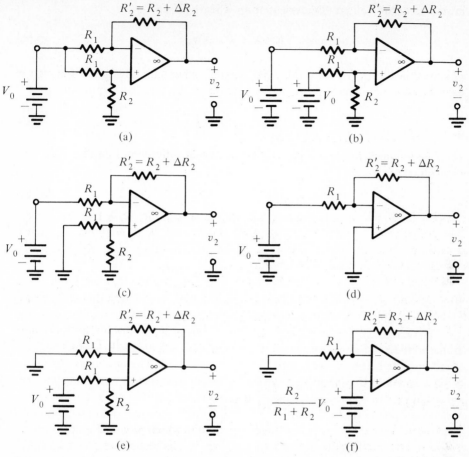

FIGURE 2.41 The steps shown in the figure illustrate the application of superposition to the analysis of the circuit shown in (a).

We make use of the principle of superposition in analyzing the circuit of Fig. 2.41a. First we observe that the insertion of a new battery of voltage V_0 as shown in Fig. 2.41b does not change any voltage in the circuit. We will consider the output voltage produced by these batteries separately, and then add these voltages. To do this, we replace one battery by a short circuit, as shown in Fig. 2.41c. Since the op-amp current i_- is zero, there is no voltage across R_1 and R_2, meaning that the + terminal of the op amp is at ground potential, as shown in Fig. 2.41d. With this circuit simplification we then have a standard inverting circuit for which we write

$$v_2' = V_0 \left(\frac{-R_2'}{R_1} \right) \qquad (2.78)$$

Next we observe the output voltage due to the other battery, with the first battery removed and so replaced by a short circuit to ground. This is shown in Fig. 2.41e, from which we see that the voltage at the + terminal of the op amp may be determined from the voltage-divider equation. Clearly, it is

$$V_0' = \frac{R_2}{R_1 + R_2} V_0 \tag{2.79}$$

as shown in Fig. 2.41f. With this simplification the circuit of Fig. 2.41f is seen to be a standard noninverting circuit, from which we see that the output voltage is

$$v_2' = \frac{R_2}{R_1 + R_2}\left(1 + \frac{R_2'}{R_1}\right) V_0 \tag{2.80}$$

Now we are prepared to use the principle of superposition and find the total output voltage:

$$v_2 = v_2' + v_2'' \tag{2.81}$$

$$= \frac{-R_2' R_1 - R_2' R_2 + R_1 R_2 + R_2 R_2'}{R_1(R_1 + R_2)} V_0 \tag{2.82}$$

$$= \frac{R_1(R_2 - R_2')}{R_1(R_1 + R_2)} V_0 = \frac{-V_0}{R_1 + R_2} \Delta R_2 \tag{2.83}$$

Since R_1, R_2, and V_0 are constants, the output voltage varies directly with ΔR_2 and so is a measure of the transduced quantity, the pressure.

The circuit shown in Fig. 2.42 is used to produce an output voltage proportional to the key that is depressed. It operates with a standard keyboard incorporated into the feedback resistor of the inverting circuit. The output of this circuit is then added to the input by means of an adder–inverter circuit. Pressing a key on the keyboard connects together points that are the horizontal and vertical projections of points A, B, C, D, E, F, G, and H from the key. Thus pushing key 1

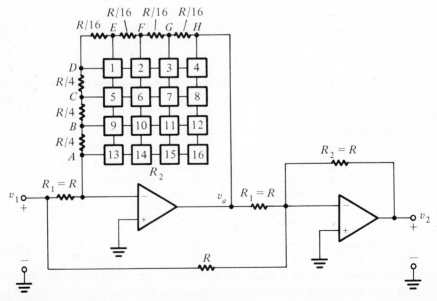

FIGURE 2.42

connects together points D and E such that $R_2 = \frac{15}{16} R$, pushing key 11 connects B and G so that $R_2 = \frac{5}{16}R$, pushing key 13 connects A and E so that $R_2 = \frac{3}{16}R$, and so on. In general we see that

$$R_2 = R\left(1 - \frac{n}{16}\right) \tag{2.84}$$

Now since $R_1 = R$, then

$$\frac{v_a}{v_1} = 1 - \frac{n}{16} \tag{2.85}$$

The second op-amp circuit adds the two inputs and inverts such that

$$v_2 = -(v_1 + v_a) \tag{2.86}$$

or, combining the last two equations,

$$v_2 = \frac{n}{16} v_1 \tag{2.87}$$

Then, as promised, the output voltage is proportional to the key pressed. The circuit might be used in component testing or as an analog voltage generator for the control of some system.

The next circuit to be studied is an instrumentation amplifier which has very low input current and is tuned to a desired gain by one resistor. The circuit is shown in Fig. 2.43, and the tuning resistor is identified as R_g. While the connection is new to us, it is composed of circuits that have been previously studied. We make use of the fact that the input voltage at the op-amp terminals is zero (being a virtual short circuit) such that v_1 appears at both nodes A and B and v_3 at nodes C and D. Then

$$v_4 = \left(1 + \frac{R_1}{R_g}\right) v_1 + \left(\frac{-R_1}{R_g}\right) v_3 \tag{2.88}$$

$$v_5 = \left(1 + \frac{R_1}{R_g}\right) v_3 + \left(\frac{-R_1}{R_g}\right) v_1 \tag{2.89}$$

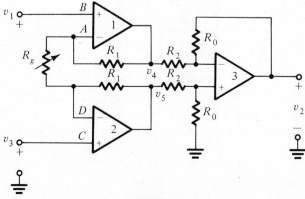

FIGURE 2.43

We see from the circuit of Fig. 2.43 that v_4 and v_5 are combined by op amp 3 to give

$$v_2 = \frac{(-R_o)}{R_2} v_4 + \frac{R_o}{R_o + R_2} \left(1 + \frac{R_o}{R_2}\right) v_5 \tag{2.90}$$

Substituting v_4 and v_5 into this equation gives v_2 as

$$v_2 = \left(1 + \frac{2R_1}{R_g}\right) \frac{R_o}{R_2} (v_3 - v_1) \tag{2.91}$$

$$v_2 = K(v_3 - v_1) \tag{2.92}$$

where K is the circuit gain. Thus the circuit is a differential amplifier and the gain K is adjusted by changing R_g. The input resistance is very high because the input voltages are applied to the + terminals of op amps 1 and 2.

As an example assume that the instrumentation amplifier has the values shown in Fig. 2.44. Then from Eq. (2.91) we determine the gain to be

$$K = \left(1 + \frac{2 \times 75}{5}\right)\left(\frac{160}{5}\right) = 992 \tag{2.93}$$

Fine adjustments on the value of gain can be made by tuning R_g.

As the last example of this chapter we study an op-amp circuit that is useful in providing a stage of high gain.* Since gain for the inverting amplifier, shown in Fig. 2.45a, is $-R_2/R_1$, we see that gain can be increased by making R_2 larger, or making R_1 smaller. But R_1 is the input resistance to the inverting amplifier, and so there is a preference for making R_2 larger. Suppose that we wish to increase the gain by α, and that we do this by increasing R_2 to $R_2' = \alpha R_2$, as shown in Fig. 2.45b. Now the circuit we will actually use to realize this increased gain is shown in Fig. 2.46, where R_2 has been replaced by a T-circuit. The circuit may be analyzed making use of Kirchhoff's current law. The circuit of Fig. 2.47 is that given

* J. I. Smith, *Modern Operational Circuit Design*, Wiley, New York, 1971, p. 137.

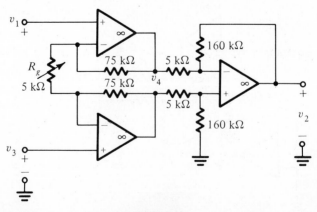

FIGURE 2.44

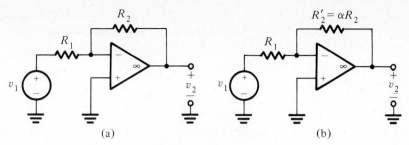

(a)   (b)

FIGURE 2.45

in Fig. 2.46 with references added for ease in analysis. At node A we have

$$\frac{v_1}{R_1} + \frac{v_B}{R_3} = 0 \tag{2.94}$$

and at node B we see that

$$-i_1 + i_2 + i_3 = 0 \tag{2.95}$$

or

$$\frac{-v_1}{R_1} + \frac{v_B}{R_4} + \frac{v_B - v_2}{R_5} = 0 \tag{2.96}$$

We solve Eq. (2.94) for v_B and substitute into Eq. (2.95) to give

$$\frac{v_2}{v_1} = \frac{R_5 + R_3(1 + R_5/R_4)}{R_1} \tag{2.97}$$

Since we have assumed that

$$\frac{v_2}{v_1} = \frac{-R_2'}{R_1} = \frac{-\alpha R_2}{R_1} \tag{2.98}$$

we see that our design equation is

$$\alpha R_2 = R_5 + R_3\left(1 + \frac{R_5}{R_4}\right) \tag{2.99}$$

There are many ways in which the design equation (2.99) can be satisfied.

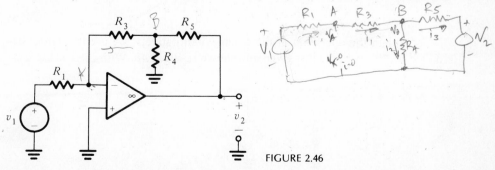

FIGURE 2.46

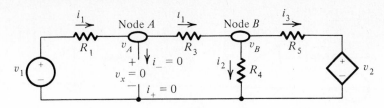

FIGURE 2.47

One method is to select two resistor values and then solve for the third. We select

$$R_3 = \frac{1}{2} R_2 \quad \text{and} \quad R_5 = \alpha \frac{1}{2} R_2 \tag{2.100}$$

Then solving Eq. (2.99) for R_4 gives

$$R_4 = \frac{\alpha}{\alpha - 1} \frac{R_2}{2} \tag{2.101}$$

As an example of circuit design using these equations, suppose that $R_1 = 10$ kΩ and $R_2 = 100$ kΩ; we wish to select $\alpha = 10$ so that the circuit designed will have a gain of 100. Then the circuit will have the following values:

$$R_3 = 50 \text{ k}\Omega, \qquad R_5 = 500 \text{ k}\Omega, \qquad R_4 = 55.56 \text{ k}\Omega \tag{2.102}$$

Another design approach begins with Eq. (2.99) for αR_2. Suppose that we select $R_3 = R_2$ as a design value. Then

$$\alpha = 1 + \frac{R_5}{R_3} + \frac{R_5}{R_4} \tag{2.103}$$

We now select R_5 so that the ratio R_5/R_3 is small, meaning much smaller than 1. Let that number be α_0. Then

$$R_4 = \frac{1}{\alpha - \alpha_0} R_5 \tag{2.104}$$

results in a large value for R_5/R_4 in Eq. (2.103) and a small value for R_4. Using the same numerical values as for the last example, we have

$$R_1 = 10 \text{ k}\Omega, \qquad R_2 = R_3 = 100 \text{ k}\Omega \tag{2.105}$$

We will select $R_5 = R_1 = 10$ kΩ, so that $R_5/R_3 = 0.1$. Then the value for R_4 is found from Eq. (2.104) as

$$R_4 = 1.124 \text{ k}\Omega \tag{2.106}$$

To explain the operation of the circuit quantitatively, we use the values determined for the example just given and shown in Fig. 2.48. With the 100 kΩ tem-

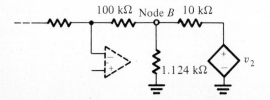

FIGURE 2.48

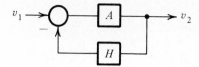

FIGURE 2.49

porarily disconnected, the remaining resistors form a voltage-divider circuit, so that the voltage at node B under this circumstance is

$$v_B = \frac{1.124}{10 + 1.124}v_2 = 0.101v_2 \qquad (2.107)$$

In other words, only about 10% of the output voltage is being fed back for addition to the input. What happens when we reduce the amount of voltage fed back? Using the block diagram representation of a feedback circuit shown in Fig. 2.49, we found in Eq. (2.20) that

$$\frac{v_2}{v_1} = \frac{1}{1/A - H} \approx \frac{-1}{H} \qquad (2.108)$$

for large A. If we reduce H by multiplying by $1/\beta$, then Eq. (2.108) becomes

$$\frac{v_2}{v_1} = -\beta \frac{1}{H} \qquad (2.109)$$

In other words, decreasing the effective value of H increases the value of the gain. We have used this principle in designing an inverting amplifier circuit that provides high gain. This action is known as *gain enhancement*, and this kind of enhancement will be used in later chapters.

PROBLEMS

2.1 In the op-amp circuit given in Fig. P2.1, it is required that

$$V_2 = \frac{V_3}{3} - 2V_1$$

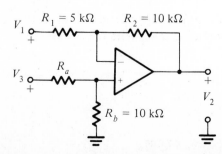

FIGURE P2.1

(a) Determine the value of R_a that gives the desired relationship.
(b) Suppose that $V_1 = -10$ V and $V_2 = +10$ V. Find the magnitude of the current in each resistor, and the average power dissipated by each resistor.
(c) Repeat (b) if $V_1 = 10$ V and $V_2 = 10$ V.

2.2 Design a circuit using a single op amp so that the output voltage is

$$V_2 = V_1 - 4V_3$$

where V_1 and V_3 are the input voltages. Select the resistor values so that no more than 0.25 W is dissipated by any resistor when V_1 and V_3 have absolute magnitudes less than 10 V.

2.3 Design an op-amp circuit to satisfy the relationship

$$V_0 = 2V_1 + V_2 - V_3$$

where V_0 is the output and V_1, V_2, and V_3 are input voltages. Using the assumption that none of the inputs will exceed 10 V in magnitude, design the circuit so that no resistor dissipates in excess of 0.1 W.

2.4 In the circuit given in Fig. P2.4, $R = 10$ kΩ. Find the value of R_1 so that $V_2 = -100 \, V_1$ and calculate the energy dissipated in each resistor.

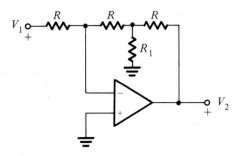

FIGURE P2.4

2.5 The circuit given in Fig. P2.4, is purely resistive.
   (a) Letting $G_j = 1/R_j$ show that,

$$\frac{V_2}{V_1} = T = \frac{(G_3/G_2 G_4)(G_1 + G_2)}{[(G_3 + G_5)/G_4] - G_1/G_2}$$

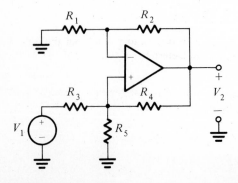

FIGURE P2.5

(b) Let all resistors be equal except for R_4 which may be tuned. Sketch T as a function of R_4.

(c) Explain the behavior of the circuit when $R_4 = R/2$.

2.6 For the circuit shown in Fig. P2.6, it is required that $R_{in} = 1\ \Omega$, and also that

$$V_3 = k_1 V_1 + k_2 V_2$$

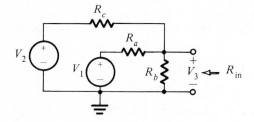

FIGURE P2.6

Show that the required resistor values are

$$R_a = \frac{1}{k_1}, \qquad R_c = \frac{1}{k_2}, \qquad \text{and} \qquad R_b = \frac{1}{[1 - (k_1 + k_2)]}$$

2.7 For the circuit shown in Fig. P2.7, determine V_2 as a function of V_1 and V_3 for the element values specified.

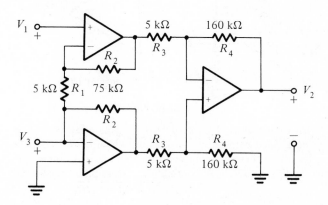

FIGURE P2.7

2.8 The circuit in Fig. P2.8 has voltage input V_1 with respect to ground, but the output load shown is "floating" with respect to ground. Determine V_a, V_b, and $V_a - V_b$ as a function of V_1.

2.9 The op-amp circuit shown in Fig. P2.9 is intended to provide any gain between -10 (when $k = 0$) and $+10$ (when $k = 1$) by adjusting the potentiometer. Let R_1 and R_2 have the scaled values shown.

(a) Determine the transfer function (gain) V_2/V_1 in terms of r_1, r_2, and k.

(b) Determine expressions for r_1 and r_2.

(c) For what value of k is the gain 0?

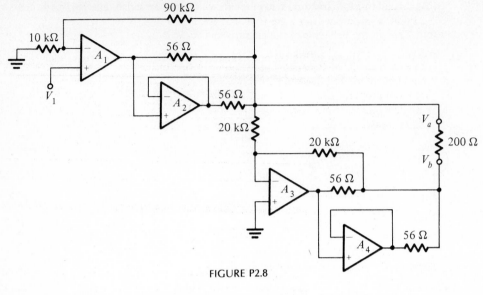

FIGURE P2.8

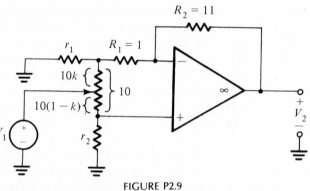

FIGURE P2.9

2.10 Piezoelectric tilt elements of the kind used to position laser beams and optical scanners can be aligned with an op-amp circuit which converts input voltages in x-y coordinates into a corresponding nonorthogonal (a, b, c) three-axis system as diagrammed in Fig. P2.10. The coordinate transformation must satisfy the relationships:

$$x = c - a, \qquad y = b - \tfrac{1}{2}(a + c), \qquad a + b + c = 0$$

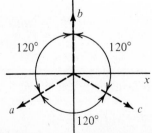

FIGURE P2.10

Design an op-amp circuit that accepts voltages x and y as inputs and yields voltages a, b, and c as outputs using the transformation:

$$a = c + x, \qquad b = -(a + c), \qquad c = -(x/2 + y/3)$$

Since it is intended that the design use a 10-kΩ R-net, all resistors in the design should be 10 kΩ, but these may be connected in series or in parallel. *Note 1:* R-nets are more properly known as "resistor arrays" and are available commercially from many companies. *Note 2:* If you need further information, see the Feb. 1, 1979 issue of *Electronics.*

2.11 For the circuit given in Fig. P2.11, show that

$$\frac{V_2}{V_1} = \frac{ab}{1 + b[(1 + a)/(1 + c)]}$$

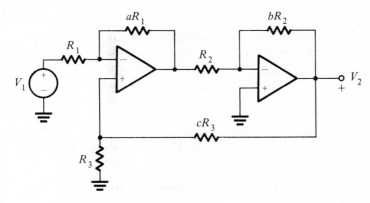

FIGURE P2.11

CHAPTER 3

Bilinear

Transfer

Functions

and

Frequency

Response

In this chapter there are two significant departures from our studies in Chapter 2. First, we now add the capacitor as a component to the resistor and the op amp. Second, beginning with this chapter we will assume that the circuits are driven by sinusoidal sources and that they are in the steady state. While we will remind ourselves from time to time of time-domain responses, the language of filter design is predominantly that of the frequency domain. This means that $v(t)$ and $i(t)$ will be replaced by the phasors V and I. In the last chapter we called the ratio v_2/v_1 the gain. Now we will refer to V_2/V_1 as the voltage ratio transfer function, and assign it the symbol T. We will also characterize the elements, alone and in combination, by their impedance $Z(s)$ or admittance $Y(s)$.

3.1 BILINEAR TRANSFER FUNCTIONS

In keeping with the changes described above, the voltage-divider circuit shown in Fig. 3.1 is characterized by impedances, phasor voltages, and current. Analysis follows the pattern of Chapter 2. Since

$$V_1 = (Z_1 + Z_2)I \quad \text{and} \quad V_2 = Z_2 I \tag{3.1}$$

then

$$\frac{V_2}{V_1} = T(s) = \frac{Z_2}{Z_1 + Z_2} \tag{3.2}$$

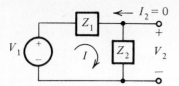

FIGURE 3.1

For the simple RC circuit of Fig. 3.2 we see that $Z_1 = R$ and $Z_2 = 1/Cs$, so that
the transfer function is

$$T(s) = \frac{1/RC}{s + 1/RC} \tag{3.3}$$

This particular transfer function is one member of a larger class of transfer functions, which we will next describe.

The familiar equation of a straight line in the form in which it is usually
taught in algebra courses is

$$y = mx + b \tag{3.4}$$

where x is the independent variable, y the dependent variable, m the slope, and b
the y intercept. We may recast this equation into a form better suited to our
study:

$$p = a_1 s + a_2 \tag{3.5}$$

This equation is described as being of first order because of the term s^1. When a
transfer function is the quotient of linear terms like Eq. (3.5), it is said to be *bilinear*. Thus a bilinear transfer function is of the form

$$T(s) = \frac{a_1 s + a_2}{b_1 s + b_2} = \frac{p(s)}{q(s)} \tag{3.6}$$

where the a and b coefficients are real constants, but may be either positive or
negative. Comparing this equation with Eq. (3.3), we see that the transfer function for the RC circuit of Fig. 3.2 is bilinear. In this chapter we study only circuits
with bilinear transfer functions.

If $T(s)$ in Eq. (3.6) is written in the form

$$T(s) = \frac{a_1}{b_1} \frac{s + a_2/a_1}{s + b_2/b_1} = K \frac{s + z_1}{s + p_1} \tag{3.7}$$

then z_1 is the *zero* of $T(s)$ and p_1 is the *pole* of $T(s)$. In the s plane these quantities
are located at

$$s = -z_1 \quad \text{and} \quad s = -p_1 \tag{3.8}$$

Since we have assumed that z_1 and p_1 are real, we see that the pole and zero of

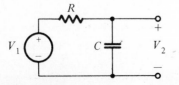

FIGURE 3.2

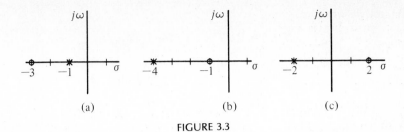

(a)   (b)   (c)

FIGURE 3.3

$T(s)$ are located on the real axis of the s plane. It will be the case that p_1 is always on the negative real axis, while z_1 may be on either the positive or the negative part of the real axis. Typical pole and zero locations in the s plane are shown in Fig. 3.3. These correspond to

$$T_a(s) = 5\,\frac{s+3}{s+1}, \qquad T_b(s) = 17\,\frac{s+1}{s+4}, \qquad T_c(s) = \frac{s-2}{s+2} \qquad (3.9)$$

It should be clear from these examples that the pole and zero locations do not specify the constant K.

If Eqs. (3.6) and (3.7) are the most general forms of a bilinear function, then we can quickly enumerate the possibilities for less general bilinear functions such as Eq. (3.3). These are shown in Table 3.1. It is a simple matter to sketch these pole and zero locations in the s plane.

3.2 PARTS OF $T(j\omega)$

In the introduction to this chapter we stated that the emphasis would be on circuits operating in the sinusoidal steady state. From elementary studies of circuits we know that this case corresponds to the condition $s = j\omega$. So Eq. (3.3), which describes the circuit of Fig. 3.2, is for the sinusoidal steady state

$$T(j\omega) = \frac{1}{RC}\,\frac{1}{j\omega + 1/RC} \qquad (3.10)$$

in which the denominator has a real and an imaginary part. Now the complex function $T(j\omega)$ may be written in either of two standard forms. In rectangular

TABLE 3.1   Bilinear functions

$T(s)$	Pole location	Zero location
$K_1 s$	$s = \infty$	$s = 0$
$K_2(s + z_1)$	$s = \infty$	$s = -z_1$
$\dfrac{K_3}{s}$	$s = 0$	$s = \infty$
$\dfrac{K_4}{s + p_1}$	$s = -p_1$	$s = \infty$
$\dfrac{K_5 s}{s + p_1}$	$s = -p_1$	$s = 0$

coordinates we write

$$T(j\omega) = \mathrm{Re}\ T(j\omega) + j\ \mathrm{Im}\ T(j\omega) \tag{3.11}$$

where Re T is the real part and Im T the imaginary part of $T(j\omega)$. In polar coordinates we express $T(j\omega)$ in terms of its magnitude and phase as follows:

$$T(j\omega) = |T(j\omega)| \angle\ \theta(j\omega) \tag{3.12}$$

Although we will need both forms, we will ordinarily deal in terms of magnitude and phase in the design of filters. The relationship between the real and imaginary parts and magnitude and phase is also familiar from elementary studies. It is

$$|T(j\omega)|^2 = [\mathrm{Re}\ T(j\omega)]^2 + [\mathrm{Im}\ T(j\omega)]^2 \tag{3.13}$$

and

$$\theta(j\omega) = \tan^{-1}\left|\frac{\mathrm{Im}\ T(j\omega)}{\mathrm{Re}\ T(j\omega)}\right| \tag{3.14}$$

We will find frequent need to plot the magnitude and phase of $T(j\omega)$ as a function of ω or f. We will make use of both linear and logarithmic scales. The logarithm of $T(j\omega)$ in Eq. (3.12) is

$$\log T(j\omega) = \log|T(j\omega)| + j\theta(j\omega) \tag{3.15}$$

We will make separate plots for the two parts of this equation. If the logarithm of the magnitude is multiplied by 20, then the unit of this quantity becomes the decibel (dB):

$$A(\omega) = 20 \log |T(j\omega)| \qquad \mathrm{dB} \tag{3.16}$$

The phase function is usually plotted in degrees, less frequently in radians. Thus the scales we will use most often are those depicted in Fig. 3.4.

Example 3.1   We may illustrate the computation of the parts of $T(j\omega)$ using Eq. (3.10). Let $1/RC = \omega_0$ so that

$$T(j\omega) = \frac{1}{1 + j\omega/\omega_0} \tag{3.17}$$

If we multiply both numerator and denominator by the conjugate of the denominator, we have

$$T(j\omega) = \frac{1 - j\omega/\omega_0}{1 + (\omega/\omega_0)^2} \tag{3.18}$$

From this we see that

$$\mathrm{Re}\ T(j\omega) = \frac{1}{1 + (\omega/\omega_0)^2} \tag{3.19}$$

and

$$\mathrm{Im}\ T(j\omega) = \frac{-\ \omega/\omega_0}{1 + (\omega/\omega_0)^2} \tag{3.20}$$

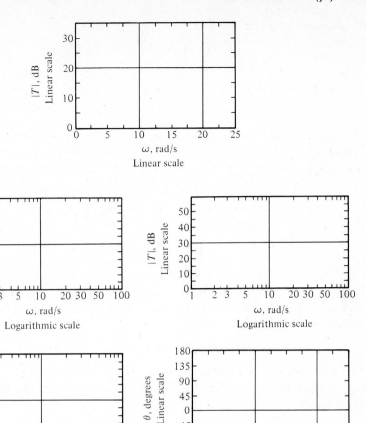

FIGURE 3.4   Various systems of coordinates to be used in showing magnitude and phase.

The magnitude function may be found directly from Eq. (3.17). It is

$$|T(j\omega)|^2 = \frac{1}{1 + (\omega/\omega_0)^2}$$ (3.21)

and the phase is

$$\theta = -\tan^{-1}\left(\frac{\omega}{\omega_0}\right)$$ (3.22)

This study of magnitude and phase serves to introduce the next section.

3.3 CLASSIFICATION OF MAGNITUDE RESPONSES

Of the choices of coordinates given in Fig. 3.4 we will select in this section that involving linear magnitude and linear frequency. For the *RC* circuit of Fig. 3.2 we have found that the magnitude squared transfer function is given by Eq.

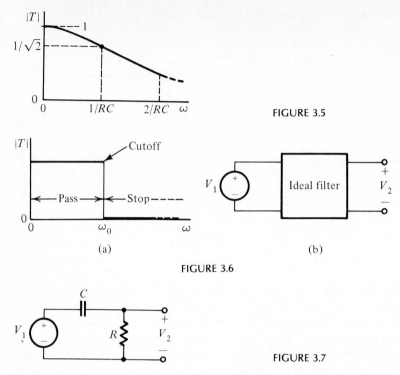

FIGURE 3.5

FIGURE 3.6

FIGURE 3.7

(3.21). Three frequencies are of special interest and correspond to these values of $|T|$:

$$|T(j0)| = 1$$

$$|T(j\omega_0)| = \frac{1}{\sqrt{2}} \approx 0.707 \tag{3.23}$$

$$|T(j\infty)| \to 0$$

The complete plot of $T(j\omega)$ is shown in Fig. 3.5. The frequency ω_0 is known as the *half-power frequency*. This plot is known as the magnitude response, meaning the response of magnitude as a function of frequency.

An idealization of the response of Fig. 3.5 is shown in Fig. 3.6a, sometimes known as a *brick wall*. The ideal filter which this response would describe if it existed has a behavior which we can visualize in terms of the circuit of Fig. 3.6b, where V_1 represents a sinusoidal voltage generator. As frequency increases in the generator, the output voltage V_2 remains fixed in amplitude until a critical value of frequency is reached, called the *cutoff frequency*, ω_0. At that frequency, and for all higher frequencies, the output V_2 is zero. The range of frequencies with output is called the *pass band*; the range with no output is called the *stop band*. The obvious classification of the filter is a *lowpass filter*. Now even though the response given by Eq. (3.21) and shown in the plot of Fig. 3.5 differs from the idealization of Fig. 3.5, it is known as a lowpass filter, and, by convention, the half-power frequency is taken as the cutoff frequency.

If the positions of the resistor and the capacitor in the circuit of Fig. 3.2 are interchanged, then the resulting circuit is that shown in Fig. 3.7. The transfer function may be found from Eq. (3.2) as

$$T(s) = \frac{R}{R + 1/Cs} = \frac{s}{s + 1/RC} \tag{3.24}$$

If we again let $1/RC = \omega_0$, then with $s = j\omega$ we obtain

$$T(j\omega) = \frac{j\omega/\omega_0}{1 + j\omega/\omega_0} \tag{3.25}$$

The corresponding magnitude function of this equation may be studied for the three frequencies, as in Eq. (3.23):

$$|T(j0)| = 0$$

$$|T(j\omega_0)| = \frac{1}{\sqrt{2}} \approx 0.707 \tag{3.26}$$

$$|T(j\infty)| = 1$$

The complete plot of the magnitude response for this circuit is shown in Fig. 3.8a and the idealized brick wall in Fig. 3.8b. Clearly, this filter is classified as a *highpass filter*, and the cutoff frequency is ω_0, as it was for the lowpass filter.

The next circuit we will study has a different appearance than those considered earlier, as shown in Fig. 3.9. In Fig. 3.9a it is classified as one section of a *lattice*. An equivalent form in which it may appear is shown in Fig. 3.9b, where it is known as a *bridge* circuit. We frequently become confused by how a circuit may be disguised through the form in which it is presented, but it is unusual that two forms are distinguished by two different names. Analysis will be carried out in terms of Fig. 3.9b. We observe that the output voltage is the difference of the voltages at nodes A and B,

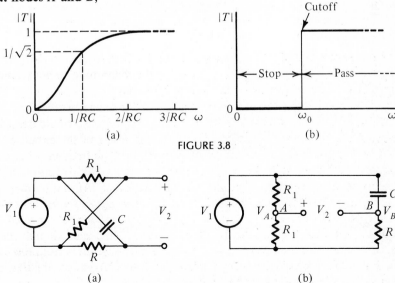

FIGURE 3.8

FIGURE 3.9

$$V_2 = V_A - V_B \tag{3.27}$$

Now clearly, with equal resistors,

$$V_A = \frac{V_1}{2} \tag{3.28}$$

The other voltage, V_B, comes directly from Eq. (3.24) of our last example so that

$$V_B = \frac{s}{s + 1/RC} V_1 \tag{3.29}$$

Substituting these two values into Eq. (3.27), we have

$$V_2 = \frac{1}{2} - \frac{s}{s + 1/RC} = -\frac{s - 1/RC}{s + 1/RC} V_1 \tag{3.30}$$

The pole–zero description of this equation is shown in Fig. 3.10, with the zero in the right half-plane and the pole in the left half-plane, on the real axis, and equal distances from the origin.

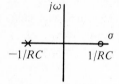

FIGURE 3.10

As before, we now let $s = j\omega$ and $1/RC = \omega_0$. Then the magnitude function becomes

$$|T(j\omega)| = \frac{|1 - j\,\omega/\omega_0|}{|1 + j\,\omega/\omega_0|} \tag{3.31}$$

Observe next that the magnitude of the numerator is identical with the magnitude of the denominator, and hence

$$|T(j\omega)| = 1 \qquad \text{for all } \omega \tag{3.32}$$

The only apt description of this characteristic, shown in Fig. 3.11, is *allpass*, and the concept of a cutoff frequency has no meaning. The circuit having this characteristic is known as an *allpass filter*.

The next simple circuit we will consider is that of Fig. 3.12, which has the voltage-divider impedances identified so that Eq. (3.2) applies. Direct substitution gives

$$T(s) = \frac{R_2}{R_2 + 1/(C_1 s + 1/R_1)} = \frac{s + 1/R_1 C_1}{s + 1/R_1 C_1 + 1/R_2 C_1} \tag{3.33}$$

FIGURE 3.11

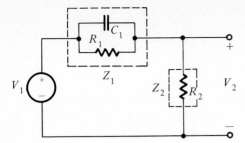

FIGURE 3.12

If we write this in the standard form

$$T(s) = \frac{s + z_1}{s + p_1} \tag{3.34}$$

then

$$z_1 = \frac{1}{R_1 C_1} \quad \text{and} \quad p_1 = z_1 + \frac{1}{R_2 C_1} \tag{3.35}$$

so that $p_1 > z_1$. The pole–zero locations for this transfer function are as shown in Fig. 3.13, with the zero always closer to the origin than the pole. From Eq. (3.34) it is clear that when s becomes very large, then T approaches the value of 1; while for $s = 0$, T has a value of z_1/p_1 which is less than 1; that is,

$$|T(j\infty)| = 1$$

$$|T(j0)| = \frac{z_1}{p_1} = \frac{R_2}{R_1 + R_2} \tag{3.36}$$

The magnitude response of this circuit is shown in Fig. 3.14. This does not approximate the ideal of Fig. 3.8b very well, but it is still known as a *highpass filter*.

The circuit of Fig. 3.15 may be analyzed using the voltage-divider equation,

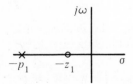

FIGURE 3.13

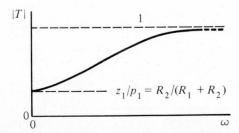

FIGURE 3.14

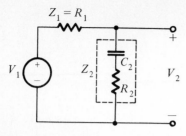

FIGURE 3.15

Eq. (3.2), to give

$$T(s) = \frac{Z_2}{Z_1 + Z_2} = \frac{R_2 + 1/C_2s}{R_1 + R_2 + 1/C_2s} \tag{3.37}$$

Algebraic manipulation permits this equation to be written in standard form as

$$T(s) = \frac{R_2}{R_1 + R_2} \frac{s + 1/R_2C_2}{s + 1/(R_1 + R_2)C_2} \tag{3.38}$$

or

$$T(s) = K \frac{s + z_1}{s + p_1} \tag{3.39}$$

For this transfer function the pole is always closer to the origin than the zero, as shown in Fig. 3.16; that is, $p_1 < z_1$. We gain insight to the magnitude response from the observation from Eq. (3.38)

$$|T(j0)| = 1$$

$$|T(j\infty)| = \frac{R_2}{R_1 + R_2} \tag{3.40}$$

The complete frequency response is shown in Fig. 3.17. Again, we compare with the kinds of responses studied thus far and see that this response is that of a *low-pass filter*.

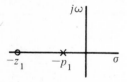

FIGURE 3.16

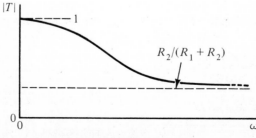

FIGURE 3.17

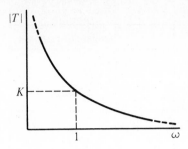

FIGURE 3.18

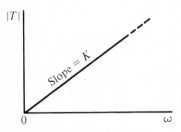

FIGURE 3.19

Two of the cases tabulated in Table 3.1 require mention. The transfer function

$$T(s) = \frac{K}{s} \tag{3.41}$$

is that of an *ideal integrator*. With $s = j\omega$, the magnitude function becomes

$$|T(j\omega)| = \frac{K}{\omega} \tag{3.42}$$

and the frequency response is that shown in Fig. 3.18. The transfer function

$$T(s) = Ks \tag{3.43}$$

is that of an *ideal differentiator*. Here

$$|T(j\omega)| = K\omega \tag{3.44}$$

The frequency response corresponding to this equation is the straight line shown in Fig. 3.19. In some sense, the integrator is a lowpass filter, and the differentiator a highpass filter.

The magnitude responses found thus far are summarized in Table 3.2.

3.4 CLASSIFICATION OF PHASE RESPONSES

If the input voltage to a circuit is

$$v_1(t) = V_1 \sin \omega t \tag{3.45}$$

TABLE 3.2

$T(s)$	Pole and zero	Magnitude response
$\dfrac{K_1}{s}$		
$K_2 s$		
$\dfrac{K_3}{s + p_1}$		
$K_4(s + z_1)$		
$K_5 \dfrac{s + z_1}{s + p_1}$		
$K_6 \dfrac{s}{s + p_1}$		
$K_7 \dfrac{s + z_1}{s + p_1}$		
$K_8 \dfrac{s - \sigma_1}{s + \sigma_1}$		

and the steady-state output is

$$v_2(t) = V_2 \sin(\omega t + \theta) \tag{3.46}$$

then we say that v_2 *leads* v_1 if θ is positive or *lags* v_1 if θ is negative. The sinusoidal variation of v_1 and v_2 for a negative or lagging angle θ is shown in Fig. 3.20a and for a positive or leading angle θ in Fig. 3.20b. In terms of the phasor representation of the sinusoids, V_1 is chosen as the reference, and so

$$V_1 = |V_1| \angle 0° \tag{3.47}$$

while the phasor representation of Eq. (3.46) is

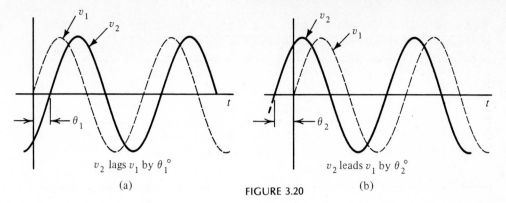

v_2 lags v_1 by θ_1°

(a)

FIGURE 3.20

v_2 leads v_1 by θ_2°

(b)

$$V_2 = |V_2| \angle \theta^{\circ} \qquad (3.48)$$

Hence we see that

$$T(j\omega) = \frac{|V_2| \angle \theta}{|V_1| \angle 0} = |T| \angle \theta \qquad (3.49)$$

so that the phase angle of $T(j\omega)$ is the phase difference between v_2 and v_1, with v_1 chosen as the reference. The phasor interpretation of these two voltages is shown in Fig. 3.21.

For the bilinear transfer function

$$T(s) = K \frac{s + z_1}{s + p_1} \qquad (3.50)$$

letting $s = j\omega$ gives

$$T(j\omega) = K \frac{z_1 + j\omega}{p_1 + j\omega} \qquad (3.51)$$

Then the phase of $T(j\omega)$ is θ where

$$\theta = \text{phase of } K + \text{phase of } (z_1 + j\omega) - \text{phase of } (p_1 + j\omega) \qquad (3.52)$$

If K is positive, its phase is $0°$, and if negative it is $180°$. Assuming that it is positive, we see that from Eq. (3.14)

$$\theta = \tan^{-1}\left(\frac{\omega}{z_1}\right) - \tan^{-1}\left(\frac{\omega}{p_1}\right) \qquad (3.53)$$

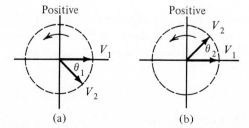

(a)

(b)

FIGURE 3.21

We use the sign of this phase angle θ as a means for classifying circuits. Those giving positive θ are known as *lead circuits*, those giving negative θ as *lag circuits*.

To apply this concept to specific circuits, we begin with the RC circuit of Fig. 3.2 for which $T(s)$ is given by Eq. (3.3). From the last result, Eq. (3.53), we see that

$$\theta = -\tan^{-1} \omega RC \qquad (3.54)$$

as shown in Fig. 3.22. Since θ is negative for all ω, this circuit is a lag circuit. It is useful to observe that when $\omega = 1/RC$, then $\theta = -\tan^{-1} 1 = -45°$.

The next circuit that we considered was given in Fig. 3.7 for which $T(j\omega)$ is given by Eq. (3.25). Applying Eq. (3.53) to Eq. (3.25), we find that the phase is

$$\theta = 90° - \tan^{-1} \omega RC \qquad (3.55)$$

This result is very similar to Eq. (3.54) with a constant angle of 90° added. Then the phase response is the same as that shown in Fig. 3.22 with the angle coordinates shifted upward by 90°, as shown in Fig. 3.23. Clearly the angle θ is positive for all values of ω, and so the circuit is a lead circuit.

The lattice circuit which gave the all-pass magnitude response is shown in Fig. 3.9 and the $T(j\omega)$ is given by Eq. (3.31). The application of Eq. (3.53) to Eq. (3.31) gives the same phase characteristic for numerator and denominator, and so

$$\theta = -2 \tan^{-1} \omega RC \qquad (3.56)$$

This phase is negative for all ω, and so the circuit is a lag circuit. The phase response is that shown in Fig. 3.22 with the angle coordinate multiplied by 2. This is shown in Fig. 3.24.

The circuits of Figs. 3.12 and 3.15 are similar in that they are both characterized by the general bilinear transfer function of Eq. (3.51) and the corresponding $\theta(j\omega)$ of Eq. (3.53). From this equation we see that θ is characterized as the difference of two angles, the first a function of the zero, the second a function of the pole. Let us write this in the form of Eq. (3.53) as

$$\theta = \theta_z + \theta_p \qquad (3.57)$$

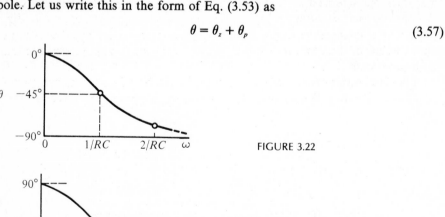

FIGURE 3.22

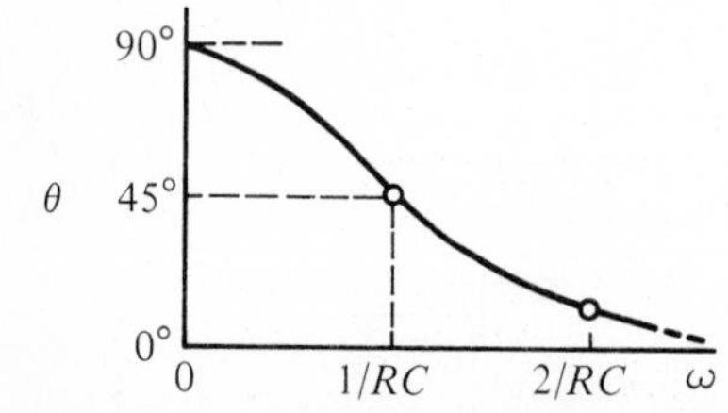

FIGURE 3.23

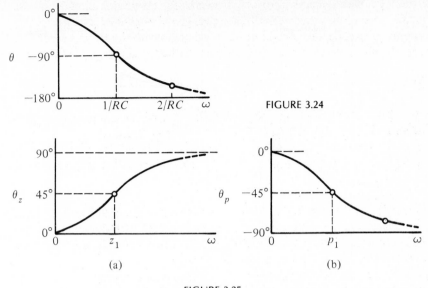

FIGURE 3.24

FIGURE 3.25

The inverse tangent form of these two phase angle functions is shown in Fig. 3.25 for θ_z and $-\theta_p$. Returning now to the two circuits under consideration, we recall that for the circuit of Fig. 3.12, $z_1 < p_1$, while for the circuit of Fig. 3.15, $p_1 < z_1$, as depicted in Fig. 3.26. This figure also reminds us that the two frequencies $\omega = z_1$ and $\omega = p_1$ are important because the phase angle has the value of $\pm 45°$ under these conditions.

The implication of the condition $z_1 < p_1$ is shown in Fig. 3.27a. The phase function θ_z reaches $+45°$ at a low frequency, while θ_p reaches $-45°$ for a higher frequency. Since θ for the circuit is the sum of θ_z and $-\theta_p$, the net phase is positive and is thus characterized as lead. Thus the circuit of Fig. 3.12 provides phase lead. The opposite situation, $p_1 < z_1$, leads to the consequences shown in Fig. 3.27b, giving a net negative angle, or phase lag. So the circuit of Fig. 3.15 is a lag circuit. Circuits of these two forms find frequent application in the compensation of control systems.

In the last section we considered the transfer function for an integrator given by Eq. (3.41), $T = K/s$. With $s = j\omega$, the phase of the function $T = -jK/\omega$ is $-90°$

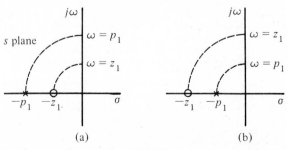

FIGURE 3.26

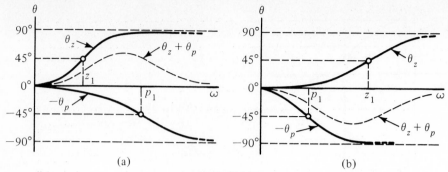

FIGURE 3.27

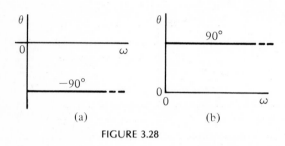

FIGURE 3.28

for all values of ω. Hence for the integrator

$$\theta = -90° \qquad \text{for all } \omega \tag{3.58}$$

A similar analysis of the differentiator transfer function of Eq. (3.43) leads to the conclusion that

$$\theta = +90° \qquad \text{for all } \omega \tag{3.59}$$

Hence the integrator is a lag circuit, while the differentiator is a lead circuit. These two phase characteristics are shown in Fig. 3.28.

Table 3.3 extends the summary of Table 3.2 to include the phase response information we have just found.

3.5 BODE PLOTS

The general bilinear $T(j\omega)$ from Eq. (3.51) is

$$T(j\omega) = K \frac{z_1 + j\omega}{p_1 + j\omega} = K \frac{z_1}{p_1} \frac{1 + j\omega/z_1}{1 + j\omega/p_1} \tag{3.60}$$

We are interested in plotting the magnitude and phase of $T(j\omega)$

$$T(j\omega) = |T(j\omega)| \angle \theta(j\omega) \tag{3.61}$$

as a function of frequency ω. If we plot the magnitude

$$A = 20 \log |T(j\omega)| \quad \text{dB} \tag{3.62}$$

as a function of ω with logarithmic coordinates, then the plot is known as a Bode

TABLE 3.3

$T_n(s)$*	Pole and zero	Magnitude response	Phase response
$\dfrac{K_1}{s}$	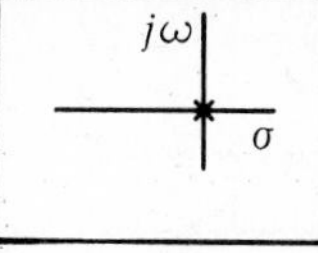	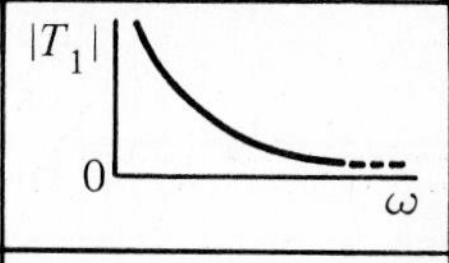	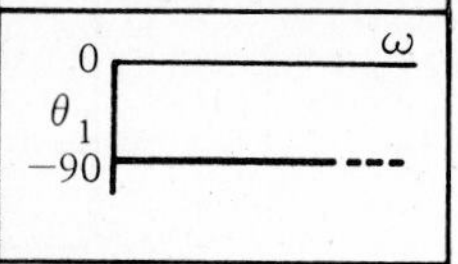
$K_2 s$	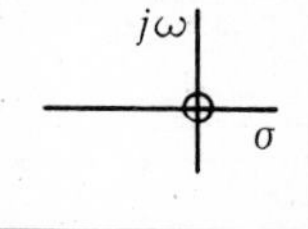	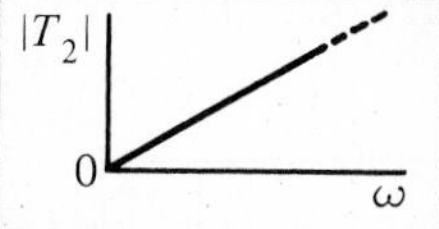	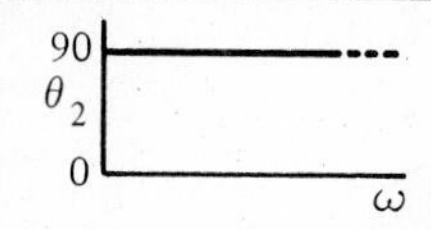
$\dfrac{K_3}{s + p_1}$	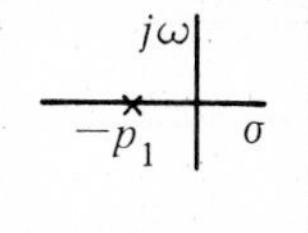	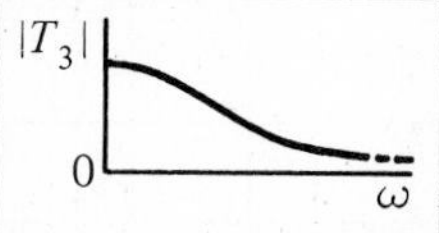	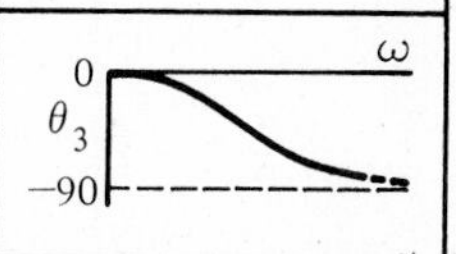
$K_4(s + z_1)$	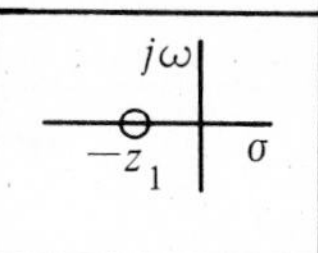	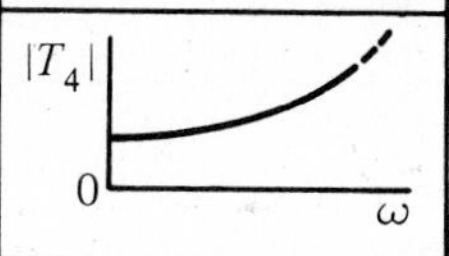	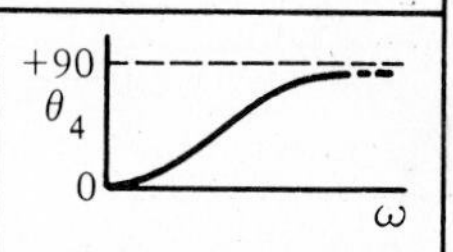
$K_5 \dfrac{s + z_1}{s + p_1}$	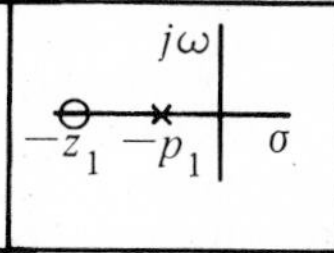	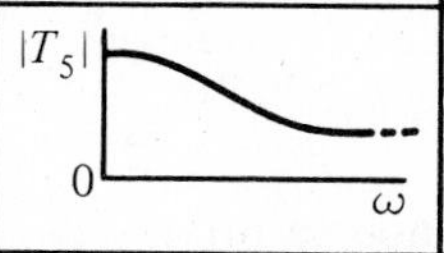	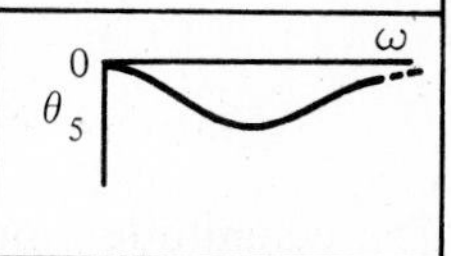
$K_6 \dfrac{s}{s + p_1}$	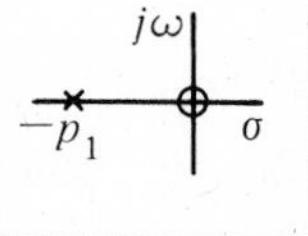	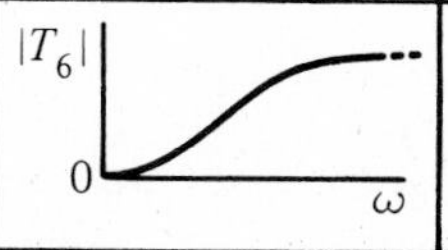	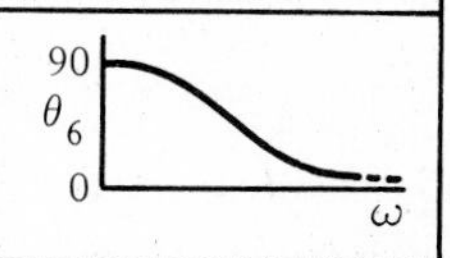
$K_7 \dfrac{s + z_1}{s + p_1}$	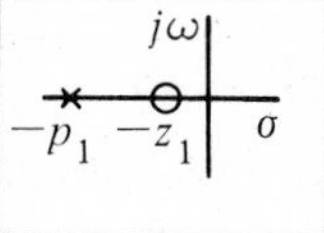	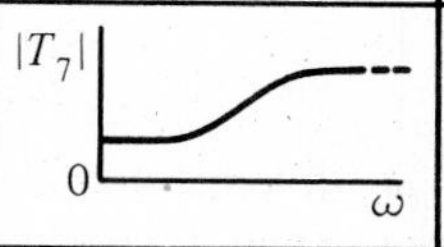	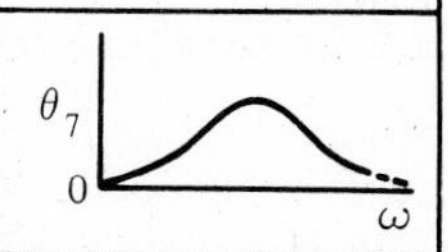
$K_8 \dfrac{s - \sigma_1}{s + \sigma_1}$	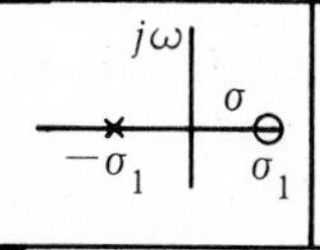	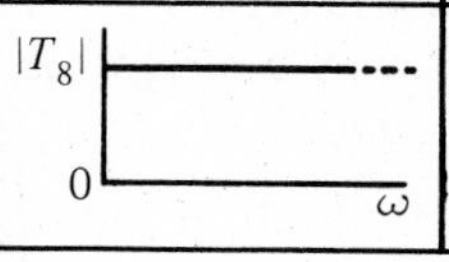	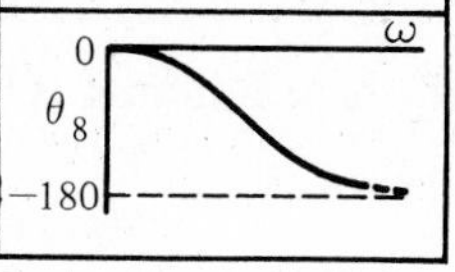

*All K_j are assumed positive.

plot after Hendrik Bode.* The plot of θ as a function of ω with the same logarithmic coordinates is also known as a Bode plot; the two together are the magnitude and phase Bode plots. This coordinate system is illustrated in Fig. 3.29. The origin of the coordinate system is difficult to define since the position of 0 on the logarithmic frequency scale is an infinite distance to the left. When $|T(j\omega)| = 1$, then $A = 0$ dB.

---

* Hendrik Bode grew up in Urbana, Illinois, where we pronounce his name *boh dee*. Purists insist on the original Dutch *boh dah*. No one uses *bohd*. Dr. Bode spent most of his professional life at Bell Laboratories and, until his second retirement, was a professor at Harvard University.

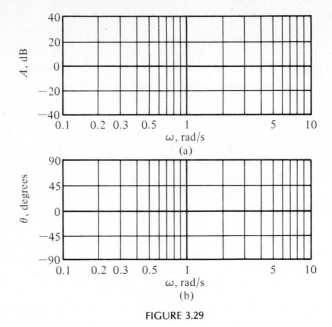

FIGURE 3.29

The magnitude function of Eq. (3.61) is, from Eq. (3.60),

$$|T(j\omega)| = \left| K \frac{z_1}{p_1} \right| \frac{|1 + j\omega/z_1|}{|1 + j\omega/p_1|} \tag{3.63}$$

From this we see that $A(\omega)$ is

$$A(\omega) = 20 \log \left| K \frac{z_1}{p_1} \right| + 20 \log \left| 1 + j \frac{\omega}{z_1} \right| - 20 \log \left| 1 + j \frac{\omega}{p_1} \right| \quad \text{dB} \tag{3.64}$$

and the phase function is

$$\theta(j\omega) = 0 \text{ or } 180° + \tan^{-1} \left(\frac{\omega}{z_1} \right) - \tan^{-1} \left(\frac{\omega}{p_1} \right) \tag{3.65}$$

The first term in Eq. (3.64) is a constant, while the second and third terms are similar, except for the sign. We will first study the second term and then consider the others in Eq. (3.64). It is

$$A_2(\omega) = 20 \log \left| 1 + j \frac{\omega}{z_1} \right| \tag{3.66}$$

or

$$A_2(\omega) = 20 \log \sqrt{1 + \left(\frac{\omega}{z_1} \right)^2} = 10 \log \left[1 + \left(\frac{\omega}{z_1} \right)^2 \right] \tag{3.67}$$

We gain insight into the form of $A_2(\omega)$ if we consider the low-frequency and high-frequency asymptotes. When $\omega \ll z_1$, then

$$A_2(\omega) = 10 \log 1 = 0 \text{ dB} \qquad (3.68)$$

At high frequencies the 1 in Eq. (3.67) can be neglected, and we obtain

$$A_2(\omega) = 20 \log \left| \frac{\omega}{z_1} \right| \qquad (3.69)$$

This is the equation of a straight line in Bode coordinates. To describe this line, we make a number of observations. First, we see that $A_2 = 0$ when $\omega = z_1$. We also see that the straight line has a positive slope. What is this slope, and how can it be described? Figure 3.30 shows two points on the frequency coordinate, marked ω_1 and ω_2. The corresponding points on a linear scale are designated as u_1 and u_2. The relationship between the linear points and the values of frequency is

$$u_1 = \log \omega_1 \qquad \text{and} \qquad u_2 = \log \omega_2 \qquad (3.70)$$

The linear distance between u_1 and u_2 is

$$u_2 - u_1 = \log \omega_2 - \log \omega_1 = \log \left| \frac{\omega_2}{\omega_1} \right| \qquad (3.71)$$

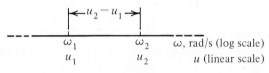

FIGURE 3.30

There are two common ratios in this equation. If $\omega_2 = 2\omega_1$, then the two frequencies are said to be separated by an *octave*, while if $\omega_2 = 10\omega_1$, then the two frequencies are separated by a *decade*. If the distance $u_2 - u_1$ corresponds to an octave, then this distance will define an octave along the entire ω scale. Thus if

$$\omega_2 = 2^n \omega_1 \qquad (3.72)$$

is substituted into Eq. (3.71), then the distance from u_1 to u_n is

$$u_n - u_1 = n \log \left| \frac{\omega_2}{\omega_1} \right| = n(u_2 - u_1) \qquad (3.73)$$

and the distance of n octaves has been described. We see that the appropriate unit for linear distance in the ω direction of the Bode plot is either the octave or the decade.

Going back to Eq. (3.69), we see that if $\omega_1 = z_1$ and $\omega_2 = 2z_1$, then A_2 will have increased from 0 dB to 6.0206 dB. It is conventional to round this down to 6 dB. If ω_2 is equal to $10z_1$, then A_2 will increase by 20 dB. Thus we see that the slope of the straight line of Eq. (3.69) can be described as either

$$6 \frac{\text{dB}}{\text{octave}} \qquad \text{or} \qquad 20 \frac{\text{dB}}{\text{decade}} \qquad (3.74)$$

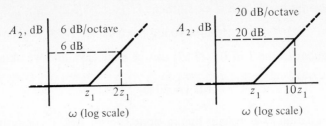

FIGURE 3.31

The line that we just described is shown in Fig. 3.31. We observe that if the low-frequency asymptote of 0 dB is extended, it intersects the high-frequency asymptote at the frequency $\omega = z_1$. These two asymptotes constitute the asymptotic plot of Eq. (3.66).

For many applications the asymptotic plot alone will suffice. If we require accuracy, then a plot may be made using Eq. (3.67) or its inverse

$$\omega = z_1 \sqrt{10^{A_2/10} - 1} \qquad (3.75)$$

Both equations are well suited for use of the calculator. However, for a simple though approximate method of plotting the actual curve, the following steps may be taken:

1. Plot the straight-line asymptotes, as in Fig. 3.31.
2. The difference between the actual and the asymptotic curves will be as follows:
   a. At $\dot{z}_1$ (called the *break frequency*) the difference is 3 dB.
   b. One octave above and one octave below the break frequency, the difference is 1 dB.

These three points allow you to rough in the actual response with fair accuracy. This is summarized in Fig. 3.32.

Now let us return to Eq. (3.64) and compare the second and third terms. The second involves z_1, while the third involves p_1, but these are simply break frequencies as identified in Fig. 3.32. The significant difference is the sign, which is negative. This simply means that the slope of the high-frequency asymptotic curve will be negative, as shown in Fig. 3.33. Otherwise everything that has been described for the second term applies to the third.

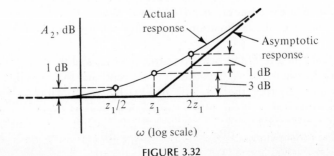

FIGURE 3.32

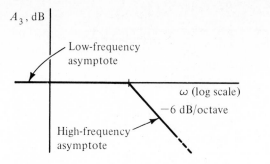

A_3, dB

Low-frequency
asymptote

ω (log scale)

-6 dB/octave

High-frequency
asymptote

FIGURE 3.33

Example 3.2 As an example of the bilinear transfer function that we have been describing, consider

$$T(s) = 6 \frac{s + 0.5}{s + 3} \tag{3.76}$$

Let $s = j\omega$ so that

$$T(j\omega) = \frac{1 + j\omega/0.5}{1 + j\omega/3} \tag{3.77}$$

Written in this form, we recognize that $z_1 = 0.5$ and $p_1 = 3$. The numerator and denominator factors are shown in asymptotic form in Fig. 3.34. Adding these two gives the total response in both asymptotic and actual forms, as identified in the figure. The corresponding phase plots are shown in Fig. 3.34c. These are also added. From the plots we conclude that the $T(s)$ of this example represents a highpass phase-lead circuit.

Example 3.3 As our next example we invert Eq. (3.76) as well as Eq. (3.77) so that

$$T(j\omega) = \frac{1 + j\omega/3}{1 + j\omega/0.5} \tag{3.78}$$

Here $z_1 = 3$ and $p_1 = 0.5$. The Bode plot for the magnitude factors is shown in Fig. 3.35a and b and for the phase factors in Fig. 3.35c. We conclude that this $T(s)$ represents a lowpass circuit which provides phase lag.

There remain the transfer functions with a single zero or pole at the origin, those representing the differentiator and the integrator. In addition, such factors may appear in other transfer functions. Consider a differentiator with the transfer function

$$T(s) = K_d s \tag{3.79}$$

For this $T(s)$ we calculate

$$A(\omega) = 20 \log \left| \frac{\omega}{1/K_d} \right| \tag{3.80}$$

This equation is the same as Eq. (3.69) given for A_2, with $1/K_d$ replacing z_1 and valid for all ω rather than only for high frequencies. This is shown in Fig. 3.36a. Note that $A = 0$ when $\omega = 1/K_d$.

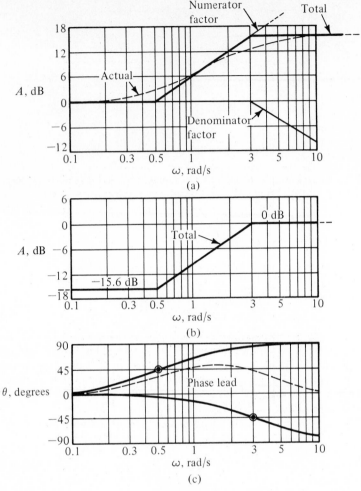

FIGURE 3.34

The transfer function of an integrator may be written

$$T(s) = \frac{1}{K_i s} \tag{3.81}$$

and for this function

$$A(\omega) = -20 \log \left| \frac{\omega}{1/K_i} \right| \tag{3.82}$$

This equation is identical with Eq. (3.80), except for the sign which indicates that the slope is negative rather than positive as it was for the differentiator. As before, $A = 0$ dB when $\omega = 1/K_i$, and the Bode plot is shown in Fig. 3.36b.

In summary, compared with linear plots, Bode plots offer a number of features that will be important in filter design. These include:

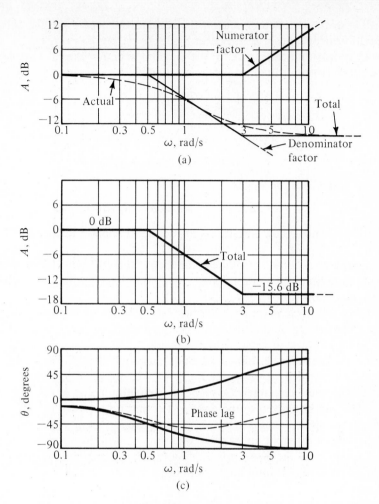

FIGURE 3.35

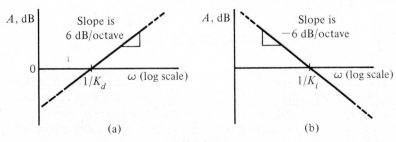

FIGURE 3.36

1. With frequency plotted on a logarithmic scale, the octave and the decade are equal linear distances, as shown in Fig. 3.37.
2. Using a logarithmic measure for the magnitude of T makes it possible to add and subtract rather than multiply and divide. For example, if

$$T = \frac{T_1 T_2 T_3}{T_4} \tag{3.83}$$

then

$$A = A_1 + A_2 + A_3 - A_4 \tag{3.84}$$

just as is the case for angles,

$$\theta = \theta_1 + \theta_2 + \theta_3 - \theta_4 \tag{3.85}$$

Figure 3.38 shows how simple it is to add lines to obtain segments of lines for $A_t = A_1 + A_2 + A_3$.

3. The slope of all lines for bilinear functions is ± 6 dB per octave or ± 20 dB per decade. In the general case all asymptotic lines are integer multiples of these two numbers.
4. In many applications we deal only with the asymptotic plots without "filling in the corners" to obtain the actual plots. In some sense the asymptotic plots are a shorthand with which you can sketch out your ideas in design.
5. Because of item 1, the shape of a Bode magnitude plot is maintained when frequency is scaled, as shown by Fig. 3.39. This makes it possible to use templates with these characteristic shapes for magnitude and phase such that the template can be moved to the right or left and up or down.
6. The phase angle plots for first-order factors are shown in Figs. 3.34 and 3.35. These have a different appearance than the linear plots, but there is nothing distinctive about them. As indicated by Eq. (3.85), angle plots are added and subtracted to obtain total angle response.

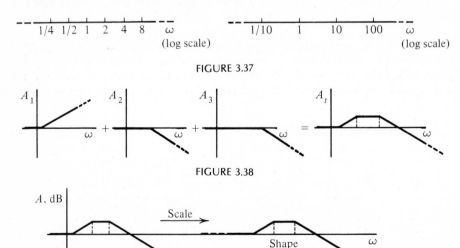

1/4 1/2 1 2 4 8 ω
(log scale)

1/10   1   10   100 ω
(log scale)

FIGURE 3.37

A_1 A_2 A_3 A_t

FIGURE 3.38

A, dB

Scale

Shape
maintained

ω
(log scale)

FIGURE 3.39

3.6 AND NOW DESIGN

We are now prepared to introduce some of the aspects of circuit design which will be amplified throughout the book. We do so in terms of some of the circuits analyzed earlier in this chapter. The RC circuit given in Fig. 3.2 and shown again in Fig. 3.40a is a lowpass lag filter having the transfer function

$$T(s) = \frac{1/RC}{s + 1/RC} = \frac{p_1}{s + p_1} \tag{3.86}$$

If a design is specified in terms of the half-power frequency p_1, then the design equation becomes

$$p_1 = \frac{1}{RC} \tag{3.87}$$

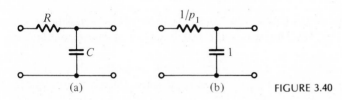

(a)   (b)   FIGURE 3.40

This equation represents a situation that is typical of design. We have one equation and two unknowns, the values of R and C. The approach which will become standard is to choose one and so determine the other. We let $C = 1$ F; it is then required that

$$R = \frac{1}{p_1} \tag{3.88}$$

These design values are given in Fig. 3.40b. To see how this works, we consider an example.

Example 3.4 We are required to design a lowpass filter with a half-power frequency of 1000 rad/s. This means that the frequency response on Bode coordinates is as shown in Fig. 3.41, and also that $p_1 = 1000$. Then the circuit elements are determined as $R = 1/1000$ and $C = 1$. These are not very practical values, and so we next resort to magnitude scaling

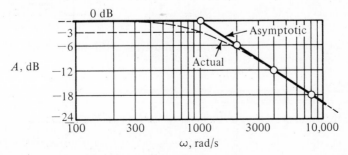

FIGURE 3.41

as described in Appendix A. For magnitude scaling the appropriate equations are Eqs. (A.15) and (A.17):

$$C_{\text{new}} = \frac{1}{k_m} C_{\text{old}} \qquad (3.89)$$

and

$$R_{\text{new}} = k_m R_{\text{old}} \qquad (3.90)$$

Any value of k_m may be selected; a good one is $k_m = 10^6$, giving the values of 1 kΩ and 1 μF shown in Fig. 3.42.

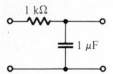

FIGURE 3.42

A second circuit to which we will apply this design strategy is the highpass phase-lead circuit of Fig. 3.12, for which the following set of design equations were given as Eq. (3.35):

$$z_1 = \frac{1}{R_1 C_1} \qquad \text{and} \qquad p_1 = z_1 + \frac{1}{R_2 C_1} \qquad (3.91)$$

Rearranging these equations,

$$R_1 C_1 = \frac{1}{z_1} \qquad (3.92)$$

and

$$R_2 C_1 = \frac{1}{p_1 - z_1} \qquad (3.93)$$

An obvious choice for the element value to select at this point is $C_1 = 1$ F. Once this choice is made, then the other two design equations become

$$R_1 = \frac{1}{z_1} \qquad (3.94)$$

and

$$R_2 = \frac{1}{p_1 - z_1} \qquad (3.95)$$

These are shown in Fig. 3.43. Again, we illustrate with an example.

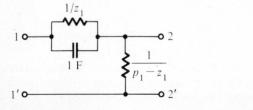

FIGURE 3.43

Example 3.5 Our design is based on the information given in Example 3.2, except that the desired break frequencies are 500 and 3000 rad/s. This difference can be accommodated by frequency scaling. If we substitute the values $z_1 = 0.5$ and $p_1 = 3$, into Eqs. (3.94) and (3.95), we obtain these design values:

$$C_1 = 1 \text{ F}, \qquad R_1 = 2 \text{ } \Omega, \qquad R_2 = \frac{2}{5} \text{ } \Omega \qquad\qquad (3.96)$$

The scaling equations

$$C_{\text{new}} = \frac{1}{k_f k_m} C_{\text{old}} \qquad\qquad (3.97)$$

and

$$R_{\text{new}} = k_m R_{\text{old}} \qquad\qquad (3.98)$$

will be used to complete the design. The requirements of the problem dictate that $k_f = 1000$. The choice of $k_m = 1000$ gives the following element values:

$$C_1 = 1 \text{ } \mu\text{F}, \qquad R_1 = 2 \text{ } k\Omega, \qquad R_2 = 400 \text{ } \Omega \qquad\qquad (3.99)$$

which are in a reasonable and practical range. These are shown in Fig. 3.44.

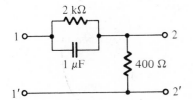

FIGURE 3.44

Example 3.6 We require a circuit that will provide 45° of phase lead at a frequency of 1000 rad/s. First we take advantage of lessons learned from the last two examples and design for $\omega = 1$, with later frequency scaling to meet the specifications. The phase relationship for a bilinear circuit is given by Eq. (3.53), which may also be written in the form

$$\theta = \tan^{-1} \left[\frac{\omega(p_1 - z_1)}{\omega^2 + p_1 z_1} \right] \qquad\qquad (3.100)$$

using a trigonometric identity. Since we require that $\theta = 45°$ at $\omega = 1$, this equation becomes

$$\frac{p_1 - z_1}{1 + p_1 z_1} = 1 \qquad\qquad (3.101)$$

which is equivalent to

$$p_1 - z_1 = p_1 z_1 + 1 \qquad\qquad (3.102)$$

Again we have a design problem with one equation and two unknowns, which we resolve by selecting a value for either p_1 or z_1 to see if a reasonable value results for the other quantity. If we let $p_1 = 5$, then Eq. (3.102) gives $z_1 = \frac{2}{3}$. This gives us the element values

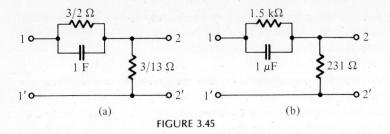

FIGURE 3.45

shown in Fig. 3.45a. We are now ready to complete the design by scaling. From the statement of the problem we know that $k_f = 1000$. One of the possible choices for k_m is 1000. This scaling gives the circuit shown in Fig. 3.45b.

The application of these techniques to the other circuits of this chapter is summarized in Table 3.4.

The design procedure described may be summarized in the following steps:

Specification Some aspect of the magnitude or phase response that is required must be given in order that the pole and zero locations may be found. For example, if a Bode plot is given, then the break frequencies and so the pole and zero locations can be found by a cut-and-try procedure.

TABLE 3.4

Pole and zero	$T(s)$	Circuit	Classification
$jω$ ✕ ⊙ $σ$	$\dfrac{s+z_1}{s+p_1}$ $p_1 > z_1$	$V_1 \circ$ —[$1/z_1$]— / $1\,F$ / $\dfrac{1}{p_1 - z_1}$ — $\circ V_2$	Lead highpass
$jω$ ✕ ⊕ $σ$	$\dfrac{s}{s+p_1},$ $p_1 = \dfrac{1}{RC}$	$V_1 \circ$ —[1]— / $1/p_1$ — $\circ V_2$	Lead highpass
$jω$ ⊙ ✕ $σ$	$T_{hi}\dfrac{s+z_1}{s+p_1},$ $z_1 > p_1$	$V_1 \circ$ —$1/p_1 - 1/z_1$— / 1 / $1/z_1$ — $\circ V_2$	Lag lowpass
$jω$ ✕ $σ$	$\dfrac{p_1}{s+p_1},$ $p_1 = \dfrac{1}{RC}$	$V_1 \circ$ —$R = 1/p_1$— / 1 — $\circ V_2$	Lag lowpass

Required   We require the complete circuit with element values in a practical range.

Procedure   The following six steps are suggested.

1. Select a circuit that seems to satisfy the requirements of magnitude or phase variation with frequency.
2. Proceed from the specifications to determine pole and zero locations.
3. Use scaling if desired to reduce the pole and zero values to small integer values.
4. Determine element values from the values of the poles and zeros.
5. Scale: frequency scale to meet specifications, magnitude scale to give convenient and practical element values.
6. Since there are never unique solutions in design, it will usually be necessary to repeat the process to see the range of possibilities available.

PROBLEMS

3.1 Prepare an asymptotic Bode plot for both magnitude and phase for the following transfer functions. In making the plot, it is useful to make use of four- or five-cycle semilog paper.

(a) $T(s) = 1000 \dfrac{(1 + 0.25s)(1 + 0.1s)}{(1 + s)(1 + 0.025s)}$

(b) $T(s) = \dfrac{(1 + 0.1s)(1 + 0.01s)}{(1 + s)(1 + 0.001s)}$

(c) $T(s) = \dfrac{1000s}{(1 + 0.01s)(1 + 0.0025s)}$

(d) $T(s) = 180 \dfrac{s(1 + 0.01s)}{(1 + 0.05s)(1 + 0.001s)}$

(e) $T(s) = 50 \dfrac{(1 + 0.025s)}{s(1 + 0.05s)}$

(f) $T(s) = \dfrac{100}{s(1 + 0.01s)(1 + 0.001s)}$

(g) $T(s) = 1000 \dfrac{s^2}{(1 + 0.17s)(1 + 0.53s)}$

3.2 Figure P3.2 shows only the asymptotes of a Bode magnitude plot. The characteristic response is used for gain enhancement—increasing the gain over a band of frequencies, but not changing either high- or low-frequency behavior. For this response, determine $T(s)$, evaluating all constants.

FIGURE P3.2

3.3 Repeat Problem 3.2 for the Bode magnitude plot shown in Fig. P3.3.

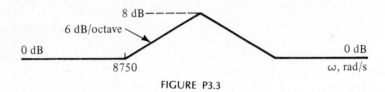

FIGURE P3.3

3.4 The bandpass response shown in Fig. P3.4 is given as a specification. As the first step in the design of the figure, determine $T(s)$, evaluating all constants.

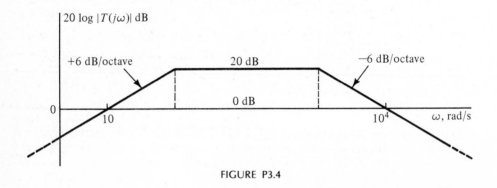

FIGURE P3.4

3.5 The problem is to design an amplifier-filter having the Bode asymptotic plot shown in Fig. P3.5. Determine $T(s)$, evaluating all constants.

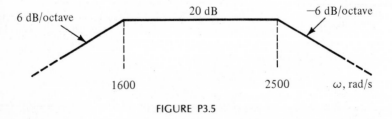

FIGURE P3.5

3.6 Given the Bode asymptotic plot for a notch filter as shown in Fig. P3.6, determine $T(s)$, evaluating all constants.

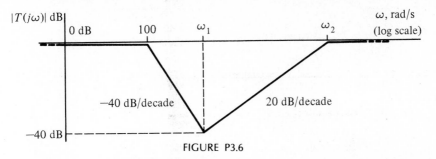

FIGURE P3.6

3.7 The transfer function corresponding to the asymptotic Bode plot shown in Fig. P3.7 has only real poles and zeros. Construct $T(s)$, determining all constants.

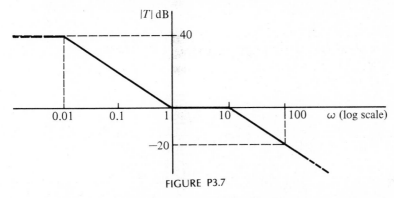

FIGURE P3.7

3.8 The asymptotic Bode plot shown in Fig. P3.8a represents a filter-amplifier having a break frequency of 1000 rad/s. Determine a transfer function $T_0(s)$ which when multiplied by $T(s)$ of Fig. P3.8a gives that specified in Fig. P3.8b.

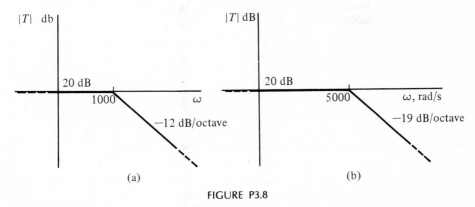

FIGURE P3.8

3.9 Figure P3.9 shows the asymptotes of a Bode plot.
  (a)   Determine the half-power frequency for this plot.
  (b)   Determine $T(s)$, evaluating all constants.

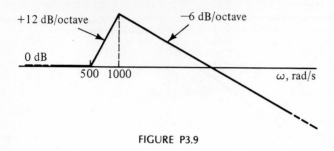

FIGURE P3.9

CHAPTER 4

Cascade

Design

with

First-Order

Circuits

In this chapter we study techniques for connecting first-order circuits in cascade such that the overall transfer function is related to the transfer functions of the individual circuits by the chain rule

$$T = T_1T_2T_3 \cdots \tag{4.1}$$

We first examine the conditions under which the chain rule is valid, and then tabulate first-order op-amp circuits that will be useful in design. This will differ from the last chapter where all realizations were passive.

4.1 WHEN MAY WE CASCADE CIRCUITS?

Figure 4.1a shows a ladder circuit made up of two resistors and two capacitors. If this circuit is routinely analyzed, the transfer function is found to be

$$\frac{V_2}{V_1} = \frac{(1/RC)^2}{s^2 + (3/RC)s + (1/RC)^2} \tag{4.2}$$

Now suppose that we break the ladder in half and insert a unit-gain voltage-controlled voltage source V_A, as shown in Fig. 4.1b. This will cause the two circuits to be isolated in the sense that the current $I_A = 0$; the second circuit does not "load" the first. Under this condition we see that the transfer function of the first section is given by Eq. (3.3):

$$T_1 = \frac{V_A}{V_1} = \frac{1/RC}{s + 1/RC} \tag{4.3}$$

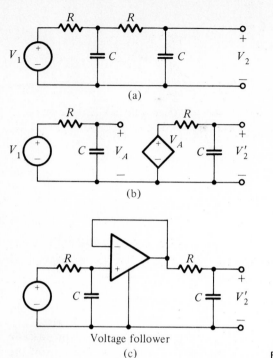

(a)

(b)

Voltage follower
(c)

FIGURE 4.1

Similarly, the transfer function of the second section is

$$T_2 = \frac{V_2'}{V_A} = \frac{1/RC}{s + 1/RC} \tag{4.4}$$

The product of these two transfer functions is

$$\frac{V_A}{V_1} \frac{V_2'}{V_A} = \frac{V_2'}{V_1} = T_1 T_2 = T \tag{4.5}$$

Multiplying the two transfer functions together, we have

$$T_1 T_2 = \frac{1/RC}{s + 1/RC} \frac{1/RC}{s + 1/RC} = \frac{(1/RC)^2}{s^2 + (2/RC)s + (1/RC)^2} \tag{4.6}$$

First we observe that this equation differs from Eq. (4.2) for the circuit without isolation of the two sections. Then we see that a practical realization of the volt-age-controlled source of Fig. 4.1b is as shown in Fig. 4.1c, which uses the voltage follower studied in Chapter 2. There we found that the input current to the volt-age follower is zero, meaning that R_{in} is infinite. At the output of the voltage fol-lower $R_{out} = 0$, and the circuit behaves as a controlled-voltage source with zero internal resistance.

　　We may generalize the observation with a simple cascade of two RC circuits to say that the chain rule applies, provided that the sections being connected in cascade are isolated in the sense that each successive circuit does not load the pre-vious circuit. When this condition is satisfied, then, with circuits cascaded as

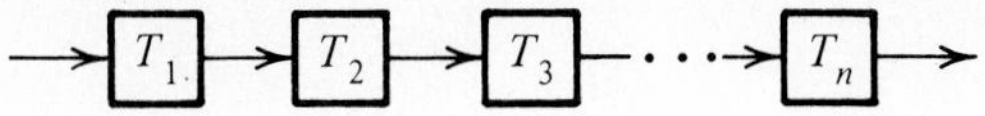

FIGURE 4.2

shown in Fig. 4.2,

$$T = T_1 T_2 T_3 T_4 \cdots T_n \qquad (4.7)$$

for n circuits (or modules) connected in cascade. Circuits of the inverting and noninverting types studied in Chapter 2 are well suited to cascade-connection design.

4.2 INVERTING OPERATIONAL-AMPLIFIER CIRCUITS

The circuits derived in Chapter 2 for op amps and resistors apply for the sinusoidal steady state with values of R replaced by $Z(s)$. Hence the transfer function of the basic inverting op-amp circuit shown in Fig. 4.3 is

$$T(s) = -\frac{Z_2}{Z_1} \qquad (4.8)$$

Since we are considering only the bilinear function for $T(s)$, we have

$$T = -\frac{Z_2}{Z_1} = -K \frac{s + z_1}{s + p_1} \qquad (4.9)$$

The problem to be considered may be formulated in terms of this equation. We assume that the specifications of the design problem are the values of K, z_1, and p_1. These may be found from a Bode plot—the break frequencies and the gain at some frequency—or obtained in any other way. The solution to the design problem involves finding a circuit and the values of the elements in that circuit. We will assume that inductors are excluded from our consideration. Hence we wish to find the values of the R's and the C's. Once found, these values can be adjusted by any necessary frequency scaling, and then by magnitude scaling to obtain convenient element values.

The procedure we will follow requires that some parts of Eq. (4.9) be assigned to Z_1 and some to Z_2. These are arbitrary assignments, each resulting in a different design strategy. Since inductors are excluded, we must avoid making the identifications $Z = Ks$ and $Y = 1/Ks$. Some of the possible assignments are identified in Table 4.1. In fact, these are all of the forms of the Z_1 and Z_2 circuits of Fig. 4.3 if we consider circuits with no more than two capacitors and two resistors.

We will illustrate the procedure by starting from the specification for a low-pass phase-lag circuit described by Eq. (3.3). From Eq. (4.9) we require that

$$\frac{Z_2}{Z_1} = \frac{K}{s + p_1} \qquad (4.10)$$

Note first that we cannot make the assignment that $Z_2 = K$ and $Z_1 = s + p_1$ for this would require that Z_1 include an inductor. However, if we write Eq. (4.10) in

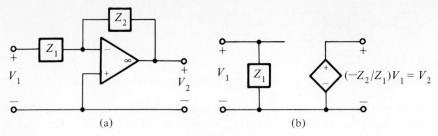

FIGURE 4.3

the form

$$\frac{Z_2}{Z_1} = \frac{1}{s/K + p_1/K} \tag{4.11}$$

then we can make the following assignments:

$$Z_1 = 1$$

$$Z_2 = \frac{1}{s/K + p_1/K} = \frac{1}{C_2 s + 1/R_2} \tag{4.12}$$

The last part of this equation was found from Table 4.1 by examining the four possibilities. We have now found the design equations

$$R_1 = 1, \qquad C_2 = \frac{1}{K}, \qquad R_2 = \frac{K}{p_1} \tag{4.13}$$

and also the circuit structure shown in Fig. 4.4. This set of element values in terms of the specification parameters completes the design procedure, except for scaling.

Example 4.1   Reconsider Example 3.4 of the last chapter which relates to the design of a lowpass circuit to satisfy the specifications of Fig. 3.41. In terms of those specifications, it

TABLE 4.1

Z_1 or Z_2 elements	Impedance	Admittance
R (resistor)	R	$\dfrac{1}{R}$
C (capacitor)	$\dfrac{1}{Cs}$	Cs
R, C in parallel	$\dfrac{1}{Cs + 1/R}$	$Cs + \dfrac{1}{R}$
R, C in series	$R + \dfrac{1}{Cs}$	$\dfrac{1}{R + 1/Cs}$

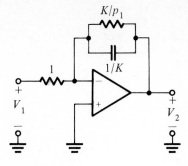

FIGURE 4.4

was found that

$$T(s) = \frac{1000}{s + 1000} \tag{4.14}$$

so that $p_1 = 1000$ and $K = 1000$. Substituting these values into Eqs. (4.13), we obtain the circuit shown in Fig. 4.5a. As in the previous case, frequency scaling is not required, and a magnitude scaling factor of $k_m = 1000$ gives the values shown in Fig. 4.5b.

Applying this same procedure to the general bilinear function, we make the following identification:

$$K\frac{s + z_1}{s + p_1} = \frac{1/(s/K + p_1/K)}{1/(s + z_1)} = \frac{Z_2}{Z_1} \tag{4.15}$$

This identification is identical to that of Eq. (4.12), from which we write the design equations as

$$R_1 = \frac{1}{z_1}, \quad C_1 = 1, \quad R_2 = \frac{K}{p_1}, \quad C_2 = \frac{1}{K} \tag{4.16}$$

The realization is shown in Fig. 4.6. This is a general solution for the bilinear transfer function and applies whether p_1 is larger or smaller than z_1.

Another strategy in making circuit identification makes use of division of both the numerator and the denominator by s. Starting with Eq. (4.9) we have,

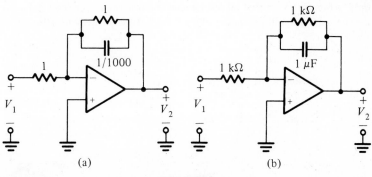

(a)

(b)

FIGURE 4.5

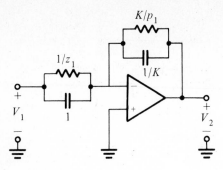

FIGURE 4.6

with $K = K_1/K_2$, an arbitrary division:

$$T = \frac{K_1(s + z_1)/s}{K_2(s + p_1)/s} = \frac{K_1 + K_1 z_1/s}{K_2 + K_2 p_1/s} = \frac{Z_2}{Z_1} \tag{4.17}$$

Comparing this result with the possible values given in Table 4.1, we see that both Z_1 and Z_2 represent series RC circuits. Then the design equations are

$$R_1 = K_2, \qquad C_1 = \frac{1}{K_2 p_1}, \qquad R_2 = K_1, \qquad C_2 = \frac{1}{K_1 z_1} \tag{4.18}$$

This realization is shown in Fig. 4.7. It is another general solution for the bilinear transfer function and applies for either relationship between p_1 and z_1.

Example 4.2   Example 3.2 of the last chapter had as specifications

$$T(s) = 6 \, \frac{s + 0.5}{s + 3} \tag{4.19}$$

Assuming that an inverting solution, with a minus sign added to $T(s)$, is a satisfactory solution (or another inverting stage can be added), we see that the specifications are

$$z_1 = 0.5, \qquad p_1 = 6, \qquad K = 6 \tag{4.20}$$

Using the design equations given as Eq. (4.16), we find that the element values are

$$R_1 = 2, \qquad C_1 = 1, \qquad R_2 = 1, \qquad C_2 = \frac{1}{6} \tag{4.21}$$

These element values are shown in Fig. 4.8a. Suppose that the $T(s)$ of this example is a scaled transfer function, and that all frequencies are to be multiplied by $k_f = 10^4$. We will arbitrarily select the magnitude scaling of $k_m = 1000$, and the element values given in Fig. 4.8b result.

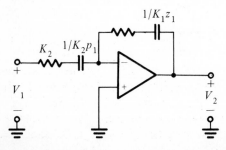

FIGURE 4.7

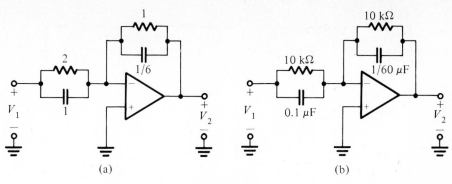

FIGURE 4.8

The approach that has been illustrated by two examples can be applied to other forms of $T(s)$ to give the entries in Table 4.2. We know that the element values in Table 4.2 are not unique, and so other entries can be found.

The circuit of the last example, shown in different form in Fig. 4.9, provides insight into the operation of the inverting amplifier. Analysis of the circuit gives

$$T(s) = -\frac{C_1}{C_2} \frac{s + 1/R_1 C_1}{s + 1/R_2 C_2} \tag{4.22}$$

From this we see that the elements in Z_1 control the position of the zero, while the elements in Z_2 control the position of the pole. Hence adjustment is very simple and routine. The pole and zero are always on the negative real axis, as shown in Fig. 4.10, but the positions of the pole and zero can easily be interchanged. For low frequencies we see from Eq. (4.22) that $T = -R_2/R_1$, while for high frequencies $T = -C_1/C_2$.

The last three entries in Table 4.2 illustrate another point. If we use a series RC connection for Z_1 but a shunt RC connection for Z_2, then the transfer function becomes one of second order. So the manner in which the capacitors are connected in the circuit determines the order of the circuit.

4.3 NONINVERTING OPERATIONAL-AMPLIFIER CIRCUITS

The noninverting op-amp circuit of Chapter 2, generalized with R's replaced by Z's, is shown in Fig. 4.11a. The transfer function is

$$T(s) = \frac{V_2}{V_1} = 1 + \frac{Z_2}{Z_1} \tag{4.23}$$

This is shown in Fig. 4.11b, which reminds us that the circuit operates as a controlled-voltage source. The figure also reminds us of the attractive feature of this connection that Z_{in} is infinite, while the Z_{in} of the inverting op-amp circuit is Z_1. We will next show that there is a disadvantage to this connection to compensate for the infinite input impedance advantage. Solving Eq. (4.23) for Z_2/Z_1 and sub-

TABLE 4.2

	Pole and zero	$T(s)$	Circuit
1	$j\omega$, σ (pole ×)	$-\dfrac{K}{s}$	V_1, $1/K$, capacitor 1, op-amp, V_2
2	$j\omega$, σ (zero ○)	$-Ks$	V_1, $1/K$, resistor 1, op-amp, V_2
3	$j\omega$, σ, $-z_1$ (zero ○)	$-K(s + z_1)$	V_1, K, $1/Kz_1$, resistor 1, op-amp, V_2
4	$j\omega$, σ, $-p_1$ (pole ×)	$\dfrac{-K}{s + p_1}$	V_1, 1, K/p_1, $1/K$, op-amp, V_2
5	$j\omega$, σ (pole ×, zero ○)	$\dfrac{-Ks}{s + p_1}$	V_1, K, K/p_1, resistor 1, op-amp, V_2

stituting the general bilinear $T(s)$, we have

$$\frac{Z_2}{Z_1} = K\,\frac{s + z_1}{s + p_1} - 1 = \frac{s(K - 1) + (Kz_1 - p_1)}{s + p_1} \tag{4.24}$$

Since our objective is to identify terms in this equation with Z_1 and Z_2 representing a passive RC circuit, we see that there are constraints on K and on the relationship for z_1 and p_1. In some cases the limitation will be on K, which can be overcome by additional stages (modules) to provide only gain. But in some cases

TABLE 4.2 (continued)

Pole and zero	T(s)	Circuit
6	$-K\dfrac{s+z_1}{s+p_1}$, any p_1 and z_1	K/p_1, $1/z_1$, $1/K$, 1
7	$-K\dfrac{s+z_1}{s+p_1}$, $K=\dfrac{K_1}{K_2}$ any p_1 and z_1	K_1, $1/K_1 z_1$, K_2, $1/K_2 p_1$
8	$\dfrac{-Ks}{(s+p_1)(s+p_2)}$	$1/K$, K/p_1, 1, $1/p_2$, ∞
9	$\dfrac{-Ks}{(s+p_1)(s+p_2)}$	$1/K$, $1/p_1$, K/p_2, 1, ∞
10	$\dfrac{-Ks}{(s+p_1)(s+p_2)}$, $K=\dfrac{K_1}{K_2}$	$1/K_1$, K_1/p_2, K_2, $1/p_1 K_2$, ∞, $K=K_1/K_2$

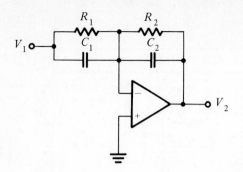

FIGURE 4.9

no realization will be possible. For example, suppose that we wish a realization in the form of Fig. 4.11 for the low-pass transfer function

$$T(s) = \frac{K}{s + p_1} \tag{4.25}$$

Then

$$\frac{Z_2}{Z_1} = \frac{-s + (K - p_1)}{s + p_1} \tag{4.26}$$

and the presence of the $-s$ term in the numerator means that no passive realization can be found. The same conclusion will hold for $T = K/s$ and $T = Ks$. However, there are some realizations possible.

Suppose that we let $K = 1$ so that Eq. (4.24) becomes

$$\frac{Z_2}{Z_1} = \frac{z_1 - p_1}{s + p_1} \tag{4.27}$$

If we divide numerator and denominator by s, we have

$$\frac{Z_2}{Z_1} = \frac{(z_1 - p_1)/s}{1 + p_1/s} \tag{4.28}$$

Next we make the identifications

$$Z_1 = 1 + \frac{p_1}{s} = R_1 + \frac{1}{C_1 s}$$

$$Z_2 = \frac{z_1 - p_1}{s} = \frac{1}{C_2 s} \tag{4.29}$$

The circuit of Fig. 4.12 is then found to be a realization, provided that $p_1 < z_1$ and $K = 1$. A different manipulation of Eq. (4.27) gives a different realization. If we

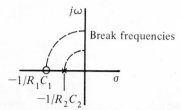

FIGURE 4.10

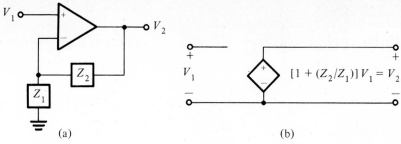

FIGURE 4.11

let

$$Z_1 = \frac{1}{z_1 - p_1} \quad \text{and} \quad Z_2 = \frac{1}{s + p_1} \tag{4.30}$$

we have the circuit shown in Fig. 4.13, which is subject to the same constraints as the circuit of Fig. 4.12.

Returning to Eq. (4.24), suppose that we next select the value of K so that the constant in the numerator vanishes. This requires that

$$K z_1 - p_1 = 0 \quad \text{or} \quad K = \frac{p_1}{z_1} \tag{4.31}$$

Substituting this value for K into Eq. (4.24) gives

$$\frac{Z_2}{Z_1} = \frac{[(p_1 - z_1)/z_1]s}{s + p_1} = \frac{(p_1 - z_1)/z_1}{1 + p_1/s} \tag{4.32}$$

From this equation we see that the identification of terms with Z_1 and Z_2 can be made such that in the circuit of Fig. 4.14a

$$R_2 = \frac{p_1 - z_1}{z_1}, \qquad R_1 = 1, \qquad C_1 = \frac{1}{p_1} \tag{4.33}$$

Another possible identification in Eq. (4.32) leads to

$$Z_1 = \frac{1}{[(p_1 - z_1)/z_1]s} = \frac{1}{C_1 s}$$

$$Z_2 = \frac{1}{s + p_1} = \frac{1}{C_2 s + 1/R_2} \tag{4.34}$$

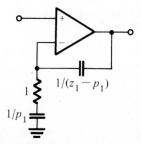

$1/(z_1 - p_1)$

1

$1/p_1$

FIGURE 4.12

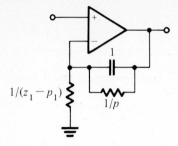

FIGURE 4.13

and from this the circuit of Fig. 4.14b is identified. Each of the two circuits found requires that $z_1 < p_1$ and $K = p_1/z_1$. These circuits are shown in Table 4.3.

Example 4.3   First we observe that the methods we have developed do not apply to the transfer function considered in Example 4.1. If we consider the $T(s)$ of Example 4.2 and the circuit of Fig. 4.14b, then the element values become those shown in Fig. 4.15a. If we scale frequency such that $k_f = 10^3$, and then scale magnitude with $k_m = 10^4$, then the circuit values of Fig. 4.15b are found. This particular realization gives us a gain of $K = p_1/z_1$ $= 12$, which is twice that specified by Eq. (4.19). To meet the specification exactly, it is necessary to reduce the gain by ½, which is done with a voltage divider at the output.

The method that has been illustrated in this section and the preceding one can be summarized in a number of steps:

1.   Reduce Bode plot information (or the equivalent) to values of K, p_1, and z_1.
2.   Look in the tables provided (a catalog) for a suitable realization in terms of pole and zero locations. Determine element values.
3.   If any element values are awkward, try another realization, or derive one not in the catalog.
4.   Frequency scale if this is called for in the specifications.
5.   Magnitude scale to get convenient element values.

Example 4.4   To illustrate these steps, we wish to design a circuit to satisfy the requirements shown in Fig. 4.16a. The circuit has the purpose of providing 6 dB of loss at low frequencies, but no loss at high frequencies. The figure shows only the asymptotic Bode plot. From the analysis of Fig. 4.16b we see that $A(\omega)$ is made up from three factors: a con-

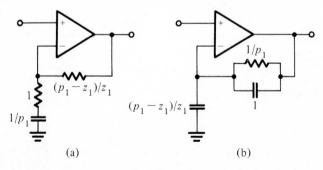

(a)   (b)

FIGURE 4.14

TABLE 4.3

Pole and zero	$T(s)$	Circuit
1	$\dfrac{s + z_1}{s + p_1}$, $z_1 > p_1$	Lag
2	$\dfrac{s + z_1}{s + p_1}$, $z_1 > p_1$	Lag
3	$K\dfrac{s + z_1}{s + p_1}$, $p_1 > z_1$ $K = \dfrac{p_1}{z_1}$	Lead
4	$K\dfrac{s + z_1}{s + p_1}$, $p_1 > z_1$ $K = \dfrac{p_1}{z_1}$	Lead
5	$K(s + z_1)$, $K = \dfrac{1}{z_1}$	Lead

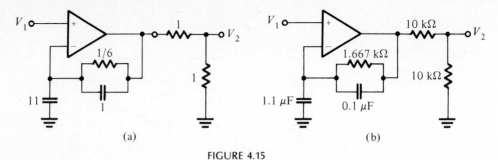

(a)   (b)

FIGURE 4.15

stant, a zero factor with a break point at $\omega = 4$, and a pole factor with a break point at $\omega = 8$. In equation form,

$$A(\omega) = A_0 + 20 \log\left|1 + j\,\frac{\omega}{4}\right| - 20 \log\left|1 + j\,\frac{\omega}{8}\right| \tag{4.35}$$

where

$$A_0 = 20 \log K \quad\text{or}\quad K = 10^{A_0/20} \tag{4.36}$$

Since $A_0 = -6$ dB from Fig. 4.16b, then $K = \frac{1}{2}$. Hence we have found that

$$T(j\omega) = \frac{1}{2}\,\frac{1 + j\,\omega/4}{1 + j\,\omega/8} \tag{4.37}$$

Letting $j\omega$ be replaced by s, we have

$$T(s) = \frac{s + 4}{s + 8} \tag{4.38}$$

The circuit realization shown in Fig. 4.17 is found in Table 4.3 and provides a gain of 2.

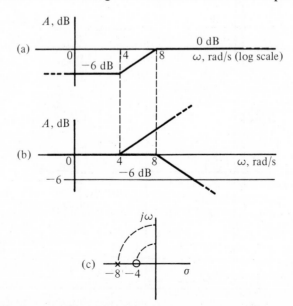

FIGURE 4.16   The decomposition of the response shown in (a) into component first-order factors in (b). The poles and zeros corresponding to (b) are shown in (c).

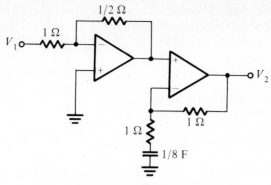

FIGURE 4.17

An inverting op-amp circuit has been used to reduce the gain by a factor of ½ to meet the specifications exactly. Frequency and magnitude scaling are accomplished as in earlier examples.

4.4 CASCADE DESIGN

In this section we make use of cascaded modules, each of first order, to satisfy specifications that are more complicated than the bilinear function. The procedure will be illustrated by three examples.

Example 4.5 The asymptotic Bode plot shown in Fig. 4.18 for $A(\omega)$ is that of a *bandstop* filter. There is no loss at low and high frequencies, but 20 dB of loss is provided in an in-

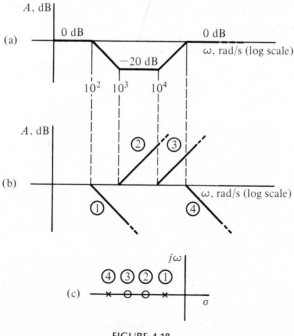

FIGURE 4.18

termediate frequency range. The composite plot may be decomposed into four first-order factors as shown in Fig. 4.18b, plus the possibility of a constant factor. Those marked 1 and 4 represent pole factors, while those marked 2 and 3 are zero factors. In other words, we see that

$$A(\omega) = A_0 - A_1 + A_2 + A_3 - A_4 \tag{4.39}$$

From the break frequencies given, we see that

$$T(j\omega) = K \frac{(1 + j\,\omega/10^3)(1 + j\,\omega/10^4)}{(1 + j\,\omega/10^2)(1 + j\,\omega/10^5)} \tag{4.40}$$

Written in this form, it is clearly seen that when $\omega = 0$, then $T(j0) = K$. From the figure we see that the low-frequency value of $A(\omega) = A_0 = 0$ dB, so that by Eq. (4.36), $K = 1$. Substituting s for $j\omega$ in Eq. (4.40) gives us the transfer function

$$T(s) = \frac{(s + 10^3)(s + 10^4)}{(s + 10^2)(s + 10^5)} \tag{4.41}$$

With experience it will be possible to write $T(s)$ in this form directly from the Bode asymptotic plot, bypassing these steps.

We next write $T(s)$ as a product of bilinear functions. The choice is arbitrary, but one possibility is

$$T(s) = T_1(s)\, T_2(s) = \frac{s + 10^3}{s + 10^2} \times \frac{s + 10^4}{s + 10^5} \tag{4.42}$$

For a circuit realization of T_1 and T_2, we next decide to use the inverting op-amp circuit 6 in Table 4.2. Using the formulas for element values given there, we obtain the realization shown in Fig. 4.19. Frequency scaling is not required for this design since we have worked directly with specified frequencies; so $k_f = 1$. The magnitude scaling of the circuit is accomplished with the equations

$$C_{\text{new}} = \frac{1}{k_m}\, C_{\text{old}} \quad \text{and} \quad R_{\text{new}} = k_m R_{\text{old}} \tag{4.43}$$

If we decide to make all capacitors have the value of 0.01 μF, this is accomplished in Eq. (4.43) by making $k_m = 10^8$. The element values that result are shown in Fig. 4.20, and the design is complete. A characteristic of design that now becomes apparent is that there is not a unique solution. If the spread of resistor sizes in the circuit of Fig. 4.20 is too large, then we begin the design process again and obtain a different solution. This process may be repeated several times before a decision is made on the circuit that will actually be used.

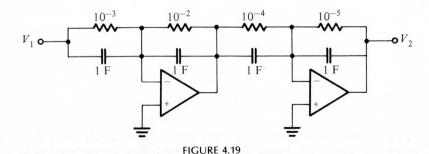

FIGURE 4.19

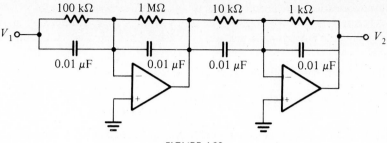

FIGURE 4.20

Example 4.6 The asymptotic Bode plot of Fig. 4.21a describes a *band-enhancement* filter. We wish to provide additional gain over a narrow band of frequencies, leaving the gain at higher and lower frequencies unchanged. We wish to design a filter to these specifications and the additional requirement that all capacitors have the value $C = 0.01\ \mu F$. This particular problem specifies that the maximum gain of the asymptotic plot is 6 dB. An asymptotic plot increases 6 dB in an octave, so that $\omega_a = 200$ rad/s. Since the plot returns to 0 dB, ω_b must be one octave greater than ω_a, or 400 rad/s. Next the first-order factors, which make up the composite Bode plot, are shown in Fig. 4.21b. As frequency increases, the first break frequency identifies a zero factor, then a double pole factor, followed by another zero factor. The pole–zero plot corresponding to these factors is shown in Fig. 4.21c.

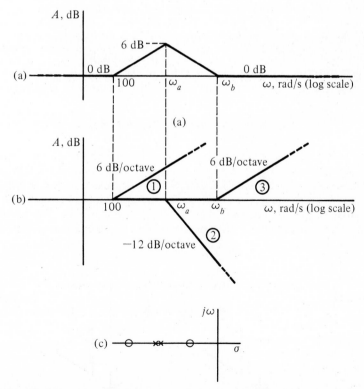

FIGURE 4.21 The decomposition of the band-enhancement filter magnitude response of (a) into component parts shown in (b), having poles and zeros for $T(s)$ as indicated in (c).

From this information we construct $T(j\omega)$ as

$$T(j\omega) = K\frac{(1 + j\,\omega/100)(1 + j\,\omega/400)}{(1 + j\,\omega/200)^2} \qquad (4.44)$$

Since $A(0) = 0$ dB, $K = 1$, and $T(s)$ is

$$T(s) = T_1(s)\,T_2(s) = \frac{s + 100}{s + 200} \times \frac{s + 400}{s + 200} \qquad (4.45)$$

If we make use of the same strategy that was used in Example 4.5, we obtain the circuit shown in Fig. 4.22. But in design there are always many possibilities. Suppose that we decide to try a design for this transfer function using the noninverting op-amp circuits of Table 4.3. The sequence of steps to be accomplished is shown in Fig. 4.23. Having identified T_1 and T_2 in Eq. (4.45), we make our choices of realizations from Table 4.3 as circuits 4 and 2. These are good choices in that all R's have the same value and all C's the value of 1. Frequency scaling is not required, and magnitude scaling is done to obtain the required C values of 0.01 μF by the choice of $k_m = 10^8$. The final circuit is shown in Fig. 4.24. Module 1 produces a gain of 2, and so a voltage divider is provided to reduce the gain by ½ to meet specifications exactly. Now the problem remaining for the designer is to decide whether to use the circuit of Fig. 4.22, that of Fig. 4.24, or whether to find other designs before a final selection is made.

Example 4.6 called for a maximum gain enhancement of 6 dB. Let us generalize this problem as follows: Given a maximum gain enhancement of h dB at a center frequency of ω_0, find the two break frequencies ω_1 and ω_2; all of these quantities are identified in Fig. 4.25. With ω_1 and ω_2 determined, then $T(s)$ may be determined following the steps of the last two examples.

We first observe that the distances AB and BC in Fig. 4.25 are equal, since the slope of one line is 6 dB per octave and that of the other is -6 dB per octave. Then

$$\log \omega_0 = \log \omega_1 + \frac{1}{2}(\log \omega_2 - \log \omega_1) \qquad (4.46)$$

$$= \frac{1}{2}\log \omega_2 + \frac{1}{2}\log \omega_1 = \frac{1}{2}\log \omega_1\omega_2 \qquad (4.47)$$

so that finally

$$\omega_0^2 = \omega_1\omega_2 \qquad (4.48)$$

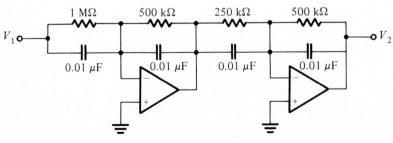

FIGURE 4.22

Item	Module 1	Module 2
Pole–zero locations	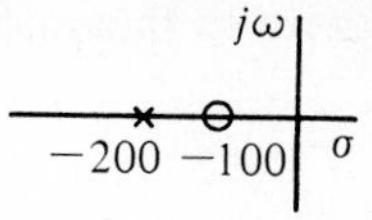 $\times$ at -200, $\bigcirc$ at -100 on σ axis; $j\omega$ axis	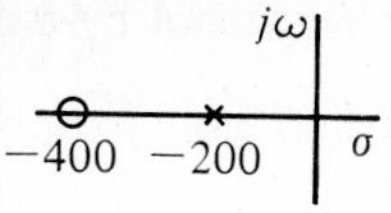 $\bigcirc$ at -400, $\times$ at -200 on σ axis; $j\omega$ axis
Transfer functions	$T_1 = \dfrac{s + 100}{s + 200}$	$T_2 = \dfrac{s + 400}{s + 200}$
Choice of circuit	Table 4.3, circuit 4	Table 4.3, circuit 2
Z_2 element values	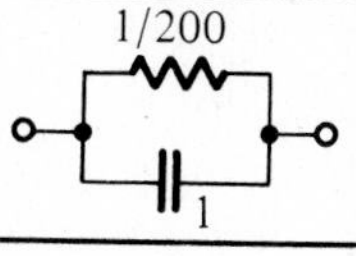 $1/200$ resistor in parallel with capacitor 1	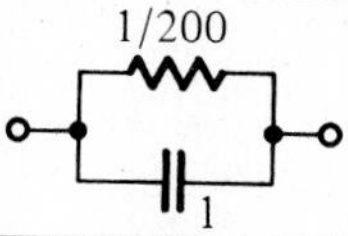 $1/200$ resistor in parallel with capacitor 1
Z_1 element value	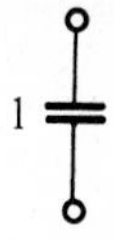 1 (capacitor)	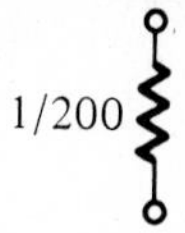 $1/200$ (resistor)

FIGURE 4.23

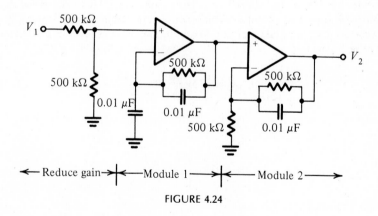

FIGURE 4.24

← Reduce gain →|← Module 1 →|← Module 2 →

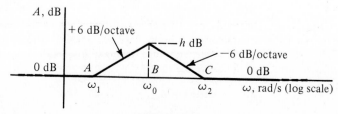

FIGURE 4.25

This is an important relationship which will be used frequently in the chapters that follow. Returning to Fig. 4.25, we make two observations. Let the frequencies ω_1 and ω_2 be separated by n octaves so that

$$\frac{\omega_2}{\omega_1} = 2^n \tag{4.49}$$

If the line of positive slope is extended to ω_2, then its value will be $A = 2h$, so that

$$n \times 6 = 2h \quad \text{dB} \tag{4.50}$$

Combining these two equations gives

$$\frac{\omega_2}{\omega_1} = 2^{h/3} \tag{4.51}$$

Combining this result with Eq. (4.48) gives us the values for the two break frequencies

$$\omega_2, \omega_1 = \omega_0 \, 2^{\pm h/6} \tag{4.52}$$

where the $+$ sign gives ω_2, the $-$ sign ω_1. As a check, note that $h = 6$ dB gives the values found for Example 4.6.

Example 4.7 The asymptotic Bode plot shown in Fig. 4.26 is that of a filter amplifier which provides bandpass filtering with a midband gain of A_{mid}. If we are given A_{mid}, ω_1, and ω_2, then we are required to find a circuit realization including element values. We proceed by decompositing the Bode plot into first-order factors, as shown in Fig. 4.26b. Observe the following factors:

$$A_1 = 20 \log \left| \frac{\omega}{\omega_1} \right| \tag{4.53}$$

which has zero value when $\omega = \omega_1$. The flat midband characteristic is provided by the factor

$$A_2 = -20 \log \left| 1 + j \, \frac{\omega}{\omega_1} \right| \tag{4.54}$$

and the high-frequency rolloff is due to

$$A_3 = -20 \log \left| 1 + j \, \frac{\omega}{\omega_2} \right| \tag{4.55}$$

The gain at frequency ω_0 is called A_{mid} and is represented by the constant term in Fig. 4.26b. Because of the way the frequency was normalized in Eq. (4.53), $A_1(\omega_1) = 0$ dB. At the break frequency ω_1, A_2 has 0 dB value; and since the break frequency ω_2 is larger than ω_1, the value of A_3 is zero at ω_1. Thus we see that

$$A(\omega_1) = A_{\text{mid}} \tag{4.56}$$

From the decomposition of $A(\omega)$, which we have just completed, we see that

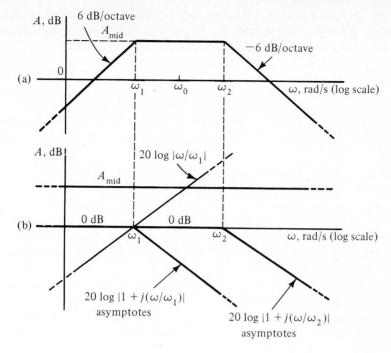

FIGURE 4.26

$$T(j\omega) = K\frac{j\omega/\omega_1}{(1 + j\omega/\omega_1)\,(1 + j\omega/\omega_2)} \qquad (4.57)$$

Letting $j\omega$ become s,

$$T(s) = K\omega_2\frac{s}{(s + \omega_1)(s + \omega_2)} \qquad (4.58)$$

where

$$K = 10^{A_{mid}/20} \qquad (4.59)$$

Let us now consider a specific example. We desire to find a circuit that will provide 20 dB of gain at a frequency of 1000 rad/s, with the break frequencies $\omega_1 = 800$ rad/s and $\omega_2 = 1250$ rad/s. We note that $\omega_1\omega_2 = \omega_0^2$, so that the specified frequencies are consistent. Referring to Table 4.2, we see that we might use a cascade connection of circuits 3 and 5. However, circuit 9 seems well suited to this problem, and its use will save one op amp. If we use the values $K = 12{,}500$, $p_1 = 800$, $p_2 = 1250$, and select $k_m = 10^3$, we obtain the element values shown in Fig. 4.27, and the circuit design is complete.

There will often be the question as to whether the specifications exclude or permit phase inversion. In this example the design of the first module of Fig. 4.27 assumes that an inverting module is permitted. If it is not, then the second module is needed.

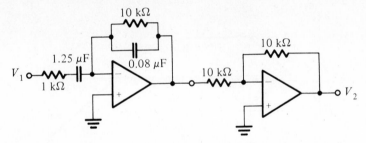

FIGURE 4.27

4.5 THE ALLPASS CIRCUIT: PHASE SHAPING

A passive RC circuit was introduced as Fig. 3.9 to attain an allpass frequency response, meaning that $|T(j\omega)|$ does not change with frequency, but the phase does. Such circuits are useful to obtain some specified phase shift at one frequency or over some band of frequencies. We will make use of allpass circuits in Chapter 18 to accomplish delay equalization.

One allpass circuit is shown in Fig. 4.28. Rather than writing node voltage equations, we employ superposition and illustrate the steps taken in terms of the circuit of Fig. 4.29. With switch S_1 in position a, and switch S_2 in position c, the circuit is a noninverting one with the voltage at the + terminal of the op amp easily determined using the voltage-divider equation. Let the voltage output for this condition be V_2', which is then

$$V_2' = \frac{R}{R + 1/Cs}\left(1 + \frac{R}{R}\right)V_1 = \frac{2R}{R + 1/Cs}V_1 \tag{4.60}$$

Next we move switch S_1 to position b, and S_2 to position d. Observe that this action connects the + terminal of the op amp to ground, since there is no current in R or C. The output voltage under this circumstance is identified as V_2'', and

$$V_2'' = -\frac{R}{R}V_1 \tag{4.61}$$

Now the output voltage is determined by superimposing V_2' and V_2'' to give

$$V_2 = V_2' + V_2'' = -V_1 + \frac{2R}{R + 1/Cs}V_1 \tag{4.62}$$

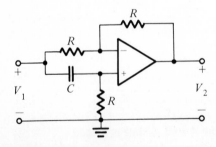

FIGURE 4.28

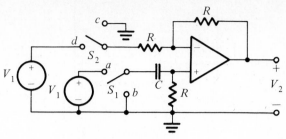

FIGURE 4.29   The switching is arranged to illustrate the application of superposition to the analysis of the circuit of Fig. 4.28.

From this we obtain

$$\frac{V_2}{V_1} = T(s) = \frac{s - 1/RC}{s + 1/RC} \tag{4.63}$$

which has the form of an allpass transfer function as discussed in Chapter 3. Another allpass circuit is shown in Fig. 4.30. Using superposition once more as the analysis tool, we see that

$$V_2 = -\left(\frac{2R}{R + 1/Cs}\right) V_1 + \left(1 + \frac{2R}{R + 1/Cs}\right) \frac{V_1}{2} \tag{4.64}$$

This simplifies to the transfer function

$$\frac{V_2}{V_1} = T(s) = -\frac{1}{2} \frac{s - 1/RC}{s + 1/RC} \tag{4.65}$$

which has the pole and zero combination of the allpass circuit, but differs from Eq. (4.63) with the negative sign indicating an inverting action, and also a magnitude factor of ½.

We will study the transfer function of Eq. (4.63) first, and then compare the result with Eq. (4.65). With $s = j\omega$ the denominator of Eq. (4.63) is

$$j\omega + \frac{1}{RC} = m_1 \angle \phi_1 \tag{4.66}$$

while the numerator is

$$j\omega - \frac{1}{RC} = M_1 \angle \theta_1 \tag{4.67}$$

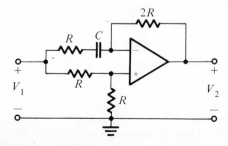

FIGURE 4.30

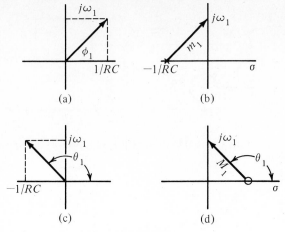

(a)   (b)

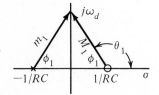

(c)   (d)

FIGURE 4.31

FIGURE 4.32

An s-plane interpretation of these factors is shown in Fig. 4.31. In Fig. 4.31a the real and imaginary parts of Eq. (4.66) are shown together with the phasor addition of the real and imaginary parts to give the polar form of Eq. (4.66). In Fig. 4.31b we show that the magnitude and phase angle are the same if the phasor is drawn from the pole position to the point $j\omega_1$. A similar analysis of Eq. (4.67) is shown in Fig. 4.31c and d, and the two factors are superimposed in Fig. 4.32 with ω_d used to indicate the design frequency. Since $m_1 = M_1$ for all points on the imaginary axis, it is clear that the magnitude $|T(j\omega)| = m_1/M_1 = 1$ for all values of frequency. The design phase angle is

$$\theta_d = \theta_1 - \phi_1 \qquad (4.68)$$

or

$$\theta_d = \tan^{-1}\left(\frac{\omega}{-1/RC}\right) - \tan^{-1}\left(\frac{\omega}{+1/RC}\right) \qquad (4.69)$$

If we let the frequency ω_d move up the imaginary axis, starting at the origin and moving toward infinity, we see that the range of the two angles will be

$$90° < \theta_1 < 180° \qquad \text{and} \qquad 0° < \phi_1 < 90° \qquad (4.70)$$

Since $\theta_d = -\phi_1 + \theta_1$, the range of θ_d will be

$$0° < \theta_d < 180° \qquad (4.71)$$

with θ_d approaching 180° for small ω and 0° for large ω. This is shown in Fig. 4.33. Since the phase angle θ_d is the phase of V_2 with V_1 as the reference, we can

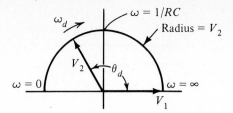

FIGURE 4.33

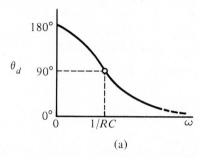

(a)

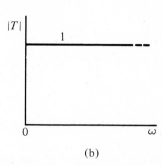

(b)

FIGURE 4.34

picture V_2 moving with constant magnitude through a range of 180° as frequency increases, as depicted in Fig. 4.33. The corresponding magnitude and phase plots on rectangular coordinates are shown in Fig. 4.34.

The transfer function for the inverting allpass circuit of Fig. 4.30, given by Eq. (4.65), differs from that just studied only by having a negative multiplier and a constant of ½. Since −1 corresponds to a constant angle of 180°, the phase angle is advanced by 180° compared to that shown in Fig. 4.33, and the result is shown in Figs. 4.35 and 4.36. The transfer function has a magnitude of ½ for all values of frequency, as shown in Fig. 4.36b. Given the two circuits, we can achieve any specified phase shift from 0° to 360°.

Given an angle θ_d required in a design, how do we determine the values of R and C for one of the circuits?

Return to Fig. 4.32 and observe that the angle from the pole to ω_d is the same as that from the zero to ω_d, the angle being ϕ_1. Then it is seen that

$$\phi_1 + \theta_1 = 180° \qquad (4.72)$$

If we combine this equation with Eq. (4.68), we have

$$\theta_1 = \frac{1}{2}(180° + \theta_d) \qquad (4.73)$$

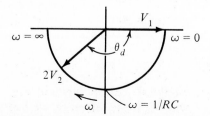

FIGURE 4.35

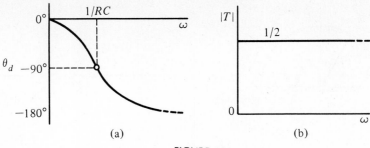

FIGURE 4.36

From Eq. (4.69) we have

$$\theta_1 = -\tan^{-1} RC\,\omega_d \qquad (4.74)$$

so that

$$RC = -\frac{\tan\theta_d}{\omega_d} \qquad (4.75)$$

As in previous design procedures, we now have one equation and two unknowns. This situation is resolved by selecting either R or C, or by a trial-and-error procedure which is continued until a suitable set of values is found.

Example 4.8 We wish to design a circuit to provide a phase shift of $\theta_d = 135°$ at a frequency $\omega_d = 10$ rad/s. We do this in the following steps:

1. Since θ_d is positive and in quadrant II, we must use the circuit of Fig. 4.28.
2. From Eq. (4.73), $\theta_1 = \frac{1}{2}(135° + 180°) = 157.5°$, and tan 157.5° = −0.4142.
3. Then from Eq. (4.75) we have the design equation

$$RC = 0.04124 \qquad (4.76)$$

4. One possible choice is

$$C = 1\ \mu F, \qquad R = 41.42\ k\Omega \qquad (4.77)$$

The required circuit is then that shown in Fig. 4.37.

Example 4.9 We wish to design a circuit to provide a set of three-phase 60-Hz voltages, each separated by 120° and equal in magnitude, as shown in Fig. 4.38. These voltages will simulate those used in ordinary three-phase transmission systems. We first design a circuit to provide $\theta_d = 120°$. Going through the steps of Example 4.8,

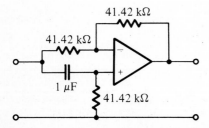

41.42 kΩ

41.42 kΩ

1 μF

41.42 kΩ

FIGURE 4.37

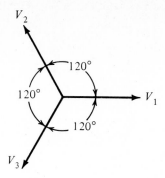

FIGURE 4.38

1. We will use the circuit of Fig. 4.28.
2. $\theta_1 = \frac{1}{2}(120° + 180°) = 150°$, and $\tan 150° = -0.57735$.
3. Then since $\omega_d = 377$, we have the design equation

$$RC = 0.001531 \tag{4.78}$$

4. The design values selected are

$$C = 1 \ \mu F, \qquad R = 1531 \ \Omega \tag{4.79}$$

Now if we use two such circuits connected in cascade, then the phase shift of the combination will be twice the phase shift of the first circuit. The final design is shown in Fig. 4.39.

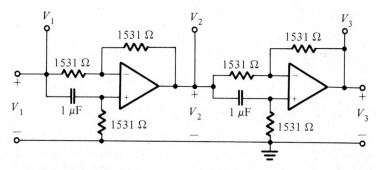

FIGURE 4.39 When a sinusoidal voltage represented by V_1 is applied to this circuit, then V_2 and V_3 satisfy the phase relationships of Fig. 4.38.

PROBLEMS

4.1 For the circuit given in Fig. P4.1, prepare an asymptotic Bode plot for the magnitude of $T(j\omega)$. Carefully identify slopes and low- and high-frequency asymptotes.

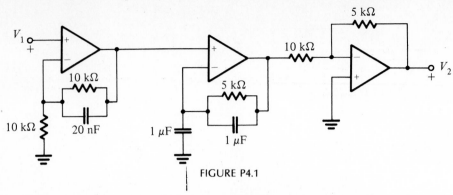

FIGURE P4.1

4.2 The circuit shown in Fig. P4.2 consists of the cascade connection of two op-amp circuits. For this circuit, determine $T(s)$ and plot the Bode asymptotic magnitude function. Identify slopes and the low- and high-frequency asymptotes.

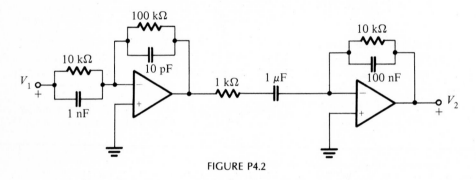

FIGURE P4.2

4.3 For the circuit given in Fig. P4.3, prepare the asymptotic Bode plot for the magnitude of $T(j\omega)$. Carefully identify all slopes and low- and high-frequency asymptotes.

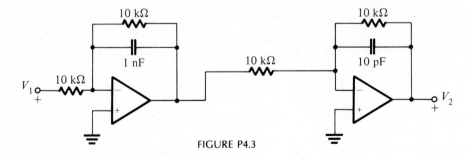

FIGURE P4.3

4.4 The circuit given in Fig. P4.4 consists of the cascade connection of three sections. The first is an inverting stage in which the input is the sum of voltages V_0 and V_1; the second stage is a passive RC circuit; and the third stage is a noninverting op-amp circuit. In the total circuit, all C's are equal, and all R's are equal.
   (a) Let $V_0 = 0$ and find the transfer function $T = V_2/V_1$.
   (b) Connect node 0 to node 2 such that $V_0 = V_2$, and determine the transfer function under this condition.

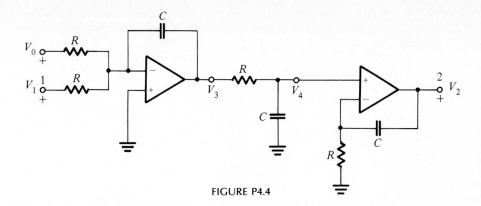

FIGURE P4.4

4.5 Design an amplifier-filter having the Bode asymptotic plot shown in Fig. P4.5. Find the circuit, give the schematic and element values. Scale so that the element values are in a practical range.

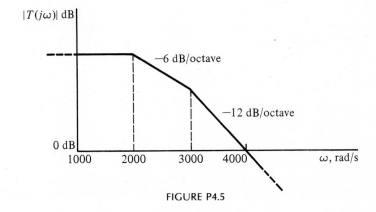

FIGURE P4.5

4.6 The accompanying Fig. P4.6 shows the asymptotic Bode plot for a desired magnitude response. Design an amplifier-filter using a minimum number of op amps. Give the schematic and indicate the element values for your design.

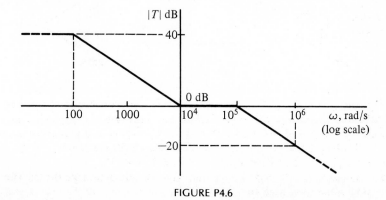

FIGURE P4.6

4.7 Design a bandpass filter having the asymptotic Bode plot shown in Fig. P4.7. Use a minimum number of op amps in your realization. Scale element values until they are in a practical range.

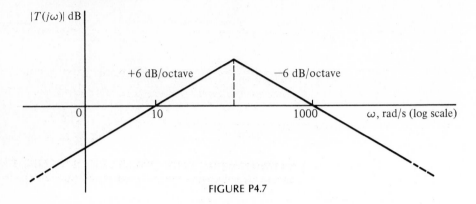

FIGURE P4.7

4.8 Figure P4.8 shows the asymptotic Bode plot for a bandpass filter. Design a circuit that realizes the given frequency response. Give the schematic, and indicate element values chosen for your design.

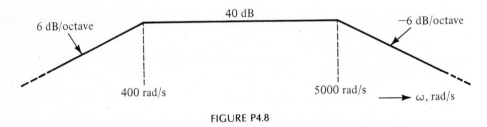

FIGURE P4.8

4.9 Design an *RC* op-amp filter to realize the bandpass response shown in Fig. P4.9. Use a minimum number of op amps in your design, and scale so that the elements are in a practical range.

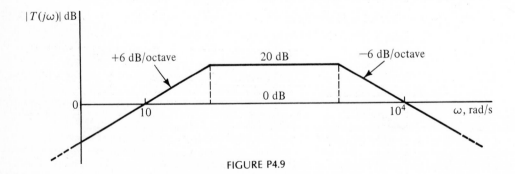

FIGURE P4.9

4.10 The accompanying Fig. P4.10 shows only the asymptotes of a Bode plot. The characteristic response is used for gain enhancement—increasing the gain over a small band of frequencies, but not changing either high- or low-frequency behavior. Find a circuit that will realize this specification characteristic. Give a schematic and the circuit element values.

FIGURE P4.10

4.11 Repeat Problem 4.10 for the specification given in Fig. 4.11.

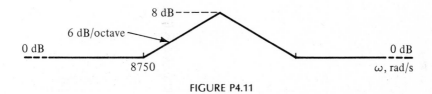

FIGURE P4.11

4.12 The asymptotes of the Bode plot shown in Fig. P4.12 represent a characteristic which has the opposite objective to that described in Problem 4.11. In this case, we wish to reduce the gain over a band of frequencies. Find a circuit that will realize the given specification characteristic. Give the schematic for the circuit chosen and also the element values.

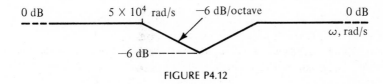

FIGURE P4.12

4.13 Repeat Problem 4.12 for the notch characteristic shown in Fig. P4.13.

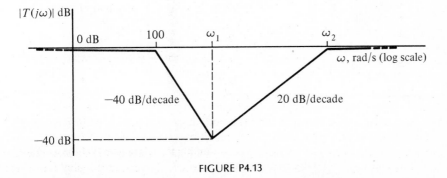

FIGURE P4.13

4.14 The asymptotic Bode plot shown in Fig. P4.14a represents a filter-amplifier with a break frequency of $\omega = 1000$ rad/s. Design a circuit to be connected in cascade with the amplifier such that the break frequency is extended to $\omega = 5000$ rad/s, but there is no change in the magnitude characteristic as far as the 20 dB of gain is concerned.

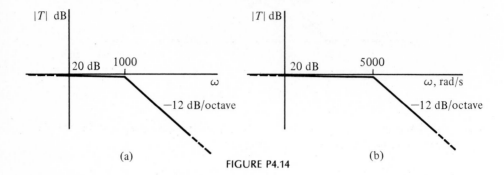

FIGURE P4.14

4.15 The asymptotic Bode plot shown in Fig. P4.15 represents a lowpass filter with gain enhancement over a range of frequencies. Design a circuit using no more than two op amps that will have this magnitude characteristic using one or both of the circuits shown in Fig. P4.15b and c.

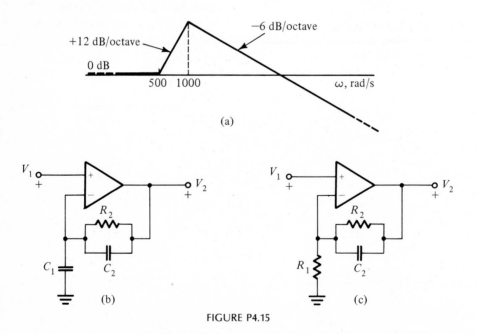

FIGURE P4.15

4.16 The characteristic shown in the asymptotic Bode plot of Fig. P4.16 indicates that considerable gain enhancement is required over a large band of frequencies, but there should be no change in gain at low and high frequencies. Design an op-amp circuit that realizes these specifications.

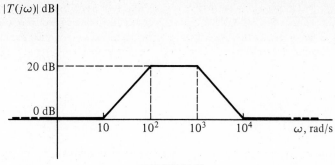

FIGURE P4.16

4.17 The circuit shown in Fig. P4.17 is known as a noninverting integrator. For the element values given, show that the transfer function is

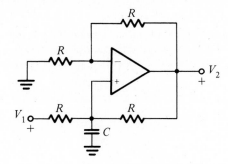

FIGURE P4.17

CHAPTER 5

The
Biquad
Circuit

The biquad circuit is one of the most useful circuits to the electrical engineer because it is a universal filter. It is widely available as a module from industrial sources. It serves as our introduction to second-order filters. In Chapters 3 and 4 we considered transfer functions with poles and zeros on the real axis of the s plane. Beginning with this chapter, poles will be complex conjugates in the left half of the s plane, and zeros will also be complex conjugates, but in either the left or the right half of the s plane.

5.1 DESIGN PARAMETERS Q AND ω_0

Jargon fills a special need for the engineer. It is a shorthand that permits the expression of ideas quickly and compactly. "Design a bandpass filter with a Q of 5 and an ω_0 of 10,000." In this section we will explore what this statement means.

We begin with the *RLC* circuit shown in Fig. 5.1, which has the now familiar form of a voltage-divider circuit. We can anticipate the form of the transfer function $T(s) = V_2(s)/V_1(s)$ by considering the behavior of the elements at low and high frequencies. At low frequencies C behaves as an open circuit; thus there is no current in R and L, and so $V_2 = V_1$ approximately. At high frequencies the capacitor C behaves as a short circuit so that V_2 approaches the value $V_2 = 0$ in the limit. From this we see that the circuit is a lowpass filter of the type familiar from studies in the last two chapters. From the voltage-divider equation,

$$T(s) = \frac{V_2(s)}{V_1(s)} = \frac{Z_2}{Z_1 + Z_2} = \frac{1/Cs}{Ls + R + 1/Cs} \tag{5.1}$$

Dividing numerator and denominator by L and multiplying by s, we obtain

$$T(s) = \frac{1/LC}{s^2 + (R/L)s + 1/LC} \tag{5.2}$$

This result may be put into a *standard form* by defining two new quantities. First

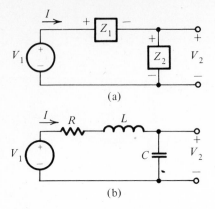

(a)

(b)   FIGURE 5.1

we observe that when the circuit is lossless with $R = 0$, then the denominator reduces to the simple form from which the pole positions may be determined:

$$s^2 + \frac{1}{LC} = 0 \tag{5.3}$$

or

$$s_1, s_2 = \pm j\sqrt{\frac{1}{LC}} = \pm j\omega_0 \tag{5.4}$$

This means that the poles are on the imaginary axis and are conjugates. The other parameter which we require originated in studies of lossy coils for which a *quality factor Q* was defined by Johnson* as

$$Q = \frac{\omega_0 L}{R} = \frac{1}{R}\sqrt{\frac{L}{C}} \tag{5.5}$$

which is the ratio of reactance at the frequency ω_0 to resistance. The historical identification of Q with a lossy coil is no longer appropriate, of course, since we will identify many kinds of circuits with the parameter Q. The last equation may be solved for the ratio R/L in Eq. (5.2):

$$\frac{R}{L} = \frac{\omega_0}{Q} \tag{5.6}$$

Substituting this equation and Eq. (5.4) for $1/LC$ into Eq. (5.2), we obtain

$$T(s) = \frac{\omega_0^2}{s^2 + (\omega_0/Q)s + \omega_0^2} = \frac{N(s)}{D(s)} \tag{5.7}$$

This is the desired standard form. Before studying $T(j\omega)$, we turn our attention to the s-plane location for the poles of $T(s)$.

The poles of $T(s)$ are the values for which $D(s) = 0$ in Eq. (5.7). Let their s-

---

* E. I. Green, "The Story of Q," Monograph 2491, Bell Laboratories, Murray Hill, NJ, 1954.

plane location be $-\alpha \pm j\beta$ so that

$$D(s) = (s + \alpha + j\beta)(s + \alpha - j\beta)$$
$$= s^2 + 2\alpha s + (\alpha^2 + \beta^2) \tag{5.8}$$

Equating like terms of this equation and $D(s)$ in Eq. (5.7), we find that

$$\alpha = \frac{\omega_0}{2Q} \tag{5.9}$$

or

$$Q = \frac{\omega_0}{2\alpha} \tag{5.10}$$

Similarly, equating the constant terms in $D(s)$ in Eqs. (5.7) and (5.8)

$$\omega_0{}^2 = \alpha^2 + \beta^2 \tag{5.11}$$

Combining this with Eq. (5.9) and solving for β, gives

$$\beta = \omega_0 \sqrt{1 - \frac{1}{4Q^2}} \tag{5.12}$$

All of these relationships are shown in Fig. 5.2. In this figure we also define the angle ψ with respect to the negative real axis as

$$\psi = \cos^{-1}\left(\frac{\alpha}{\omega_0}\right) = \cos^{-1}\left(\frac{1}{2Q}\right) \tag{5.13}$$

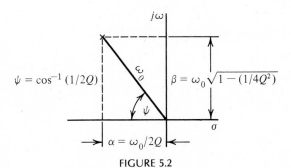

FIGURE 5.2

Figure 5.3 shows important contours in the s plane. Contours of constant ω_0 are circles of radius ω_0 with their centers at the origin, as shown in Fig. 5.3a. From Eq. (5.13) we see that lines of constant Q are lines of constant angle ψ, as shown in Fig. 5.3b. Finally, lines of constant ratio $\omega_0/2Q$ are lines parallel to the imaginary axis, as shown in Fig. 5.3c.

In circuit design we will ordinarily deal with Q values greater than 1. This has implications with respect to pole positions. From Eq. (5.13) we make the following tabulation:

Q	ψ (degrees)
0.707	45
1	60
2	75.52
5	84.3
20	88.5
100	89.7

Hence we conclude that we will be interested in a small sector of the s plane, which is shaded in Fig. 5.4. Observe that when Q is greater than 5, then Eq. (5.12) for β simplifies to $\beta = \omega_0$ with an error less than 1%.

Example 5.1   The two poles of a given $T(s)$ are located in the s plane on lines of slope ± 2, as shown in Fig. 5.5. (a) Determine an expression for the Q of these poles. (b) Express the pole locations $-\alpha \pm j\beta$ in terms of ω_0.

Since the slope of the s-plane line is 2, then tan $\psi = 2$. For this angle, shown in Fig. 5.5b, cos $\psi = 1/\sqrt{5}$. Combining this result with Eq. (5.13), we see that

$$\cos \psi = \frac{1}{2Q} = \frac{1}{\sqrt{5}} \tag{5.14}$$

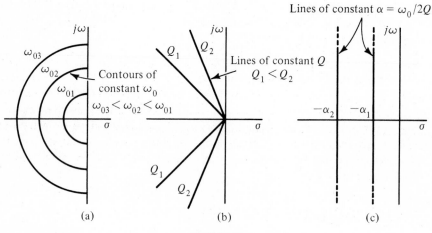

FIGURE 5.3

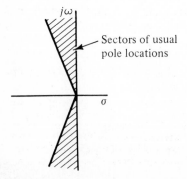

Sectors of usual pole locations

FIGURE 5.4

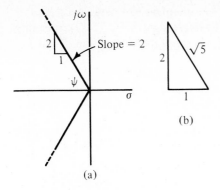

(b)

(a)

FIGURE 5.5

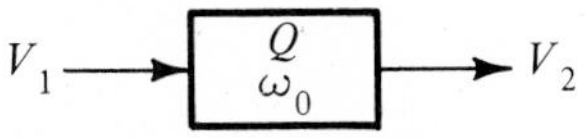

$V_1 \longrightarrow \boxed{\begin{matrix} Q \\ \omega_0 \end{matrix}} \longrightarrow V_2$

FIGURE 5.6

so that

$$Q = \frac{\sqrt{5}}{2} \tag{5.15}$$

Once Q is determined, then Eqs. (5.9) and (5.12) give the pole location as

$$\alpha = \frac{1}{2Q}\omega_0 = \frac{\omega_0}{\sqrt{5}}, \qquad \beta = \omega_0\sqrt{1 - \frac{1}{4Q^2}} = \frac{2}{\sqrt{5}}\omega_0 \tag{5.16}$$

If ω_0 is also specified, then the pole locations are fixed.

Returning to the circuit of Fig. 5.1a and the associated transfer function which described it, Eq. (5.2), we see that the specification of the three element values R, L, and C completely specifies $T(s)$. But we have now shown that the specification of the two parameters Q and ω_0 also specifies $T(s)$, as given by Eq. (5.7). We now have to relate Q and ω_0 to the magnitude and phase responses which we will in turn relate to specifications. In doing so, we now assume that the second-order circuit can be specified by the parameters Q and ω_0, as suggested by the block diagram of Fig. 5.6.

5.2   THE BIQUAD CIRCUIT*

The transfer function for the low-pass filter derived as Eq. (5.7) was written in a normalized form such that $T(j0) = 1$. A more general form for $T(s)$ will recognize the possibility of gain and also that the associated circuit may be inverting or noninverting. Such a transfer function is

$$T(s) = \frac{\pm H\omega_0^2}{s^2 + (\omega_0/Q)s + \omega_0^2} \tag{5.17}$$

* The name biquad for this circuit was first suggested by J. Tow, "Active RC Filters—a State-space Realization," *Proc. IEEE*, vol. 56, pp. 1137–1139, 1968 and by L. C. Thomas in two papers, "The Biquad: Part I—Some Practical Design Considerations," *IEEE Trans. Circuits and Syst.*, vol. CAS-18, pp. 350–357, 1971, and "The Biquad: Part II—A Multipurpose Active Filtering System," *IEEE Trans. Circuits and Syst.*, vol. CAS-18, pp. 358–361, 1971. It is also sometimes called the ring of 3 circuit.

Next we do what will be done frequently in the chapters to follow: we scale frequency so that $\omega_0 = 1$. We also choose the negative sign in Eq. (5.17), meaning that we anticipate an inverting realization from the transfer function. Then Eq. (5.17) becomes

$$T(s) = \frac{-H}{s^2 + (1/Q)s + 1} = \frac{V_2}{V_1} \tag{5.18}$$

We wish to manipulate this equation until it has a form that can be identified with simple circuits which have been studied in past chapters. We rewrite Eq. (5.18) as

$$\left(s^2 + \frac{1}{Q}s + 1\right)V_2 = -HV_1 \tag{5.19}$$

If we divide this equation by the factor $s(s + 1/Q)$, it becomes

$$\left[1 + \frac{1}{s(s + 1/Q)}\right]V_2 = \frac{-H}{s(s + 1/Q)}V_1 \tag{5.20}$$

We may now manipulate this equation to the form

$$V_2 = \left[\frac{-1}{s + 1/Q}V_2 + \frac{-H}{s + 1/Q}V_1\right] \cdot \left(-\frac{1}{s}\right) \cdot (-1) \tag{5.21}$$

Starting at the right-hand side of this equation, we recognize that the (-1) term may be realized by an inverting circuit of gain 1. Similarly, the factor $(-1/s)$ is realized by an inverting integrator. Two operations are indicated by the remaining factor. The circuit realization must produce a sum of voltages, and it must have a transfer function of the form $1/(s + 1/Q)$. The three circuits that provide for these three operations are shown in Fig. 5.7. The circuit marked T_1 sums voltages V_1 and V_2 with appropriate multiplication, and also realizes the first-order transfer function with a circuit that is sometimes called a *lossy integrator*. The circuit marked T_2 is the standard inverting integrator circuit, and the circuit marked T_3 is an inverting circuit of unity gain. If we connect the three circuits together, including a *feedback* connection of the output V_2 to the input, the result is the circuit shown in Fig. 5.8. This is a scaled version of the circuit called the *biquad circuit* or the *ring of 3* circuit, or sometimes the *Tow—Thomas biquad*.

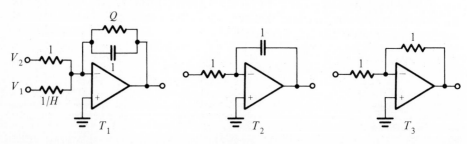

FIGURE 5.7

As a different approach to the study of this circuit, suppose that we start with the circuit itself, with the elements identified by R's and C's as in Fig. 5.9. Thinking of this circuit would be quite natural for an engineer with a background in analog computers since the three modules involved are familiar in analog computers. Routine analysis of the circuit gives us

$$T(s) = \frac{V_2}{V_1} = \frac{-1/R_3 R_4 C_1 C_2}{s^2 + (1/R_1 C_1)s + 1/R_2 R_4 C_1 C_2} \tag{5.22}$$

We may identify this result with the standard form of the low-pass filter transfer function by equating the appropriate coefficients here to those of Eq. (5.17):

$$\omega_0^2 = \frac{1}{R_2 R_4 C_1 C_2} \tag{5.23}$$

$$Q = \sqrt{\frac{R_1^2 C_1}{R_2 R_4 C_2}} \tag{5.24}$$

$$H = \frac{R_2}{R_3} \tag{5.25}$$

To the design parameters introduced in Section 5.1 we now add a third, which is identified with the low-frequency gain. We wish to design to satisfy the parameters ω_0, Q, and H, and we have six circuit elements to adjust to satisfy these parameters. This is a typical situation in design, and our approach will always be to arbitrarily select any three and then examine the consequences on the remaining three. Since we intend to use both frequency and magnitude scaling, we have no hesitation in selecting unit values for the circuit elements. Let us make the following choices:

$$C_1 = C_2 = 1 \quad \text{and} \quad R_4 = 1 \tag{5.26}$$

Let us also decide that we will scale frequency so that $\omega_0 = 1$. Then we may solve Eqs. (5.23)–(5.25) to obtain the values

$$R_1 = Q, \quad R_2 = 1, \quad R_3 = \frac{1}{H} \tag{5.27}$$

This choice gives us exactly the circuit previously derived and shown as Fig. 5.8.

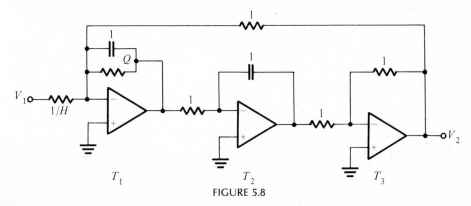

FIGURE 5.8

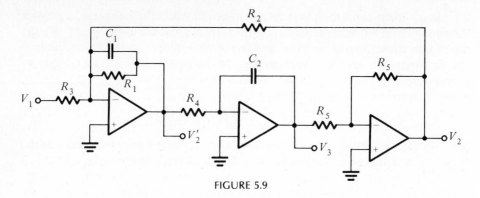

FIGURE 5.9

An important property of the biquad circuit is that it can be *orthogonally* tuned. By this we mean that

1. R_2 can be adjusted to a specified value of ω_0.
2. R_1 can then be adjusted to give the specified value of Q without changing ω_0, which has already been adjusted.
3. Finally R_3 can be adjusted to give the desired value of H or gain for the circuit, without affecting either ω_0 or Q which have already been set.

These steps are often called the *tuning algorithm*. This algorithm provides for orthogonal tuning. If this tuning is not possible, then the tuning is called *iterative*, meaning that we try to adjust successively each of the tuning elements until all specifications are met. Orthogonal tuning is always much preferred, especially when the filter is to be produced on a production line with a laser used to adjust each circuit element value.

One other voltage in the biquad circuit is of special interest. Referring to Fig. 5.9, observe that $V_2/V_3 = -1$, so that

$$\frac{V_3}{V_1} = \frac{V_2}{V_1} \times \frac{V_3}{V_2} = -\frac{V_2}{V_1} \tag{5.28}$$

This tells us that the transfer function V_3/V_1 represents a *noninverting* low-pass filter. Moving back one module in the circuit, we see that V_3 is related to V_2' by the transfer function

$$\frac{V_3}{V_2'} = \frac{1}{R_4 C_2 s} \tag{5.29}$$

From the chain rule

$$\frac{V_2'}{V_3} \times \frac{V_3}{V_2} \times \frac{V_2}{V_1} = \frac{V_2'}{V_2} \tag{5.30}$$

we obtain the result

$$\frac{V_2'}{V_1} = \frac{(-1/R_3 C_1)s}{s^2 + (1/R_1 C_1)s + 1/R_2 R_4 C_1 C_2} \tag{5.31}$$

We show in the next section that this is the transfer function of a *bandpass filter*. To emphasize that the filter is a bandpass filter, the schematic of Fig. 5.9 may be redrawn as shown in Fig. 5.10.

Referring to Fig. 5.9, let us consider the function of the unit-gain inverting section of the biquad circuit, which was referred to as T_3 in Fig. 5.7. This biquad circuit operates with *negative feedback*. Since each section in the biquad is inverting, there must be an odd number of sections, for otherwise the feedback would be positive. So the circuit marked T_3 has the function of inverting the output marked V_3 to provide negative feedback. The same would be accomplished by any odd number of sections in cascade. An alternative approach, which accomplishes the same objective, is to have one inverting stage and one non-inverting stage, as shown in Fig. 5.11. The second stage will be recognized as a noninverting integrator (see Problem 5.5), so that it accomplishes the same objective as T_2 and T_3 together.

Example 5.2 We require a circuit that will provide poles at $-577 \pm j816.5$ and a dc ($\omega \to 0$) gain of 2. Using the equations of Section 5.1, we find that these pole locations correspond to $\omega_0 = 1000$ rad/s and $Q = \sqrt{3}/2$, and that the gain of 2 requirement means that $H = 2$. First we set $\omega_0 = 1$, and then use the biquad circuit of Fig. 5.8, noting that H and Q values are specified. To do the necessary scaling, we set $k_f = 1000$, and then note that selecting $k_m = 10,000$ gives convenient element sizes in the circuit shown in Fig. 5.12.

5.3 FREQUENCY RESPONSE OF THE BIQUAD CIRCUIT

We have found that the biquad circuit shown in Fig. 5.9 is described by two transfer functions, depending on our selection of the output. To begin, let the output be V_2 so that the transfer function that applies is that given by Eq. (5.18). We are interested in the magnitude and phase of this $T(j\omega)$. For simplicity, let $H = 1$

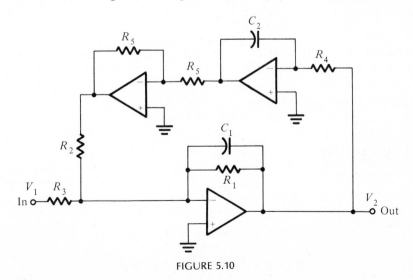

FIGURE 5.10

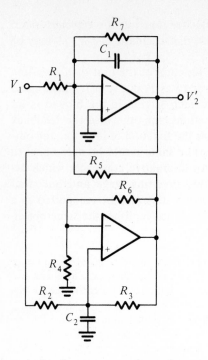

FIGURE 5.11

and also frequency scale so that $\omega_0 = 1$. Then

$$T(j\omega) = \frac{1}{1 - \omega^2 + j\,\omega/Q} \qquad (5.32)$$

From this complex quantity we find that the magnitude is

$$|T(j\omega)| = \frac{1}{\sqrt{(1 - \omega^2)^2 + (\omega/Q)^2}} \qquad (5.33)$$

and the phase is

$$\theta = -\tan^{-1}\left(\frac{\omega/Q}{1 - \omega^2}\right) \qquad (5.34)$$

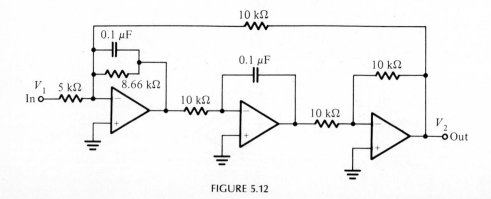

FIGURE 5.12

The magnitude and phase functions are plotted in Fig. 5.13 for a given value of Q. For the magnitude function we see from Eq. (5.33) that

$$|T(j0)| = 1, \qquad |T(j1)| = Q, \qquad |T(j\infty)| \to 0 \qquad (5.35)$$

and that for large $\omega |T(j\omega)| \approx 1/\omega^2$. Similarly for the phase

$$\theta(j0) = 0°, \qquad \theta(j1) = -90°, \qquad \theta(j\infty) \to -180° \qquad (5.36)$$

The magnitude plot on Bode coordinates is shown in Fig. 5.14 for a range of values of Q from 0.707 to 10. The asymptotic Bode plot decreases at the rate of -12 dB per octave, and this is sometimes described as *two-pole rolloff*.

These responses can be visualized in terms of the pole locations of the transfer function. Starting with Eq. (5.18),

$$T(s) = \frac{1}{s^2 + (1/Q)s + 1} \qquad (5.37)$$

The poles of this function are located on a circle of radius 1 and at an angle with respect to the negative real axis of

$$\psi = \cos^{-1}\left(\frac{1}{2Q}\right) \qquad (5.38)$$

This equation can be combined with the previous one to give an alternative representation:

$$T(s) = \frac{1}{s^2 + 2\cos\,\psi\,s + 1} \qquad (5.39)$$

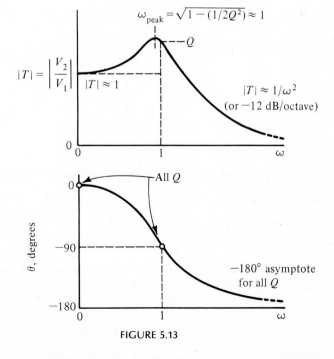

FIGURE 5.13

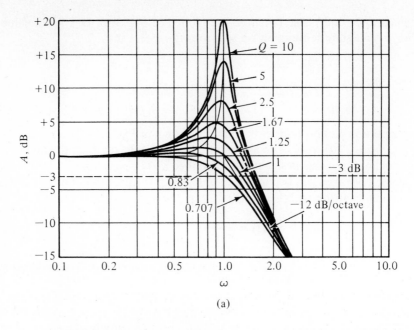

(a)

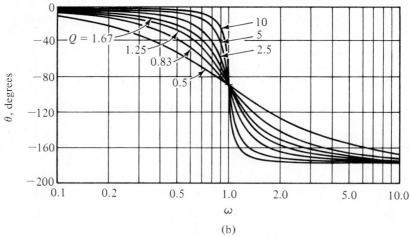

(b)

FIGURE 5.14

In terms of the poles shown in Fig. 5.15,

$$T(s) = \frac{1}{(s + p_1)(s + \bar{p}_1)} \qquad (5.40)$$

where $\bar{p}_1$ is the conjugate of p_1. With $s = j\omega$, the two factors in this equation become

$$j\omega + p_1 = m_1 \angle \phi_1 \qquad \text{and} \qquad j\omega + \bar{p}_1 = m_2 \angle \phi_2 \qquad (5.41)$$

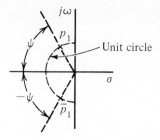

FIGURE 5.15

In terms of these quantities, the magnitude and phase are

$$|T(j\omega)| = \frac{1}{m_1 m_2} \tag{5.42}$$

and

$$\theta = -(\phi_1 + \phi_2) \tag{5.43}$$

Phasors representing Eq. (5.41) are shown in Fig. 5.16, as was done in Fig. 4.31. The figure shows the values computed using these last two equations for three different values of frequency—one below ω_0, one at ω_0, and one above ω_0. From this construction we see that the short length m_2 near the frequency ω_0 is the reason why the magnitude function reaches a peak near ω_0. These plots are useful in

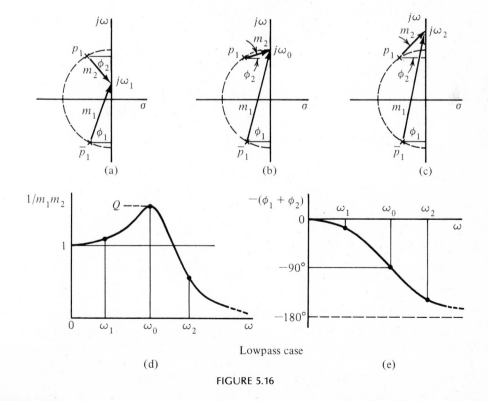

Lowpass case

FIGURE 5.16

visualizing the behavior of the circuit. In solving problems, Eqs. (5.33) and (5.34) may be evaluated using a hand-held calculator.

For a lowpass filter the usual specifications will be the half-power frequency and the value of $|T|_{peak}$; these quantities are identified in Fig. 5.17. The term *half-power* comes from the equation for power $P = I^2R$, from which we see that if P is to be reduced by one-half, then it is necessary that I be reduced by $1/\sqrt{2}$. We are not dealing with current, but with $|T(j\omega)|$; however, the name *half-power* is now applied to most response curves. So we see that the half-power frequency ω_{hp} corresponds to the value of $|T|$ of 0.707. The relationship between ω_0 and ω_{hp} can be estimated from Fig. 5.17. Since $|T|_{peak}$ is approximately equal to Q, the specification of a relatively flat response in the pass band implies a low value of Q. Better methods for designing filters with flat pass-band characteristics will be considered in Chapter 6.

We now return to the biquad circuit of Fig. 5.9 and consider the case where the output voltage is taken to be V_2' for which Eq. (5.31) was found to apply. This transfer function differs from that for the low-pass case in that it has a zero at the origin. The denominators of the two transfer functions are identical, of course. The general transfer function for which Eq. (5.31) is a special case must have a form patterned after Eq. (5.17), but with a different numerator. It must be

$$T(s) = \frac{Ks}{s^2 + (\omega_0/Q)s + \omega_0^2} \tag{5.44}$$

with K to be determined. Suppose that we require that $|T(j\omega_0)|$ have the value H, which will be analogous to dc gain for the lowpass case. Setting $s = j\omega_0$ in Eq. (5.44) and then setting the equation to H, we see that the first and last terms in the denominator cancel and

$$K = H\frac{\omega_0}{Q} \tag{5.45}$$

so that Eq. (5.44) is

$$T(s) = \frac{H(\omega_0/Q)s}{s^2 + (\omega_0/Q)s + \omega_0^2} \tag{5.46}$$

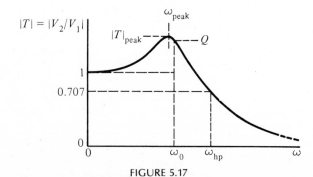

FIGURE 5.17

where H may be either positive or negative. If we now return to Eq. (5.31), we see that

$$H = -\frac{R_1}{R_3} \qquad (5.47)$$

and the equations for ω_0 and Q given by Eqs. (5.23) and (5.24) apply to this case as well as to the lowpass filter.

Now we scale frequency by letting $\omega_0 = 1$ and then let $s = j\omega$ in Eq. (5.46). From this we find that

$$|T(j\omega)| = \frac{H\,\omega/Q}{\sqrt{(1 - \omega^2)^2 + (\omega/Q)^2}} \qquad (5.48)$$

and

$$\theta = 90° - \tan^{-1}\frac{\omega/Q}{1 - \omega^2} \qquad (5.49)$$

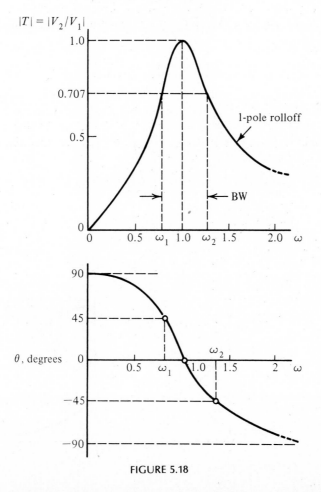

FIGURE 5.18

Plots of these two functions for one value of Q are given in Fig. 5.18 with $H = 1$. From the magnitude response we see that it starts with zero value at $\omega = 0$, increases to unit value at $\omega = 1$, and then decreases at the rate of -6 dB per octave, or with *one-pole rolloff*. The phase response differs from that found for the lowpass case only in that $90°$ is added to all values. Hence it has a value of $90°$ when $\omega = 0$, $0°$ when $\omega = 1$, and then approaches $-90°$ as ω increases.

An analysis similar to that given in terms of the pole locations in Fig. 5.16 is presented for this case in Fig. 5.19. The difference here and in the lowpass case is that a zero at the origin has been added, which contributes the factor $M_1 \angle \theta_1$ to the magnitude and phase characteristics.

We next compute the half-power frequencies for the bandpass response. If we let $H = 1$ and impose the requirement that $|T|^2 = \frac{1}{2}$ in Eq. (5.48), then we obtain an equation for which there are four solutions. Selecting only those that are positive, we have

$$\omega_2, \omega_1 = \sqrt{1 + \left(\frac{1}{2Q}\right)^2} \pm \frac{1}{2Q} \tag{5.50}$$

These frequencies are identified in Fig. 5.18. The product and difference of these two frequencies are

$$\omega_1 \omega_2 = 1 \quad \text{and} \quad \omega_2 - \omega_1 = \frac{1}{Q} \tag{5.51}$$

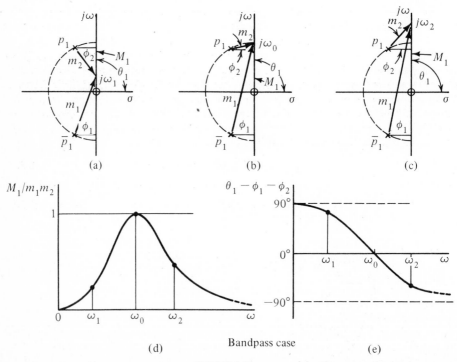

(a)   (b)   (c)

(d)   Bandpass case   (e)

FIGURE 5.19

Restoring the ω_0 gives

$$\omega_1\omega_2 = \omega_0{}^2 \quad \text{and} \quad \omega_2 - \omega_1 = \frac{\omega_0}{Q} \qquad (5.52)$$

This frequency difference is defined as the bandwidth (BW); so we see that

$$\text{BW} = \frac{\omega_0}{Q} \qquad (5.53)$$

or

$$Q = \frac{\omega_0}{\text{BW}} = \frac{\omega_0}{\omega_2 - \omega_1} \qquad (5.54)$$

These equations tell us that Q and BW are inversely related, as shown in Fig. 5.20. We can use these results to show that there is symmetry in this response in that

$$|T(j\omega_1)| = |T(j\omega_2)| \qquad (5.55)$$

and

$$\theta(j\omega_1) = -\theta(j\omega_2) \qquad (5.56)$$

We also observe that Eq. (5.46) with $H = 1$ can be written

$$T(s) = \frac{-\text{BW}\, s}{s^2 + \text{BW}\, s + \omega_0{}^2} \qquad (5.57)$$

However, this applies only for the bandpass case.

Example 5.3 We wish to design a bandpass filter with a center frequency at $\omega_0 = 1000$ rad/s, a bandwidth of 200 rad/s, and a maximum gain of 1, using the biquad circuit. From the given facts we see that $Q = 5$ from Eq. (5.54) and $H = 1$ in the biquad circuit of Fig. 5.8, which was based on $\omega_0 = 1$. Hence it is necessary to frequency scale to match specifications, and this is done by letting $k_f = 1000$. The choice of the magnitude scale of $k_m = 10,000$ gives the circuit elements shown in Fig. 5.21. It is interesting to note that we can find the half-power frequencies by solving the two equations

$$\omega_2 - \omega_1 = 200 \quad \text{and} \quad \omega_1\omega_2 = 10^6 \qquad (5.58)$$

to give

$$\omega_2 = 1105 \text{ rad/s} \quad \text{and} \quad \omega_1 = 905 \text{ rad/s} \qquad (5.59)$$

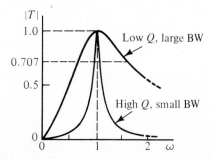

FIGURE 5.20

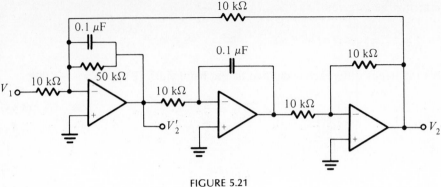

FIGURE 5.21

5.4 The FOUR-OP-AMP BIQUAD CIRCUIT

The addition of a fourth op amp to the biquad circuit gives it versatility in the kinds of filtering it can provide. Since it is common to manufacture op amps with four units on a chip—the *quad op amp*—the fourth unit is available to the designer. We use the extra op amp to add voltages taken from the biquad circuit. In terms of the quantities in Fig. 5.22 we see that

$$V_2'' = -(V_2' + V_1) \tag{5.60}$$

Dividing by V_1, we have a new transfer function:

$$\frac{V_2''}{V_1} = -\left(\frac{V_2'}{V_1} + \frac{V_1}{V_1}\right) \tag{5.61}$$

If we merge the circuit of Fig. 5.22 with that for the biquad circuit given in Fig. 5.9, the result is the circuit of Fig. 5.23. If we substitute Eq. (5.31) for V_2'/V_1 into the last equation, there results

$$\frac{V_2''}{V_1} = -\frac{(-1/R_3C_1)s}{s^2 + (1/R_1C_1)s + 1/R_2R_4C_1C_2} + \frac{s^2 + (1/R_1C_1)s + 1/R_2R_4C_1C_2}{s^2 + (1/R_1C_1)s + 1/R_2R_4C_1C_2} \tag{5.62}$$

Combining the two equations gives us

$$\frac{V_2''}{V_1} = -\frac{s^2 + (1/R_1C_1 - 1/R_3C_1)s + 1/R_2R_4C_1C_2}{s^2 + (1/R_1C_1)s + 1/R_2R_4C_1C_2} \tag{5.63}$$

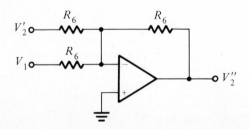

FIGURE 5.22

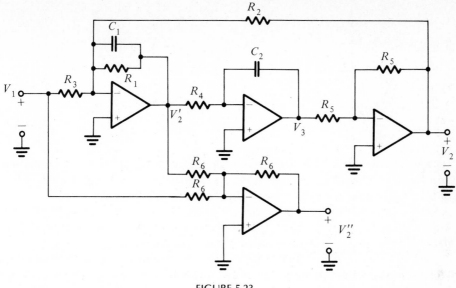

FIGURE 5.23

This equation reduces to an especially simple form if we let $R_1 = R_3$. The consequence of this choice is seen from Eqs. (5.27) to be that the gain is specified

$$H = \frac{1}{Q} \qquad (5.64)$$

Under this condition, Eq. (5.63) becomes

$$\frac{V_2''}{V_1} = -\frac{s^2 + 1/R_2 R_4 C_1 C_2}{s^2 + (1/R_1 C_1)s + 1/R_2 R_4 C_1 C_2} \qquad (5.65)$$

The denominator has already been identified in terms of the parameters ω_0 and Q, and from the equation we see that the constant of the numerator is ω_0^2. The minus sign indicates that the circuit is an inverting one and will be omitted until later. Then Eq. (5.65) becomes

$$\frac{V_2''}{V_1} = \frac{s^2 + \omega_0^2}{s^2 + (\omega_0/Q)s + \omega_0^2} \qquad (5.66)$$

The poles are those associated with the lowpass and bandpass filter operation of the biquad circuit, but the zeros are located on the imaginary axis, as shown in Fig. 5.24. Both the poles and the zeros are located on a circuit of radius ω_0. Let the transfer function of Eq. (5.66) be designated as T_{BE}, which has a magnitude response

$$|T_{BE}(j\omega)| = \frac{|N(j\omega)|}{|D(j\omega)|} \qquad (5.67)$$

We have already studied $|D(j\omega)|$ in Eq. (5.33). From Eq. (5.66), we see that

$$|N(j\omega)| = |\omega_0^2 - \omega^2| \qquad (5.68)$$

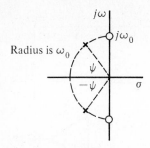

FIGURE 5.24

such that $|N(j\omega_0)| = 0$, and there is no output at frequency ω_0. If $T_{BE}(j\omega)$ is plotted as a function of ω for some Q, then the result is that shown in Fig. 5.25. Because of the particular shape of this magnitude response, the filter giving this response is known as a *notch filter* and also as a *band-elimination* or a *bandstop filter*. This kind of filter is useful in applications where a specific frequency must be eliminated. For example, instrumentation systems require that the power-line frequency of 60 Hz be eliminated. But it reduces the output voltage for a fixed input voltage over a band of frequencies, and in this sense it is band elimination in nature. The relationship between the band of frequencies and the frequencies at which there is half-power (or 3 dB of loss) is the same as that found for the bandpass case. For the frequencies ω_1 and ω_2 and bandwidth BW, as identified in Fig. 5.25, then

$$\omega_1\omega_2 = \omega_0^2 \quad \text{and} \quad \omega_2 - \omega_1 = \text{BW} = \frac{\omega_0}{Q} \tag{5.69}$$

It is also useful to note that there is the following symmetry:

$$|T(j\omega_1)| = |T(j\omega_2)| \tag{5.70}$$

and

$$\theta(j\omega_1) = -\theta(j\omega_2) \tag{5.71}$$

This symmetry exists for all frequencies which are related by the equation

$$\omega_a\omega_b = \omega_0^2 \tag{5.72}$$

Sometimes the specifications for a notch filter are given in terms of the depth

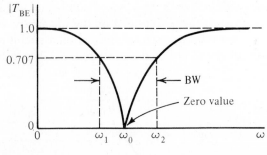

FIGURE 5.25

of the notch and the band of frequencies to be eliminated. Let it be required to provide α_x dB of loss over a bandwidth bw_x. Then it may be shown (see Problem 5-30) that

$$Q = \frac{\omega_0}{\mathrm{bw}_x \sqrt{10^{0.1\alpha_x} - 1}} \tag{5.73}$$

is required. Note that when $\alpha_x = 3$ dB, then $\mathrm{bw}_3 = \omega_0/Q$, as in Eq. (5.69).

We next return to Eq. (5.63) and select the value of $R_3 = R_1/2$. This causes the multiplier of s in the numerator to be the negative of the multiplier of s in the denominator. Then the general form of the transfer function of Eq. (5.63) in terms of ω_0 and Q is

$$T_{\mathrm{AP}}(s) = \frac{s^2 - (\omega_0/Q)s + \omega_0^2}{s^2 + (\omega_0/Q)s + \omega_0^2} \tag{5.74}$$

If we follow Eq. (5.67) and let $T_{\mathrm{AP}}(j\omega) = N(j\omega)/D(j\omega)$, then we see that

$$|N(j\omega)| = |D(j\omega)| \quad \text{and} \quad |T_{\mathrm{AP}}(j\omega)| = 1 \quad \text{for all } \omega \tag{5.75}$$

Thus the circuit has an allpass frequency response as found in Chapters 3 and 4 for first-order transfer functions. From Eq. (5.74) we also see that the positions of the poles and zeros of $T_{\mathrm{AP}}(s)$ differ in the sign of the real part. The pole–zero configuration shown in Fig. 5.26 is known as a *quad*. We postpone consideration of the phase characteristic associated with $T_{\mathrm{AP}}(s)$ until the next section. Allpass circuits find application in the design of delay compensation systems to be studied in Chapter 18.

Following the same pattern in modifying the biquad circuit that was used to obtain the notch and allpass circuits, we modify the circuit by adding a connection to V_2, as shown in Fig. 5.27, such that

$$V_2'' = -(V_1 + V_2' + V_2) \tag{5.76}$$

Dividing by V_1 gives the required transfer function

$$\frac{V_2''}{V_1} = -\left(1 + \frac{V_2'}{V_1} + \frac{V_2}{V_1}\right) \tag{5.77}$$

To Eq. (5.63) we add the transfer function V_2/V_1 of Eq. (5.22) so that

$$T_{\mathrm{HP}} = \frac{V_2''}{V_1} = \frac{s^2 + (1/R_1C_1 - 1/R_3C_1)s + (1/r_2R_4C_1C_2 - 1/R_3R_4C_1C_2)}{s^2 + (1/R_1C_1)s + 1/R_2R_4C_1C_2} \tag{5.78}$$

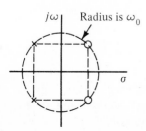

FIGURE 5.26

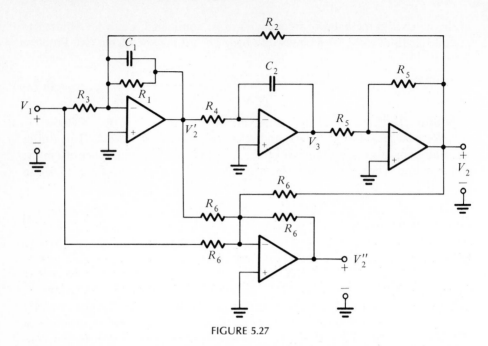

FIGURE 5.27

Now the choices $R_1 = R_3$ and $R_2 = R_3$ cause the second and third terms in the numerator to vanish, leaving only the s^2 term. Writing this result in general for ω_0 and Q, we have

$$T_{\text{HP}}(s) = \frac{s^2}{s^2 + (\omega_0/Q)s + \omega_0^2} \tag{5.79}$$

For $s = j\omega$ the numerator has a magnitude of ω^2 and the denominator has a magnitude of the same form as in earlier cases. For $\omega_0 = 1$, it is

$$|T_{\text{HP}}(j\omega)| = \frac{\omega^2}{\sqrt{(1 - \omega^2)^2 + (\omega/Q)^2}} \tag{5.80}$$

From this equation we see the following:

$$|T_{\text{HP}}(j0)| = 0, \qquad |T_{\text{HP}}(j1)| = Q, \qquad |T_{\text{HP}}(j\infty)| = 1 \tag{5.81}$$

which verifies the highpass filter nature of this response. A plot of this magnitude function is given in Fig. 5.28. We note the similarity of this response and that found earlier for the lowpass case with the behavior at 0 and ∞ frequencies interchanged. We will show later that with $\omega_0 = 1$, the relationship between responses is

$$\omega_{\text{HP}} = \frac{1}{\omega_{\text{LP}}} \tag{5.82}$$

The locations of the poles and zeros for the highpass case are given by Eq. (5.79). We see in Fig. 5.29 that there is a double zero at the origin of the s plane, with poles in the same position as has been the case previously.

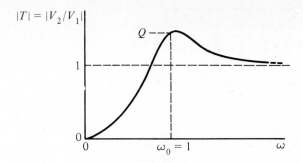

FIGURE 5.28

We now see the great advantage to the four-op-amp biquad circuit. Starting with the most general case, which we have just considered, with connections made as shown in Fig. 5.27, the simple disconnecting of certain resistors from the circuit makes it possible to also realize the allpass, notch or bandstop, bandpass, or lowpass filter. In this sense the biquad circuit is a truly *universal* filter. It is manufactured by a number of companies and is widely applied in solving practical filtering problems.

Example 5.4 A notch filter is required to remove an objectionable 60-Hz hum associated with a power supply in an audio application. The filter must pass frequencies below 55 Hz and above 65 Hz with at most 3 dB of loss and the dc loss must be 0 dB. We are required to design the notch filter. The biquad circuit with a notch frequency characteristic is that given in Fig. 5.23 with the condition that $R_1 = R_3$, meaning that both must have the value of Q. In addition, this will cause $H = 1/Q$ by Eq. (5.64), for which we must compensate. This can be accomplished by making the feedback R_6 for the fourth op amp have the value $R_6' = QR_6$. From the statement of the problem we see that $f_0 = 60$ Hz and BW = 10 Hz, so that we require a Q of 6. Now the design equations for the biquad circuit of Fig. 5.23 were given on the basis that $\omega_0 = 1$. Since we require $\omega_0 = 377$ rad/s, we require that the frequency scaling constant be $k_f = 377$, and we may choose k_m to give appropriate element sizes. No single value of k_m will give convenient element sizes for both resistors and capacitors. In this application suppose that we decide to make the resistors have the value of an integer times 10 kΩ. This will make it possible to use a resistor network such as that shown in Fig. 5.30 with some resistors connected in series as required. Since

$$R_{\text{new}} = k_m R_{\text{old}} \quad \text{and} \quad C_{\text{new}} = \frac{1}{k_f k_m} C_{\text{old}} \tag{5.83}$$

we must select $k_m = 10^4$, making R in Fig. 5.30 the required value of 10 kΩ and the C's have the value of 0.2653 μF. The complete design is shown in Fig. 5.31.

Radius is ω_0

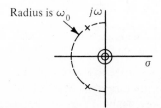

FIGURE 5.29

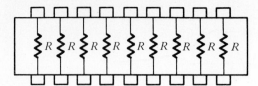

FIGURE 5.30

Example 5.5   For some applications, a deep notch is not required, but a *dip* for gain equalization such as that shown in Fig. 5.32. In a particular cable transmission system it is desired that $|T(j\omega_0)| = \frac{1}{2}|T(j0)|$, which corresponds to 6 dB of loss at ω_0. We are to design a filter to achieve this response with $\omega_0 = 100$ rad/s.

We will select $Q = 5$ and base the design on the biquad circuit of Fig. 5.23. Then

$$T(s) = \frac{s^2 + as + 10^4}{s^2 + 20s + 10^4} \tag{5.84}$$

When $\omega_0 = 100$, then the first and last terms of both numerator and denominator vanish so that

$$T(j\omega_0) = \frac{a\,j\omega_0}{20\,j\omega_0} = \frac{a}{20} \tag{5.85}$$

Then $a = 10$ meets the specifications of Fig. 5.32. For the notch circuit the cancellation was achieved by making V_2' cancel part of V_1. For this problem we wish only partial cancellation. If instead of the condition of Eq. (5.60) we make

$$V_2'' = -\left(\frac{1}{2}V_2' + V_1\right) \tag{5.86}$$

then we achieve the desired modification of Eq. (5.63). This is realized by changing the summing circuit of the fourth op amp, as shown in Fig. 5.33, where $2R_6$ has replaced R_6. $R_7 = QR_6$ achieves the 0-dB requirement at low frequencies, as it did in Example 5.4.

The procedure we have followed has had the result shown in the pole and zero plot of

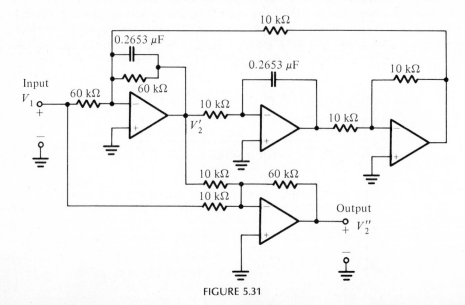

FIGURE 5.31

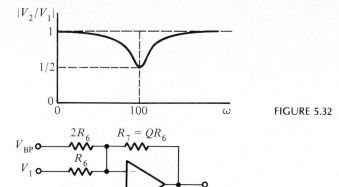

FIGURE 5.32

FIGURE 5.33

Fig. 5.34 for Eq. (5.84) with $a = 10$. The zeros on the imaginary axis for the notch frequency response have moved into the left half-plane and remain on the circuit of radius ω_0. Clearly, they could have moved into the right plane with a different value chosen to multiply R_6 in Fig. 5.33. To complete the design, we frequency and magnitude scale using Eq. (5.84). The choice $k_m = 10{,}000$ along with the required $k_f = 100$ gives the element values shown in Fig. 5.35, which is the final design.

In summary, we can write the general form of the biquadratic transfer function as

$$T(s) = \frac{k_1 s^2 + k_2(\omega_0/Q)s + k_3\omega_0^2}{s^2 + (\omega_0/Q)s + \omega_0^2} \tag{5.87}$$

in which the constants k_1, k_2, and k_3 are ± 1 or 0. The possibilities and names associated with the cases are given in Table 5.1. For each of the first five cases in Table 5.1, the magnitude response $|T(j\omega)|$ and the pole–zero locations in the s plane are shown in Table 5.2. We note that since the denominators of the five forms of transfer function are the same, the pole locations are also the same. The biquad circuit with the five kinds of responses studied and identified as to output location is shown in Fig. 5.36 (on page 146).

5.5   PHASE RESPONSE OF THE BIQUAD CIRCUIT

The phase angle for the lowpass filter function was found in Eq. (5.34) and is shown in Fig. 5.14b. This angle is

$$\theta_{LP} = -\tan^{-1}\left(\frac{\omega/Q}{1 - \omega^2}\right) \tag{5.88}$$

for $\omega_0 = 1$, or, in general,

$$\theta_{LP} = -\tan^{-1}\left[\frac{(1/Q)(\omega/\omega_0)}{1 - (\omega/\omega_0)^2}\right] \tag{5.89}$$

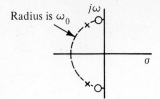

FIGURE 5.34

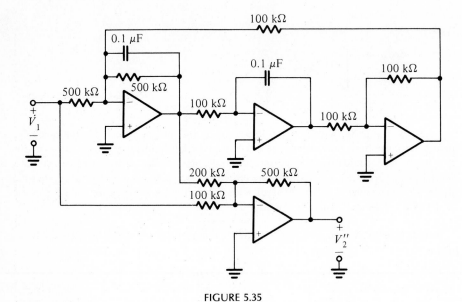

FIGURE 5.35

The plot of Fig. 5.14b is repeated for several values of Q in Fig. 5.37. In this section we wish to tabulate the phase responses for the five magnitude responses studied, the first being θ_{LP}. For the transfer function with $s = j\omega$,

$$T(j\omega) = \frac{N(j\omega)}{D(j\omega)} \qquad (5.90)$$

We have let θ_1 be the angle of $N(j\omega)$, ϕ_1 be the angle of $D(j\omega)$, and θ be the angle

TABLE 5.1

Case	k_1	k_2	k_3	Name
a	0	0	1	Lowpass
b	1	0	0	Highpass
c	0	1	0	Bandpass
d	1	0	1	Bandstop
e	1	−1	1	Allpass
f*	0	1	1	Lowpass
g*	1	1	0	Bandpass

* No realizations were considered.

TABLE 5.2   Standard forms of second-order responses

	Frequency response	*Poles/zeros*	*Name*
$T_{\mathrm{LP}} = \dfrac{\omega_0{}^2}{s^2 + \dfrac{\omega_0}{Q}s + \omega_0{}^2}$			Lowpass
$T_{\mathrm{BP}} = \dfrac{\dfrac{\omega_0}{Q}s}{s^2 + \dfrac{\omega_0}{Q}s + \omega_0{}^2}$			Bandpass
$T_{\mathrm{BE}} = \dfrac{s^2 + \omega_0{}^2}{s^2 + \dfrac{\omega_0}{Q}s + \omega_0{}^2}$			Bandstop "notch"
$T_{\mathrm{HP}} = \dfrac{s^2}{s^2 + \dfrac{\omega_0}{Q}s + \omega_0{}^2}$			Highpass
$T_{\mathrm{AP}} = \dfrac{s^2 - \dfrac{\omega_0}{Q}s + \omega_0{}^2}{s^2 + \dfrac{\omega_0}{Q}s + \omega_0{}^2}$			Allpass

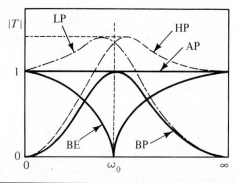

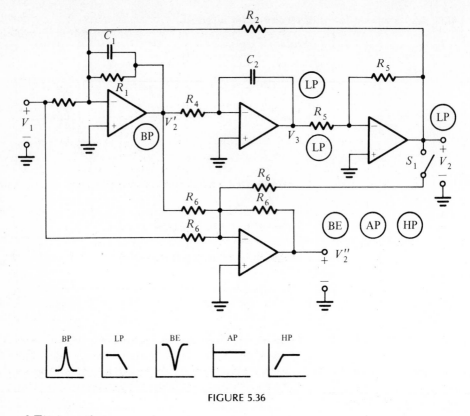

FIGURE 5.36

of $T(j\omega)$, so that

$$\theta = \theta_1 - \phi_1 \tag{5.91}$$

Since the poles of all five kinds of filter responses studied are the same, ϕ_1 will be the same for all of the responses, namely, $-\theta_{LP}$. Then we must find θ_1 for the other four responses.

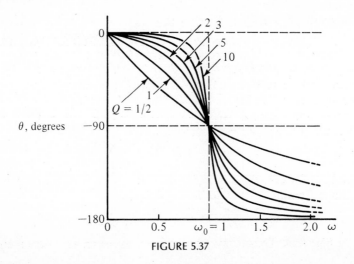

FIGURE 5.37

TABLE 5.3

Name	N(s)	N(jω)	Plot of $\theta_1(\omega)$
Lowpass	ω_0^2	ω_0^2	
Bandpass	$\dfrac{\omega_0}{Q}s$	$j\dfrac{\omega_0\omega}{Q}$	
Bandstop	$s^2 + \omega_0^2$	$-\omega^2 + \omega_0^2$	
Highpass	s^2	$-\omega^2$	

The different forms for $N(s)$ are given in Table 5.3. For the bandpass case we found that $N(s) = (\omega_0/Q)s$, for which $N(j\omega) = j(\omega_0\omega/Q)$ and the j signifies a phase angle of $+90°$. Then

$$\theta_{BP} = \theta_{LP} + 90° \qquad (5.92)$$

Similarly, for the highpass case $N(s) = s^2$ and $N(j\omega) = -\omega^2$, which means a phase angle of $180°$, and

$$\theta_{HP} = \theta_{LP} + 180° \qquad (5.93)$$

For the allpass circuit the transfer function was such that the angle of $N(j\omega)$ is the same as the angle of $D(j\omega)$, and these relate simply to the lowpass case as

$$\theta_{AP} = 2\theta_{LP} \qquad (5.94)$$

From these equations and from the plot of Fig. 5.38 it is apparent that the phase angles for these three cases—bandpass, highpass, and allpass—are the same as that for the lowpass case in appearance, with an angle added for the highpass and bandpass cases, and the angle simply doubled for the allpass case.

The bandstop filter function has an $N(j\omega)$

$$N(j\omega) = -\omega^2 + \omega_0^2 \qquad (5.95)$$

which is positive for $\omega < \omega_0$ and negative for $\omega > \omega_0$. In other words, the phase abruptly jumps from $0°$ to $180°$ when $\omega = \omega_0$. Hence

$$\theta_{BE} = \theta_{LP} + 0° \text{ or } 180° \qquad (5.96)$$

as shown in Fig. 5.39. Thus all of the phase responses for the biquad circuit are

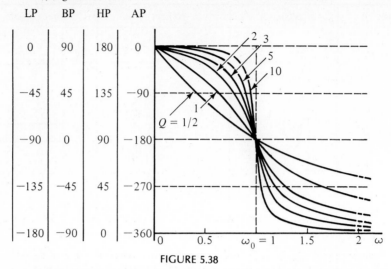

FIGURE 5.38

basically inverse tangent functions, with the allpass case being special in that there is an abrupt shift of angle when $\omega = \omega_0$.

Finally we should note that all of the rules just given apply for the non-inverting circuit. For an inverting circuit an additional phase shift of 180° is required.

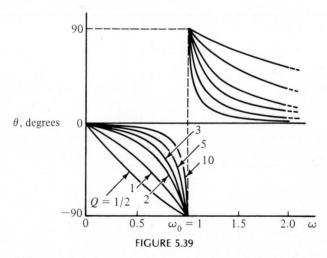

FIGURE 5.39

PROBLEMS

5.1 For the circuit given in Fig. P5.1, let $R_1 = QR$ and $\omega_0 = 1/RC$.
(a) Show that the transfer function then becomes

$$\frac{V_2}{V_1} = \frac{2s(\omega_0/Q)}{s^2 + (\omega_0/Q)s + \omega_0^2}$$

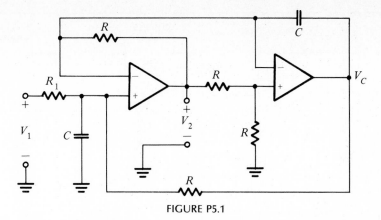

FIGURE P5.1

(b) Design a bandpass filter for which $f_0 = 10$ kHz, and BW = 1 kHz, scaling element sizes to be in a practical range.

5.2 In the circuit of Fig. P5.2, the capacitors have the same value C and all resistors except one have the common value of R. Analyze the circuit to determine the transfer functions

$$T_a = \frac{V_2}{V_1}, \qquad T_b = \frac{V_3}{V_1}, \qquad \text{and} \qquad T_c = \frac{V_4}{V_1}$$

and classify each as lowpass, bandpass, bandstop, etc.

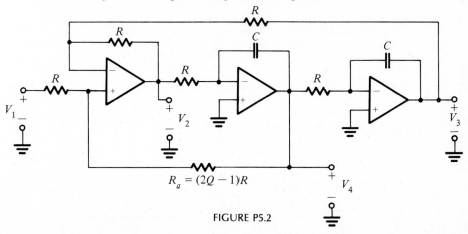

FIGURE P5.2

5.3 Using one of the connections described in Problem 5.2, design a bandpass filter for which $\omega_0 = 10,000$ rad/s with a bandwidth of 1,000 rad/s. Scale the circuit so that all element values are in a practical range.

5.4 For the op-amp filter shown in Fig. P5.4, determine sizes for C_A and C_B so that the bandpass filter has a center frequency of 1 kHz and a Q of 10.

5.5 For the circuit shown in Fig. P5.5, show that

$$\frac{V_2}{V_1} = \frac{1}{RCs}$$

Show that the use of this circuit in the biquad circuit permits us to reduce the number of op amps required to two.

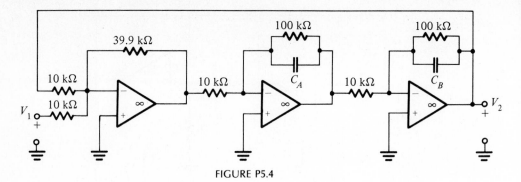

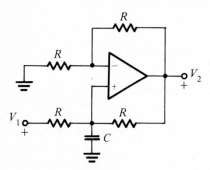

FIGURE P5.4

FIGURE P5.5

5.6 Consider the three-op-amp circuit shown in Fig. P5.6. Writing node equations at nodes a, b, and c, determine the transfer function $T = V_2/V_1$. Show that when $R_2 = R_3$, the filter becomes a notch filter, while when $R_2 = 2R_3$, the filter is an allpass one. Show that design can be accomplished by the choices $C_1 = C_2 = 1$, $R_1 = 1/Q$, and $R_2 = Q$.

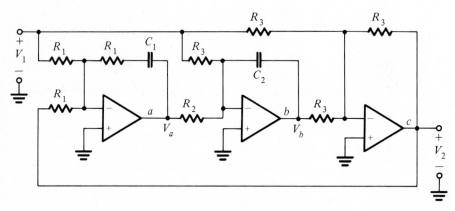

FIGURE P5.6

5.7 Using the results of Problem 5.6, design a filter such that

$$T(s) = \frac{s^2 + 10s + 10,000}{s^2 + 20s + 10,000}$$

and scale your design to practical element values.

5.8   (a)   For the circuit shown in Fig. P5.8, show that with $K = 1 + (R_7/R_6)$,

$$\frac{V_2}{V_1} = K\frac{R_4}{R_5}\frac{s^2 + R_5/R_1R_2R_4C_1C_2}{s^2 + (1/R_2C_4)s + K/R_2R_3C_1C_2}$$

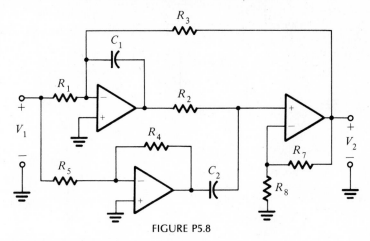

FIGURE P5.8

(b)   Devise a tuning algorithm for the circuit to meet Q, ω_0, ω_z, and gain specifications.

(c)   Devise a design algorithm for the circuit to match the specifications given in (b).

5.9   The circuit shown in Fig. P5.9 is a generalization of that given in Fig. 5.23. We wish to use it to study the lowpass notch and highpass notch filters illustrated by Fig. 12.39, and also the gain-equalizer filter illustrated by Fig. 5.32. Let $R_2 = R_4 = R_5 = R_7 = R_{10} = R$, $C_1 = C_2 = C$, $R_1 = QR$, and $R_9 = R/K$.

(a)   Show that $\omega_0 = 1/RC$ and

$$\frac{V_2}{V_1} = -K\frac{s^2 + (1/Q - R/KR_3)\,\omega_0 s + (1 \pm R^2/KR_3R_8)\omega_0^2}{s^2 + (\omega_0/Q)\,s + \omega_0^2}$$

where $\pm$ is $-$ when Sw_1 is closed, $+$ when Sw_2 is closed. Let the denominator of this equation be $D(s)$ and $V_2/V_1 = T$.

Find the values for R_3, R_8, and the switch positions that will provide the following values of $T(s)$:

(b) $T = \dfrac{s^2 + \omega_z^2}{D}$ $\omega_z < \omega_0$,

(c) $T = \dfrac{s^2 + \omega_z^2}{D}$ $\omega_z > \omega_0$

(d) $T = \dfrac{s^2 + (\omega_0/Q_z)s + \omega_0^2}{D}$

5.10   Using the results of Problem 5.9, design a so-called "universal filter" for which $Q = 5$ and $f_0 = 10$ kHz. The preferred value of C is 0.01 μF. Devise a switching arrangement so that you may obtain bandpass, lowpass, highpass, and allpass, characteristics.

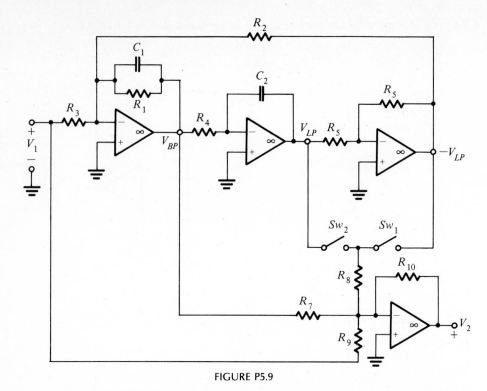

FIGURE P5.9

5.11 Figure P5.11 shows the pole and zero plot for $T(s)$. Sketch the magnitude and phase for $T(j\omega)$. Make clear the $\omega = 0$ value, the $\omega = \infty$ value, and the general shape of the response.

5.12 Repeat Problem 5.11 for the pole zero plot for $T(s)$ which has only one finite zero rather than two (Fig. P5.12).

5.13 A bandpass filter has a Q of 2. Find the pole locations in the s plane as a function of the bandwidth, BW.

Problems 5.14 to 5.24 are intended to give experience in the design of filters.

	Design ω_0	Q	Kind of filter	Preferred C
5.14	10,000	5	LP + AP	0.001 μF
5.15	10,000	5	LP + BE	0.001 μF
5.16	10,000	5	LP + HP	0.001 μF
5.17	5,000	8	LP + AP	0.01 μF
5.18	5,000	8	LP + BE	0.01 μF
5.19	5,000	8	LP + HP	0.01 μF
5.20	7,500	10	BP + AP	0.1 μF
5.21	7,500	10	BP + BE	0.1 μF
5.22	7,500	10	BP + HP	0.01 μF
5.23	10,000	5	BP + BE	0.01 μF
5.24	10,000	5	BP + HP	0.01 μF

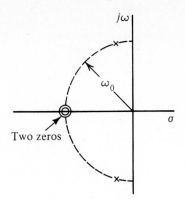

FIGURE P5.11

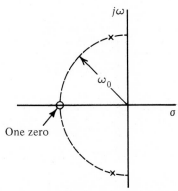

FIGURE P5.12

5.25 Figure P5.25a shows a passive RLC circuit for which $C = 1/4$ F, $L = 1$ H, and $R = 1/2$ Ω. Design an inductorless circuit to fit in the box of Fig. P5.25b which is the analog of that shown in Fig. P5.25a with respect to ω_0 and Q. Give schematic and element values.

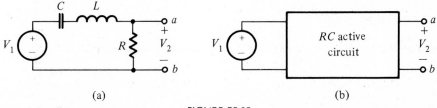

(a)   (b)

FIGURE P5.25

5.26 Consider an RLC circuit as shown in Fig. P5.26 in which one element is varied over its range of values from 0 to ∞. Verify that the locus of the poles of $T(s) = V_2/V_1$ are those shown in the figure.

5.27 Consider the RLC circuit shown in Fig. P5.27.
   (a) Show that the transfer function $T(s) = V_2/V_1$ has the form

$$T(s) = \frac{s^2 + \omega_z^2}{s^2 + (\omega_0/Q)\,s + \omega_0^2}$$

   (b) Express ω_0, ω_z, Q and the low- and high-frequency asymptotic gains in terms of the element values.

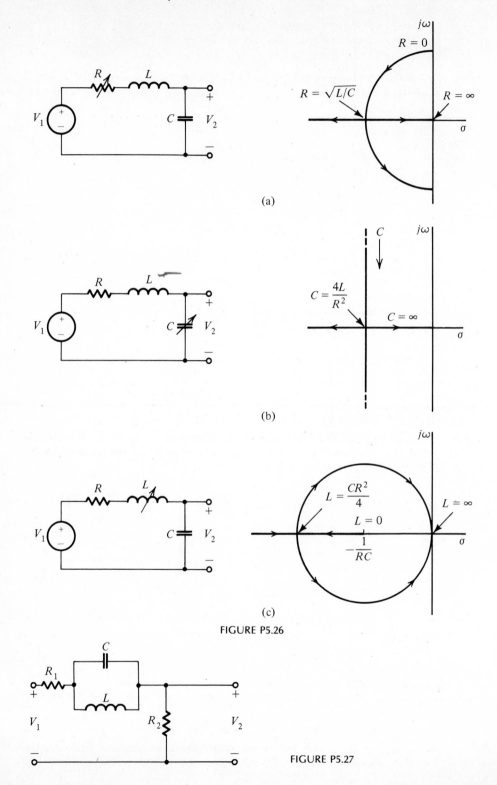

(a)

(b)

(c)

FIGURE P5.26

FIGURE P5.27

(c) Determine expressions for $|T(j\omega)|$ and the phase of $T(j\omega)$.

(d) In the expressions determined for Part (c), let $\omega_0 = 1$ and $Q = 5$. Sketch the magnitude and phase as a function of frequency from $\omega = 0$ to $\omega = 5$ rad/s.

5.28 Repeat the steps outlined in Problem 5.27 for the circuit shown in Fig. P5.28 which is a lowpass notch circuit.

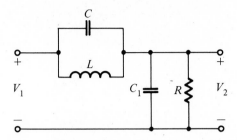

FIGURE P5.28

5.29 Repeat the steps outlined in Problem 5.27 for the circuit shown in Fig. P5.29 which is a highpass notch circuit.

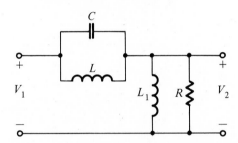

FIGURE P5.29

5.30 As a generalization of Eq. (5.73), let α_{dc} be the loss at $\omega = 0$, and α_x be the loss in excess of α_{dc} corresponding to the bandwidth frequencies ω_{1x} and ω_{2x}, as shown in the accompanying Fig. P5.30. Show that the required Q is

$$Q = \frac{\omega_0}{bw_x \sqrt{10^{0.1\alpha_x} - 1}}$$

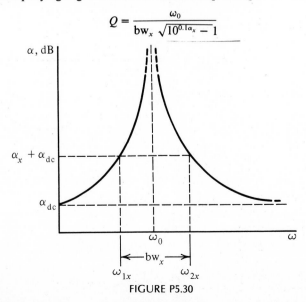

FIGURE P5.30

CHAPTER 6

Butterworth
Lowpass
Filters

This chapter is concerned with the design of lowpass filters of the general class realized by the biquad circuit of the last chapter. In the biquad circuit the parameter ω_0 fixed the transition from pass band to stop band, leaving only Q to shape the magnitude response. Here our objective is to approximate the lowpass filter characteristic through the cascade connection of a number of circuits, each tuned to a different Q (said to be stagger tuned), together contributing to achieve the required overall response.

6.1 THE IDEAL LOWPASS FILTER

The input voltage v_1 shown in Fig. 6.1 contains a low-frequency signal plus *hash*, a term we apply to unwanted high-frequency signals such as shrill tones, scratching sounds, or chirps. To remove the hash, leaving only the low-frequency signal, requires that we have a lowpass filter capable of passing low frequencies and rejecting high frequencies. Had voltage v_1 contained several low-frequency signal components, we would like the filter design to be such that each was transmitted without change in amplitude. This would not be the case if we had used the biquad circuit of the last chapter with a moderately high value of Q, for, as shown by Fig. 6.2, signals near $\omega_0 = 1$ rad/s would be multiplied by as much as Q, in contrast with lower frequency signals which pass through the filter without multiplication.

From this discussion it is clear that the ideal filter characteristic we seek is that shown in Fig. 6.3. Below the normalized frequency of $\omega_0 = 1$, the amplitude of $T(j\omega)$ is a constant; above that frequency the value of T is 0. The pass band and stop band are clearly separated at $\omega = 1$. Because of its shape, this characteristic is called a *brick wall*; it is the ideal lowpass filter characteristic. While we recognize that we will not be able to achieve the ideal, it provides a basis on which we can rate an approximation. As shown in Fig. 6.4, we desire that $|T|$ be as nearly constant as possible in the pass band. In the stop band we require n-pole

FIGURE 6.1

rolloff, where n is a large number, in contrast to the $n = 2$ rolloff for the biquad circuit. We want the transition from pass band to stop band to be as abrupt as possible.

The method we will use in our approach to this problem is illustrated in Figs. 6.5 and 6.6. Suppose that we connect three modules in cascade such that the overall transfer function T is equal to the product $T_1 T_2 T_3$. The product of the magnitudes is shown in Fig. 6.6 as the dashed line, which is of the form required in Fig. 6.4. The large values of $|T_1|$ are just overcome by the small values of $|T_2|$ and $|T_3|$ to achieve the approximation to the brick wall. The transfer functions have the same value of ω_0, but different values of Q. How do we determine the required values of Q? To answer this question will be our first objective.

6.2   BUTTERWORTH RESPONSE

We first review a topic in the algebra of complex numbers. If we denote the real and imaginary parts of the complex transfer function as

$$T(j\omega) = \text{Re } T(j\omega) + j \text{ Im } T(j\omega) \tag{6.1}$$

then we may enumerate some of the properties of $T(j\omega)$. Now the real part of Eq. (6.1) is an *even* function, while the imaginary part is an *odd* function. This means that replacing $j\omega$ by $-j\omega$ will change the sign of the imaginary part, but not that of the real part. Hence

$$T(-j\omega) = \text{Re } T(j\omega) - j \text{ Im } T(j\omega) \tag{6.2}$$

This function is also known as the *conjugate* of $T(j\omega)$, so that

$$T(-j\omega) = T^*(j\omega) \tag{6.3}$$

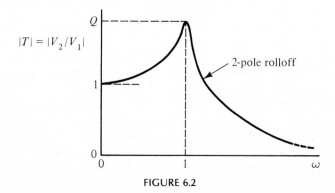

FIGURE 6.2

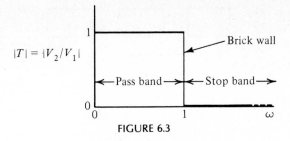

FIGURE 6.3

Since

$$T(j\omega)T^*(j\omega) = (\text{Re } T)^2 + (\text{Im } T)^2 = |T(j\omega)|^2 \qquad (6.4)$$

we have the important relationship

$$|T(j\omega)|^2 = T(j\omega)T(-j\omega) \qquad (6.5)$$

In the past we have frequently replaced s by $j\omega$ or $j\omega$ by s, so that

$$|T(j\omega)|^2 = T(s)T(-s)|_{s=j\omega} \qquad (6.6)$$

which is an important relationship in our study. Now the magnitude-squared function is an even function in that $|T(j\omega)|^2 = |T(-j\omega)|^2$. If we represent the magnitude-squared function as a quotient of polynomials, then both the numerator and the denominator polynomial must be even. Let this quotient be

$$|T_n(j\omega)|^2 = \frac{A(\omega^2)}{B(\omega^2)} \qquad (6.7)$$

We choose a simple form for $A(\omega^2)$ by letting it be a constant A_0. Then

$$|T_n(j\omega)|^2 = \frac{A_0}{B_0 + B_2\omega^2 + B_4\omega^4 + \cdots + B_{2n}\omega^{2n}} \qquad (6.8)$$

The reason for this choice is that we wish to make the rolloff of $|T_n(j\omega)|$ large for large ω, which is accomplished by making the difference of the degree of A and the degree of B as large as possible. This choice will give a $|T_n(j\omega)|$ with n-pole rolloff and a $T_n(s)$ that will be known as an all-pole function. The special case in which all B coefficients except B_0 and B_{2n} have zero value, $A_0 = B_0$ such that $T_n(j0) = 1$, and

$$B_{2n} = \left(\frac{1}{\omega_0}\right)^{2n} \qquad (6.9)$$

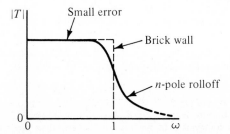

FIGURE 6.4

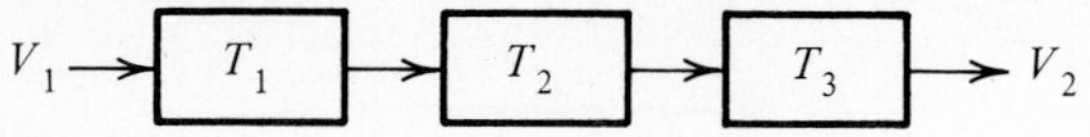

$$V_1 \rightarrow \boxed{T_1} \rightarrow \boxed{T_2} \rightarrow \boxed{T_3} \rightarrow V_2$$

$$V_2 = T_1 T_2 T_3 V_1$$

FIGURE 6.5

gives us the simple form of Eq. (6.8):

$$|T_n(j\omega)|^2 = \frac{1}{1 + (\omega/\omega_0)^{2n}} \tag{6.10}$$

This response is known as the *Butterworth response*.* We may follow our usual procedure and let the frequency be normalized such that $\omega_0 = 1$, giving

$$|T_n(j\omega)| = \frac{1}{\sqrt{1 + \omega^{2n}}} \tag{6.11}$$

From this equation we may observe some interesting properties of the Butterworth response:

1. $|T_n(j0)| = 1$ for all n; the consequence of normalization.
2. $|T_n(j1)| = 1/\sqrt{2} \cong 0.707$ for all n.
3. For large ω, $|T_n(j\omega)|$ exhibits n-pole rolloff.
4. The derivatives of $|T_n(j\omega)|$ for small ω are of interest. If we express Eq. (6.10) by a Taylor series,

$$|T_n(j\omega)| = (1 + \omega^{2n})^{-1/2} = 1 - \frac{1}{2}\omega^{2n} + \frac{3}{8}\omega^{4n} - \cdots \tag{6.12}$$

then it follows that

$$\frac{d^k |T_n(j\omega)|}{d\omega^k}\bigg|_{\omega=0} = 0, \qquad k = 1, 2, \ldots, 2n - 1 \tag{6.13}$$

while

$$\frac{d^{2n} |T_n(j\omega)|}{d\omega^{2n}}\bigg|_{\omega=0} = -\frac{1}{2} \tag{6.14}$$

Since this form of response has all derivatives but one equal to zero near $\omega = 0$, the response is also known as *maximally flat*. These properties are shown in Fig. 6.7. Observe the maximally flat property, and also that the case $n = 10$ comes close to our brick-wall ideal response.

6.3 BUTTERWORTH POLE LOCATIONS

Our next objective is to determine the location of the poles for the transfer function with a Butterworth response. We begin by combining Eq. (6.6) with Eq.

---

*S. Butterworth's original paper appears in the collection of papers in M. E. Van Valkenburg, *Circuit Theory: Foundations and Classical Contributions*, Dowden, Hutchinson & Ross, Stroudsburg, Pa., 1974. This form of response was used by other earlier contributors to the field, but its association with the name Butterworth is now secure.

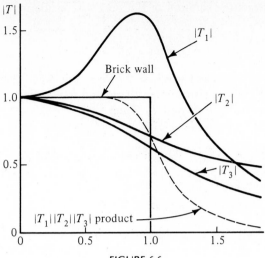

FIGURE 6.6

(6.10) modified by letting $\omega_0 = 1$ and $\omega = s/j$:

$$T_n(s)T_n(-s) = \frac{1}{1 + (s/j)^{2n}} \tag{6.15}$$

$$= \frac{1}{1 + (-1)^n s^{2n}} \tag{6.16}$$

The poles of Eq. (6.16) are the roots of the equation

$$B_n(s)B_n(-s) = 1 + (-1)^n s^{2n} = 0 \tag{6.17}$$

where B_n has been introduced to designate the *Butterworth polynomial.*

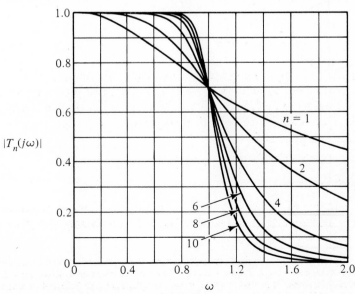

FIGURE 6.7

We will illustrate the solution of Eq. 6.17 by considering several examples. Let $n = 1$ so that

$$1 - s^2 = (1 + s)(1 - s) = 0 \qquad (6.18)$$

Thus the poles are located as $s = \pm 1$, as shown in Fig. 6.8a. The pole in the right half-plane corresponds to an unstable system, and so we select the pole in the left half-plane to associate with B_1 and T_1. Then

$$B_1 = s + 1 \qquad \text{and} \qquad T_1 = \frac{1}{s + 1} \qquad (6.19)$$

If we let $n = 2$, then Eq. (6.17) becomes

$$s^4 + 1 = 0 \qquad \text{or} \qquad s^4 = -1 \qquad (6.20)$$

If we write $-1 + j0$ in the polar form,

$$-1 = 1 \angle (180° + k360°) \qquad (6.21)$$

for integer values of k and $k = 0$, then we see that the angles of this equation are

$$\theta_k = \frac{180° + k360°}{4} = 45°, 135°, 225°, 315° \qquad (6.22)$$

as shown in Fig. 6.8b. As we did for the $n = 1$ case, we select the roots in the left half-plane to assign to $T_2(s)$.

$$B_2(s) = (-0.707 + j0.707)(-0.707 - j0.707) = s^2 + \sqrt{2}s + 1 \qquad (6.23)$$

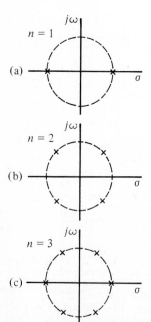

FIGURE 6.8

and

$$T_2 = \frac{1}{s^2 + \sqrt{2}s + 1} \tag{6.24}$$

For $n = 3$ the form of Eq. (6.17) is

$$1 - s^6 = 0 \quad \text{or} \quad s^6 = 1 \tag{6.25}$$

The angles corresponding to Eq. (6.22) are

$$\theta_k = \frac{k360°}{6} = 0°, 60°, 120°, 180°, 240°, 300° \tag{6.26}$$

and all roots of Eq. (6.25) are on a unit circle. If we generalize the two angle relationships of Eqs. (6.22) and (6.26), we have

$$\theta_k = 90° \left| \frac{2k + n - 1}{n} \right| \qquad k = 1, 2, \ldots, 2n \tag{6.27}$$

We will seldom use this, since a different form is better suited to our needs.

All of the poles of $T_n(s)$ will be located on a unit circle because of the frequency normalization in which we set $\omega_0 = 1$. In addition, we will always select the poles in the left half-plane, since only these correspond to a stable circuit. Now θ_k in Eq. (6.27) is the angle measured from the positive real axis. Since our concern is in the left half-plane, let us measure the angle with respect to the negative real axis, designating it as ψ_k, as in Chapter 5. Using this approach, we return to the case $n = 3$ and Eq. (6.25) and see that the angles of the poles are

$$\psi_k = 0°, +60°, -60° \tag{6.28}$$

Knowing the sine and cosine of 60°, we see that

$$B_3(s) = (s + 1)\left(s + \frac{1}{2} + j\frac{\sqrt{3}}{2}\right)\left(s + \frac{1}{2} - j\frac{\sqrt{3}}{2}\right) = (s + 1)(s^2 + s + 1) \tag{6.29}$$

This form is better suited to our needs, especially if we use the form of Eq. (5.29), from which

$$B_n = \frac{1}{s + 1} \quad \text{or} \quad \prod_k (s^2 + 2\cos\psi_k\, s + 1) \tag{6.30}$$

Two simple rules permit us to determine ψ_k:

1. If n is odd, then there is a pole at $\psi = 0°$; if n is even, then there are poles at $\psi = \pm 90°/n$.
2. Poles are separated by $\psi = 180°/n$.

The consequences of these rules are that there are never poles on the imaginary axis, and there is always symmetry with respect to both the real and the imaginary axes when the poles for both $T(s)$ and $T(-s)$ are included.

There are various ways in which information about the Butterworth response can be presented. The angles for each value of n can be presented, as is

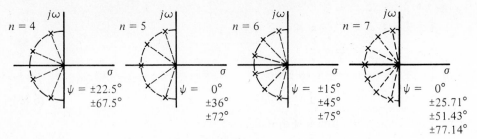

FIGURE 6.9

done in Fig. 6.9, as derived from the rules given following Eq. (6.30). Table 6.1 tabulates the pole locations for $n = 2$ to $n = 10$, and Table 6.2 gives the coefficients of the Butterworth polynomials $B_n(s)$. The value of Q for each of the pole locations is routinely found using the result given in Eq. (5.39) and apparent in Eq. (6.30), since $\omega_0 = 1$,

$$Q = \frac{1}{2 \cos \psi} \tag{6.31}$$

Such values are given in Table 6.3. Finally, the phase angle associated with the Butterworth response is found, once $B_n(s)$ is known, as

$$\theta_B = -\tan^{-1} \frac{\operatorname{Im} B_n(j\omega)}{\operatorname{Re} B_n(j\omega)} \tag{6.32}$$

These angles for $n = 1$ to $n = 10$ are given in Fig. 6.10.

Example 6.1 We wish to tabulate information concerning the fifth-order Butterworth response and calculate the phase angle at the frequency $\omega = 1$.

Since $n = 5$ indicates an odd Butterworth function, we know that one pole is located at $\psi = 0$ and that the others are separated from it by multiples of $180°/5 = 36°$. Thus

$$\psi_k = 0°, \pm 36°, \pm 72° \tag{6.33}$$

The pole locations are

$$p_i, \bar{p}_i = -\cos \psi \pm j \sin \psi \tag{6.34}$$

They are then at

$$-1.0000000$$
$$-0.3090170 \pm j\,0.9510565$$
$$-0.8090170 \pm j\,0.5877852$$

The values of Q for the poles are found from Eq. (6.31) as 0.500, 0.618 and 1.618. The fifth-order Butterworth function is

$$B_5(s) = (s + 1)(s^2 + 2 \cos 36° \, s + 1)(s^2 + 2 \cos 72° \, s + 1) \tag{6.35}$$

which can be compared with values tabulated in Table 6.3. The phase angle at $\omega = 1$ can be calculated from Eq. (6.35) by simply determining the phase of each of the terms. Here we see that

$$\theta_B(j1) = 45° + 90° + 90° = 225° \tag{6.36}$$

which agrees with an estimated value taken from Fig. 6.10.

TABLE 6.1 Pole locations for Butterworth Responses

$n=2$	$n=3$	$n=4$	$n=5$	$n=6$	$n=7$	$n=8$	$n=9$	$n=10$
-0.7071068	-0.5000000	-0.3826834	-0.8090170	-0.2588190	-0.9009689	-0.1950903	-0.9396926	-0.1564345
$\pm j0.7071068$	$\pm j0.8660254$	$\pm j0.9238795$	$\pm j0.5877852$	$\pm j0.9659258$	$\pm j0.4338837$	$\pm j0.9807853$	$\pm j0.3420201$	$\pm j0.9876883$
	-1.0000000	-0.9238795	-0.3090170	-0.7071068	-0.2225209	0.5555702	-0.1736482	-0.4539905
		$\pm j0.3826834$	$\pm j0.9510565$	$\pm j0.7071068$	$\pm j0.9749279$	$\pm j0.8314696$	$\pm j0.9848078$	$\pm j0.8910065$
			-1.0000000	-0.9659258	0.6234898	-0.8314696	-0.5000000	-0.7071068
				$\pm j0.2588190$	$\pm j0.7818315$	$\pm j0.5555702$	$\pm j0.8660254$	$\pm j0.7071068$
					-1.0000000	-0.9807853	-0.7660444	-0.8910065
						$\pm j0.1950903$	$\pm j0.6427876$	$\pm j0.4539905$
							-1.0000000	-0.9876883
								$\pm j0.1564345$

TABLE 6.2 Coefficients of the Butterworth polynomial $B_n(s) = s^n + \sum_{i}^{n-1} a_i s^i$

n	a_0	a_1	a_2	a_3	a_4	a_5	a_6	a_7	a_8	a_9
2	1.0000000	1.4142136								
3	1.0000000	2.0000000	2.0000000							
4	1.0000000	2.6131259	3.4142136	2.6131259						
5	1.0000000	3.2360680	5.2360680	5.2360680	3.2360680					
6	1.0000000	3.8637033	7.4641016	9.1416202	7.4641016	3.8637033				
7	1.0000000	4.4939592	10.0978347	14.5917939	14.5917939	10.0978347	4.4939592			
8	1.0000000	5.1258309	13.1370712	21.8461510	25.6883559	21.8461510	13.1370712	5.1258309		
9	1.0000000	5.7587705	16.5817187	31.1634375	41.9863857	41.9863857	31.1634375	16.5817187	5.7587705	
10	1.0000000	6.3924532	20.4317291	42.8020611	64.8823963	74.2334292	64.8823963	42.8020611	20.4317291	6.3924532

TABLE 6.3 Q of Butterworth Poles

n even

2	4	6	8	10	12	14	16
0.71	0.54	0.52	0.51	0.51	0.50	0.50	0.50
	1.31	0.71	0.60	0.56	0.54	0.53	0.52
		1.93	0.90	0.71	0.63	0.59	0.57
			2.56	1.10	0.82	0.71	0.65
				3.20	1.31	0.94	0.79
					3.83	1.51	1.06
						4.47	1.72
							5.10

*n odd**

3	5	7	9	11	13	15
1.00	0.62	0.55	0.53	0.52	0.51	0.51
	1.62	0.80	0.65	0.59	0.56	0.55
		2.24	1.00	0.76	0.67	0.62
			2.88	1.20	0.88	0.75
				3.51	1.41	1.00
					4.15	1.62
						4.78

* For *n* odd there is also a real pole for which $Q = 0.5$.

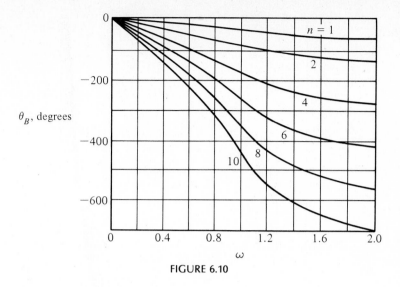

θ_B, degrees

FIGURE 6.10

6.4 LOWPASS FILTER SPECIFICATIONS

Since the early 1920s it has been traditional for those who design electronic amplifiers to think in terms of gain decibels being positive, while those who design filters think of loss in decibels as being positive. With the advent of the op amp and thus active filters, we need both concepts. Rather than adopt one point of view or the other, we resolve the problem by using two symbols as explained in Chapter 1. Thus

$$\alpha = -A \quad \text{dB} \tag{6.37}$$

where

$$A = 20 \log |T(j\omega)| \quad \text{dB} \tag{6.38}$$

In doing so we are simply introducing another coordinate system which reverses the direction of our plots up or down. The Butterworth response of Fig. 6.7 is shown on linear coordinates. The corresponding plot of α in decibels as a function of linear ω is shown in Figs. 6.11 and 6.12, one for the pass band and the other for the stop band. Such plots are useful for visualizing magnitudes, but design values will always be found using a calculator.

The manner in which specifications for a filter will be given to the engineer is illustrated by the plot of Fig. 6.13. For the pass band extending from $\omega = 0$ to $\omega = \omega_p$, the attenuation should not exceed α_{max}. From ω_p to ω_s we have a transition band. Then the specifications indicate that from ω_s and for all higher frequencies the attenuation should not be less than α_{min}. Given this information, we need to find n and ω_0 as applied to the Butterworth response, from which the design can proceed. We begin with Eq. (6.10) for the Butterworth response, retaining ω_0 since it is now one of the unknowns. Substituting this equation into Eqs. (6.37)

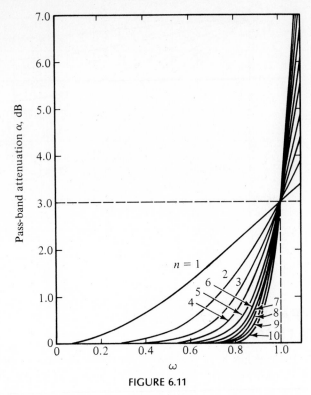

FIGURE 6.11

and (6.38) gives us

$$\alpha = 10 \log \left[1 + \left(\frac{\omega}{\omega_0} \right)^{2n} \right] \quad \text{dB} \tag{6.39}$$

Dividing by 10 and then finding the antilogarithm gives us

$$10^{\alpha/10} = 1 + \left(\frac{\omega}{\omega_0} \right)^{2n} \tag{6.40}$$

and from this equation,

$$\omega_0 = \frac{\omega}{[10^{\alpha/10} - 1]^{1/2n}} \tag{6.41}$$

Thus if we are given corresponding values of α and ω, then ω_0 is determined. If we select $\alpha_{\max}$ and ω_p as defined in Fig. 6.13, then Eq. (6.41) becomes

$$\omega_0 = \frac{\omega_p}{[10^{\alpha_{\max}/10} - 1]^{1/2n}} \tag{6.42}$$

which expresses ω_0 in terms of specified quantities.

To determine n, we start with Eq. (6.40) and substitute values of α and ω that go together, as indicated in Fig. 6.13. Then

$$\left(\frac{\omega_p}{\omega_0} \right)^{2n} = 10^{\alpha_{\max}/10} - 1 \tag{6.43}$$

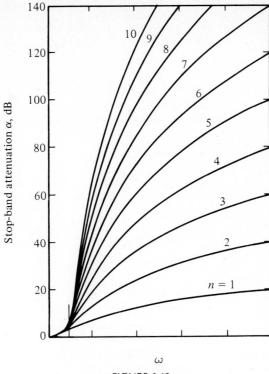

Stop-band attenuation α, dB

FIGURE 6.12

and

$$\left(\frac{\omega_s}{\omega_0}\right)^{2n} = 10^{\alpha_{min}/10} - 1 \qquad (6.44)$$

Dividing these equations gives us

$$\left(\frac{\omega_s}{\omega_p}\right)^{2n} = \frac{10^{\alpha_{min}/10} - 1}{10^{\alpha_{max}/10} - 1} \qquad (6.45)$$

Taking the logarithm of this equation and solving for n gives the desired result:

$$n = \frac{\log\,[(10^{\alpha_{min}/10} - 1)/(10^{\alpha_{max}/10} - 1)]}{2\,\log\,(\omega_s/\omega_p)} \qquad (6.46)$$

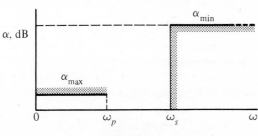

FIGURE 6.13

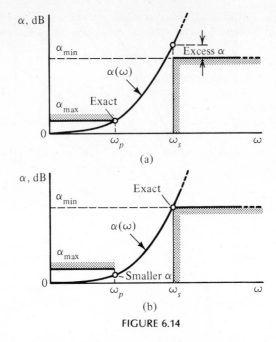

FIGURE 6.14

This is the second required equation to be used in design. A calculator is useful in carrying out the operations indicated by this equation.

A design procedure is carried out in two steps:

1. Using Eq. (6.46), find n. This will ordinarily be a noninteger, so we *round up* to the next integer value and assign it to n.
2. Using this integer n, we find ω_0. We cannot meet the specifications exactly now because we are not using the noninteger n. However, we have two choices:
   a. If we use the value of ω_0 given by Eq. (6.42), then we meet one specification exactly, as shown in Fig. 6.14a, but there is excess α at ω_s.
   b. If we compute ω_0 using Eq. (6.41) matched to the other specification point,

$$\omega_0 = \frac{\omega_s}{(10^{\alpha_{min}/10} - 1]^{1/2n}} \qquad (6.47)$$

as shown in Fig. 6.14b, this will result in meeting one specification exactly with a smaller value of attenuation in the pass band than is required.

In a given design problem we can try both possibilities with the aid of a calculator to see which offers an advantage.

Example 6.2 Suppose that we are required to realize the following specifications with a Butterworth response:

$$\alpha_{max} = 0.5 \text{ dB}, \qquad \alpha_{min} = 20 \text{ dB},$$

$$\omega_p = 1000 \text{ rad/s}, \qquad \omega_s = 2000 \text{ rad/s}$$

For these specifications we wish to determine the transfer function $T(s)$ from which a realization can be found. Substituting the required values into Eq. (6.46) we find that

$$n = 4.83209, \qquad \text{round up to } n = 5 \tag{6.48}$$

Suppose that we decide to use Eq. (6.47) to determine ω_0. It is found to be $\omega_0 = 1263.2$. If we had used Eq. (6.42), it would have been found to be $\omega_0 = 1234$. The Butterworth case $n = 5$ has been considered earlier in this chapter, and it was found that the required values of Q are 0.5, 0.618, and 1.618. Hence the realization of circuits to meet these specifications will be in the block diagram form shown in Fig. 6.15, where each block could be realized using the biquad circuit of Chapter 5, for example, but other alternatives will be given in

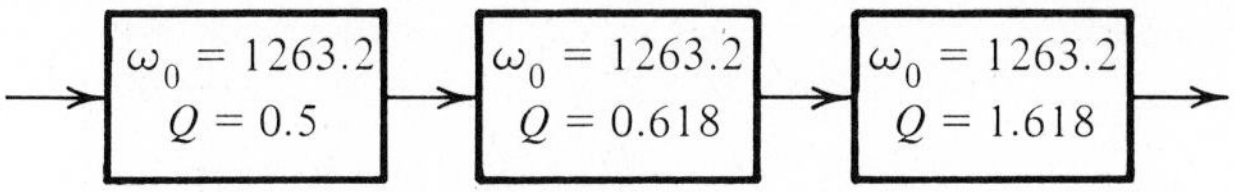

FIGURE 6.15

the next section. We will do our design by letting $\omega_0 = 1$ initially and then frequency scale to the required ω_0 by using the scaling constant $k_f = 1263.2$. Finally we should check to see what the attenuation is at ω_p. Using Eq. (6.39), we find that

$$\alpha(1000) = 10 \log \left[1 + \left(\frac{1000}{1263.2} \right)^{10} \right] = 0.4007 \text{ dB} \tag{6.49}$$

which is less than the specified 0.5 dB as predicted.

6.5 SALLEN AND KEY CIRCUIT

The circuit given in Fig. 6.16 is one of a class of circuits that were described in 1955 by Sallen and Key,* then at MIT's Lincoln Laboratory. In the circuit the noninverting op-amp circuit provides a constant relationship between V_2 and V_a, which is

$$\frac{V_2}{V_a} = 1 + \frac{R_B}{R_A} = K \tag{6.50}$$

The controlled-source representation of the Sallen and Key circuit is given in Fig. 6.17. This circuit may be routinely analyzed using Kirchhoff's current law. At node a the currents directed out of the node must sum to zero, or

$$\frac{1}{R_2} \left(\frac{V_2}{K} - V_b \right) + \left(\frac{V_2}{K} - 0 \right) C_2 s = 0 \tag{6.51}$$

Similarly, the sum of the currents out of node b is

$$\frac{1}{R_2} \left(V_b - \frac{V_2}{K} \right) + C_1 s (V_b - V_2) + \frac{1}{R_1} (V_b - V_1) = 0 \tag{6.52}$$

We next rearrange this equation in a form for solution:

$$\left(\frac{1}{R_1} + \frac{1}{R_2} + C_1 s \right) V_b - \frac{1}{R_2} \frac{V_2}{K} = \frac{V_1}{R_1} + C_1 s V_2 \tag{6.53}$$

* R. P. Sallen and E. L. Key, "A Practical Method of Designing RC Active Filters," *IRE Trans. Circuit Theory,* vol. CT-2, pp. 74–85, 1955.

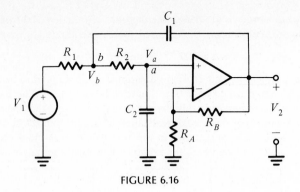

FIGURE 6.16

and

$$-\frac{1}{R_2}V_b + \left(\frac{1}{R_2} + C_2s\right)\frac{V_2}{K} = 0 \tag{6.54}$$

We now eliminate the voltage V_b and solve for the ratio $V_2/V_1 = T$. The result is, after some algebraic simplification,

$$T(s) = \frac{V_2}{V_1} = \frac{K\,1/R_1R_2C_1C_2}{s^2 + (1/R_1C_1 + 1/R_2C_1 + 1/R_2C_2 - K/R_2C_2)s + 1/R_1R_2C_1C_2} \tag{6.55}$$

This transfer function is recognized as being of the general form

$$T(s) = \frac{K\,\omega_0^2}{s^2 + (\omega_0/Q)s + \omega_0^2} \tag{6.56}$$

which is that of a lowpass filter. As was the case in Chapter 5, our objective is to find a design strategy to determine K and the four circuit elements, given the design parameters ω_0 and Q. Before doing this, we will examine the role of K with respect to pole placement. If we let $C_1 = C_2 = 1$ and also $R_1 = R_2 = 1$, then Eq. (6.55) reduces to

$$\frac{V_2}{V_1} = \frac{K}{s^2 + (3 - K)s + 1} \tag{6.57}$$

The locus of poles for the upper left half of the s plane is shown in Fig. 6.18. Since $\omega_0 = 1$ in Eq. (6.57), the locus is on a circle of radius 1. When $K = 1$, then the poles of Eq. (6.57) are both at -1. As K increases, the poles move into the complex plane, and when $K = 3$, they are on the imaginary axis. Thus we see that K alone can place the poles in a position to satisfy any Q requirement. This is also seen from the relationship in Eq. (6.57):

$$Q = \frac{1}{3 - K} \tag{6.58}$$

(Since K may have any value simply by adjusting R_A and R_B, we are left with the question of the meaning of a circuit with negative Q.)

Returning to the general equation for $T(s)$ for the Sallen–Key circuit, we

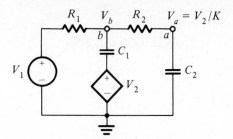

FIGURE 6.17

now outline design procedures to permit choices of element sizes. We have already made the decision that frequency will be scaled such that $\omega_0 = 1$, which reduces the specification parameters to simply the value of Q. A few of the large number of possible choices will be outlined as different design strategies. In general we will select most of the elements to have unit value, knowing that these will be changed to a practical range of values by frequency and magnitude scaling.

Design 1

For this design we will use the element values that led to Eq. (6.57), $R_1 = R_2 = 1$, and $C_1 = C_2 = 1$. Then from Eq. (6.58),

$$K = 3 - \frac{1}{Q} = 1 + \frac{R_B}{R_A} \tag{6.59}$$

If we make the further choice that $R_A = 1$, then R_B is determined:

$$R_B = 2 - \frac{1}{Q} \tag{6.60}$$

The resulting circuit is given in Fig. 6.19a. Only Q need be specified to complete the design to which frequency and magnitude scaling can be applied.

Design 2

We make the choice $K = 1$, which requires that the noninverting op-amp circuit be replaced by a voltage follower, as shown in Fig. 6.19b. We also make the decision that $R_1 = R_2 = 1$ and $\omega_0 = 1$. Applying these choices to Eq. (6.55), gives the

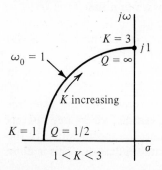

FIGURE 6.18

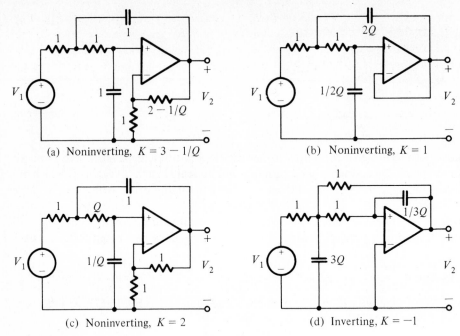

(a) Noninverting, $K = 3 - 1/Q$ (b) Noninverting, $K = 1$

(c) Noninverting, $K = 2$ (d) Inverting, $K = -1$

FIGURE 6.19

two conditions

$$\frac{2}{C_1} = \frac{1}{Q} \quad \text{and} \quad C_1 C_2 = 1 \tag{6.61}$$

From these equations we find that

$$C_1 = 2Q \quad \text{and} \quad C_2 = \frac{1}{2Q} \tag{6.62}$$

which become the design equations, as shown in Fig. 6.19b.

Design 3

Since using equal resistors has an advantage in design, suppose that we let $R_A = R_B = 1$ so that $K = 2$. As always, $\omega_0 = 1$, and as assumptions we let $C_1 = 1$ and $R_1 C_1 = R_2 C_2$. Applying these conditions to Eq. (6.55), we find that

$$R_1 = 1, \quad R_2 = Q, \quad C_1 = 1, \quad C_2 = \frac{1}{Q} \tag{6.63}$$

as shown in Fig. 6.19c. A fourth realization, which differs from the other three in that it is inverting, is given in Fig. 6.19d.

Returning now to the Butterworth response, we recall that the zero-frequency response is $T(j0) = 1$, or 0 dB, while we see that the Sallen–Key circuits give us a zero-frequency gain $T(j0) = K$. If we must meet the specifications of the filter exactly, then we must reduce the gain of our circuit realization. This is done with a resistive voltage divider, which we consider next. Observe from Fig. 6.19 that all of the realizations have a resistor $R_1 = 1$ in series with the input voltage V_1, as shown in Fig. 6.20a. The proposed voltage divider is shown in Fig. 6.20b, for which we require that $R_{in} = 1$ and that the voltage be reduced by the amount H, which is

$$\frac{V_2}{V_1} = H = \frac{R_b}{R_a + R_b}, \qquad H < 1 \qquad (6.64)$$

If the circuit of Fig. 6.20b is such that $R_{in} = 1$, then it can replace the circuit of Fig. 6.20a without changing the overall transfer function except for the gain constant. So in addition to satisfying Eq. (6.64), we require that

$$\frac{R_a R_b}{R_a + R_b} = 1 \qquad (6.65)$$

If we divide this equation by Eq. (6.64), we find that

$$R_a = \frac{1}{H} \qquad (6.66)$$

Then solving Eq. (6.64) gives

$$R_b = \frac{1}{1 - H} \qquad (6.67)$$

These values are shown in the circuit of Fig. 6.20c.

We next apply this result to two different situations. The transfer functions for the Sallen–Key circuit given by Eq. (6.55) have a zero-frequency gain of

$$T(j0) = K \qquad (6.68)$$

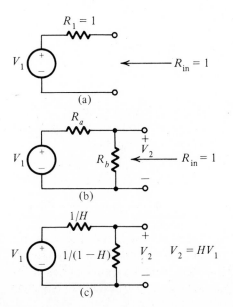

(a)

(b)

(c)

FIGURE 6.20

Hence for gain equalization it is necessary that a resistive voltage divider be provided such that $H = 1/K$, meaning that

$$R_a = K \quad \text{and} \quad R_b = \frac{1}{K-1} \tag{6.69}$$

Let us apply this result to the Sallen–Key circuit of Fig. 6.19a for which

$$K = 3 - \frac{1}{Q} \tag{6.70}$$

Then combining Eqs. (6.69) and (6.70) gives the required values of the voltage-divider resistors

$$R_a = 3 - \frac{1}{Q} \quad \text{and} \quad R_b = \frac{Q}{2Q-1} \tag{6.71}$$

Then the circuit shown in Fig. 6.21a has a gain of 0 dB at zero frequency for all values of Q. Rather than accepting this requirement, we wish the gain adjusted so that it has the value of 0 dB at the normalized frequency $\omega = 1$. Since

$$T(s) = \frac{3 - 1/Q}{s^2 + (1/Q)s + 1} \tag{6.72}$$

we have from this

$$|T(j1)] = 3Q - 1 \tag{6.73}$$

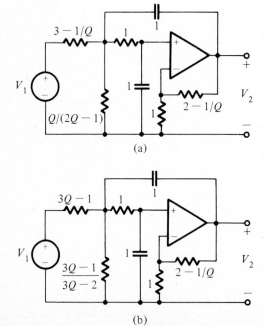

(a)

(b)

FIGURE 6.21

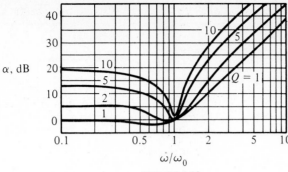

FIGURE 6.22

so that we require that

$$H = \frac{1}{3Q - 1} \tag{6.74}$$

This gives the circuit of Fig. 6.21b. The response function $\alpha(\omega)$ is shown in Fig. 6.22 for several values of Q.

Example 6.3 We now return to the design of the fifth-order Butterworth filter specified in Fig. 6.15 as to Q and ω_0. For the realization we will select the circuit of Fig. 6.19b because of its simplicity and due to the fact that it requires no gain adjustment with a voltage-divider circuit. The cascade connection of the three modules is shown in Fig. 6.23. The responses of the individual stages and the overall response of the circuit are shown in Fig. 6.24. To complete the design requires scaling. We had found that $k_f = 1263.2$, and we will make the choice that $k_m = 10^4$, so that all resistors will have the value of 10 kΩ. The capacitor values are found from

$$C_{\text{new}} = \frac{1}{k_f k_m} C_{\text{old}} \tag{6.75}$$

where $1/k_f k_m = 0.79164 \times 10^{-9}$. The capacitor values are given in the following tabulation:

	Values of C_{new}		
C_{old}	Stage 1 $Q_0 = 0.5$	Stage 2 $Q_1 = 0.618$	Stage 3 $Q_2 = 1.618$
1	0.79164 nF	—	—
$2Q$	—	0.9784 nF	2.5617 nF
$1/2Q$	—	0.6405 nF	0.2446 nF

6.6 RESISTIVE GAIN ENHANCEMENT

In studying the Sallen–Key circuit we have found that the gain K is adjusted to control Q. Any excess gain is compensated by a resistive voltage-divider circuit.

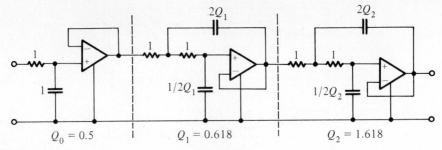

FIGURE 6.23

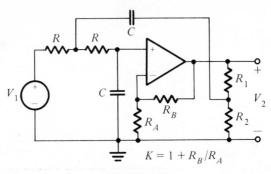

FIGURE 6.24

In this section we consider a different problem. Suppose that we want the lowpass Butterworth response, but we also want more gain than the Sallen–Key circuit provides. While such gain is realized with additional stages, it can also be attained by a modification of the Sallen–Key circuit, or other circuits operating on the same principle. In the circuit shown in Fig. 6.25 the Sallen–Key circuit is modified in that only a fraction of the output voltage V_2 is fed back through the capacitor, the amount being kV_2, where, from the voltage-divider equation,

$$k = \frac{R_2}{R_1 + R_2} \tag{6.76}$$

The circuit used for analysis as shown in Fig. 6.17 need only be changed with kV_2 substituting V_2, the controlled source. If we let $R_1 = R_2 = R$ and $C_1 = C_2 = C$, then it is found that the transfer function becomes

$$T(s) = \frac{K/R^2 C^2}{s^2 + [(3 - kK)/RC]\, s + 1/(RC)^2} \tag{6.77}$$

From this

$$Q = \frac{1}{3 - kK} \tag{6.78}$$

This compares with the value given in Eq. (6.58), which is

$$Q = \frac{1}{3 - K} \tag{6.79}$$

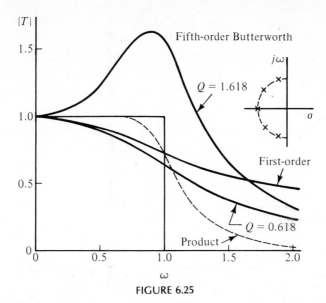

FIGURE 6.25

Comparing these two equations we see that for a Q that is given as a result of specifications, K can be made larger provided that k is made smaller to compensate. This permits a much larger value of K to be employed in the circuit realization. This is known as *resistive gain enhancement*.

6.7 RC–CR TRANSFORMATION

The general subject of frequency transformations is covered in Chapter 11. Postponing this detailed study, we now present a simple and very useful transformation known as the *RC–CR* transformation. The consequence of it is that if we can design lowpass filters, then we can design highpass filters by a simple change of the kind of element. This objective is indicated in Fig. 6.26.

If we divide the frequency scale of Fig. 6.27a into two parts, from 0 to 1 and from 1 to ∞, then the portion of the frequency scale of interest is seen to have a reciprocal relationship for the lowpass and the highpass filter cases. This may be written in equation form as

$$\omega = \frac{1}{\Omega} \tag{6.80}$$

We will later generalize this to

$$S = \frac{\omega_0}{s} \tag{6.81}$$

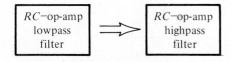

FIGURE 6.26

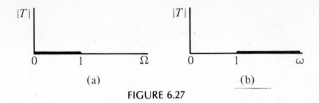

FIGURE 6.27

where the normalized frequency 1 has been scaled to ω_0. It was initially shown by Mitra* that this transformation is accomplished if the circuit is modified as follows:

$$R_i \text{ is replaced by } C_i = \frac{1}{R_i}$$

$$C_j \text{ is replaced by } R_j = \frac{1}{C_j} \tag{6.82}$$

as shown in Fig. 6.28. We have modeled the op amp as a voltage-controlled source which is not affected by this transformation, and so K, a gain factor used earlier in this chapter, is not changed. Note also that it is not necessary to apply this transformation to the resistors used to set the gain for the noninverting op-amp circuit, R_A and R_B.

To illustrate the procedure, we return to Design 2 of the Sallen–Key circuit shown in Fig. 6.19b for which the design equations were given in Eq. (6.62). With $K = 1$ and $R_1 = R_2 = 1$, then Eq. (6.55) was

$$T(s) = \frac{1/C_1 C_2}{s^2 + (2/C_1)\,s + 1/C_1 C_2} \tag{6.83}$$

From the requirement that $C_1 C_2 = 1$ and $2/C_1 = 1/Q$, we found the design equa-

* S. K. Mitra, "A Network Transformation for Active *RC* Networks," *Proc. IEEE,* vol. 55, pp. 2021–2022, 1967.

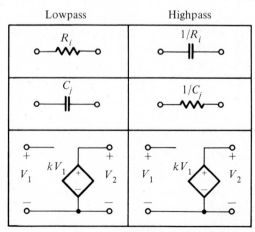

FIGURE 6.28

tions to be

$$C_1 = 2Q \quad \text{and} \quad C_2 = \frac{1}{2Q} \qquad (6.84)$$

The resulting circuit of Fig. 6.19b is also shown in Fig. 6.29a. If we apply the element transformations specified in Eqs. (6.82), then we obtain the highpass filter shown in Fig. 6.29b which has the magnitude response shown in Fig. 6.30. Substituting

$$R_1 = \frac{1}{2Q} \quad \text{and} \quad R_2 = 2Q \qquad (6.85)$$

along with replacing s by $(1/s)$ gives us

$$T(s) = \frac{s^2}{s^2 + (2/R_2)\,s + 1/R_1R_2} \qquad (6.86)$$

which is the transfer function for the highpass circuit of Fig. 6.29b.

As a second example of the use of the *RC–CR* transformation we make use

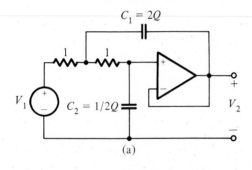

(a)

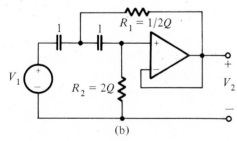

(b)

FIGURE 6.29

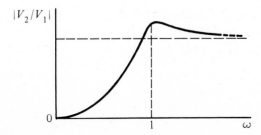

FIGURE 6.30

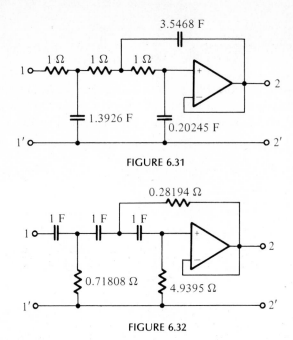

FIGURE 6.31

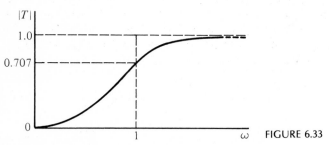

FIGURE 6.32

FIGURE 6.33

of the circuit due to Geffe,* shown in Fig. 6.31, which realizes a third-order But-terworth frequency response with $\omega_0 = 1$. Applying the RC–CR transformation, we obtain the circuit shown in Fig. 6.32, which will have the frequency response given in Fig. 6.33. Such a circuit will substitute for the cascade connection of a first-order and a second-order circuit.

Example 6.4   We require a highpass filter having the attenuation characteristics shown in Fig. 6.34, which indicate an attenuation of at least 30 dB at 60 Hz and 3 dB or less attenu-ation for all frequencies in excess of 200 Hz.

   For our design we first test the suitability of the Geffe circuit of Fig. 6.32. This is a third-order Butterworth circuit. The normalizing frequency is 200 Hz or 1256.6 rad/s, which will be made to correspond to $\omega_0 = 1$ in the Geffe circuit. In terms of this frequency normalization, 60 Hz corresponds to 0.3 rad/s. To compute the attenuation at that fre-quency, we use Eq. (6.39) which was derived for the lowpass case and recognize that 0.3

* P. R. Geffe, "How to Build High-Quality Filters out of Low-Quality Parts," *Electronics,* pp. 111–113, Nov. 11, 1976.

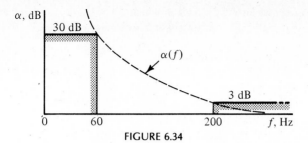

FIGURE 6.34

rad/s for the highpass case is equivalent to $1/0.3 = 3.333$ for the lowpass case. Then from Eq. (6.39)

$$\alpha = 10 \log (1 + 3.333^6) = 31.35 \text{ dB} \qquad (6.87)$$

which means that the Geffe circuit provides more than the needed attenuation at 60 Hz.

From the specifications we must frequency scale 1 rad/s to 1256.6 rad/s (200 Hz), meaning that $k_f = 1256.6$. Suppose that we decide to use 0.1-μF capacitors. This then fixes the value of k_m in Eq. (6.75) at

$$k_m = \frac{10^7}{k_f} = \frac{10^7}{1256.6} = 7958 \qquad (6.88)$$

We may now scale the resistors of the Geffe circuit, Fig. 6.32, by multiplying each value by 7958. This gives the final design shown in Fig. 6.35.

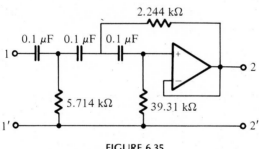

FIGURE 6.35

PROBLEMS

6.1 In the circuit shown in Fig. P6.1, it is given that $R_1 = 1$. The elements L_1 and C_2 are to be determined such that V_2/V_1 gives a Butterworth frequency response.

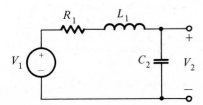

FIGURE P6.1

6.2 Figure P6.2 shows an *RLC* circuit in which $R_2 = 1$ and L_1 and C_2 are to be determined so that V_2/V_1 gives a Butterworth frequency response.

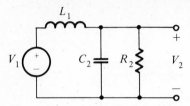

FIGURE P6.2

6.3 Figure P6.3 shows an *RLC* circuit driven by a current source I_1. It is given that $R_2 = 1$. You are to find the values of C_1 and L_2 such that V_2/I_1 gives a Butterworth frequency response.

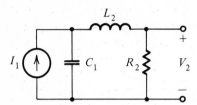

FIGURE P6.3

6.4 Consider the following three sets of specifications:

	α_{max}, *dB*	α_{min}, dB	ω_p, rad/s	ω_s, rad/s
(a)	0.25	15	10,000	14,000
(b)	0.50	30	750	1,750
(c)	1.00	25	1,250	4,375

For each of the three cases, do the following:
  (i)   Determine *n*, the required order of the Butterworth LP filter.
  (ii) Determine the half-power frequency, ω_0.
  (iii) Determine the actual attenuation at the edge of the passband and the edge of the stop band, $\alpha(\omega_p)$ and $\alpha(\omega_s)$.
  (iv) Determine the attenuation at the frequencies $2 \times \omega_s$ and $10 \times \omega_s$

6.5 Repeat Problem 6.4 for the following three specifications:

	α_{max}, dB	α_{min}, dB	f_p, Hz	f_s, Hz
(a)	0.25	18	1000	1400
(b)	0.50	20	2000	2800
(c)	0.50	30	1000	1400

6.6 A fifth-order lowpass Butterworth filter characteristic has the values $\omega_p = 1000$ rad/s and $\alpha_{max} = 0.35$ dB. What will be the attenuation of the Butterworth response when $\omega = 2000$ rad/s?

In this series of problems, a lowpass filter is to be studied having specified loss character-istics as shown in Fig. P6.4. For each set of specifications, determine the following:
(a) Determine n, the required order of the Butterworth response.
(b) Determine the s plane location of the poles, and the Q of each pole.
(c) Determine the actual loss at the edge of the passband and the edge of the stopband, $\alpha(\omega_p)$ and $\alpha(\omega_s)$.
(d) Determine the half-power frequency, ω_0.

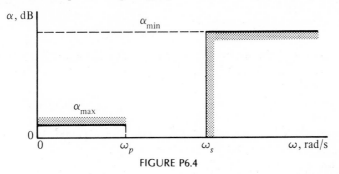

FIGURE P6.4

	α_{max}, dB	α_{min}, dB	ω_p, rad/s	ω_s, rad/s
6.7	0.5	30	1000	2330
6.8	0.5	20	1000	2000
6.9	1.0	35	1000	3500
6.10	0.5	20	1000	1725

6.11 Repeat Problem 6.9 if the frequencies specified are in Hz rather than rad/s: $f_p = 1000$ Hz and $f_s = 3500$ Hz.

We wish to design a lowpass filter to satisfy the loss specifications shown in Fig. P6.12. Thus we wish a filter with a flat loss of α_1 dB and a Butterworth response specified by α_2, α_3, ω_1, and ω_2. Our final design might better be described as a filter-attenuator. For uniformity, we will specify that the Sallen–Key circuit of Fig. 6.19c be used and that the capacitors in the realization should have the value of 1 μF.

	α_1, dB	α_2, dB	α_3, dB	ω_1, rad/s	ω_2, rad/s
6.12	8	9	23	1000	2300
6.13	6	8	32	1000	3000
6.14	10	10.5	40	100	800

6.15 It is required that we design a lowpass filter. However, only one op-amp is available and the stockroom has only 0.1-μF capacitors. It does have a good stock of resistors. You are to design a filter to meet the following specifications:
(a) The response is to be Butterworth.
(b) You are to obtain the maximum value of n that is possible.
(c) The half-power frequency is to be $\omega_0 = 2000$ rad/s.
(d) In addition to the filtering action, we wish a low-frequency gain of 14 dB.

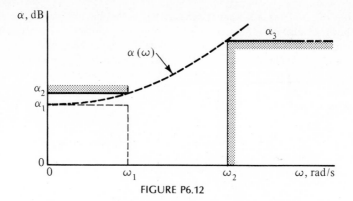

FIGURE P6.12

6.16 You are required to design an amplifier-filter using two stages of Sallen–Key circuits. The gain at dc is to be $A = 20$ dB, and a fourth-order Butterworth response is required with $A = 17$ dB at $\omega = 10{,}000$ rad/s. Use magnitude scaling to obtain elements in a practical range.

The next series of problems relates to the specifications given in Problems 6.7 through 6.11. For each set of specifications, do the following:
(a) Find a Sallen–Key realization and magnitude scale to obtain elements in a practical range.
(b) Modify the circuit found in (a) to obtain gain enhancement of 20 dB (flat for all frequencies).

6.17 Design using the specifications in Problem 6.7.
6.18 Design using the specifications in Problem 6.8.
6.19 Design using the specifications in Problem 6.9.
6.20 Design using the specifications in Problem 6.10.
6.21 Design using the specifications in Problem 6.11.

6.22 Design a lowpass filter with a Butterworth response to meet the specifications: $\alpha_{max} = 0.50$ dB, $\alpha_{min} = 30$ dB, $\omega_p = 750$ rad, $\omega_s = 1750$ rad/s.
(a) Make use of Sallen–Key circuits with $K = 1$, and magnitude scale to obtain elements in a practical range.
(b) To the filter designed in part (a), we wish to add gain enhancement of 20 dB (flat for all frequencies). Modify the design of part (a) to accomplish this objective.

6.23 Consider the Sallen–Key lowpass circuit with the following choice made for the design of the fixed elements:

$$K = 2, \qquad R_1 C_1 = R_2 C_2, \qquad C_1 = 1$$

Determine design equations which express the values of R_1, R_2, and C_2 in terms of ω_0 and Q.

6.24 Consider the Sallen–Key lowpass circuit for $K = 1$ which is to be designed to realize a pair of poles located at the angles $\pm\psi$ with respect to the negative real axis of the s plane. Using the assumption that $R_1 = R_2 = 1$, show that $\cos\psi = \sqrt{C_2/C_1}$.

6.25 Show that the circuit given in Fig. P6.25 provides a Butterworth lowpass response with 60 dB of gain at dc and with a half-power frequency of 1.577 kHz.

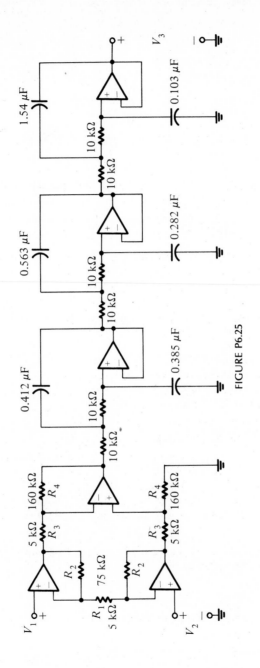

FIGURE P6.25

187

6.27 Consider the *RC* op-amp circuit shown in Fig. P6.27. What value of R_1 and R_2 will give the transfer function

$$T(s) = \frac{s^2}{s^2 + (1/Q)s + 1}$$

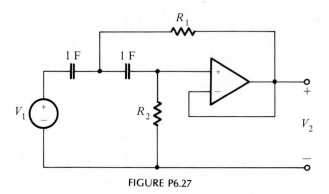

FIGURE P6.27

6.28 A highpass filter is required to meet the specifications shown in Fig. P6.28. Make use of the circuit given in Fig. 6.32 to design the filter, and scale so that all capacitors have the value of $C = 0.1\ \mu\text{F}$.

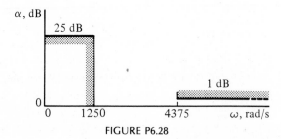

FIGURE P6.28

CHAPTER 7

Butterworth
Bandpass
Filters

This chapter is a continuation of the preceding one with the range of frequencies over which $|T|$ is maximally flat transformed from the lowpass case to the bandpass case, as shown in Fig. 7.1. The main question to be answered is how to locate the poles and zeros of $T(s)$. This must be known so that poles can be assigned to cascaded modules from which a circuit can be designed to meet specifications.

7.1 A FREQUENCY TRANSFORMATION

One of the outputs for the biquad circuit studied in Chapter 5 made it a bandpass filter having a frequency response as shown in Fig. 7.2a. The sharpness of this response depended on the Q of the circuit, but for all values of Q the response had one-pole rolloff. The response was produced by a pair of complex conjugate poles and a zero at the origin, as shown in Fig. 7.2b. For a bandpass filter that is an approximation to the brick wall studied in the last chapter, we anticipate the advantages of a maximally flat response in the pass band and also n-pole rolloff outside the pass band. For this response we may anticipate that the zeros of $T(s)$ will be located at or near the origin, and that the poles will be in the cross-hatched regions shown in Fig. 7.2d. The exact location of the poles will next be determined through the use of a frequency transformation.

Figure 7.3 shows the lowpass brick wall with the frequency marked Ω, normalized to extend from -1 to $+1$. We wish to transform values of Ω to the corresponding values of ω that will define the pass bands and stop bands shown in Fig. 7.3b. Only frequency is to be transformed, with $|T|$ unchanged in going from Fig. 7.3a to Fig. 7.3b. Such a transformation may be written

$$\Omega = X(\omega) \tag{7.1}$$

Frequency transformations play an important role in filter design, and Chapter 11 will be devoted to this subject in its entirety. In that chapter we will find that the appropriate transformation from the lowpass to the bandpass case has the

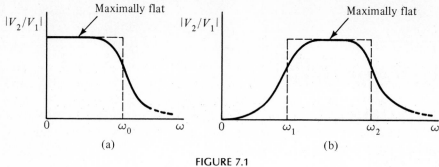

FIGURE 7.1

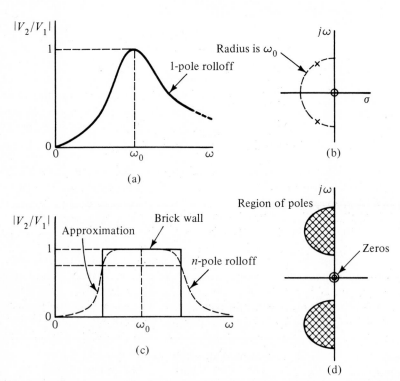

FIGURE 7.2   The bandpass magnitude characteristic of *a* corresponds to the poles and zero of *b*. The figure suggests by analogy the magnitude response of *c* will correspond to poles located in the crosshatched region and zeros at the origin as in *d*.

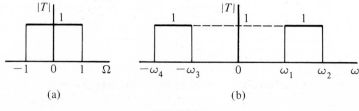

FIGURE 7.3

general form

$$\Omega(\omega) = K \frac{1}{\omega} (\omega^2 - \omega_0^2) \tag{7.2}$$

where K is an unknown, and ω_0^2 would normally be an unknown, but we have anticipated its value. This function is one of a class of functions known as *Foster reactance functions*, to be studied extensively in Chapter 16. To match it to the problem we are studying, we make use of the fact, shown in Fig. 7.4, that we wish $\Omega = +1$ to transform into ω_2 and $\Omega = -1$ to transform to ω_1. Substituting these facts into Eq. (7.1) gives

$$\Omega(\omega_1) = -1 = K \frac{\omega_1^2 - \omega_0^2}{\omega_1} \tag{7.3}$$

$$\Omega(\omega_2) = +1 = K \frac{\omega_2^2 - \omega_0^2}{\omega_2} \tag{7.4}$$

Solving these two equations, we obtain

$$\omega_0^2 = \omega_1 \omega_2 \tag{7.5}$$

and

$$K = \frac{1}{\omega_2 - \omega_1} \tag{7.6}$$

Thus ω_0 is the geometrical mean of the two frequencies ω_1 and ω_2, and K is the reciprocal of a bandwidth. In Fig. 7.4 this bandwidth is defined as the ends of the brick wall, while in earlier studies of the Butterworth response it has been the frequency at which the response had decreased to 0.707 of its maximum value. To maintain the generality of bandwidth, we designate the half-power or 3-dB bandwidth as BW, and a bandwidth in general as bw. Hence Eq. (7.2) becomes

$$\Omega(\omega) = \frac{1}{bw} \frac{\omega^2 - \omega_1 \omega_2}{\omega} \tag{7.7}$$

If we replace ω by s/j and similarly Ω by S/j, then this equation has the general

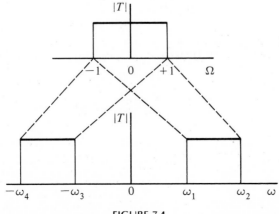

FIGURE 7.4

form

$$S(s) = \frac{1}{bw} \frac{s^2 + \omega_0^2}{s} \tag{7.8}$$

Before continuing, we examine Eq. (7.7) by considering the special case $\omega_0 = 1$ and bw = 1. For these values Eq. (7.7) becomes

$$\Omega = \frac{\omega^2 - 1}{\omega} = \omega - \frac{1}{\omega} \tag{7.9}$$

or

$$\omega^2 - \omega\Omega - 1 = 0 \tag{7.10}$$

We make use of the quadratic equation to solve for the four values of ω:

$$\omega_a, \omega_b, \omega_c, \omega_d = \frac{\pm\Omega \pm \sqrt{\Omega^2 + 4}}{2} \tag{7.11}$$

Let ω_b and ω_a be the two positive values of the solution (with corresponding negative values for ω_c and ω_d). These are tabulated in Table 7.1. From this table and Fig. 7.5 we see that one value moves toward the origin and the other toward infinity for increasing values of Ω. From the example, we see that the transformation indeed "maps" all values of Ω into the corresponding values for ω, as was intended.

7.2 GEFFE ALGORITHM

The example used in the last section provides a good starting point for explaining the problem of calculating the pole positions. From Eq. (7.8) with bw = 1 and $\omega_0 = 1$, the transformation is

$$S = s + \frac{1}{s} \tag{7.12}$$

We wish to apply this frequency transformation to the third-order Butterworth function given in Table 6.1 with pole positions as shown in Fig. 7.6. The transfer function is

$$T(s) = \frac{1}{s^3 + 2s^2 + 2s + 1} = \frac{1}{(s + 1)(s^2 + s + 1)} \tag{7.13}$$

TABLE 7.1

Ω	ω_b	ω_a
0	1	1
±0.5	1.28	0.78
±1	1.618	0.618
±1.5	2	0.5
±2	2.414	0.414
±10	10.1	0.1

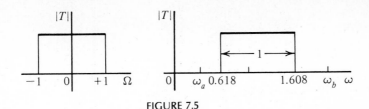

FIGURE 7.5

The second form of the equation reminds us that the Butterworth polynomial $B_n(s)$ contains the factor $(s + 1)$ for n odd, but otherwise is the product of second-order factors $(s^2 + (1/Q_i)s + 1)$ with $\omega_0 = 1$. For the example we are studying, let $T = T_1 T_2$. Using the lowpass variable $S = \Sigma + j\Omega$, then

$$T_1 = \frac{1}{S + 1} \quad \text{and} \quad T_2 = \frac{1}{S^2 + S + 1} \tag{7.14}$$

We will call our study of T_1 case I. Substituting Eq. (7.12) into T_1's denominator,

$$S + 1 = s + \frac{1}{s} + 1 = \frac{s^2 + s + 1}{s} \tag{7.15}$$

Then the case I pole on the negative real axis has transformed into two conjugate poles at $\pm 60°$ with respect to the negative real axis and a zero at the origin, as shown in Fig. 7.7. The same operation for T_2 gives

$$S^2 + S + 1 = \left(s + \frac{1}{s}\right)^2 + \left(s + \frac{1}{s}\right) + 1 = \frac{s^4 + s^3 + 3s^2 + s + 1}{s^2} \tag{7.16}$$

Then for case II the two poles transform into four poles and two zeros at the origin. The location of the poles is determined by finding the roots of the equation

$$s^4 + s^3 + 3s^2 + s + 1 = 0 \tag{7.17}$$

When we find them, the poles and zeros in the s plane are as shown in the case II part of Fig. 7.7. How do we solve Eq. (7.17)?

If you have ready access to a computer, the roots of Eq. (7.17) can be found. Or you might use an iterative procedure, guessing the location of two roots and then testing to see if it is a solution. Fortunately there is a routine procedure for accomplishing the factoring of Eq. (7.17) due to Geffe* which has been arranged to give the answer in terms of ω_{0i} and Q_i. It is known as *Geffe's algorithm* and is of great value to filter designers.

* P. R. Geffe, "Designers' Guide to Active Bandpass Filters," *EDN*, pp. 46–52, Apr. 5, 1974. The name is pronounced "geffie," with the g sounded as in "golley."

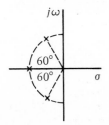

FIGURE 7.6

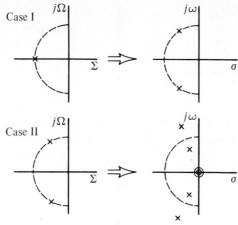

FIGURE 7.7

We next generalize the simple example of Eqs. (7.14) for which the frequency transformation of Eq. (7.8) was specialized to the case of bw = 1 and ω_0 = 1. In particular we wish to examine the quantity bw. If Eq. (7.8) is written in the form

$$S = \frac{\omega_0}{\text{bw}}\left(\frac{s}{\omega_0} + \frac{\omega_0}{s}\right) = q_c\left(\frac{s}{\omega_0} + \frac{\omega_0}{s}\right) \qquad (7.18)$$

then we have defined a new quantity:

$$q_c = \frac{\omega_0}{\text{bw}} = \frac{\omega_0}{\omega_2 - \omega_1} \qquad (7.19)$$

If the bandwidth is defined as the half-power or 3-dB frequencies, then bw becomes BW and q_c becomes Q, the design quantity introduced in Chapter 5. In a more general sense bw is defined in terms of $\alpha_{\max}$ as shown in Fig. 7.8, as the frequency difference $\omega_2 - \omega_1$. Now $\alpha_{\max}$ may have any design value (0.1 dB, 1 dB, etc.), and so the quantity q_c is one related to the specifications. On the other hand, the 3-dB bandwidth relates to the design value Q because the circuits we will use in our realizations relate to this Q. This important difference will become clear when we give design examples.

The denominator of the first-order transfer function of Eq. (7.14) will be written in terms of the pole location being at $-\Sigma_1$, as shown in Fig. 7.9a. Then

$$S + \Sigma_1 = q_c\left(\frac{s}{\omega_0} + \frac{\omega_0}{s}\right) + \Sigma_1 \qquad (7.20)$$

$$= \frac{s^2 + (\Sigma_1\omega_0/q_c)s + \omega_0^2}{s} \cdot \frac{1}{(\omega_0/q_c)} \qquad (7.21)$$

The standard form of the second-order factor is

$$s^2 + \frac{\omega_0}{Q}s + \omega_0^2 \qquad (7.22)$$

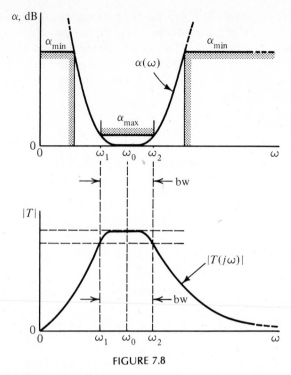

FIGURE 7.8

and so we see that the design Q is related to the defined q_c as

$$Q = \frac{q_c}{\Sigma_1} \qquad (7.23)$$

The fact that the constant term in the numerator of Eq. (7.21) is ω_0^2, which is the same as the standard form Eq. (7.22), may be indicated by saying that the radius to which the simple pole of Fig. 7.9a is transformed is

$$\omega_T = \omega_0 \qquad (7.24)$$

The angle of the poles with respect to the negative real axis is

$$\psi = \cos^{-1}\left|\frac{\Sigma_1}{2q_c}\right| \qquad (7.25)$$

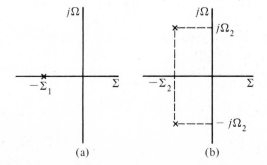

(a)   (b)   FIGURE 7.9

These quantities are shown in Fig. 7.10. We note that it may be convenient to let $\omega_0 = \omega_T = 1$ and later frequency scale. Then we see that the simple pole on the negative real axis transforms to a pair of complex poles and a zero at the origin, and that the pole locations are fixed by the radius of the circle and the angle ψ.

The second-order poles of $T_2(S)$ are shown in Fig. 7.9b and are located at $-\Sigma_2 \pm j\Omega_2$. Then

$$T_2(S) = \frac{1}{S^2 + 2\Sigma_2 S + \Sigma_2{}^2 + \Omega_2{}^2} \tag{7.26}$$

If we let $\omega_0 = 1$, then the frequency transformation is

$$S = q_c\left(s + \frac{1}{s}\right) \tag{7.27}$$

When this is substituted into Eq. (7.26), the denominator becomes

$$q_c{}^2\left(\frac{s^2 + 1}{s}\right)^2 + 2\Sigma_2 q_c\left(\frac{s^2 + 1}{s}\right) + \Sigma_2{}^2 + \Omega_2{}^2 \tag{7.28}$$

and this reduces to the fourth-order equation

$$s^4 + \left(\frac{2\Sigma_2}{q_c}\right)s^3 + \left(2 + \frac{\Sigma_2{}^2 + \Omega_2{}^2}{q_c{}^2}\right)s^2 + \left(\frac{2\Sigma_2}{q_c}\right)s + 1 = 0 \tag{7.29}$$

for which we wish to find the four roots.

Leaving this result, we recognize that the transfer function of a standard bandpass filter section was given in Chapter 5 as

$$T_1(s) = \frac{(\omega_{01}/Q_1)s}{s^2 + \omega_{01}/Q_1 + \omega_{01}{}^2} \tag{7.30}$$

where ω_{01} and Q_1 are the design values from which a circuit may be found. Further, to realize a fourth-order specification such as Eq. (7.29), we will require two modules connected in cascade, each having the transfer function of the form of Eq. (7.30). Letting the second module have the transfer function of Eq. (7.30), but with the subscript 1 replaced by the subscript 2, then the denominator of $T_1(s)T_2(s)$ will be

$$\left(s^2 + \frac{\omega_{01}}{Q_1}s + \omega_{01}{}^2\right)\left(s^2 + \frac{\omega_{02}}{Q_2}s + \omega_{02}{}^2\right) \tag{7.31}$$

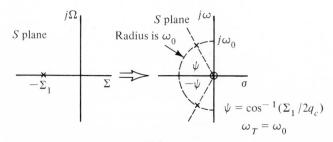

FIGURE 7.10

Carrying out the multiplication and rearranging the terms in an orderly fashion, we obtain

$$s^4 + \left(\frac{\omega_{02}}{Q_2} + \frac{\omega_{01}}{Q_1} \right) s^3 + \left(\omega_{02}{}^2 + \omega_{01}{}^2 + \frac{\omega_{01}\omega_{02}}{Q_1 Q_2} \right) s^2$$

$$+ \left(\frac{\omega_{01}\omega_{02}{}^2}{Q_1} + \frac{\omega_{02}\omega_{01}{}^2}{Q_2} \right) s + \omega_{01}{}^2 \omega_{02}{}^2 = 0 \qquad (7.32)$$

The strategy that we will follow is to equate coefficients of this equation and Eq. (7.29) and thereby determine the conditions for a solution in the cascade module form assumed.

The most obvious equality is that

$$\omega_{01}{}^2 \, \omega_{02}{}^2 = 1 \qquad (7.33)$$

or

$$\omega_{01}\omega_{02} = 1 \qquad \text{or} \qquad \omega_{01} = \frac{1}{\omega_{02}} \qquad (7.34)$$

Furthermore, since the 1 in Eq. (7.29) arose because we required that $\omega_0 = 1$, the general form of Eq. (7.34) will be

$$\omega_{01}\omega_{02} = \omega_0{}^2 \qquad (7.35)$$

In Eq. 7.29 we note that the coefficients of s^3 and s are identical. For this to be the case in Eq. (7.32), we must require that

$$\frac{\omega_{02}}{Q_2} + \frac{\omega_{01}}{Q_1} = \frac{\omega_{01}\omega_{02}{}^2}{Q_1} + \frac{\omega_{02}\omega_{01}{}^2}{Q_2} \qquad (7.36)$$

Since $\omega_{01}\omega_{02} = 1$, this condition is satisfied only if

$$Q_1 = Q_2 = Q \qquad (7.37)$$

an important result!

There remains only the coefficient of s^3, which we know is the same as that for s, plus the coefficient of s^2. The first of these is now simplified to

$$\frac{2\Sigma_2}{q_c} = \frac{1}{Q} \left(\omega_{02} + \frac{1}{\omega_{02}} \right) \qquad (7.38)$$

The second, after adding 2 to both sides of the equation, is

$$\frac{\Sigma_2{}^2 + \Omega_2{}^2}{q_c{}^2} + 4 = \left(\omega_{02} + \frac{1}{\omega_{02}} \right)^2 + \frac{1}{Q^2} \qquad (7.39)$$

Let

$$C = \Sigma_2{}^2 + \Omega_2{}^2 \qquad (7.40)$$

$$D = \frac{2\,\Sigma_2}{q_c} \qquad (7.41)$$

$$E = 4 + \frac{C}{q_c{}^2} \qquad (7.42)$$

Geffe showed that the conditions for satisfying Eqs. (7.38) and (7.39) could be expressed in algorithmic form in terms of the defined quantities C, D, and E. The Geffe algorithm requires that the following calculations be made:

$$G = \sqrt{E^2 - 4D^2} \tag{7.43}$$

$$Q = \frac{1}{D}\sqrt{\frac{1}{2}(E + G)} \tag{7.44}$$

$$K = \frac{\Sigma_2 Q}{q_c} \tag{7.45}$$

$$W = K + \sqrt{K^2 - 1} \tag{7.46}$$

Then finally,

$$\omega_{02} = W\,\omega_0 \quad \text{and} \quad \omega_{01} = \frac{1}{W}\,\omega_0 \tag{7.47}$$

The form of these equations makes calculation particularly easy with a calculator. The result is that Q is given by Eq. (7.44), and the two pole locations are specified in terms of ω_0 and a factor W. The transformation also gives two zeros at the origin. All of this is shown in Fig. 7.11.

To illustrate the application of this important algorithm, let us return to the unfinished business of factoring Eq. (7.17). This equation arose from the conjugate poles for which

$$\Sigma_2 = -0.5 \tag{7.48}$$

and

$$\Omega_2 = \pm 0.866$$

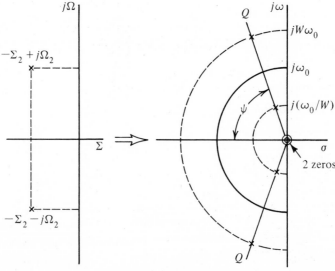

FIGURE 7.11

and the further requirements that bw = 1 and ω_0 = 1. We begin the algorithm with Eq. (7.40) and then carry out the calculations in a step-by-step manner as follows:

$$C = 1$$

$$D = 1$$

$$E = 5$$

$$G = 4.5826$$

$$Q = 2.189 \tag{7.49}$$

$$K = 1.0945$$

$$W = 1.5393$$

$$\omega_{02} = 1.5393 \text{ rad/s}$$

$$\omega_{01} = 0.6497 \text{ rad/s}$$

Thus the three quantities we will need for design are now known; namely, Q, ω_{01}, and ω_{02}. These pole locations for the upper half of the s plane, together with that found previously arising from the pole on the negative real axis, are shown in Fig. 7.12, together with the two zeros at the origin. If we are specifically interested in pole location, as we are in order to factor Eq. (7.17), then we may find the s-plane locations of the poles from the equations

$$\beta_i = \omega_{0i} \sin \cos^{-1}\left(\frac{1}{2Q_i}\right) \tag{7.50}$$

$$\alpha_i = \omega_{0i} \cos \cos^{-1}\left(\frac{1}{2Q_i}\right) = \frac{\omega_{0i}}{2Q_i} \tag{7.51}$$

Using these equations and a calculator, we find that the roots of Eq. (7.17) are

$$-0.148 \pm j0.6325$$

$$-0.3516 \pm j1.4985 \tag{7.52}$$

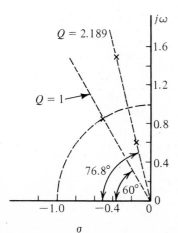

FIGURE 7.12

The corresponding quadratic terms are

$$(s^2 + 0.296s + 0.422)(s^2 + 0.7032s + 2.3691) \qquad (7.53)$$

Finally, if this multiplication is carried out, we find that (with tolerance for round-off errors)

$$s^4 + s^3 + 3s^2 + s + 1 = 0 \qquad (7.54)$$

While this algorithm is very useful for design, it is also useful as a method for factoring fourth-order algebraic equations, assuming only that the roots are complex.

Returning to Fig. 7.12 we now ask the question concerning the loci of the poles: might they be on a circle? Stated in other terms, does a circle in the S plane transform into a circle in the s plane? The answer is no; the property we have just described applies for bilinear transformations, but the transformation of Eq. (7.8) is not bilinear. Actually the poles in the s plane are on a curve that might be called a distorted ellipse. This is illustrated in Fig. 7.13. The circle of radius 1 in the S plane, shown in Fig. 7.13a, maps into the s plane as shown in Fig. 7.13b. Notice also the way in which lines of constant Q are distorted by the transformation.

7.3 FINDING *n*:
BANDPASS BACK TO LOWPASS

In the example of the last section, we assumed that $n = 3$ was specified. This assumption is not necessary, of course, since n is obtained directly from the specification. The strategy for finding the required value of n is very simple. We use the lowpass to bandpass transformation in reverse, and convert the bandpass specification into equivalent lowpass specifications which we studied in Chapter 6.

The bandpass specifications are given in Fig. 7.14. The two frequencies ω_1 and ω_2 which define the bandwidth bw are identical to those shown in Fig. 7.8.

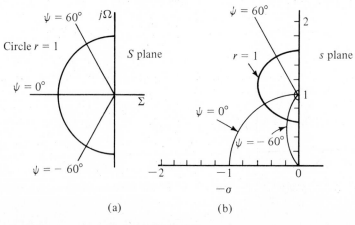

(a)   (b)

FIGURE 7.13

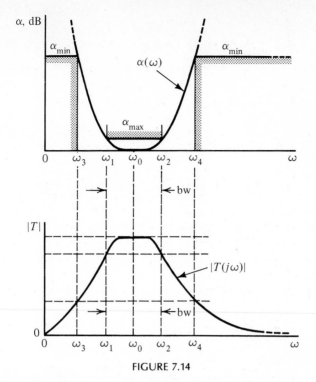

FIGURE 7.14

We assume that the low frequency and high frequency attenuation specifications are identical and specified by α_{min}. The two frequencies at which the attenuation is α_{min} are identified as ω_3 and ω_4. Just as

$$\omega_1\omega_2 = \omega_0{}^2 \tag{7.55}$$

so also

$$\omega_3\omega_4 = \omega_0{}^2 \tag{7.56}$$

Since bw $= \omega_2 - \omega_1$, the equation for the frequency transformation of Eq. (7.7) becomes

$$\Omega = -\; \frac{-\omega^2 + \omega_0{}^2}{\omega(\omega_2 - \omega_1)} \tag{7.57}$$

This transformation will be applied to the lowpass frequency characteristics of Fig. 7.15. Note that α_{min} and α_{max} are identical in Figs. 7.14 and 7.15; we transform only frequencies and not attenuation. Since we derivated the last equation, Eq. (7.57), assuming that $\Omega = 1$ transformed into ω_1 and ω_2, we would be surprised if substituting $\omega = \omega_2$ into Eq. 7.57 did not give the value 1. So

$$\Omega_p = -\; \frac{-\omega_2{}^2 + \omega_0{}^2}{\omega_2(\omega_2 - \omega_1)} = 1 \tag{7.58}$$

Similarly, substituting ω_4 into Eq. (7.57) and making use of Eq. (7.56), we have

$$\Omega_s = \frac{\omega_4 - \omega_3}{\omega_2 - \omega_1} \tag{7.59}$$

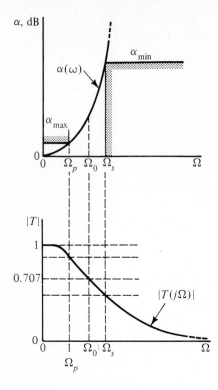

FIGURE 7.15

where the subscript p was chosen to indicate the end of the pass band, and s the beginning of the stop band (as far as specifications are concerned). So we have started with the bandpass specifications and reduced these to equivalent normalized lowpass specifications:

$$\alpha_{min}, \qquad \alpha_{max}, \qquad \Omega_p = 1, \qquad \Omega_s = \frac{\omega_4 - \omega_3}{\omega_2 - \omega_1} \qquad (7.60)$$

Now we may return to Chapter 6 and our analysis for the lowpass filter. The number n is given by Eq. (6.46). In terms of Ω_s and Ω_p,

$$n = \frac{\log\left[(10^{\alpha_{min}/10} - 1)/(10^{\alpha_{max}/10} - 1)\right]}{2 \log (\Omega_s/\Omega_p)} \qquad (7.61)$$

Once this value of n is obtained, including the usual rounding up, then the half-power frequency is given by Eq. (6.42):

$$\Omega_0 = \frac{1}{(10^{\alpha_{max}/10} - 1)^{1/2n}} \qquad (7.62)$$

where we have substituted $\Omega_p = 1$ in the numerator. From this lowpass specification we may make use of the Geffe algorithm to proceed to find the pole locations.

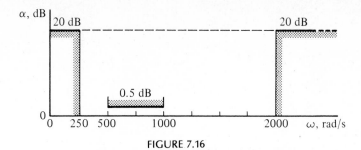

FIGURE 7.16

Example 7.1 Let the bandpass specifications be those given in Fig. 7.16, from which we see that $\omega_0 = 707$ rad/s and bw $= 500$ rad/s. Then from Eq. (7.59),

$$\Omega_s = \frac{2000 - 250}{1000 - 500} = 3.5 \tag{7.63}$$

From Eq. (7.61)

$$n = \frac{\log\left[(10^2 - 1)/(10^{0.05} - 1)\right]}{2 \log 3.5} = 2.6735 \tag{7.64}$$

which we must round up to $n = 3$. Finally, from Eq. (7.62) we have

$$\Omega_0 = \frac{1}{(10^{0.05} - 1)^{1/6}} = 1.42 \text{ rad/s} \tag{7.65}$$

The next step in a design would be to reverse the transformation and make use of the Geffe algorithm to determine the pole locations for the bandpass case.

7.4 DELYIANNIS–FRIEND CIRCUIT[*,†]

We next introduce a useful bandpass circuit which we will use as the module to be connected in cascade to realize the Butterworth bandpass specifications. The circuit is based on a bridged-T *RC* circuit with an op amp to provide negative feedback. It has been used in many applications. Two of those to first exploit the circuit were Delyiannis and Friend, whose circuits will be covered in later chapters. The filter designer will find that there are many circuits available to realize a second-order bandpass response. From the large number of possibilities, most designers have a favorite based on experience. We will adopt the Delyiannis–Friend circuit as our favorite bandpass circuit.

To analyze the circuit shown in Fig. 7.17, we will employ a different strategy than has been used in the past. Observe that the part of the circuit enclosed within the dashed lines is a stage of inverting op-amp circuit, which has a controlled-source equivalent. That part of the circuit and its equivalent are shown in

[*] T. Delyiannis, "High-Q Factor Circuit with Reduced Sensitivity," *Electron. Lett.,* 4, p. 577, Dec. 1968.
[†] J. J. Friend, "A Single Operational-Amplifier Biquadratic Filter Section," *1970 IEEE ISCT Digest Technical Papers,* 1970, p. 189.

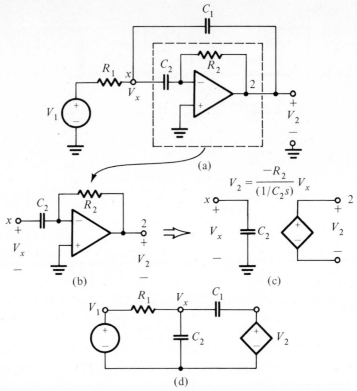

FIGURE 7.17 The figure illustrates the analysis of the circuit of (a) by transforming part of it shown in (b) to the equivalent circuit of (c). This results in the simplified circuit of (d) which is analyzed with one node equation.

Fig. 7.17b and c. Using the relationship for the inverting op-amp stage which is $T = -Z_2/Z_1$, we see that

$$V_2 = -\frac{R_2}{1/C_2 s} V_x = -R_2 C_2 s\, V_x \qquad (7.66)$$

With this simplification the circuit is as shown in Fig. 7.17d, and we may apply Kirchhoff's current law at node x. Summing the currents into the node gives the equation

$$\frac{1}{R_1}(V_x - V_1) + C_2 s\, V_x + (V_x - V_2)\, C_1 s = 0 \qquad (7.67)$$

If we eliminate the voltage V_x from the last two equations and solve for the ratio of V_2 to V_1, we obtain

$$\frac{V_2}{V_1} = T(s) = \frac{(-1/R_1 C_1)\, s}{s^2 + (1/R_2)(1/C_1 + 1/C_2)s + 1/R_1 R_2 C_1 C_2} \qquad (7.68)$$

As the first simplification, we let $C_1 = C_2 = C$, and the equation has a simplified form:

$$T(s) = \frac{(-1/R_1 C)s}{s^2 + (2/R_2 C)s + 1/R_1 R_2 C^2} \qquad (7.69)$$

Comparing this equation with the standard form of the bandpass filter response,

$$T(s) = \frac{-(\omega_0/Q)s}{s^2 + (\omega_0/Q)s + \omega_0^2} \qquad (7.70)$$

we equate coefficients in the last two equations and obtain the following equations:

$$\omega_0 = \frac{1}{C\sqrt{R_1 R_2}}, \qquad Q = \frac{1}{2}\sqrt{\frac{R_2}{R_1}}, \qquad BW = \frac{\omega_0}{Q} = \frac{2}{R_2 C} \qquad (7.71)$$

From these equations we see that the circuit may be orthogonally tuned. Having assumed that $C_1 = C_2$ to obtain Eqs. (7.71), we must gang the capacitors together (for example, connect them physically on the same shaft). Then one of the resistors can be used to tune for Q or BW. Once this is set, then the capacitor may be adjusted to match the ω_0 specifications.

A normalized form of the circuit is obtained by making the choice that $R_1 = 1$ and $\omega_0 = 1$. Then Eqs. (7.71) may be solved to give the remaining element values:

$$R_2 = 4Q^2 \qquad \text{and} \qquad C = \frac{1}{2Q} \qquad (7.72)$$

This circuit is shown in Fig. 7.18. It reminds us that the bandpass circuit is specified by the one value, Q. Once this is set, then frequency scaling can be employed to match the ω_0 requirement, and magnitude scaling can be used to obtain reasonable element values. Under the assumptions just stated, the transfer function for the Friend circuit is

$$\frac{V_2}{V_1} = \frac{-2Qs}{s^2 + (1/Q)s + 1} \qquad (7.73)$$

Comparing this equation with the standard form of Eq. (7.70), with $\omega_0 = 1$,

$$\frac{V_2}{V_1} = \frac{-(1/Q)s}{s^2 + (1/Q)s + 1} \qquad (7.74)$$

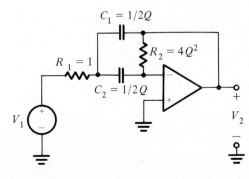

FIGURE 7.18

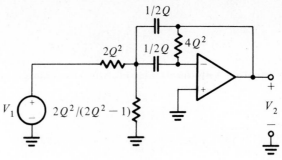

FIGURE 7.19

for which

$$\left|\frac{V_2}{V_1}(j1)\right| = 1 \tag{7.75}$$

we see that the circuit described by Eq. (7.73) gives

$$\left|\frac{V_2}{V_1}(j1)\right| = 2Q^2 \tag{7.76}$$

If it is important that the circuit provide unit gain at the resonant peak at $\omega = 1$, then the gain of the circuit may be reduced by a voltage-divider circuit for the Sallen–Key circuit in Chapter 6 using the circuit elements of Eqs. (6.66) and (6.67), also given in Fig. 6.20c. In terms of that notation we require that

$$H = \frac{1}{2Q^2} \tag{7.77}$$

giving the element sizes shown in Fig. 7.19. The magnitude response of this normalized circuit is then such that Eq. (7.75) is satisfied (or $A = 0$ dB) for all values of Q, as shown in Fig. 7.20.

As an example of the Friend circuit design, suppose that we wish the frequency response shown in Fig. 7.20 for $Q = 5$ and at the frequency $f_0 = 10{,}000$ Hz. In addition we have a preference for 0.01-μF capacitors. The normalized element values are determined by Q alone, and are those shown in Fig. 7.21a. Frequency

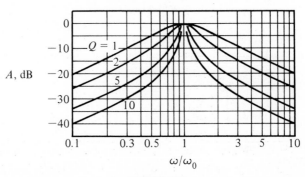

FIGURE 7.20

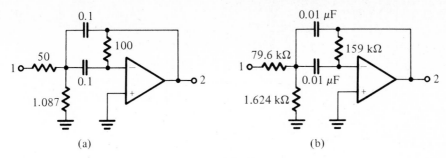

FIGURE 7.21

scaling is required from $\omega = 1$ to $f = 1000$, requiring that $k_f = 2\pi \times 10{,}000$. Using the scaling equation for capacitors,

$$C_{new} = \frac{1}{k_f k_m} C_{old} = 10^{-8} = \frac{1}{20{,}000\pi k_m} \qquad (7.78)$$

we determine k_m to be 1591.6. Then $R_{new} = k_m R_{old}$, and from this we find the element values shown in Fig. 7.21b, which completes the circuit design. This circuit provides the frequency response that was specified by $Q = 5$ in Fig. 7.20. Filter design using the Friend circuit is very simple and takes very little time.

7.5 STAGGER-TUNED BANDPASS FILTER DESIGN

Thus far in this chapter we have been collecting pieces needed for the complete bandpass filter design. Now we are ready to assemble the pieces and give a design procedure. We will do so in terms of an example. Let the specifications be those shown in Fig. 7.22, which are identical to those given for Example 7.1, except that the frequency is in hertz rather than in radians per second. The analysis given in Eqs. (7.63)–(7.65) applies here and permits us to find the lowpass equivalent of the bandpass specifications. Then we know that $n = 3$ is the required order of the Butterworth response, and that the poles are required to be on a circle of radius $F_0 = 1.42$ Hz. This is shown in Fig. 7.23. To these poles we apply the Geffe algorithm to find the location of the bandpass poles and zeros. The poles are located at

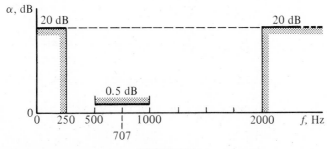

FIGURE 7.22

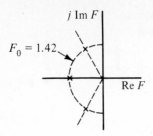

FIGURE 7.23

$$-1.42 + j0$$
$$-0.710 + j1.230 \tag{7.79}$$
$$-0.710 - j1.230$$

The first listed gives the information for case I of the Geffe algorithm. It is

$$f_0 = 707$$

$$q_c = \frac{707}{500} = 1.414 \tag{7.80}$$

$$\Sigma_1 = 1.42$$

From this information the transformed values are

$$f_0 = f_T = 707 \text{ Hz}$$

$$Q = \frac{q_c}{\Sigma_1} = 0.9958 \tag{7.81}$$

For case II of the Geffe algorithm we see that

$$\text{Re } F_2 = 0.710 \quad \text{and} \quad \text{Im } F_2 = 1.230 \tag{7.82}$$

We follow the steps of the algorithm beginning with Eq. (7.40):

$$C = (0.710)^2 + (1.230)^2 = 2.017$$

$$D = \frac{2 \times (0.710)}{1.414} = 1.0042$$

$$E = 4 + \frac{2.017}{(1.414)^2} = 5.009$$

$$G = [(5.009)^2 - 4(1.0042)^2]^{1/2} = 4.589$$

$$Q = \frac{1}{1.0042} \left[\frac{1}{2} (5.009 + 4.589) \right]^{1/2} = 2.181 \tag{7.83}$$

$$K = \frac{0.710 \times 2.181}{1.414} = 1.095$$

$$W = 1.095 + [(1.095)^2 - 1]^{1/2} = 1.541$$

$$f_{01} = 707 \times 1.541 = 1089.8 \text{ Hz}$$

$$f_{02} = \frac{707}{1.541} = 458.8 \text{ Hz}$$

Recapping our results, the following f_0 and Q are required:

Stage	f_0(Hz)	Q	
1	707	0.9958	
2	1089.8	2.181	(7.84)
3	458.8	2.181	

This result is shown in terms of s-plane location in Fig. 7.24. The dashed lines show the two values of Q, and the distance from the origin to the pole is the value of f_0.

The strategy of cascade design is illustrated by Fig. 7.25. Each pole pair and one of the zeros at the origin are assigned to one of the three cascade-connected stages. The order of assignment is not important; the three indicated are as good as any. In this way the specification for each stage is given in terms of the required Q and the frequency to which that stage must be scaled.

The implementation of the block diagram shown in Fig. 7.25b using the Friend circuit is shown in Fig. 7.26. The element values with the exception of R_1 = 1 are specified in terms of Q. The circuit has yet to be scaled in frequency and magnitude. Since the Friend circuit was determined in normalized form under the assumption that $\omega_0 = 1$, it is necessary to multiply the values in Eq. (7.84) by 2π since they are given in hertz. Then the frequency scaling for each stage is accomplished as

$$k_{fi} = 2\pi \times f_{0i} \tag{7.85}$$

Because each stage is scaled to a different value of frequency, the combined circuit is said to be *stagger tuned*. Had all of the circuits been tuned to the same frequency, the overall circuit would be described as synchronously tuned.

In terms of design it is important to recognize that a different value of the magnitude scaling factor k_m may be used for each of the three stages. This is possible because no coupling exists between the three stages. Thus, for example, a different value of k_m can be selected for each stage to make all capacitors in the

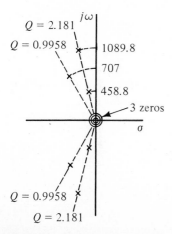

FIGURE 7.24

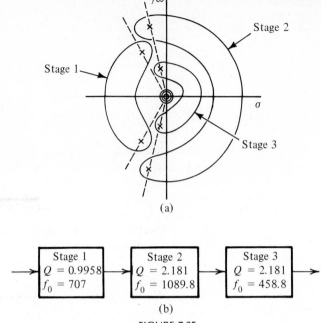

(a)

Stage 1
$Q = 0.9958$
$f_0 = 707$

Stage 2
$Q = 2.181$
$f_0 = 1089.8$

Stage 3
$Q = 2.181$
$f_0 = 458.8$

(b)

FIGURE 7.25

circuit have the same value. For this example suppose that we decide that all capacitors will have the value of 0.1 μF.

Calculations for the determination of final element values can conveniently be carried out in the form shown in Table 7.2. The resulting scaled circuit is shown in Fig. 7.27.

We next wish to plot the response we have achieved through this design. If we restore the normalizing frequency ω_{0i} to the transfer function, we have for each of the three stages

$$T_i(s) = \frac{-2\,Q_i\omega_{0i}s}{s^2 + (\omega_{0i}/Q_i)s + \omega_{0i}{}^2}, \qquad i = 1, 2, 3 \qquad (7.86)$$

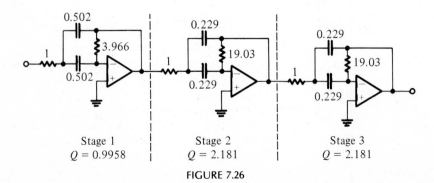

Stage 1
$Q = 0.9958$

Stage 2
$Q = 2.181$

Stage 3
$Q = 2.181$

FIGURE 7.26

TABLE 7.2

Stage	Element		
	$R_1 = 1$	$R_2 = 4Q^2$	$C_1 = C_2 = \dfrac{1}{2Q}$
1. $Q = 0.9958$	1	3.966 Ω	0.502 F
$k_f = 2\pi \times 707$. $k_m = 1130$	1.13 kΩ	4.48 kΩ	0.1 μF
2. $Q = 2.181$	1	19.0 Ω	0.229 F
$k_f = 2\pi \times 1089.8$ $k_m = 334.4$	334 Ω	6.35 kΩ	0.1 μF
3. $Q = 2.181$	1	19.0 Ω	0.229 F
$k_f = 2\pi \times 458.8$ $k_m = 794.4$	794 Ω	15 kΩ	0.1 μF

If we let $s = j\omega$, we obtain

$$|T_i(j\omega)|^2 = \frac{(2\,Q_i\omega_{0i}\omega)^2}{(\omega_{0i}^2 - \omega^2)^2 + (\omega_{0i}\omega/Q_i)^2} \tag{7.87}$$

If we let ω_{0i} have the three values given in Eq. (7.84), then we may plot the magnitude response for each of the three stages, as is done in Fig. 7.28, along with the composite magnitude response found by summing the responses of each of the three stages. From the figure it is clear that the magnitude function reaches a maximum at the center frequency $f_{01} = 707$. The overall response at that frequency is

$$|T_1(jf_{01})|\,|T_2(jf_{01})|\,|T_3(jf_{01})| = 1.98 \times (4.35)^2 = 37.5 \tag{7.88}$$

This value of gain corresponds to 31.5 dB, as shown in Fig. 7.28. If we wish the combined circuit to behave as a filter and not as a filter-amplifier, then we must compensate for this gain. This was considered in Chapter 6, and the results are given in Eqs. (6.66) and (6.67) for a voltage-divider attenuator. The circuit of Fig. 7.27 is modified to include the attenuator as shown in Fig. 7.29. We note that we have chosen to reduce the gain at the input to the filter. It would also have been

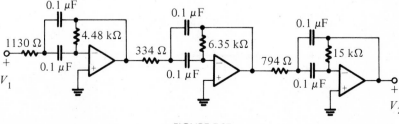

FIGURE 7.27

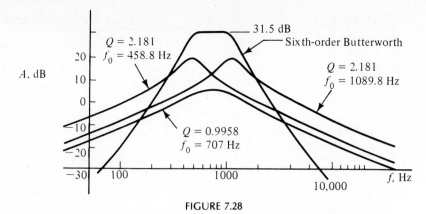

FIGURE 7.28

possible to distribute the attenuation throughout the circuit by reducing the gain of each of the three circuits.

In summary, a stagger-tuned bandpass filter may be designed by a simple step-by-step procedure as described in this section. The steps are the following:

1. The following are given as defined in Fig. 7.14: the four frequencies ω_1, ω_2, ω_3, ω_4, and α_{max} and α_{min}. It is assumed that the frequencies satisfy the relationship

$$\omega_1\omega_2 = \omega_3\omega_4 = \omega_0^2 \qquad (7.89)$$

We also assume that the two values of α_{min}, one for low frequencies and one for high frequencies, are equal. If any of these assumptions are violated, then a "best solution" may be found by trying all of the possibilities. Usually there are only a few such possibilities.

2. We next find the lowpass equivalent to the bandpass filter. This means that we calculate Ω_p and Ω_s from the equations

$$\Omega_p = 1 \qquad \text{and} \qquad \Omega_s = \frac{\omega_4 - \omega_3}{\omega_2 - \omega_1} \qquad (7.90)$$

3. We determine n and Ω_0. The degree of the Butterworth function required is found from

$$n = \frac{\log\left[(10^{\alpha_{min}/10} - 1)/(10^{\alpha_{max}/10} - 1)\right]}{2 \log \Omega_s} \qquad (7.91)$$

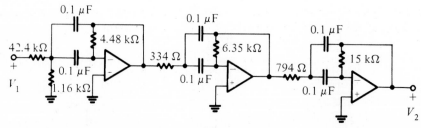

FIGURE 7.29

rounded up to the next integer, and then the radius of the Butterworth circle is

$$\Omega_0 = (10^{\alpha_{max}/10} - 1)^{-1/2n} \qquad (7.92)$$

4. From the given n we determine the location of the Butterworth poles and scale them by Ω_0.
5. Apply the Geffe algorithm and so determine the pole locations for the bandpass filter.
6. Each pole pair and one zero are assigned to a stage of filter circuit, and these stages are to be connected in cascade.
7. Using the Friend circuit (or any other bandpass circuit for that matter), find a realization.
8. If adjustment of gain is required, introduce a resistive voltage-divider circuit with resistor values determined by Eqs. (6.66) and (6.67).
9. Frequency scale to match specifications, and magnitude scale to achieve convenient element values.

These steps are illustrated in Table 7.3.

7.6 Q ENHANCEMENT OF THE FRIEND CIRCUIT

The pole locations of the transfer function of the Friend circuit may be adjusted by the addition of a summing circuit, as shown in Fig. 7.30. In particular it is possible to achieve higher values of Q by the use of this circuit addition. This is called Q enhancement: the overall circuit achieves a high value of Q, while the Friend circuit is operating with a moderate value of Q.

The circuit shown in Fig. 7.30a with the voltage-divider circuit added to the normal Friend circuit is described by the transfer function

$$T = \frac{(1/Q)s}{s^2 + (1/Q)s + 1} \qquad (7.93)$$

If the fraction of the output βV_2 is fed back such that the input voltage to the circuit of Fig. 7.30a is $V_1 + \beta V_2$, then the transfer function becomes

$$T_2 = \frac{(1/Q)s}{s^2 + (1/[Q/(1-\beta)])s + 1} \qquad (7.94)$$

From this equation we see that the resonant frequency is not changed, remaining at the normalized value of $\omega_0 = 1$, but that there is a new value of Q:

$$Q_{new} = \frac{Q}{1 - \beta} \qquad (7.95)$$

where

$$\beta = 1 - \frac{Q}{Q_{new}} \qquad (7.96)$$

TABLE 7.3   Step-by-step design of Butterworth bandpass filter

Step 1: These are specifications with $\alpha_{\min_1} = \alpha_{\min_2}$, $\omega_1\omega_2 = \omega_3\omega_4 = \omega_0^2$.

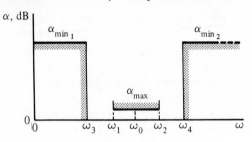

Step 2:
$$\Omega_p = 1, \qquad \Omega_s = \frac{\omega_4 - \omega_3}{\omega_2 - \omega_1}$$

Note that $\alpha_{\max}$ and $\alpha_{\min}$ are the same as bandpass specifications.

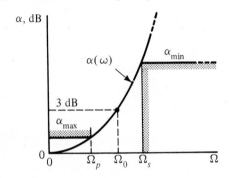

Step 3:

$$n = \frac{\log\,[(10^{\alpha_{\min}/10} - 1)/(10^{\alpha_{\max}/10} - 1)]}{2\,\log\,(\Omega_s/\Omega_p)} \qquad \text{and round up to integer.}$$

$$\Omega_0 = [10^{\alpha_{\max}/10} - 1]^{-1/2n}$$

Step 4: Find poles using Butterworth equations.

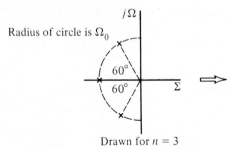

Drawn for $n = 3$

TABLE 7.3 (continued)

Step 5: Use Geffe algorithm.

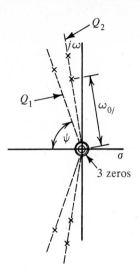

Step 6: Each pole pair and a zero is assigned to a stage. Here ω_{oj} is the distance to pole and the angle is

$$\psi = \cos^{-1}\left[\frac{1}{2Q_j}\right]$$

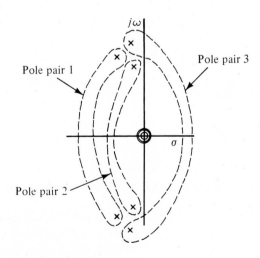

Step 7: Realize as a cascade of Friend circuits

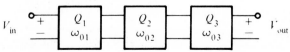

TABLE 7.3 (*continued*)

The bandpass Friend circuit for each Q_j is:

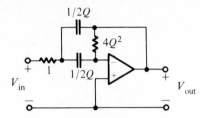

Step 8: Attenuation if needed. Make $\alpha(\omega_0) = 0$ dB:

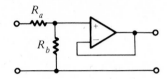

Step 9: Scale $k_f = \omega_{0j}$ and select any k_m to give a proper range of element sizes.

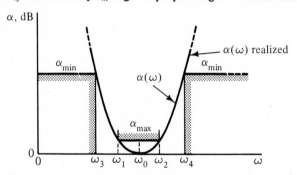

In addition we observe that

$$T_{2_{max}} = \frac{1}{1 - \beta} \tag{7.97}$$

Thus the Friend circuit operates at a low value of Q, but the overall circuit operates at a Q that may be much higher.

To illustrate the design of the circuit of Fig. 7.30b, suppose that we require that $Q_{new} = 60$, and we selected the value $Q = 10$ for the Friend circuit. The bandpass filter is to operate such that $f_0 = 3600$ Hz. From these values we calculate

$$\beta = \frac{60 - 10}{60} = 0.833 \quad \text{or} \quad \frac{1}{\beta} = 1.2 \tag{7.98}$$

In order to achieve the required resonance frequency, it is necessary that

$$k_f = 22,619 \tag{7.99}$$

If we make the decision that the capacitors have the value of 0.01 μF, then

$$k_m = 221.1 \tag{7.100}$$

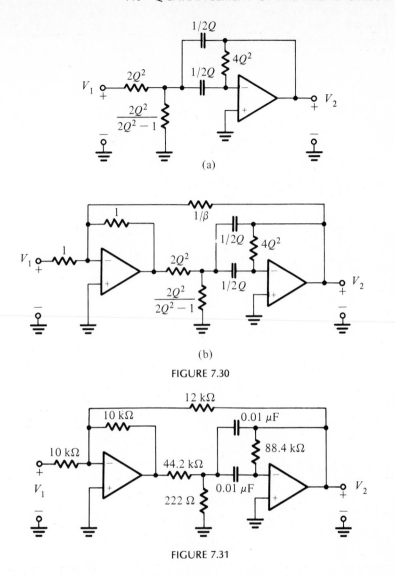

FIGURE 7.30

FIGURE 7.31

The resulting circuit, which meets the design specifications, is shown in Fig. 7.31. When $f = 3600$ Hz, then $T_{max} = 6$.

PROBLEMS

7.1 Consider a bandpass filter with the half-power (3 dB) passband extending from 905 to 1105 Hz. The lowpass prototype for the filter is to be a three-pole Butterworth characteristic.

(a) Determine the ω_0 and Q for each of the passband poles.

(b) Determine the s-plane location for each of the passband poles.

7.2 Repeat Problem 7.1 with the same specifications except the passband is to extend from 940 to 1064 Hz.

7.3 For the bandpass specifications shown in Fig. P7.3, a Butterworth response is required. As an intermediate stage in the design, we are required to find the number of poles and the radius of the circle on which the lowpass prototype poles are located.

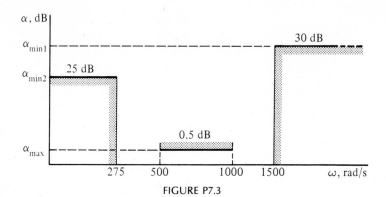

FIGURE P7.3

7.4 For a given filter, we require that α satisfy the specifications shown in Fig. P7.4 and that it be Butterworth in form. Find the pole locations that correspond to this response.

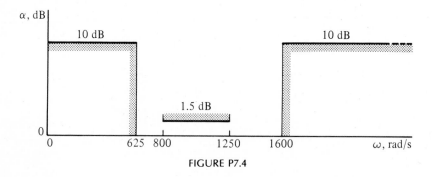

FIGURE P7.4

7.5 Design a bandpass filter to meet the specifications shown in Fig. P7.5, with the added requirement that the response be Butterworth. It is required that $\alpha(2000) = 0$. Show steps in the Geffe algorithm carefully. In your final design, at least one capacitor should have the value of 0.1 μF.

7.6 Figure P7.6 shows a response that is maximally flat (Butterworth) in the passband. It is given that $\alpha(500) = \alpha(1000) = 1$ dB. It is also given that the filter realization has two op-amps. Determine the loss (attenuation) at a frequency of $\omega = 3000$ rad/s.

7.7 From the specifications given on Fig. P7.7, we require 12 dB of attenuation in the two stop bands, and as little attenuation as possible in the passband. You are permitted only two op amps for the filter and you also are required to use only 0.01-μF capacitors. Design a filter to meet the specifications and indicate the value of α_{max} that your design achieves.

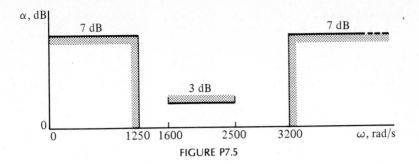

FIGURE P7.5

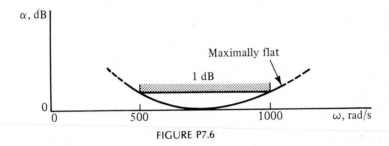

FIGURE P7.6

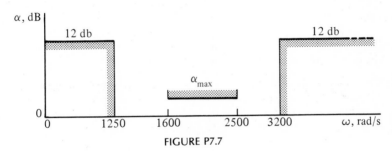

FIGURE P7.7

7.8   Design a filter to satisfy the bandpass specifications shown in Fig. P7.8. Adjust the gain of the filter such that $\alpha(707) = 0$ dB, and use only $0.1\text{-}\mu\text{F}$ capacitors in your realization.

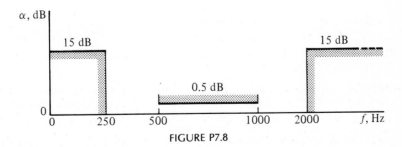

FIGURE P7.8

7.9   From the specifications given in Fig. P7.9, design a Butterworth filter with the gain adjusted such that $\alpha(707) = 0$ dB. Magnitude scale so that all capacitors in the filter have the same value.

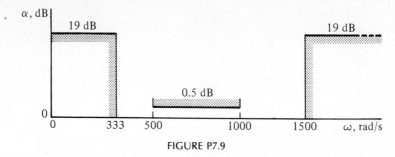

FIGURE P7.9

7.10 From the specifications given in Fig. P7.10, design a Butterworth filter with the gain adjusted such that $\alpha(707) = 0$ dB. Magnitude scale so that all capacitors in the filter have the same value.

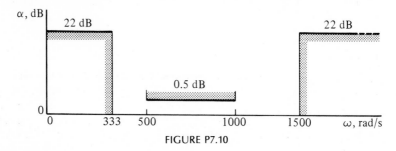

FIGURE P7.10

7.11 Repeat Problem 7.10 for the specifications given in Fig. P7.11 with the addition that the gain be adjusted to 0 dB at the frequency ω_0.

7.12 The specifications of Problem 7.10 apply to this problem except that $\alpha_{min} = 28$ dB. Design the filter as outlined in Problem 7.10.

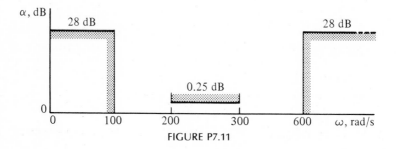

FIGURE P7.11

7.13 A passive *RLC* circuit is shown in Fig. P7.13a for which $C = 1/4$ F, $L = 1$ H, and $R = 1/2 \, \Omega$. Find an inductorless circuit to place in the box in Fig. P7.13b which is the analog of that in Fig. P7.13a with respect to ω_0, Q and $|T(j\omega_0)|$.

7.14 With the switch open, the circuit shown in Fig. P7.14 is found to have the voltage-ratio transfer function

$$\frac{V_2}{V_1} = \frac{-k_1 s}{s^2 + (k_2/C_1)\,s + (k_3/C_1)}$$

where the k's are constants. The switch is then closed connecting the capacitor into the circuit.

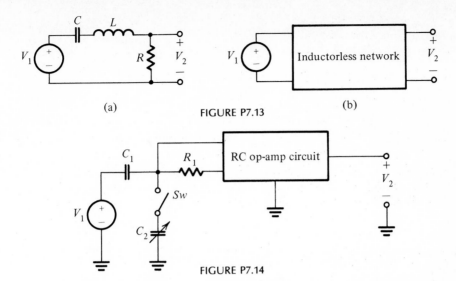

FIGURE P7.13

FIGURE P7.14

(a) Write an expression for the following quantities and indicate whether connecting C_2 into the circuit has caused them to increase or decrease: ω_0, BW, $[V_2/V_1]_{max}$.

(b) The capacitor C_2 is tuned to a specific value C_0, and then to a second value $2C_0$. Using the results of part (a), sketch $|V_2/V_1|$ as a function of ω for the two cases.

7.15 The circuit of Fig. P7.15 is the Friend circuit with the capacitors ganged (connected) together. Give an algorithm for the orthogonal tuning of ω_0 and the bandwidth BW.

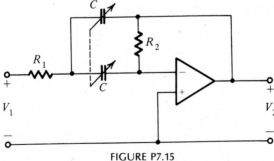

FIGURE P7.15

7.16 In the circuit of Fig. P7.16, the capacitor C_1 is tuned. What is the effect of C_1 on ω_0 and BW?

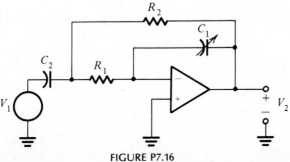

FIGURE P7.16

CHAPTER 8

The Chebyshev Response

The two preceding chapters have dealt with the Butterworth response which has a magnitude-squared form:

$$|T_n(j\omega)|^2 = \frac{1}{1 + (\omega^n)^2} \tag{8.1}$$

A generalization of this response may be written as

$$|T_n(j\omega)|^2 = \frac{1}{1 + [F_n(\omega)]^2} \tag{8.2}$$

where the function F_n is to be found. Suppose that we wish the response function to be confined to the cross-hatched strip shown in Fig. 8.1a, where the strip may be made as narrow as we wish. What does this requirement made on T_n imply about the form of F_n?

The graphical construction of Fig. 8.1 provides us with the answer. In Fig. 8.1b we show the denominator of Eq. 8.2. We then subtract 1 from this figure to obtain the required limits of $F_n(\omega)^2$ shown in Fig. 8.1c. Finally we extract the square root and scale the magnitude and obtain the cross-hatched strip shown in Fig. 8.1d. This has the dimensions of a square, two units on each side; the limits of the amplitude are ± 1 and the limits of ω are ± 1. We seek a function $F_n(\omega)$ which may be scaled to fit into this box. Once this is found, then a response of the form of Fig. 8.1a and Eq. (8.2) will result. This, it turns out, is of great importance in the design of filters.

8.1 LISSAJOUS FIGURES

A familiar experiment in the laboratory is performed with two sinusoidal oscillators, each connected to the deflecting plates of a cathode-ray oscillograph, as shown in Fig. 8.2. The oscillator connected to the horizontal deflecting plates is maintained at a fixed frequency, that connected to the vertical deflecting plates

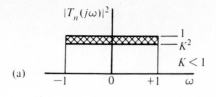

(a)

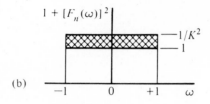

(b)

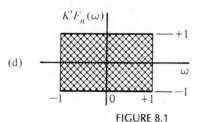

(c)

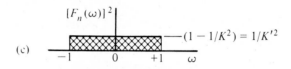

(d)

FIGURE 8.1

has adjustable frequency. As frequency is increased, a variety of figures appear on the screen of the cathode-ray oscillograph having the appearance of a cosine wave painted on an invisible tuna can, whirling rapidly, gradually slowing down, and then standing still when one frequency is the exact multiple of the other. The stationary figures reappear each time another multiple frequency is reached. A record of the figures is made in Fig. 8.3 for $n = 1$ to $n = 4$. These figures are known as *Lissajous figures* after the French mathematician* who first described them in 1857. If electrical engineers had been concerned with approximation in the 19th century, we might be writing about Lissajous functions in this chapter. Instead, they are attributed to Pafnuti L. Chebyshev[†] who used these functions in 1899 in studying the construction of steam engines.

Let the deflection due to the voltage on the horizontal plates be

$$x = \cos kt, \qquad k = \frac{2\pi}{T} \tag{8.3}$$

* Jules Antoine Lissajous (1822–1880).
[†] P. L. Chebyshev (1821–1894). His paper was published in *Oeuvres,* vol. I, St. Petersburg, 1899. Some authors use the German transliteration Tschebyscheff.

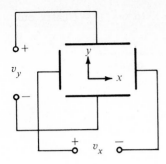

FIGURE 8.2   The figure represents the two deflection plates of a cathode-ray oscillograph.

and that due to the voltage on the vertical plates,

$$y = \cos nkt \qquad (8.4)$$

where n is an integer. Solving Eq. (8.3) for kt and substituting this value into Eq. (8.4), we have the equation of the curves of Fig. 8.3:

$$y = \cos n \cos^{-1} x = C_n(x) \qquad (8.5)$$

which is the equation of the stationary Lissajous figure. In other words, we have found that

$$\text{vertical deflection} = C_n \text{ (horizontal deflection)} \qquad (8.6)$$

We observe that these last two equations are not functions of time t.

To study Eq. (8.5) in more detail, let kt in Eqs. (8.3) and (8.4) be θ, so that Eq. (8.3) becomes

$$\theta = \cos^{-1} x \qquad (8.7)$$

This is shown in Fig. 8.4. If we select the value $n = 4$, then Eq. (8.5) becomes

$$y = \cos 4\theta \qquad (8.8)$$

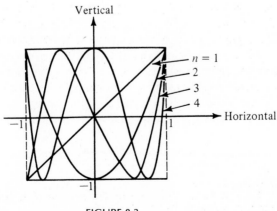

FIGURE 8.3

which is also shown in Fig. 8.4 for the range from $x = 0$ to $x = 1$. We carry out the following calculations:

θ^0	x	$4\theta^0$	$\cos 4\theta$
0	1.00	0	1
22.5	0.924	90	0
45	0.707	180	−1
67.5	0.383	270	0
90	0	360	1

$$(8.9)$$

From these values, as marked in Fig. 8.4, we see that we have identified values of x at which y has the values 1, 0, or −1, so that $y = 0$ at $x = 0.383$ and 0.924. In some sense this is a distorted cosine wave, bunched toward the $x = 1$ value. This distortion will be noted throughout our studies of these functions. Similar constructions for $n = 1$ to $n = 6$ with the range of values of x extended to include $x = -1$ to $x = 0$ are shown in Fig. 8.5. It is clear from this figure that the function $C_n(x)$ satisfies our need for a function that always fits into the box of Fig. 8.1d. Furthermore, the ripples within the box are always equal, the upper limit always being 1 and the lower limit being −1. This is also true for the square of this function C_n^2, and for this reason the functions we are discussing are also known as *equal-ripple functions*.

Return to the rotating tuna can with the cosine wave painted on it, as pictured on the screen of the cathode-ray oscillograph. In Fig. 8.6 the upper part shows half of the can as viewed from the top. As we move around the rim of the can, we take steps of 15°, chosen for the case $n = 6$ for which $90°/6 = 15°$ and marked $\theta_1, \theta_2, \ldots$ The projection of the value to the $\theta = 0°$ line is the one we

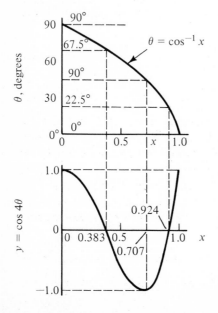

FIGURE 8.4

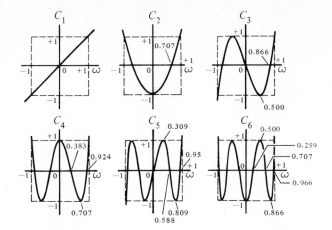

FIGURE 8.5   Plots of the Chebyshev functions $C_n(\omega)$, $n = 1$ to 6, in the range $-1 < \omega < 1$. The numbers on the figure represent the value of ω at which the function is either zero or a maximum.

would view looking head on at the can. The bottom of Fig. 8.6 shows an edge view of the can. The cosine wave, which was painted on the can, is seen in this view and is identical to the case $n = 6$ in Fig. 8.5. If n is not an integer, then the can appears to be rotating, as viewed on the screen of the cathode-ray oscillograph.

8.2   CHEBYSHEV MAGNITUDE RESPONSE

The function F_n we sought for Eq. (8.2) is clearly the Lissajous function with x replaced by ω, scaled by the constant ε, which is less than or equal to 1. Then the magnitude response is

$$|T_n(j\omega)|^2 = \frac{1}{1 + \varepsilon^2 C_n^{\,2}(\omega)} \tag{8.10}$$

where

$$C_n(\omega) = \cos n \cos^{-1} \omega, \qquad |\omega| \le 1 \tag{8.11}$$

which is defined as the *Chebyshev magnitude response*. This function satisfies the requirements cited earlier in the frequency range from $\omega = -1$ to $\omega = 1$ as being inside of the square box of Fig. 8.1d. We know that Eq. (8.10) must also apply for ω larger than 1, and so we do have the problem with the cosine function and its inverse for $|\omega| > 1$. There are two approaches to resolve this problem. One is to express C_n as a polynomial which is valid for any ω. A procedure better suited to our purposes is to observe that when ω becomes larger than 1, the inverse cosine function becomes imaginary.* Thus

$$\cos^{-1} \omega = jz \qquad \text{or} \qquad \omega = \cos jz \tag{8.12}$$

* R. W. Daniels, *Approximation Methods for Electronic Filter Design*, McGraw-Hill, New York, 1974, p. 33.

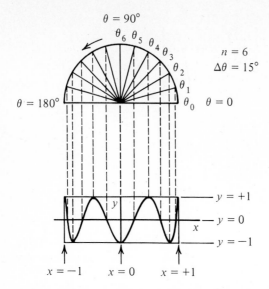

FIGURE 8.6

From the Euler form of the cosine function,

$$\cos jz = \frac{e^{j(jz)} + e^{-j(jz)}}{2} = \cosh z = \omega \qquad (8.13)$$

or

$$z = \cosh^{-1} \omega \qquad (8.14)$$

Substituting this result into Eq. (8.12), we have

$$\cos^{-1} \omega = j \cosh^{-1} \omega \qquad (8.15)$$

Finally we substitute this result into Eq. (8.11) and obtain

$$C_n(\omega) = \cos nj \cosh^{-1} \omega = \cosh n \cosh^{-1} \omega, \qquad |\omega| > 1 \qquad (8.16)$$

Thus we will use either Eq. (8.11) or Eq. (8.16), depending on the value of ω. Some of the properties of the magnitude response may be found by relating the behavior of $|T_n(j\omega)|$ to $C_n^2(\omega)$. From the square root of Eq. (8.10),

$$|T_n(j\omega)| = \frac{1}{\sqrt{1 + \varepsilon^2 C_n^2(\omega)}} \qquad (8.17)$$

which is plotted for one value of ε in Fig. 8.7, and from the plots of $C_n(\omega)$ from $\omega = -1$ to $\omega = +1$ in Fig. 8.5, we see the following:

1. Behavior at $\omega = 0$: Since

$$C_n(0) = 0, \qquad n \text{ odd}$$
$$C_n(0) = 1, \qquad n \text{ even} \qquad (8.18)$$

then

$$|T_n(j0)| = 1, \qquad\qquad n \text{ odd}$$
$$|T_n(j0)| = \frac{1}{\sqrt{1 + \varepsilon^2}}, \qquad n \text{ even} \qquad (8.19)$$

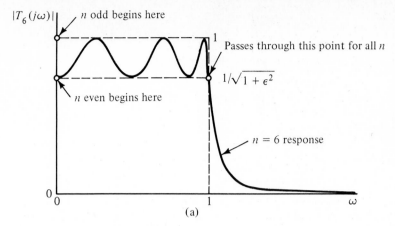

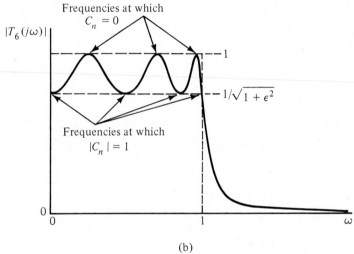

(b)

FIGURE 8.7

2. Similarly, the behavior at $\omega = 1$ follows from the fact that

$$C_n(1) = 1, \qquad \text{for all } n \tag{8.20}$$

and so

$$|T_n(j1)| = \frac{1}{\sqrt{1 + \epsilon^2}}, \qquad \text{for all } n \tag{8.21}$$

This is shown in Fig. 8.7a for the case $n = 6$.

Returning to the plots of $C_n(\omega)$ in Fig. 8.5, observe that we must first square these functions before they are substituted into Eq. (8.17). The minima of $|T_n|$ will occur when $C_n^2 = 1$, and the maxima of $|T_n|$ will occur when $C_n^2 = 0$. These are shown in Fig. 8.5 and also in Fig. 8.7b. We also observe that in the range of frequencies from $\omega = -1$ to $\omega = 1$, there will be n ripples. Since the magnitude function is even, there will be n half-ripples or half-cycles (from maximum to minimum or from minimum to maximum) in the range from $\omega = 0$ to $\omega = 1$. In general there will be n maxima from $\omega = -1$ to $\omega = 1$, and $n + 1$ minima.

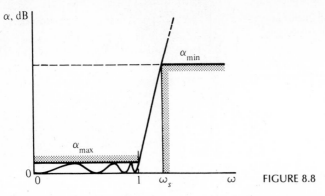

FIGURE 8.8

Applying these concepts to the $n = 6$ response of Fig. 8.7, we see from Fig. 8.5 that all maxima and minima will occur when ω has the value $\cos n15°$, where $n = 0\text{-}6$. Then $|T_6|$ will be a maximum when

$$\omega = 0.259, \qquad 0.707, \qquad 0.966 \tag{8.22}$$

and a minimum when

$$\omega = 0, \qquad 0.500, \qquad 0.866, \qquad 1 \tag{8.23}$$

The definition of attenuation is

$$\alpha_n = -20 \log |T_n(j\omega)| \quad \text{dB} \tag{8.24}$$

so that the attenuation for the Chebyshev response is

$$\alpha_n = 10 \log [1 + \varepsilon^2 C_n^2(\omega)] \quad \text{dB} \tag{8.25}$$

A plot of this response for the case $n = 7$ is given in Fig. 8.8 and for $n = 1$ to 10 in Fig. 8.9a (page 232). These show the rippling nature of the Chebyshev response from $\omega = 0$ to $\omega = 1$. For larger values of ω Eq. (8.25) becomes

$$\alpha_n = 10 \log [1 + \varepsilon^2(\cosh n \cosh^{-1} \omega)^2] \quad \text{dB} \tag{8.26}$$

This attenuation in the stop band is shown in Fig. 8.9 for the case $\varepsilon = 0.1526$ (0.1 dB).

Two values of α are of special interest to us. The maximum value of attenuation in the pass band $\alpha_{\max}$ occurs when $C_n^2(\omega) = 1$. In that case Eq. (8.25) simplifies to the form

$$\alpha_{\max} = 10 \log (1 + \varepsilon^2) \tag{8.27}$$

which may be solved for ε as

$$\varepsilon = \sqrt{10^{\alpha_{\max}/10} - 1} \tag{8.28}$$

This is simple to solve using a calculator. The case $\alpha_{\max} = 0.1$ dB used in Fig. 8.9 is seen to be

$$\varepsilon = \sqrt{10^{0.01} - 1} = \sqrt{1.0233 - 1} = 0.15262 \tag{8.29}$$

Again referring to Eq. (8.25), we see that when

$$\varepsilon^2 C_n^2(\omega) = 1 \tag{8.30}$$

then $\alpha = 3.01$, which is the attenuation that defines the half-power frequency ω_{hp}. Then this frequency is found from Eq. (8.30) to be

$$C_n(\omega_{hp}) = \frac{1}{\varepsilon} = \cosh n \cosh^{-1}\omega_{hp} \qquad (8.31)$$

Then

$$n \cosh^{-1} \omega_{hp} = \cosh^{-1}\left(\frac{1}{\varepsilon}\right) \qquad (8.32)$$

$$\cosh^{-1} \omega_{hp} = \frac{1}{n} \cosh^{-1}\left(\frac{1}{\varepsilon}\right) \qquad (8.33)$$

and finally,*

$$\omega_{hp} = \cosh \left(\frac{1}{n}\right) \cosh^{-1}\left(\frac{1}{\varepsilon}\right) \qquad (8.34)$$

Here we see that $\omega_{hp} > 1$ since $0 < \varepsilon < 1$. We may combine this last result with Eq. (8.28) to give

$$\omega_{hp} = \cosh \left(\frac{1}{n}\right) \cosh^{-1} (10^{\alpha_{max}/10} - 1)^{-1/2} \qquad (8.35)$$

Let us now return to Fig. 8.8 and examine the specifications for a filter with a Chebyshev response. The pass band extends from $\omega = 0$ to $\omega = 1$. Unlike the Butterworth case,[†] the end of the pass band (or ripple band) is always $\omega = 1$. At some higher frequency ω_s the stop band begins, and there the attenuation is required to have the minimum value it attains in the stop band α_{min}. Hence the specifications for a Chebyshev response are values of

$$\alpha_{max}, \quad \alpha_{min}, \quad \omega_s$$

We now turn to a determination of the value of n that will satisfy these specifications. We begin with α_{min} and note that

$$\alpha_{min} = \alpha(\omega_s) = 10 \log[1 + \varepsilon^2 C_n^2(\omega_s)] \qquad (8.36)$$

If we divide both sides of this equation by 10, take the antilogarithm, and subtract 1 from both sides of the equation, we have

$$\varepsilon^2 \cosh^2 (n \cosh^{-1} \omega_s) = 10^{\alpha_{min}/10} - 1 \qquad (8.37)$$

Now ε^2 is given by Eq. (8.28). When we take the square root of the resulting equation, we have

$$\cosh (n \cosh^{-1}\omega_s) = \left[\frac{10^{\alpha_{min}/10} - 1}{10^{\alpha_{max}/10} - 1}\right]^{1/2} \qquad (8.38)$$

---

* For some calculators it is convenient to make use of the identities

$$\cosh^{-1} x = \ln (x + \sqrt{x^2 - 1}), \qquad \sinh^{-1} x = \ln (x + \sqrt{x^2 + 1})$$

† In the Butterworth case $\omega = 1$ is always the half-power frequency.

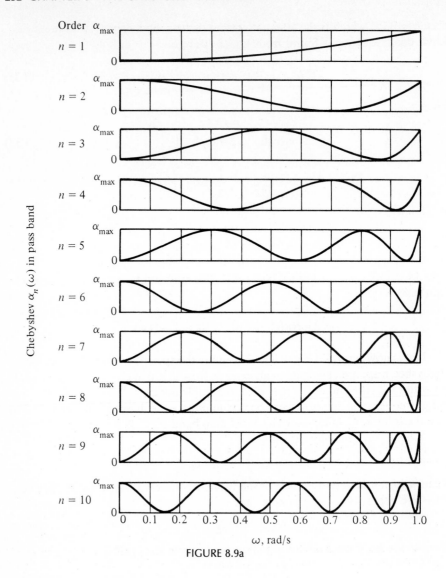

FIGURE 8.9a

Finally this equation is solved for n:

$$n = \frac{\cosh^{-1}[(10^{\alpha_{min}/10} - 1)/10^{\alpha_{max}/10} - 1)]^{1/2}}{\cosh^{-1} \omega_s} \qquad (8.39)$$

In the days before the widespread availability of inexpensive calculators, this was a difficult calculation, and various kinds of approximations were proposed. There is no need to worry about approximation any longer, since all of the calculations involved in the equation are very routinely carried out.

Example 8.1 The following specifications are given for a Chebyshev lowpass filter:

$$\omega_p = 1, \qquad \omega_s = 2.33, \qquad \alpha_{max} = 0.5 \text{ dB}, \qquad \alpha_{min} = 22 \text{ dB} \qquad (8.40)$$

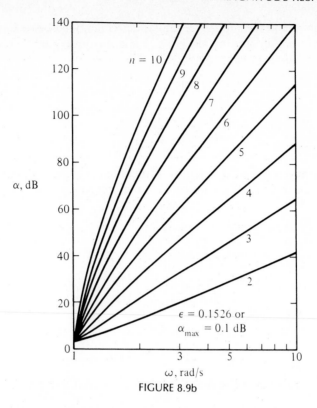

FIGURE 8.9b

Substituting these values into Eq. (8.39) gives

$$n = \frac{\cosh^{-1} 35.926}{\cosh^{-1} 2.33} = \frac{4.274}{1.489} = 2.87 \text{ (round up to 3)} \tag{8.41}$$

The half-power frequency is found from Eq. (8.35) as

$$\omega_{hp} = \cosh\left(\frac{1}{3} \cosh^{-1} \frac{1}{0.3493}\right) = \cosh 0.571 = 1.167 \tag{8.42}$$

It is interesting to calculate the value of n that is necessary to meet the specifications that are given in Eqs. (8.40) for a Butterworth response. From Eq. (6.46) we have the value

$$n = \frac{\log (35.926)^2}{2 \log 2.33} = 4.234 \text{ (round up to 5)} \tag{8.43}$$

Hence a fifth-order Butterworth response is required to meet the specifications that led to a third-order Chebyshev. We will further compare the Butterworth and Chebyshev responses in Section 8.5.

8.3   LOCATION OF THE CHEBYSHEV POLES

In Chapter 6 we found that the poles of the Butterworth function were located on a circle of radius ω_0, at angles with respect to the negative real axis that were easy to determine. We next undertake a similar determination of pole locations for the Chebyshev case. Our starting place is to replace ω by s/j in Eq. (8.10) and then recall from Chapter 6 that

$$|T_n(j\omega)|^2|_{\omega=s/j} = T(s)T(-s) \tag{8.44}$$

so that

$$T(s)T(-s) = \frac{1}{1 + \varepsilon^2 C_n^{\ 2}(s/j)} \tag{8.45}$$

The pole locations are determined by setting the denominator of this equation to zero, and then solving the equation. This requires that

$$C_n\left(\frac{s}{j}\right) = 0 \pm j\frac{1}{\varepsilon} \tag{8.46}$$

For $\omega < 1$,

$$C_n\left(\frac{s}{j}\right) = \cos n \cos^{-1}\left(\frac{s}{j}\right) = \cos nw \tag{8.47}$$

where w is a complex number:

$$\cos^{-1}\left(\frac{s}{j}\right) = w = u + jv \tag{8.48}$$

We next calculate $\cos nw$ so that this may be equated to Eq. (8.46). Then

$$\cos nw = \cos(nu + jnv)$$

$$= \cos nu \cosh nv - j \sin nu \sinh nv \tag{8.49}$$

$$= 0 \pm j\frac{1}{\varepsilon}$$

Equating the real and imaginary parts of these equations gives us two relationships:

$$\cos nu \cosh nv = 0 \tag{8.50}$$

$$\sin nu \sinh nv = \pm j\frac{1}{\varepsilon} \tag{8.51}$$

From the first of these equations we observe that $\cosh uv$ can never be zero since its smallest value is 1, and so it is necessary that $\cos nu$ be zero. The values of u for which this is true are

$$u_k = \frac{\pi}{n}, \frac{3\pi}{n}, \frac{5\pi}{n}, \ldots \tag{8.52}$$

or generally,

$$u_k = \frac{\pi}{2n}(2k + 1) \qquad k = 0, 1, \ldots, 2n - 1 \tag{8.53}$$

For these values of u_k

$$\sin nu_k = \pm 1 \qquad (8.54)$$

so that

$$v_k = \pm \frac{1}{n} \sinh^{-1}\left(\frac{1}{\varepsilon}\right) = \pm a \qquad (8.55)$$

Having determined the necessary values of u_k and v_k, we return to Eq. (8.48), $w = \cos^{-1}(s/j)$. For each value of $w_k = u_k + jv_k$ the corresponding value of s is

$$s_k = j \cos w_k = j \cos\left[\frac{\pi}{2n}(2k+1) + ja\right] \qquad (8.56)$$

Then if $s_k = \sigma_k + j\omega_k$, we have from this equation

$$\sigma_k = \pm \sinh a \sin \frac{2k+1}{2n} \pi \qquad (8.57)$$

$$\omega_k = \cosh a \cos \frac{2k+1}{2n} \pi \qquad (8.58)$$

which completes our solution, for these are the s-plane locations of the poles of $T(s)T(-s)$. We can routinely assign those in the left half-plane to $T(s)$, as we did for the Butterworth case.

We gain insight to the pole locations if we square the last two equations and add them. Since $\sin^2 x + \cos^2 x = 1$ for any x, we obtain

$$\left(\frac{\sigma_k}{\sinh a}\right)^2 + \left(\frac{\omega_k}{\cosh a}\right)^2 = 1 \qquad (8.59)$$

If we drop the index k and let σ and ω have any values, then we recognize this to be the equation of an ellipse for which:

$$\text{major semiaxis is } \cosh a$$

$$\text{minor semiaxis is } \sinh a \qquad (8.60)$$

$$\text{foci are at } \pm j1$$

These features are illustrated in Fig. 8.10.

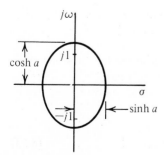

FIGURE 8.10

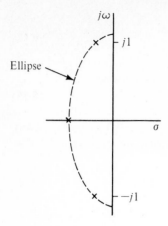

FIGURE 8.11

Example 8.2   We will determine the location of the Chebyshev poles using Eqs. (8.57) and (8.58). Assume that we wish to determine these for $n = 3$ and $A_{max} = 1$ dB. We first determine a from Eq. (8.55):

$$a = \frac{1}{3} \sinh^{-1} (10^{1/10} - 1)^{-1/2} = 0.476 \tag{8.61}$$

From this, $\sinh a = 0.494$ and $\cosh a = 1.115$. The angles are found from Eq. (8.53) to be $30°, 90°, 150°, 210°$, etc. When we locate those in the left half-plane, the poles are

$$s_1 = -0.494$$
$$s_2, s_3 = -0.247 \pm j0.966 \tag{8.62}$$

These are shown in Fig. 8.11 as well as the ellipse that passes through these three points. The polynomial corresponding to these poles is

$$_3(s) = (s + 0.494)(s^2 + 0.494s + 0.997) \tag{8.63}$$

where we have used the Cyrillic letter Ч$_n$ (Chia) to distinguish this polynomial from C_n. For the second-order factor we see that $\omega_0 = 0.999$ and $Q = 2.02$. These values will be useful in finding a circuit realization.

The determination of the poles as in Eqs. (8.62) and of the polynomial Ч$_3$ as in Eq. (8.63) is important in filter design. We next give a simplified method for their determination.

8.4   GUILLEMIN'S ALGORITHM

An alternative method for calculating Chebyshev pole positions is due to Guillemin.* His method has the conceptual advantage of relating the Butterworth circle to the Chebyshev ellipse, and the practical advantage that it makes the computation of pole positions more routine. We return to Eqs. (8.57) and (8.58) and define the angle involved in the trigonometric functions as

$$\phi_k = \frac{2k + 1}{n} \frac{\pi}{2} \qquad k = 0, 1, \ldots, 2n - 1 \tag{8.64}$$

* E. A. Guillemin, *Synthesis of Passive Networks*, Wiley, New York, 1957, p. 598.

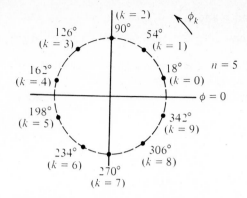

FIGURE 8.12

This family of angles is shown for the case $n = 5$ in Fig. 8.12. We observe a symmetry with respect to the angle $\phi = \pm 90°$. With the angle defined, then Eqs. (8.57) and (8.58) have the simplified forms

$$-\sigma_k = \sinh a \sin \phi_k \qquad (8.65)$$

$$\pm \omega_k = \cosh a \cos \phi_k \qquad (8.66)$$

The forms of these two equations suggest the right-angle construction shown in Fig. 8.13. With ϕ_k in the position shown, the triangle of Eq. 8.13a corresponds to Eq. (8.66) and that of Fig. 8.13b to Eq. (8.65). The complementary angle to ϕ_k

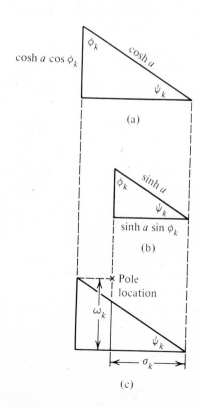

FIGURE 8.13

is identified as ψ_k in both Fig. 8.13a and b and is

$$\psi_k = 90° - \phi_k \tag{8.67}$$

We wish to show that if ϕ_k is the Chebyshev angle of Eq. (8.64) measured with respect to the positive real axis, then ψ_k is the Butterworth angle as measured from the negative real axis. That this is indeed the case is suggested from angles taken from Fig. 8.12:

k	θ_k	$\psi_k = 90° - \theta_k$
0	18°	72°
1	54°	36°
2	90°	0°
3	126°	−36°
4	162°	−72°

where ψ_k are the Butterworth angles for the case $n = 5$. In terms of the angles shown in Fig. 8.12 we see that the effect of Eq. (8.67) is to rotate the dots indicating the angle by 90°, so that the symmetry exists with respect to the real axis rather than the imaginary axis. This is shown in Fig. 8.14.

To generalize this relationship, we substitute Eq. (8.64) into Eq. (8.67) to obtain

$$\psi_k = \frac{\pi}{2} - \left(\frac{2k+1}{n}\right)\frac{\pi}{2} = \frac{n-2k-1}{n}\frac{\pi}{2} \tag{8.68}$$

This equation is to be compared with Eq. (8.27) for the Butterworth case. However, this equation gives θ_k as measured from the $\theta = 0°$ line while ψ_k is measured from the $\theta = 180°$ line. In addition, Eq. (8.27) gives the index $k = 1, 2, \ldots, 2n$, indicating that we should change the index. To accomplish this, we find

$$\psi_k = \pi - \theta_k \tag{8.69}$$

Also let $k' = k = 1$. The result is

$$\psi_k = \frac{n-2k'-1}{n}\frac{\pi}{2} \tag{8.70}$$

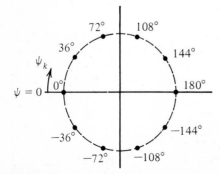

FIGURE 8.14

which is identical with Eq. (8.68). Thus ψ_k shown in Fig. 8.13 is the familiar Butterworth angle discussed at length in Chapter 6.

Returning next to Fig. 8.13 we see that

$$\sin \phi_k = \cos \psi_k \tag{8.71}$$

$$\cos \phi_k = \sin \psi_k \tag{8.72}$$

Substituting these equations into Eqs. (8.65) and (8.66), we obtain

$$-\sigma_k = \sinh a \cos \psi_k \tag{8.73}$$

$$\pm \omega_k = \cosh a \sin \psi_k \tag{8.74}$$

Since the pole position is $s_k = \sigma_k \pm j \omega_k$, it is located as shown in Fig. 8.13c in which Fig. 8.13a and b has been superimposed. We may use these equations directly, or we may observe a geometrical construction due to Guillemin.

We assume that we are given α_{max}, the ripple width, and the required value of n. Then we may follow these steps in determining the Chebyshev pole locations:

1. Determine a from the equation

$$a = \frac{1}{n} \sinh^{-1} (10^{\alpha_{max}/10} - 1)^{-1/2} \tag{8.75}$$

2. Compute $\cosh a$ and $\sinh a$.
3. On the s plane draw a semicircle of radius $\cosh a$ and a smaller semicircle of radius $\sinh a$. These are shown in Fig. 8.15.
4. Draw the Butterworth angles ψ_k. In Fig. 8.15 these are at the angles $0°$, $\pm 36°$, and $\pm 72°$.
5. As shown in Fig. 8.15, these lines from step 4 intersect the semicircles at two points, marked A and B. The horizontal projection of the line $0A$ is $\sinh a \cos \psi_k$, which is the real part of the Chebyshev pole location, $0D$. The vertical projection of the line $0B$, which is BC, is $\cosh a \sin \psi_k$, which is the imaginary part of the Chebyshev pole location.
6. By construction, point E is the actual Chebyshev pole location, since it has the Chebyshev values for its real and imaginary parts. When all such points are located, these are the Chebyshev poles. They are located on the Chebyshev ellipse.

Example 8.3 For this example we wish to determine the pole locations and the polynomials for the specifications that $n = 5$ and $\alpha_{max} = 0.5$ dB. The following calculations are routinely made with a calculator. From Eq. (8.75) $a = 0.3548$, and as illustrated by Fig. 8.15, the Butterworth angles are 0, 36°, and 72°. Then from Eqs. (8.73) and (8.74) we find the following:

$$\text{for } \psi = 72°, \quad p_1, p_2 = -0.1120 \pm j1.0116$$

$$\text{for } \psi = 36°, \quad p_3, p_4 = -0.2931 \pm j0.6252 \tag{8.76}$$

$$\text{for } \psi = 0°, \quad p_5 = -0.3623$$

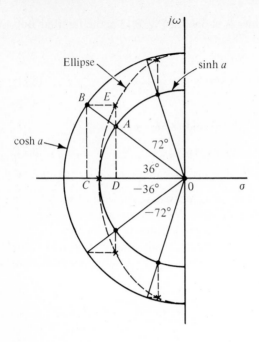

FIGURE 8.15 This illustrates a graphical procedure to locate the poles of a Chebyshev response, shown here at point E.

For filter design we frequently wish the denominator of the transfer function $T_n(s)$ to be the product of first- and second-order factors. If these poles are identified as $p = -\alpha \pm j\beta$, then the second-order factors are

$$(s + \alpha + j\beta)(s + \alpha - j\beta) = s^2 + 2\,\alpha s + \alpha^2 + \beta^2 \tag{8.77}$$

In addition we usually normalize $T_n(s)$ such that $T_n(0) = 1$. Then for the pole locations given in Eq. (8.76) we have

$$T_5 = \frac{0.1789}{(s^2 + 0.2240s + 1.0358)(s^2 + 0.5863s + 0.4768)(s + 0.3623)} \tag{8.78}$$

For some purposes we may wish to multiply the denominator factors of this equation to give the polynomial*

$$\mathrm{Ч}_5 = s^5 + 1.1725s^4 + 1.9374s^3 + 1.3096s^2 + 0.7525s + 0.1789 \tag{8.79}$$

Finally we may wish to recognize that the second-order factors can be classified in terms of ω_0 and Q from the standard form

$$s^2 + \frac{\omega_0}{Q}s + \omega_0{}^2 \tag{8.80}$$

---

* The function $C_n(\omega)$ can be expressed in polynomial form and is then called a *Chebyshev polynomial*. The denominator polynomial corresponding to a Chebyshev response is different. To avoid confusion, it is here called a *Chia polynomial*.

For T_5 given in Eq. (8.78), this classification gives the following values.

ω_0 or σ_0	Q	
0.3623	0.5	(8.81)
0.6905	1.18	
1.0178	4.54	

Extensive tables are available to give the pole locations or the polynomials Ψ_n. However, using the Guillemin relationships of Eqs. (8.73) and (8.74), the determination of the values is simple and routine so that tables seem hardly necessary for anyone with an inexpensive calculator.

8.5 COMPARISON OF THE BUTTERWORTH AND CHEBYSHEV RESPONSES

We now have two response functions, the Butterworth and the Chebyshev, from which to choose when designing a filter, and other responses will be introduced in later chapters. How shall we make a choice of the response to be used? We first observe that in making comparisons we should identify frequencies that have the same significance in the two responses. And thereby we describe our first problem: in the Butterworth response $\omega = 1$ identifies the half-power frequency, while in the Chebyshev response $\omega = 1$ identifies the end of the ripple band. These two frequencies are different unless $\varepsilon = 1$ ($\alpha_{max} = 3$ dB). As shown on Fig. 8.16, the half-power frequency for the Chebyshev response is given by Eq. (8.35):

$$\omega_{hp} = \cosh \frac{1}{n} \cosh^{-1}\left(\frac{1}{\varepsilon}\right)$$

(8.82)

and this compares with $\omega = 1$, which is the half-power frequency for the Butterworth response.

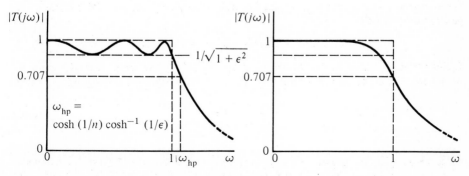

FIGURE 8.16

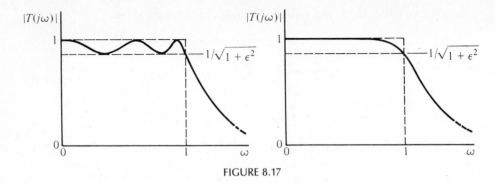

FIGURE 8.17

Rather than use these two frequencies, we introduce a new Butterworth function:

$$|T_n(j\omega)|^2 = \frac{1}{1 + \varepsilon^2 \omega^{2n}} \tag{8.83}$$

by frequency scaling by a factor of $\varepsilon^{1/n}$. Then when $\omega = 1$,

$$|T_n(j1)| = \frac{1}{\sqrt{1 + \varepsilon^2}} \tag{8.84}$$

for both the new Butterworth function and the old Chebyshev function. Then $\omega = 1$ plays the same role in both functions, as shown in Fig. 8.17. We can now make comparisons.

The attenuation corresponding to Eq. (8.83) is

$$\alpha(\omega) = 10 \log (1 + \varepsilon^2 \omega^{2n}) \tag{8.85}$$

For large values of ω with respect to 1, this function has the simplified form

$$\alpha(\omega) = 20 \log \varepsilon\omega^n = 20 \log \varepsilon + 20n \log \omega \tag{8.86}$$

The attenuation for the Chebyshev case that corresponds to Eq. (8.85) is

$$\alpha(\omega) = 10 \log [1 + \varepsilon^2 C_n^2(\omega)] \tag{8.87}$$

For $\omega \gg 1$, $C_n(\omega)$ has the approximate form*

$$C_n(\omega) = 2^{n-1}\omega^n \tag{8.88}$$

so that

$$\alpha(\omega) = 20 \log 2^{n-1} \varepsilon\omega^n = 6(n - 1) + 20 \log \varepsilon + 20n \log \omega \tag{8.89}$$

Comparing this equation with Eq. (8.86), we see that

$$\alpha_{Ch} - \alpha_{Bu} = 6(n - 1) \quad dB \tag{8.90}$$

This is a significant difference. For $n = 5$, for example, the Chebyshev response

* R. W. Daniels, *Approximation Methods for Electronic Filter Design*, McGraw-Hill, New York, 1974, p. 32.

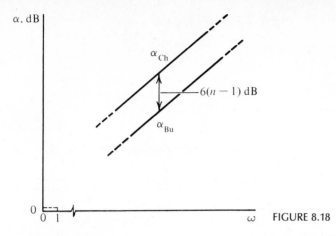

FIGURE 8.18

has 24 dB greater attenuation than the Butterworth response! This difference is illustrated by Fig. 8.18.

What price do we pay for this increased attenuation? Part of the answer is that the Q of the poles is larger. If the position of the Chebyshev poles is $p = -\alpha \pm j\beta$, then the Q is determined by comparing Eqs. (8.77) and (8.80), namely,

$$Q = \frac{1}{2\alpha} \sqrt{\alpha^2 + \beta^2} \tag{8.91}$$

For the Butterworth case

$$Q_k = \frac{1}{2 \cos \psi_k} \tag{8.92}$$

For the $n = 5$ Butterworth case, the values of Q are found to be 0.5, 0.62, and 1.62. The Q's for the Chebyshev case $n = 5$ and $A_{max} = 0.5$ dB were given in Eq. (8.81) as found from Table 8.1. For the two complex poles a comparison is as follows:

ψ_k	Q_{Bu}	Q_{Ch}
72°	1.62	4.54
36°	0.62	1.18

Indeed the Q's are higher for the Chebyshev case. This might seem intuitive from the Guillemin construction of Fig. 8.15. The Butterworth poles are on a circle of radius 1, which is just inside the circle of radius cosh a. The construction moves the poles toward the imaginary axis and so increases the value of Q.

An equation for comparison with Eq. (8.92) is found by substituting the values of α and β from Eqs. (8.73) and (8.74) into Eq. (8.91). Thus

$$Q_{Ch} = \frac{1}{2 \cos \psi_k} \sqrt{\frac{\sinh^2 a \cos^2 \psi_k + \cosh^2 a \sin^2 \psi_k}{\sinh^2 a}} \tag{8.93}$$

After some manipulation this equation reduces to

$$Q_{Ch} = Q_{Bu} \sqrt{1 + \left(\frac{\sin \psi_k}{\sinh (1/n) \sinh^{-1} (1/\varepsilon)} \right)^2} \tag{8.94}$$

Since ψ_k are the Butterworth angles and

$$\frac{1}{\varepsilon} = (10^{\alpha_{max}/10} - 1)^{-1/2} \tag{8.95}$$

it is simple to compute the Chebyshev values of Q using Eq. (8.94). For the case $\alpha_{max} = 0.5$ dB, the following table is easily found:

n	2	3	4	5	6	7	8	9	10
Q	0.86	1.71	0.71	1.18	0.68	1.09	0.68	1.06	0.67
		2.94	4.54	1.81	2.58	1.61	2.21	1.53
				6.51	8.84	3.47	4.48	2.89
						11.53	14.58	5.61
								17.99

The value $Q = 1/2$ for poles on the negative real axis has not been included in the table. The Q values are higher than those for the Butterworth case. As a practical matter, it is often more difficult to adjust stages that will make up the filter for high values of Q.

Another price we pay for the increased attenuation for the Chebyshev case is a phase characteristic of increased nonlinearity. The phase of $T(j\omega)$, which is $\theta(\omega)$, and the derivative of θ with respect to ω will be studied in detail in Chapter 10. Until then the two graphs of phase θ for the Butterworth and Chebyshev cases are shown in Fig. 8.19. The phase in the pass band from $\omega = 0$ to $\omega = 1$ is more nearly linear in the Butterworth case than in the Chebyshev case, and this will result in delay distortion.

It is instructive to examine the case of the Chebyshev response with $\varepsilon = 1$ when the half-power frequencies are identical. This is not a typical situation. When the Chebyshev response is chosen, it is usually done to secure a small ripple width. Plots for the $n = 2$ and $n = 4$ Chebyshev and Butterworth responses are shown in Fig. 8.20. The figure illustrates the feature that the Chebyshev response does decrease more rapidly beyond the pass band. The cutoff frequency $\omega = 1$ is of special interest. For this special case with $\varepsilon = 1$, the Chebyshev response is n times steeper at cutoff than the Butterworth response*:

$$\left. \frac{dT}{d\omega} \right|_{\omega=1} \text{(Chebyshev)} = n \left. \frac{dT}{d\omega} \right|_{\omega=1} \text{(Butterworth)} \tag{8.96}$$

In summary, the choice between a Butterworth and a Chebyshev response will require trade-off between the conflicting requirements of (1) high attenuation in the stop band and steeper rolloff near the cutoff frequency, and (2) higher values of Q leading to difficulties in achieving circuit realizations and a nonlinear

---

* A. Budak, *Passive and Active Network Analysis and Synthesis*, Houghton Mifflin, Boston, 1974, p. 516.

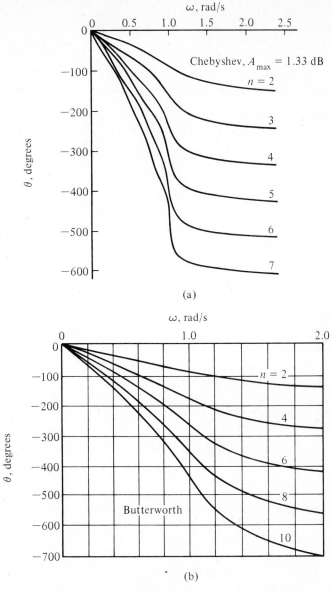

(a)

(b)

FIGURE 8.19

phase characteristic. The actual choice will depend on the weight that the designer assigns to each of the conflicting requirements.

8.6   CHEBYSHEV FILTER DESIGN

We are now prepared to go through step-by-step procedures in designing lowpass and bandpass Chebyshev filters. This will be done with examples. Table 8.2 compares the steps required for Butterworth and Chebyshev cases.

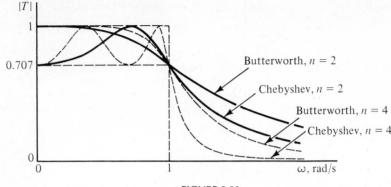

FIGURE 8.20

Example 8.4   We require a lowpass filter to satisfy specifications given in Fig. 8.21. The loss in the pass band to 10 kHz is to be 0.3 dB or less, and beyond 24.58 kHz the loss is to be at least 22 dB. The normalized frequencies are also shown in the figure with the end of the ripple band indicated as $\Omega = 1$. We first compute the degree of the approximation that will be required using Eq. (8.39). Then

$$n = \frac{\cosh^{-1}[(10^{2.2} - 1)/(10^{0.03} - 1)]^{1/2}}{\cosh^{-1} 2.458} = 2.933 \qquad (8.97)$$

which we round up to 3. Next we compute a using Eq. (8.55):

$$a = \frac{1}{3} \sinh^{-1} (10^{0.03} - 1)^{-1/2} = 0.6765 \qquad (8.98)$$

from which we find $\cosh a = 1.2377$ and $\sinh a = 0.7293$. The location of the poles is given by Eqs. (8.57) and (8.58) from which we find

$$p_1, p_2 = -0.3646 \pm j1.072$$
$$p_3 = -0.7293 \qquad (8.99)$$

For the complex poles we use Eq. (8.91) (or the values given in Table 8.1) to find that $Q = 1.553$ and $\omega_0 = 1.1323$. Finally the half-power frequency is given by Eq. (8.35) as

$$\Omega_{\text{hp}} = \cosh \left[\frac{1}{3} \cosh^{-1} (10^{0.03} - 1)^{-1/2} \right]$$
$$= 1.22906 \qquad (8.100)$$

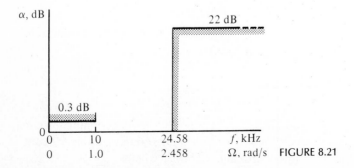

FIGURE 8.21

The attenuation actually realized at the edge of the stop band is found using Eq. (8.26).

$$\alpha_n(2.458) = 10 \log[1 + \varepsilon^2(\cosh 3 \cosh^{-1} 2.458)^2] = 22.89 \text{ dB} \qquad (8.101)$$

The first circuit realization will be made with a cascade connection to two stages having the properties summarized in Fig. 8.22. We will use a simple RC voltage divider for the first-order section and a Sallen–Key circuit to realize the second order section. The particular Sallen–Key circuit chosen is that given in Fig. 6.19b with $Q = 1.553$. This circuit is matched to the requirements of the complex pair of poles shown in Fig. 8.23 for which we require that $Q = 1.553$ and $\omega_0 = 1.1323$.

The design equation for stage 1 is

$$\sigma_0 = \frac{1}{RC} = 0.7293 \qquad (8.102)$$

If we select $R = 1$, then $C = 1.3712$. The last step is scaling. Since the edge of the ripple band must be scaled to 10 kHz, we require that $k_f = 2\pi \times 10^4$, and we may select $k_m = 1500$, giving the values shown in Fig. 8.24 for stage 1. Stage 2 requires additional frequency scaling due to the requirement that $\Omega_0 = 1.1323$. If we select $k_m = 1500$, then Eq. (A.26) becomes

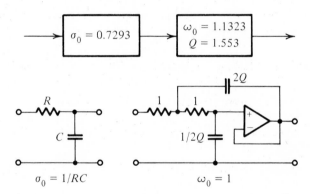

FIGURE 8.22

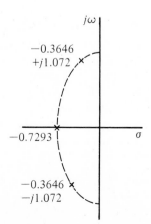

FIGURE 8.23

$$C_{\text{new}} = \frac{1}{2\pi \times 10,000 \times 1500 \times 1.1323} C_{\text{old}} \qquad (8.103)$$

Using this scaling, we obtain the element values given for stage 2 in Fig. 8.24, and the design is complete. The specifications realized are given in Fig. 8.25. The way this is accomplished in terms of the frequency responses of the first- and second-order sections in cascade is shown in Fig. 8.26. The product of the two responses gives the required ripple in the pass band extending from $\Omega = 0$ to $\Omega = 1$.

TABLE 8.1   Chebyshev pole locations

n	$\alpha_{\max} = 0.5$ dB α	β	$\alpha_{\max} = 1$ dB α	β	$\alpha_{\max} = 2$ dB α	β	$\alpha_{\max} = 3$ dB α	β
1	2.8628	0	1.9652	0	1.3076	0	1.0024	0
2	0.7128	1.0040	0.5489	0.8951	0.4019	0.8133	0.3224	0.7772
3	0.3132	1.0219	0.2471	0.9660	0.1845	0.9231	0.1493	0.9038
	0.6265	0	0.4942	0	0.3689	0	0.2986	0
4	0.1754	1.0163	0.1395	0.9834	0.1049	0.9580	0.0852	0.9465
	0.4233	0.4209	0.3369	0.4073	0.2532	0.3968	0.2056	0.3920
5	0.1120	1.0116	0.0895	0.9901	0.0675	0.9735	0.0549	0.9659
	0.2931	0.6252	0.2342	0.6119	0.1766	0.6016	0.1436	0.5970
	0.3623	0	0.2895	0	0.2183	0	0.1775	0
6	0.0777	1.0085	0.0622	0.9934	0.0470	0.9817	0.0382	0.9764
	0.2121	0.7382	0.1699	0.7272	0.1283	0.7187	0.1044	0.7148
	0.2898	0.2702	0.2321	0.2662	0.1753	0.2630	0.1427	0.2616
7	0.0570	1.0064	0.0457	0.9953	0.0346	0.9866	0.0281	0.9827
	0.1597	0.8071	0.1281	0.7982	0.0969	0.7912	0.0789	0.7881
	0.2308	0.4479	0.1851	0.4429	0.1400	0.4391	0.1140	0.4373
	0.2562	0	0.2054	0	0.1553	0	0.1265	0
8	0.0436	1.0050	0.0350	0.9965	0.0625	0.9898	0.0216	0.9868
	0.1242	0.8520	0.0997	0.5448	0.0754	0.8391	0.0614	0.8365
	0.1859	0.5693	0.1492	0.5644	0.1129	0.5607	0.0920	0.5590
	0.2193	0.1999	0.1760	0.1982	0.1332	0.1969	0.1055	0.1962
9	0.0345	1.0040	0.0277	0.9972	0.0209	0.9919	0.0171	0.9896
	0.0992	0.8829	0.0797	0.8769	0.0603	0.8723	0.0491	0.8702
	0.1520	0.6553	0.1221	0.6509	0.0924	0.6474	0.0753	0.6459
	0.1864	0.3487	0.1497	0.3463	0.1134	0.3445	0.0923	0.3437
	0.1984	0	0.1593	0	0.1206	0	0.0983	0
10	0.0279	1.0033	0.0224	0.9978	0.0170	0.9935	0.0138	0.9915
	0.0810	0.9051	0.1013	0.7143	0.0767	0.7113	0.0401	0.8945
	0.1261	0.7183	0.0650	0.9001	0.0493	0.8962	0.0625	0.7099
	0.1589	0.4612	0.1277	0.4586	0.0967	0.4567	0.0788	0.4558
	0.1761	0.1589	0.1415	0.1580	0.1072	0.1574	0.0873	0.1570

In terms of the α and β values given, the second-order factor is $s^2 + 2\alpha s + \alpha^2 + \beta^2$, and $\omega_0 = \sqrt{\alpha^2 + \beta^2}$, $Q = \sqrt{\alpha^2 + \beta^2}/2\alpha$.

TABLE 8.2 Comparison of steps for Butterworth and Chebyshev cases

Butterworth	Step	Chebyshev
$$n = \dfrac{\log \left[(10^{\alpha_{min}/10} - 1)/(10^{\alpha_{max}/10} - 1) \right]}{2 \log (\Omega_s/\Omega_p)}$$ Round up to an integer	1. Find n.	$$n = \dfrac{\cosh^{-1}\left[(10^{\alpha_{min}/10} - 1)/(10^{\alpha_{max}/10} - 1) \right]^{1/2}}{\cosh^{-1}(\Omega_s/\Omega_p)}$$ Round up to an integer
Not used	2. Find ϵ.	$\epsilon = [10^{\alpha_{max}/10} - 1]^{1/2}$
$\Omega_{hp} = (10^{\alpha_{max}/10} - 1)^{-1/2n}$	3. Find half-power (3-dB) frequency Ω_{hp}.	$\Omega_{hp} = \cosh \left[\dfrac{1}{n} \cosh^{-1}\left(\dfrac{1}{\epsilon} \right) \right]$
a. If n is odd, $\theta_k = 0°, \quad k \dfrac{180°}{n}, \quad k$ is an integer b. If n is even, $\theta_k = \pm \dfrac{180°}{2n}, \quad k \dfrac{180°}{n} \pm \dfrac{180°}{2n}$ Radius $= \Omega_0$ $-\sigma_k = \Omega_0 \cos \theta_k, \quad \pm \omega_k = \Omega_0 \sin \theta_k$	4. Find pole locations. 	a. Find θ_k for Butterworth case b. Find $a = \dfrac{1}{n} \sinh^{-1}\left(\dfrac{1}{\epsilon} \right)$ c. Then $-\sigma_k = \sin \theta_k \sinh a$ $\pm \omega_k = \cos \theta_k \cosh a$

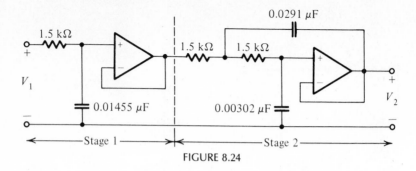

FIGURE 8.24

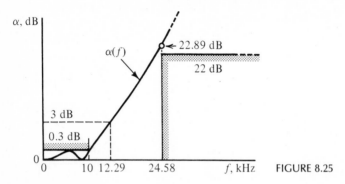

FIGURE 8.25

A second solution to the same problem makes use of a special third-order circuit described by Geffe* and shown in Fig. 8.27. The circuit is a lowpass filter, and the position of the three poles may be fixed by determining the values for the three capacitors. Geffe gives an algorithm for finding the capacitor values in general and tabulates the following values for the Chebyshev case:

α_{max} (dB)	C_1 (F)	C_2 (F)	C_3 (F)
0.03	0.097357	3.3128	1.0325
0.10	0.096911	4.7921	1.3145
0.30	0.085819	7.4077	1.6827
1.00	0.05872	14.784	2.3444

Using the values given for $\alpha_{max} = 0.30$ dB, there remains only the operation of scaling. To scale the end of the ripple band from $\omega = 1$ to $f = 10$ kHz requires that $k_f = 2\pi \times 10^4$. Selecting the magnitude scaling factor k_m to be 1500 gives the element values given in Fig. 8.28.

Example 8.5 We are required to design a bandpass filter to the attenuation specifications given in Fig. 8.29 with $\alpha_{min} = 20$ dB and $\alpha_{max} = 0.5$ dB. The response is to be Chebyshev in the pass band, which extends from 500 to 1000 rad/s. The circuit employed in the design is to make use of 0.1-μF capacitors only.

We first make use of Eq. (7.59) to determine the lowpass equivalent of the bandpass

---

* P. R. Geffe, "How to Build High-Quality Filters out of Low-Quality Parts," *Electronics*, pp. 111-113, Nov. 11, 1976.

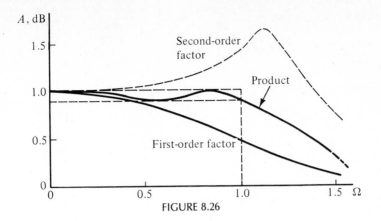

FIGURE 8.26

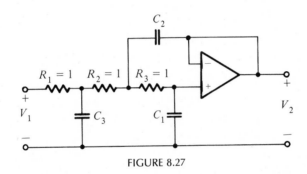

FIGURE 8.27

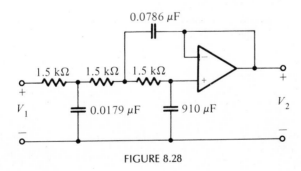

FIGURE 8.28

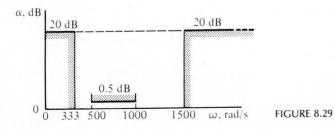

FIGURE 8.29

specifications. This gives us

$$\Omega_p = 1 \quad \text{and} \quad \Omega_s = \frac{1500 - 333}{1000 - 500} = 2.334 \tag{8.104}$$

as shown in Fig. 8.30. From Eq. (8.39) we determine the required value of n:

$$n = \frac{\cosh^{-1}[(10^2 - 1)/(10^{0.05} - 1)]^{1/2}}{\cosh^{-1} 2.334} = 2.71 \quad (n = 3) \tag{8.105}$$

We require the value of $1/\varepsilon = (10^{0.05} - 1)^{-1/2} = 2.8628$, from which we use Eq. (8.35) to determine

$$\Omega_{\text{hp}} = \cosh\left(\frac{1}{3} \cosh^{-1} 2.8628\right) = 1.167 \tag{8.106}$$

and from Eq. (8.55),

$$a = \frac{1}{3} \sinh^{-1} 2.8628 = 0.5914 \tag{8.107}$$

From this we determine the pole locations to be as given by Eqs. (8.57) and (8.58):

$$p_1, p_2 = -0.3132 \pm j1.022$$
$$p_3 = -0.6265 \tag{8.108}$$

as shown in Fig. 8.31. With this we have completed our analysis of the lowpass filter, and we use the Geffe algorithm to determine the bandpass pole locations. Starting with Eq.

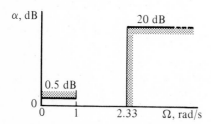

FIGURE 8.30

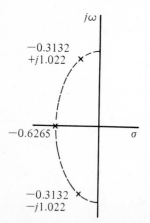

FIGURE 8.31

(7.20), the real pole is transformed as follows:

$$\Sigma_1 = 0.6265, \qquad q_c = \frac{707}{500} = 1.414, \qquad \omega_0 = 707, \qquad Q = \frac{q_c}{\Sigma_1} = 2.257 \qquad (8.109)$$

For the complex poles:

$$
\begin{aligned}
\alpha &= 0.3132 & Q &= 4.804 \\
\beta &= 1.1022 & K &= 1.064 \\
C &= 1.143 & W &= 1.428 \\
D &= 0.443 & \omega_{01} &= 1009.4 \\
E &= 4.572 & \omega_{02} &= 495.2 \\
G &= 4.48
\end{aligned}
\qquad (8.110)
$$

The location of the bandpass poles together with the three zeros at the origin is shown in Fig. 8.32. The strategy we will employ in finding a circuit realization is shown in Fig. 8.33 where each of the three stages will be realized using the Friend circuit of Fig. 7.18, which is shown in Fig. 8.34 with a voltage-divider circuit at the input for gain adjustment, should this prove necessary.

Let us next examine the Chebyshev response we have achieved as shown by Fig. 8.35. We observe the following:

1. We have satisfied the requirement that $\alpha_{\max} = 0.5$ dB exactly from 500 to 1000 rad/s.
2. Concerning the $\alpha_{\min}$ specification, we determine $\alpha(333) = \alpha(1500)$ using Eq. (8.26) with $\Omega_s = 2.334$. Then

$$(2.334) = 10 \log [1 + \varepsilon^2 (\cosh 3 \cosh^{-1} 2.334)^2] = 23.723 \text{ dB} \qquad (8.111)$$

which is larger than the 22-dB specification.

3. The half-power frequencies may be found if needed. The numerical transformation from lowpass to bandpass is

$$\Omega = \frac{1}{500} \frac{\omega^2 - (707)^2}{\omega} \qquad (8.112)$$

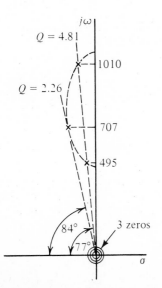

FIGURE 8.32

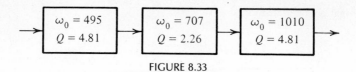

FIGURE 8.33

The lowpass half-power frequency was found in Eq. (8.106). When this value is substituted into Eq. (8.112), the polynomial that results is

$$\omega^2 \pm 583.74\omega \pm 500{,}000 = 0 \tag{8.113}$$

The two positive solutions of this quadratic equation are

$$\omega_{hp} = 473 \text{ and } 1057 \text{ rad/s} \tag{8.114}$$

4. From Fig. 8.35 and the property of odd Chebyshev polynomials in the pass band, we see that

$$\alpha(707) = 0 \text{ dB} \tag{8.115}$$

which has implications with respect to the circuit to be used for a realization.

To achieve the results required by Eq. (8.115), we turn to Eq. (7.87), which gives the gain of the Friend circuit:

$$|T(j\omega)| = \sqrt{\frac{(2\,Q_i\,\omega\omega_{0i})^2}{(\omega^2 - \omega_{0i}{}^2)^2 + (\omega\omega_{0i}{}^2)/Q_i}} \tag{8.116}$$

where i is the number of the stage being considered. If we let $\omega = 707$ in this equation, then we may calculate the value of Eq. (8.116) for each of the three stages to obtain

$$|T(j\,707)| = 12.69,\ 10.19,\ \text{and } 12.69 \tag{8.117}$$

To achieve these results, we consider the voltage-divider circuit shown in Fig. 8.34, which was originally considered in Eq. (6.64). The circuit shown in Fig. 8.36a achieves the result

$$V_0 = HV_1 \tag{8.118}$$

with the element values given for R_a and R_b. Since the values of $|T(j707)|$ are larger than 1, it is necessary to reduce the gain of each of the Friend circuit stages by

$$H = \frac{1}{|T(j\,707)|} \tag{8.119}$$

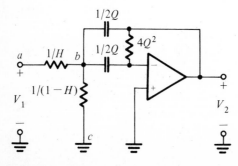

FIGURE 8.34

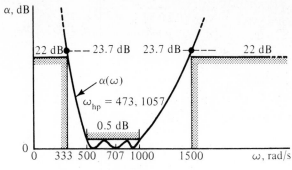

FIGURE 8.35

Then for the purposes of this problem it is necessary that

$$R_a = |T(j707)| \quad \text{and} \quad R_b = \frac{|T(j707)|}{|T(j707)| - 1} \quad (8.120)$$

as shown in Fig. 8.36b.

There remains the problem of achieving the specification that all capacitors have the value of 0.1 μF. This is accomplished by the choice of a different value of k_m for each of the three stages, chosen according to Eq. (A.25) as

$$k_m = \frac{C_{\text{old}}}{k_f \times 10^{-7}} \quad (8.121)$$

We may summarize the results found thus far in the following table:

	Stage 1	Stage 2	Stage 3
ω_0	459.2	707	1009.4
Q	4.81	2.26	4.81
k_f	459.2	707	1009.4
k_m	2103	3133	1031
$\|T(j707)\|$	12.69	10.19	12.68

Using this table, all element values are determined to give the circuit shown in Fig. 8.37.

Example 8.6 In this problem we are interested in the detail of the variation of attenuation in the pass band of a bandpass filter with a Chebyshev response. The specifications require that the pass band extend from 800 to 1250 rad/s, meaning that $\omega_0 = 1000$ rad/s, with $\alpha_{\text{max}} = 0.5$ dB and $n = 5$. The attenuation is

$$\alpha(\Omega) = 10 \log [1 + \varepsilon^2 C_5^2(\Omega)] \quad \text{dB} \quad (8.122)$$

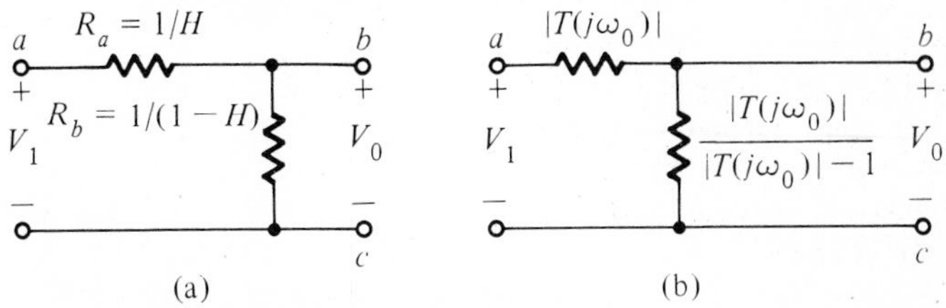

FIGURE 8.36

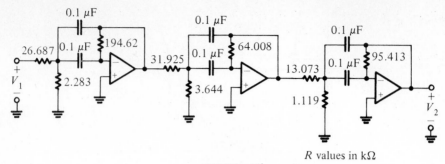

FIGURE 8.37

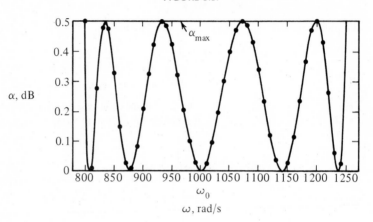

FIGURE 8.38

where

$$\epsilon^2 = 10^{0.05} - 1 = 0.12202 \tag{8.123}$$

The frequency transformation may be written

$$\Omega = \frac{\omega_0}{bw}\left(\frac{\omega}{\omega_0} - \frac{\omega_0}{\omega}\right) = 2.22\left(\frac{\omega}{10^3} - \frac{10^3}{\omega}\right) \tag{8.124}$$

Hence

$$\alpha(\omega) = 10\log\left[1 + 0.122\cos^2 5\cos^{-1} 2.22\left(\frac{\omega}{10^3} - \frac{10^3}{\omega}\right)\right] \tag{8.125}$$

This is plotted in Fig. 8.38 over the range of frequencies for which it is applicable. From the figure we make two observations:

1. There are two distorting factors in the plot. The Chebyshev response itself and, in addition, the nonlinear nature of the frequency transformation given by Eq. (8.124).
2. The specification $n = 5$ pertained to the lowpass filter. The frequency transformation from lowpass to bandpass doubles the degree of the transfer function so that there are twice as many poles. Thus the filter for which the response of Fig. 8.38 applies is a tenth-order Chebyshev response. In terms of cycles of ripple, we observe that Fig. 8.38 shows five full cycles or ten half-cycles of ripple in the pass band.

PROBLEMS

8.1 Consider the attenuation characteristic shown in Fig. P8.1 where $\alpha_{max} = 0.5$ dB. Find the attenuation, α, at the frequency which is twice the frequency at the end of the ripple band, $\omega = 2000$ rad/s.

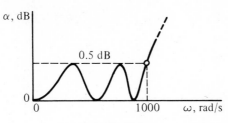

FIGURE P8.1

8.2 Equation (8.39) gives the value of n, the order of the Chebyshev function required for a given set of specifications. It is sometimes useful to use an approximate form of this equation. Show that

$$n \approx \frac{\ln [2(10)^{\alpha_{min}/20}/\varepsilon]}{\ln (\omega_s + \sqrt{\omega_s^2 - 1})}$$

8.3 Design a lowpass filter with a Chebyshev response satisfying the following specifications: $\alpha_{max} = 0.25$ dB, $\alpha_{min} = 18$ dB, $\omega_p = 1000$ rad/s, $\omega_s = 1400$ rad/s. Adjust the gain so the minimum value of $\alpha(\omega)$ is 0 dB. Magnitude scale so that the element values in your circuit realization are in a practical range (Fig. P8.3).

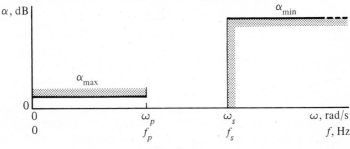

FIGURE P8.3

8.4 Repeat Problem 8.3 for the specifications: $\alpha_{max} = 0.5$ dB, $\alpha_{min} = 25$ dB, $\omega_p = 1000$ rad/s, $\omega_s = 1800$ rad/s.

8.5 Repeat the lowpass filter design described in Problem 8.3 for the following specifications: $\alpha_{max} = 1.0$ dB, $\alpha_{min} = 60$ dB, $\omega_p = 1000$ rad/s, $\omega_s = 3000$ rad/s.

8.6 From the specifications given in Fig. P8.6, we see that 12 dB of attenuation is required in the two stop bands. It is prescribed that the filter which will satisfy these specifications will have only two op amps.

(a) Determine the value of α_{max} that is achieved by a Butterworth response; by a Chebyshev response.

(b) For each of the two responses, determine the two half-power frequencies.

8.7 Design a filter with a Chebyshev response that satisfies the specifications given in Fig. P8.7. Adjust the gain of the filter so that the attenuation is always positive, and use magnitude scaling to obtain element sizes for your design in a practical range.

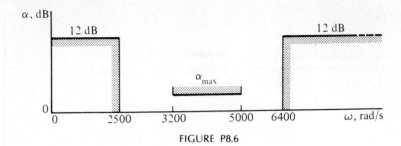

FIGURE P8.6

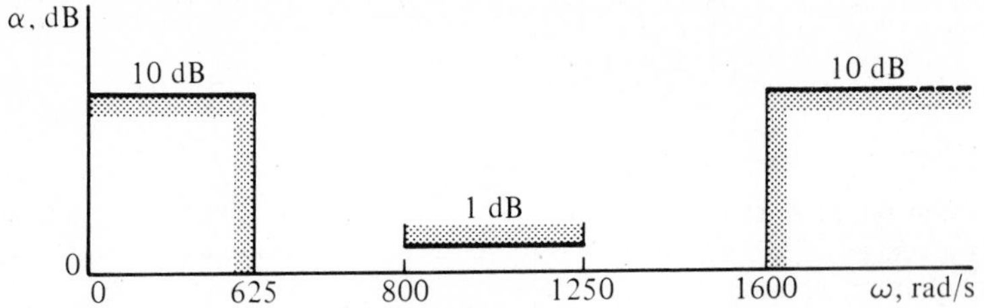

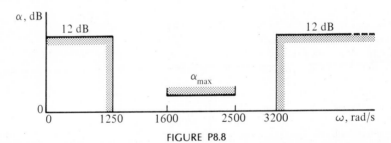

FIGURE P8.7

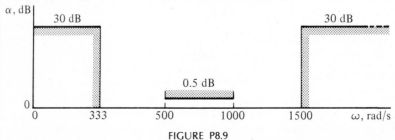

FIGURE P8.8

8.8 The specifications for a Chebyshev filter require 12 dB of attenuation in the stop bands and as little attenuation as possible in the pass bands. For the design of the filter for this problem, you are permitted to use only two op amps, and all capacitors must have the value of 0.01 μF. The gain of the filter must be adjusted so that the attentuation is always positive. For your design, give all element sizes (Fig. P8.8).

8.9 Design a filter with a Chebyshev response that satisfies the specifications given in Fig. P8.9. Adjust the gain such that the attenuation is always positive, and magnitude scale so that element sizes are in a practical range.

FIGURE P8.9

8.10 Find the order *n* of the Butterworth and Chebyshev responses needed to meet the specifications given in Fig. P8.10.

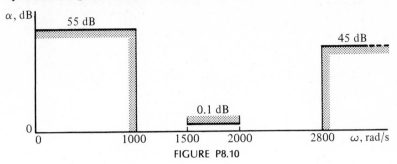

FIGURE P8.10

CHAPTER 9
Sensitivity

In earlier chapters we have assumed that circuits are constructed with ideal components. With this assumption, different circuits that produce a given response are interchangeable. The Sallen–Key circuit of Chapter 6, for example, has a different form depending upon the value of K chosen, as shown in Fig. 6.19a–c. Is there any reason to select one circuit over another? In practice the components used in circuits will deviate from their nominal values due to tolerances related to their manufacture. Similar deviations will be due to chemical changes because of component aging and changes caused by the temperature and humidity environment in which they operate. How do such changes relate to specifications such as ω_0, Q, and gain? How do such changes relate to our preference for one circuit over another? One answer to such questions is provided by a *sensitivity measure* first introduced by Bode in the early 1940s, which we study in this chapter. We will find that the sensitivity of a circuit may be far more important than the quality or expense of components used in constructing the circuit.

9.1 DEFINITION OF BODE SENSITIVITY

Figure 9.1 represents a circuit in which one component has been singled out for study, R_1. We assume that R_1 changes by an amount ΔR_1 and as a result the transfer function $T = V_2/V_1$ changes. We are interested in the frequency ω_0 which changes by an amount $\Delta\omega_0$. This change may manifest itself in different ways in $T(s)$. For example, the pole position may change, as shown in Fig. 9.2a, or the magnitude response may change as shown in Fig. 9.2b. If the circuit of Fig. 9.1 is one stage in a stagger-tuned circuit, such as those designed in Chapters 7 and 8, then this may result in a failure of the circuit to satisfy specifications, as illustrated in Fig. 9.3.

Now we are not as interested in the value of $\Delta\omega_0$ as we are in the percentage change in ω_0. Thus

$$\frac{\Delta\omega_0}{\omega_0} \quad \text{is the per unit change in } \omega_0 \qquad (9.1)$$

$$\frac{\Delta R_1}{R_1} \quad \text{is the per unit change in } R_1 \qquad (9.2)$$

If these quantities are multiplied by 100, then they become the percent changes in

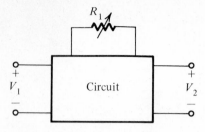

FIGURE 9.1

ω_0 and R_1. Now the ratio

$$\frac{\% \text{ change in } \omega_0}{\% \text{ change in } R_1} = \frac{(\Delta\omega_0/\omega_0) \times 100\%}{(\Delta R_1/R_1) \times 100\%} \tag{9.3}$$

is an interesting characterization of a circuit, for a small value of the ratio would imply that the circuit would be useful for design, while a large value would identify a circuit that should be avoided.

The relationships that we will find convenient to use are based on an incremental change in R_1 and ω_0 so that in the limit

$$\Delta R_1 \to \partial R_1 \quad \text{and} \quad \Delta\omega_0 \to \partial\omega_0 \tag{9.4}$$

A partial derivative is used because element values other than R_1 may also be changing. The incremental equivalent of Eq. (9.3) defines the sensitivity

$$S_{R_1}^{\omega_0} = \frac{\partial\omega_0/\omega_0}{\partial R_1/R_1} \tag{9.5}$$

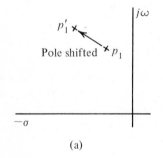
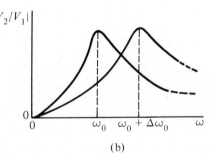

(a)

(b)

FIGURE 9.2

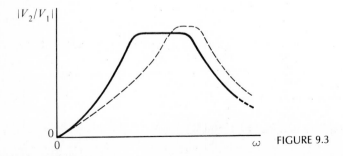

FIGURE 9.3

In the form

$$\frac{\partial \omega_0}{\omega_0} = S_{R_1}^{\omega_0} \frac{\partial R_1}{R_1} \tag{9.6}$$

we see that the sensitivity function $S_{R_1}^{\omega_0}$ is the number that multiplies the incremental change in R_1 to indicate the change that will occur in ω_0. Two other forms of the defining equation, Eq. (9.5), will prove to be useful in our studies. The first is

$$S_{R_1}^{\omega_0} = \frac{R_1}{\omega_0} \frac{\partial \omega_0}{\partial R_1} \tag{9.7}$$

The second comes from differential calculus; recall that

$$\frac{d}{dx} \ln u = \frac{1}{u} \frac{du}{dx} \tag{9.8}$$

or

$$d(\ln u) = \frac{du}{u} \tag{9.9}$$

Then from Eq. (9.5) we see that

$$S_{R_1}^{\omega_0} = \frac{\partial \ln \omega_0}{\partial \ln R_1} \tag{9.10}$$

While we have formulated the concept of sensitivity in terms of the example of ω_0 and R_1, the results are general, and the sensitivity function can be defined in terms of general variables x and Y as

$$S_x^Y = \frac{x}{Y} \frac{\partial Y}{\partial x} = \frac{\partial \ln Y}{\partial \ln x} \tag{9.11}$$

The problem of interest to Bode when he first introduced the concept of sensitivity* was the changes in the transfer function T resulting from large changes in one of the elements in the transmission system, the vacuum tube. If the system is represented as shown in Fig. 9.4, where T_2 represents the vacuum tube and the overall transfer function is

$$T = T_1 T_2 \tag{9.12}$$

it is clear that changes in T_2 directly affect T. Studied in terms of the sensitivity function,

* H. W. Bode, *Network Analysis and Feedback Amplifier Design*, Van Nostrand Reinhold, 1945, 551 pp.

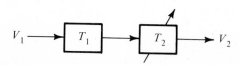

FIGURE 9.4

$$S_{T_2}{}^T = \frac{T_2}{T}\frac{\partial T}{\partial T_2} = \frac{T_1 T_2}{T_1 T_2} = 1 \qquad (9.13)$$

From this we see that a 1% change in T_2 will cause a 1% change in T, as anticipated. Now if feedback is introduced as shown in Fig. 9.5, then as studied in Chapter 2 in Eq. (2.19),

$$T = \frac{T_1 T_2}{1 + H T_1 T_2} \qquad (9.14)$$

The sensitivity function is found by differentiating a quotient:

$$S_{T_2}{}^T = \frac{T_2}{T}\frac{\partial T}{\partial T_2} = \frac{T_2}{T}\frac{(1 + H T_1 T_2)T_1 - H T_1 T_2 T_1}{(1 + H T_1 T_2)^2} \qquad (9.15)$$

or

$$S_{T_2}{}^T = \frac{1}{1 + H T_1 T_2} \qquad (9.16)$$

If we select a system such that $T_1 T_2 \gg 1$, then Eq. (9.14) reduces to

$$T = \frac{1}{H} \qquad (9.17)$$

as found in Eq. (2.21), and the sensitivity function becomes

$$S_{T_2}{}^T = \frac{1}{H T_1 T_2} \qquad (9.18)$$

For example, if we select $H = 1$, and $T_1 T_2 = 1000$, then the sensitivity function becomes

$$S_{T_2}{}^T = 10^{-3} \qquad (9.19)$$

compared to the value of 1 for the open-loop system. Such an analysis led to the widespread use of feedback in transmission systems.

Applying this method of sensitivity analysis to the inverting amplifier of Chapter 2, shown in Fig. 9.6, we start with the transfer function adapted from Eq. (2.42):

$$T = \frac{-A R_2}{R_2 + (1 + A)R_1} \qquad (9.20)$$

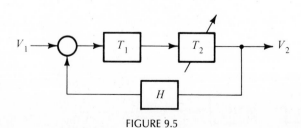

FIGURE 9.5

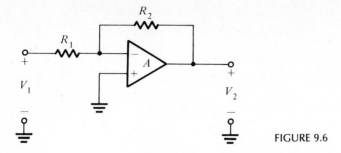

FIGURE 9.6

The sensitivity function is found by again differentiating a quotient:

$$S_A{}^T = \frac{1 + R_2/R_1}{1 + R_2/R_1 + A} \tag{9.21}$$

Since A is normally very large compared to $1 + R_2/R_1$, the sensitivity is approximately

$$S_A{}^T = \frac{1}{A}\left(1 + \frac{R_2}{R_1}\right) \tag{9.22}$$

which is a very small number. However, if R_2 is removed so that there is no feedback, then $T = -A$, and the sensitivity function is

$$S_A{}^T = \frac{A}{T}\frac{\partial T}{\partial A} = 1 \tag{9.23}$$

for $A = 10^5$, and $1 + R_2/R_1 = 10$, Eq. (9.22) gives a sensitivity of the closed-loop system of 10^{-4}, compared to the open-loop value of 1 just found.

We will next formulate some general properties of the sensitivity function which follow from Eq. (9.11). We assume that Y is a function of x or several values of x, such as x_1, x_2, x_3. If Y is not a function of some x, then $S_x{}^Y = 0$. For the simple form $Y = kx$, where k is a constant, we see that $S_x{}^Y = 1$, as was the case in Eq. (9.23). One of the forms we will encounter most often is

$$Y = x_1{}^a x_2{}^b x_3{}^c \tag{9.24}$$

The natural logarithm of this function is

$$\ln Y = a \ln x_1 + b \ln x_2 + c \ln x_3 \tag{9.25}$$

If we differentiate this expression with respect to $\ln x_1$, then from Eq. (9.11),

$$S_{x_1}{}^Y = \frac{\partial \ln Y}{\partial \ln x_1} = a \tag{9.26}$$

and similarly,

$$S_{x_2}{}^Y = b \quad \text{and} \quad S_{x_3}{}^Y = c \tag{9.27}$$

If Y may be expressed as a product of two other functions Y_1 and Y_2, we see that

$$S_x{}^Y = \frac{\partial \ln Y_1 Y_2}{\partial \ln x} = \frac{\partial \ln Y_1}{\partial \ln x} + \frac{\partial \ln Y_2}{\partial \ln x} \tag{9.28}$$

or

$$S_x^{\ Y} = S_x^{\ Y_1} + S_x^{\ Y_2} \tag{9.29}$$

In like manner, if $Y = Y_1/Y_2$, then

$$S_x^{\ Y} = S_x^{\ Y_1} - S_x^{\ Y_2} \tag{9.30}$$

The general procedure may be followed to establish other simple relationships. For example,

$$S_x^{\ 1/Y} = \frac{\partial \ln(1/Y)}{\partial \ln x} = -\frac{\partial \ln Y}{\partial \ln x} = -S_x^{\ Y} \tag{9.31}$$

Another form that will be found useful occurs when $Y = Y_1 + Y_2$ for which the sensitivity function is

$$S_x^{\ Y_1+Y_2} = \frac{Y_1 S_x^{\ Y_1} + Y_2 S_x^{\ Y_2}}{Y_1 + Y_2} \tag{9.32}$$

Example 9.1   In Chapter 5 we studied a series RLC circuit shown as Fig. 5.1 and repeated as Fig. 9.7. For this circuit we found that the design equations were

$$\omega_0 = \frac{1}{\sqrt{LC}} \quad \text{and} \quad Q = \frac{1}{R}\sqrt{\frac{L}{C}} \tag{9.33}$$

given in Chapter 5 as Eqs. (5.4) and (5.5). If we write these equations in the form of Eq. (9.24), then we see that since

$$\omega_0 = L^{-1/2}C^{-1/2} \quad \text{and} \quad Q = R^{-1}L^{1/2}C^{-1/2} \tag{9.34}$$

Eq. (9.26) gives us the sensitivity values

$$S_L^{\ \omega_0} = -\frac{1}{2}, \quad S_C^{\ \omega_0} = -\frac{1}{2}, \quad S_R^{\ \omega_0} = 0 \tag{9.35}$$

and

$$S_R^{\ Q} = -1, \quad S_L^{\ Q} = \frac{1}{2}, \quad S_C^{\ Q} = -\frac{1}{2} \tag{9.36}$$

This analysis shows that a 1% change in any of the three elements shown in Fig. 9.7 results in either $\frac{1}{2}$% or 1% change in ω_0 or Q, with the sign of the value designating whether the change is increasing or decreasing. With more experience in finding sensitivity values we will find that these values are *low* and that, in general, a passive ladder structure is characterized by low values of sensitivity.

Example 9.2   As our second example consider the biquad circuit given in Fig. 5.9 for which the design equations were given in Chapter 5 as Eqs. (5.23), (5.24), and (5.25). We

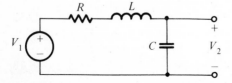

FIGURE 9.7

may rewrite these equations in a form suited to sensitivity analysis as

$$\omega_0 = \frac{1}{\sqrt{R_2 R_4 C_1 C_2}} = R_2^{-1/2} R_4^{-1/2} C_1^{-1/2} C_2^{-1/2} \tag{9.37}$$

$$Q = \sqrt{\frac{R_1^2 C_1}{R_2 R_4 C_2}} = R_1^1 C_1^{1/2} R_2^{-1/2} R_4^{-1/2} C_2^{-1/2} \tag{9.38}$$

$$H = \frac{R_2}{R_3} = R_2^1 R_3^{-1} \tag{9.39}$$

Since these equations are now in the form of Eq. (9.24), we may write the sensitivities directly as

$$S_{R_2}^{\omega_0} = -\frac{1}{2}, \quad S_{R_4}^{\omega_0} = -\frac{1}{2}, \quad S_{C_1}^{\omega_0} = -\frac{1}{2}, \quad S_{C_2}^{\omega_0} = -\frac{1}{2} \tag{9.40}$$

meaning that a change of 1% in any of the four circuit parameters would cause a decrease in ω_0 of $\frac{1}{2}\%$. Further, a change in R_3 does not change ω_0 at all. The Q sensitivities have a range of values which are the exponents of Eq. (9.38):

$$S_{R_1}{}^Q = 1, \quad S_{C_1}{}^Q = \frac{1}{2}, \quad S_{R_2}{}^Q = -\frac{1}{2}, \quad S_{R_4}{}^Q = -\frac{1}{2}, \quad S_{C_2}{}^Q = -\frac{1}{2} \tag{9.41}$$

Finally from Eq. (9.39) the gain is determined only by R_2 and R_3, and the sensitivities are

$$S_{R_2}{}^H = 1 \quad \text{and} \quad S_{R_3}{}^H = -1 \tag{9.42}$$

This analysis gives the reasons for the popularity of the biquad circuit: the sensitivity values are about the same as the passive *RLC* ladder circuit, and much lower than related circuits which accomplish the same objective.

Example 9.3 Another circuit that was introduced in Chapter 7 and has been used in circuit design is due to Delyiannis and Friend, shown in Fig. 7.17. With $C_1 = C_2 = C$, the design equations for the circuit are given in Eqs. (7.71), which are repeated here:

$$\omega_0 = \frac{1}{C\sqrt{R_1 R_2}}, \quad Q = \frac{1}{2}\sqrt{\frac{R_2}{R_1}} \tag{9.43}$$

or

$$\omega_0 = C^{-1} R_1^{-1/2} R_2^{-1/2}, \quad Q = \frac{1}{2} R_2^{1/2} R_1^{-1/2} \tag{9.44}$$

We see from these equations that

$$S_C^{\omega_0} = -1, \quad S_{R_1}^{\omega_0} = -\frac{1}{2}, \quad S_{R_2}^{\omega_0} = -\frac{1}{2}, \quad S_{R_1}{}^Q = -\frac{1}{2}, \quad S_{R_2}{}^Q = \frac{1}{2} \tag{9.45}$$

Again, all values are low, meaning that the Friend circuit has a range of sensitivities comparable to a passive ladder circuit. The bandwidth BW for a bandpass circuit is given by

$$BW = \frac{\omega_0}{Q} \tag{9.46}$$

so that from Eq. (9.30),

$$S_x{}^{BW} = S_x{}^{\omega_0} - S_x{}^Q \tag{9.47}$$

We may use this equation and the values found in Eqs. (9.45) to find the bandwidth sensitivities:

$$S_C^{BW} = -1, \qquad S_{R_1}^{BW} = 0, \qquad S_{R_2}^{BW} = -1 \tag{9.48}$$

9.2 SENSITIVITY ANALYSIS OF THE SALLEN–KEY CIRCUIT

We have made use of the Sallen–Key circuit given in Fig. 6.16 and repeated here as Fig. 9.8 as a module in lowpass filter applications. Three different realizations have been studied and are given in Fig. 6.19. The sensitivity analysis of the circuit will prove to be more difficult than the three examples of the last section, and the fact that different forms of realization are available will provide comparisons of sensitivity values. The transfer function for Fig. 9.8 was obtained in Chapter 6 as Eq. (6.55), which is

$$T(s) = \frac{V_2}{V_1} = \frac{K/R_1R_2C_1C_2}{s^2 + (1/R_1C_1 + 1/R_2C_1 + 1/R_2C_2 - K/R_2C_2)s + 1/R_1R_2C_1C_2} \tag{9.49}$$

where

$$K = 1 + \frac{R_B}{R_A} \tag{9.50}$$

From the denominator we relate to the standard form of the lowpass transfer function and see that

$$\omega_0 = \frac{1}{\sqrt{R_1R_2C_1C_2}} = R_1^{-1/2}R_2^{-1/2}C_1^{-1/2}C_2^{-1/2} \tag{9.51}$$

The coefficient that multiplies s is

$$\frac{\omega_0}{Q} = Y = \frac{1}{R_1C_1} + \frac{1}{R_2C_1} + \frac{1-K}{R_2C_2} \tag{9.52}$$

where Y has been defined for convenience in calculating Q sensitivities. Since

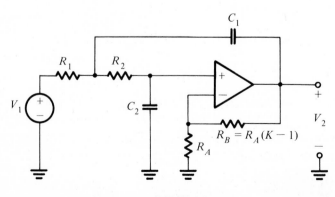

FIGURE 9.8

$Q = \omega_0 / Y$, then for any element x we have

$$S_x{}^Q = S_x{}^{\omega_0} - S_x{}^Y \tag{9.53}$$

where

$$S_x{}^Y = \frac{x}{Y}\frac{\partial Y}{\partial x} = x\frac{Q}{\omega_0}\frac{\partial Y}{\partial x} \tag{9.54}$$

The ω_0 sensitivities are calculated from Eq. (9.51) and are

$$S_{R_1,R_2,C_1,C_2}^{\omega_0} = -\frac{1}{2}, \qquad S_K{}^{\omega_0} = 0 \tag{9.55}$$

These values are as low as those for the passive circuit studied in the last section. They will later be substituted into Eq. (9.53) to find the Q sensitivities.

To determine expressions for the Q sensitivities, we begin with Eq. (9.54) with $x = R_1$. We differentiate Eq. (9.52) with respect to R_1 to give

$$\frac{\partial Y}{\partial R_1} = \frac{-1}{R_1{}^2 C_1} \tag{9.56}$$

Then Eq. (9.54) becomes

$$S_{R_1}{}^Y = R_1 Q (R_1 R_2 C_1 C_2)^{1/2}\left(\frac{-1}{R_1{}^2 C_1}\right)$$

$$= -Q\sqrt{\frac{R_2 C_2}{R_1 C_1}} \tag{9.57}$$

Finally from Eq. (9.53) we have

$$S_{R_1}{}^Q = -\frac{1}{2} - Q\sqrt{\frac{R_2 C_2}{R_1 C_1}} \tag{9.58}$$

A similar procedure is followed in calculating the Q sensitivity with respect to R_2. We begin by differentiating Y with respect to R_2:

$$\frac{\partial Y}{\partial R_2} = \frac{-1}{R_2{}^2 C_1} + \frac{-(1 - K)}{R_2{}^2 C_2} \tag{9.59}$$

Then

$$S_{R_2}{}^Y = \frac{R_2 Q}{\omega_0}\frac{\partial Y}{\partial R_2} = -Q\left[\sqrt{\frac{R_1 C_2}{R_2 C_1}} + (1 - K)\sqrt{\frac{R_1 C_1}{R_2 C_2}}\right] \tag{9.60}$$

Combining this with the first term of Eq. (9.53) we have finally

$$S_{R_2}{}^Q = -\frac{1}{2} + Q\left[\sqrt{\frac{R_1 C_2}{R_2 C_1}} + (1 - K)\sqrt{\frac{R_1 C_1}{R_2 C_2}}\right] \tag{9.61}$$

The pattern we have followed in this derivation may be followed in determining the Q sensitivities due to C_1 and C_2 and also K. The result is

$$S_{C_1}{}^Q = -\frac{1}{2} + Q\left[\sqrt{\frac{R_1 C_2}{R_2 C_1}} + \sqrt{\frac{R_2 C_2}{R_1 C_1}}\right] \tag{9.62}$$

$$S_{C_2}{}^Q = -\frac{1}{2} + (1-K)Q\sqrt{\frac{R_1 C_1}{R_2 C_2}} \tag{9.63}$$

$$S_K{}^Q = KQ\sqrt{\frac{R_1 C_1}{R_2 C_2}} \tag{9.64}$$

Additional relationships for the Sallen–Key circuit come from the equation for the gain of the noninverting op-amp stage:

$$K = 1 + \frac{R_B}{R_A} \tag{9.65}$$

or

$$\frac{R_B}{R_A} = K - 1 \tag{9.66}$$

Using this last relationship, the expression for Y of Eq. (9.52) becomes

$$Y = \frac{1}{R_1 C_1} + \frac{1}{R_2 C_1} + \frac{-R_B/R_A}{R_2 C_2} \tag{9.67}$$

Following the usual procedure, we determine the sensitivity functions:

$$S_{R_B}{}^Q = -S_{R_A}{}^Q = -(1-K)\sqrt{\frac{R_1 C_1}{R_2 C_2}} \tag{9.68}$$

Finally, from Eq. (9.65), we apply Eq. (9.11) to obtain

$$-S_{R_A}{}^K = S_{R_B}{}^K = \frac{K-1}{K} \tag{9.69}$$

All of these sensitivity equations have been derived for general values of the circuit parameters and the gain K. Once a specific circuit is selected so that these values are known, and the design value of Q is determined, the numerical values of the sensitivity functions can be determined.

9.3 SENSITIVITY COMPARISONS OF THREE CIRCUITS

In Chapter 6, in studying the Sallen–Key circuit, three different circuits were determined depending on the design strategy adopted. These were given in Fig. 6.19 and are repeated here as Fig. 9.9. All of these circuits are realizations of the normalized lowpass transfer function

$$T = \frac{K}{s^2 + (1/Q)s + 1} \tag{9.70}$$

for the three values of K, namely, $K = 1$, $K = 2$, and $K = 3 - 1/Q$. If we substitute the values of R_1, R_2, C_1, C_2, and K into the sensitivity equations of the last section, we may determine the sensitivity functions. From these calculations Table 9.1 may be computed. A comparison will be more meaningful for a specific

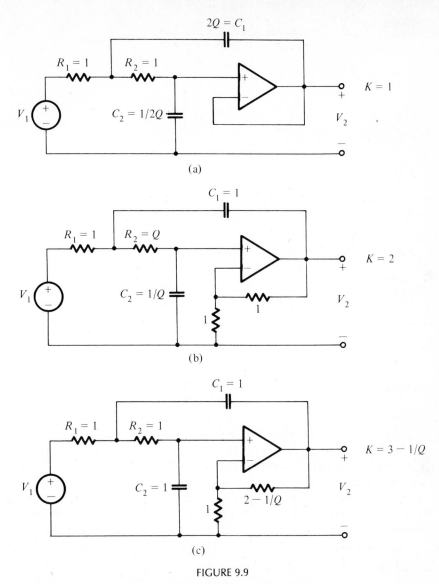

FIGURE 9.9

value of Q. The choice $Q = 10$ is made, and the values of sensitivity shown in Table 9.2 are found. It is seen that the sensitivities of circuit a are smallest in each case, so that it is the clear choice in terms of sensitivity minimization.

Each circuit has advantages and disadvantages, and sensitivity is only one factor in the trade-offs that must be considered. We next discuss each circuit with the identification of trade-offs as our objective.

Circuit a

The sensitivity values shown in Table 9.2 are remarkably low and independent of Q. Furthermore, fewer elements are required for realization of the circuit since

TABLE 9.1

	Circuit a $K = 1$	Circuit b $K = 2$	Circuit c $K = 3 - 1/Q$
$S_{R_1}^Q$	0	$-\dfrac{1}{2} + Q$	$-\dfrac{1}{2} + Q$
$S_{R_2}^Q$	0	$\dfrac{1}{2} - Q$	$\dfrac{1}{2} - Q$
$S_{C_1}^Q$	$\dfrac{1}{2}$	$\dfrac{1}{2} + Q$	$-\dfrac{1}{2} + 2Q$
$S_{C_2}^Q$	$-\dfrac{1}{2}$	$-\dfrac{1}{2} - Q$	$\dfrac{1}{2} - 2Q$
S_K^Q	$\dfrac{1}{2}$	$2Q$	$3Q - 1$
$S_{R_A}^Q$	0	-1	$-(2Q - 1)$
$S_{R_B}^Q$	0	1	$2Q - 1$
$S_{R_A}^K$	0	$-\dfrac{1}{2}$	$-\dfrac{2Q - 1}{3Q - 1}$
$S_{R_B}^K$	0	$\dfrac{1}{2}$	$\dfrac{2Q - 1}{3Q - 1}$

with $K = 1$ the op-amp connection is that of a voltage follower. The main disadvantage of this circuit realization is the required spread of element sizes. The ratio of C_1/C_2 is $4Q^2$, so that with $Q = 10$ the ratio of capacitor sizes becomes 400. This is ordinarily considered to be a large spread of values and may be difficult to realize in practice.

Circuit b

The advantage of making $K = 2$ is that $R_B = R_A$, which is also equal to R_1. If matched resistors are available, then this is an advantage. The ratio of elements is smaller than for circuit a since $R_2/R_1 = Q$ and $C_1/C_2 = Q$. For this example with $Q = 10$ this is not as severe a limitation as for circuit a. The sensitivity figures are much higher than for circuit a, especially the Q sensitivity due to K, S_K^Q meaning that the gain of the noninverting amplifier would have to be adjusted carefully.

Circuit c

This circuit has the advantage that the resistors are equal and the capacitors are equal, so that both can be scaled to the same value in a design problem. The price

TABLE 9.2 Q = 10

	Circuit a $K = 1$	Circuit b $K = 2$	Circuit c $K = 3-1/Q$
$S_{R_1}^Q$	0	9.5	9.5
$S_{R_2}^Q$	0	−9.5	−9.5
$S_{C_1}^Q$	$\dfrac{1}{2}$	10.5	19.5
$S_{C_2}^Q$	$-\dfrac{1}{2}$	−10.5	−19.5
S_K^Q	$\dfrac{1}{2}$	20	29
$S_{R_A}^Q$	0	−1	−19
$S_{R_B}^Q$	0	1	19
$S_{R_A}^K$	0	$-\dfrac{1}{2}$	$-\dfrac{19}{29}$
$S_{R_B}^K$	0	$\dfrac{1}{2}$	$\dfrac{19}{29}$

paid for this advantage is that all sensitivities are relatively high, expecially the sensitivity of Q to the adjustment of K, S_K^Q. This means that the adjustment of K would be very sensitive in attaining a specified value of Q.

9.4 MAGNITUDE OF $T(j\omega)$ SENSITIVITY

Another sensitivity figure of interest relates $T(j\omega)$ to some circuit parameter, such as element values, Q, or ω_0. The approach is the same as that followed in earlier sections. We begin with the definition

$$S_x^{T(s)} = \frac{x}{T(s)} \frac{\partial T(s)}{\partial x} \tag{9.71}$$

If we let $s = j\omega$ such that

$$T(j\omega) = |T(j\omega)|e^{j\theta(\omega)} \tag{9.72}$$

then the sensitivity function becomes

$$S_x^{T(j\omega)} = \frac{x}{|T(j\omega)|e^{j\omega(\omega)}} \frac{\partial}{\partial x} [|T(j\omega)|e^{j\theta(\omega)}] \tag{9.73}$$

We make use of the product rule for the differentiation of a product so that

$$S_x{}^{T(j\omega)} = \frac{x}{|T(j\omega)|} \frac{\partial}{\partial x} |T(j\omega)| + jx \frac{\partial \theta(\omega)}{\partial x} \tag{9.74}$$

From this equation, we see that

$$S_x|T(j\omega)| = \operatorname{Re} S_x{}^{T(j\omega)} \tag{9.75}$$

$$S_x{}^{\theta(\omega)} = \frac{1}{\theta(\omega)} \operatorname{Im} S_x{}^{T(j\omega)} \tag{9.76}$$

We are particularly interested in Eq. (9.75), since it gives a convenient way of computing the sensitivity function for a magnitude $|T(j\omega)|$.

In Eq. (9.75) x may be any quantity of interest for the circuit under study, such as any element value, ω_0, or Q. We will apply this equation with $x = \omega_0$ and $x = Q$ for the second-order lowpass and bandpass transfer functions. For the lowpass transfer function

$$T(s) = \frac{\omega_0{}^2}{s^2 + (\omega_0/Q) + \omega_0{}^2} \tag{9.77}$$

we compute the sensitivity function

$$S_Q{}^{T(s)} = \frac{Q}{T} \frac{\partial T}{\partial Q} = \frac{-(\omega_0/Q)s}{s^2 + (\omega_0/Q)s + \omega_0{}^2} \tag{9.78}$$

We next let $s = j\omega_0$ and determine the real part of the resulting function to obtain

$$S_Q{}^{|T(j\omega_0)|} = \operatorname{Re} S_Q{}^{T(j\omega_0)} = -1 \tag{9.79}$$

For the bandpass case

$$T(s) = \frac{(\omega_0/Q)s}{s^2 + (\omega_0/Q)s + \omega_0} \tag{9.80}$$

We proceed as with Eq. (9.78) and find that

$$S_Q{}^{T(s)} = \frac{s^2 + \omega_0{}^2}{s^2 + (\omega_0/Q)s + \omega_0{}^2} \tag{9.81}$$

If we let $s = j\,\omega_0$ and then determine the real part of the result, we find that

$$S_Q{}^{|T(j\omega_0)|} = 0 \tag{9.82}$$

Comparisons have been made only at the frequency ω_0, although we have determined general equations that could be plotted as a function of ω if desired. The results obtained should be examined in terms of Fig. 5.13 for the lowpass case and Fig. 5.20 for the bandpass case. In these figures we see that $|T(j\omega_0)| = Q$ for the lowpass case, while $|T(j\omega_0)| = 1$ for the bandpass case. So it makes sense that the sensitivity should be zero for the bandpass case since the value is the same for all Q. In the lowpass case, however, the value depends directly on Q, and so the sensitivity should not be zero.

We next determine the sensitivity of $|T(j\omega)|$ due to variations in ω. Starting

CHAPTER 10

Delay

Filters

One of the basic operations in signal processing is time delay. Such delay occurs naturally in the transmission of a signal through space, such as on a coaxial cable or an optical fiber. The same delay can be provided approximately by circuits known as delay filters, basically the same circuits that we have already studied but tuned differently.

10.1 TIME DELAY AND TRANSFER FUNCTIONS

Delay is a time-domain quantity, but filters are designed from frequency-domain specifications such as magnitude and phase. Hence it is necessary that we relate the two quantities. We begin with the input signal v_1 shown in Fig. 10.1a. This signal is introduced into a circuit which provides a delay of D seconds. The output is a delayed replica of the input signal and is shown in Fig. 10.1c as v_2. In equation form we are given $v_1(t)$, and we require that

$$v_2(t) = v_1(t - D) \tag{10.1}$$

From Fourier analysis we know that any signal can be decomposed into an infinite summation of sinusoidal signals. Let one of these be

$$v_1 = A \sin (\omega t + \phi) \tag{10.2}$$

Then from Eq. (10.1) the output will be

$$v_2 = A \sin [\omega(t - D) + \phi] \tag{10.3}$$

or

$$v_2 = A \sin (\omega t - \omega D + \phi) \tag{10.4}$$

Then we see that the input and output signals differ only by a phase angle, which is

$$\theta = -\omega D \tag{10.5}$$

If all Fourier components of the input signal are delayed by the same amount D

279

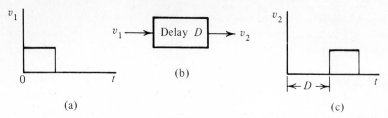

FIGURE 10.1

and not changed in amplitude, then the output will indeed be a delayed replica of the input.

Now the phasor representation of Eq. (10.2) is

$$V_1 = A \angle \phi \qquad (10.6)$$

while that for Eq. (10.4) is

$$V_2 = A \angle (\phi - \omega D) \qquad (10.7)$$

Hence the ratio V_2/V_1 is

$$\frac{V_2}{V_1} = 1 \angle -\omega D \qquad (10.8)$$

or in exponential form,

$$\frac{V_2(j\omega)}{V_1(j\omega)} = 1 e^{-j\omega D} \qquad (10.9)$$

It is usual to employ time scaling, discussed in Appendix A, and let $D = 1$ as a normalized time delay. If we make this simplification and generalize by letting $j\omega = s$, then the transfer function of Eq. (10.9) becomes

$$T(s) = \frac{V_2(s)}{V_1(s)} = e^{-s} \qquad (10.10)$$

The magnitude and phase corresponding to this transfer function, as given by Eq. (10.9), are shown in Fig. 10.2. It tells us that if the phase is linear with negative slope, the magnitude is constant, and then the delay will be constant. Under such circumstances a signal will be delayed without distortion.

Unfortunately it is not possible to realize the transfer function $T = e^{-s}$ with

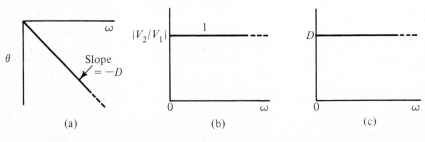

FIGURE 10.2

lumped elements. The best we can do is to approximate Eq. (10.10) with a rational quotient of polynomials:

$$T(s) = \frac{N(s)}{D(s)} \tag{10.11}$$

polynomials $N(s)$ and $D(s)$ being of relatively small degree, perhaps between 2 and 10. Our next task is to find such a polynomial quotient and then to study how well it does approximate the requirement of Eq. (10.10).

If we let $s = j\omega$ in Eq. (10.11), then $T(j\omega)$ becomes complex and may be written

$$T(j\omega) = R(\omega) + jX(\omega) \tag{10.12}$$

The phase of $T(j\omega)$ is

$$\theta = \tan^{-1}\left[\frac{X(\omega)}{R(\omega)}\right] \tag{10.13}$$

Substituting this into Eq. (10.5) and solving for the delay, gives

$$D = \frac{-1}{\omega} \tan^{-1}\left(\frac{R}{X}\right) \tag{10.14}$$

This is a transcendental function and not well suited to our purposes. If we define the delay to be the derivative of Eq. (10.5),

$$D = \frac{-d\theta}{d\omega} \tag{10.15}$$

then we may differentiate Eq. (10.13) to obtain

$$D = \frac{-R\,dX/d\omega + X\,dR/d\omega}{R^2 + X^2} \tag{10.16}$$

which is a rational quotient of polynomials in ω. Fortunately measurement techniques have been developed which make time delay as defined by Eq. (10.15) easy to measure,* and so it is the definition most often used. To distinguish the two, delay as defined by Eq. (10.5) is known as *phase delay*, while that defined by Eq. (10.15) is known as *group delay* or sometimes as *signal delay* or *envelope delay*. In the remainder of our discussion we will make use of the definition of Eq. (10.15) exclusively, and so will refer to it simply as *delay*.

10.2 BESSEL–THOMSON RESPONSE

Our objective is to find a family of transfer functions $T_f(s)$ that will give approximately constant time delay for as large a range of ω as possible. The strategy to be followed is to (1) assume a form for the transfer function, (2) compute the corresponding delay, (3) expand in the form of a Taylor series about $\omega = 0$, and (4)

* G. C. Temes and J. W. LaPatra, *Introduction to Circuit Synthesis and Design*, McGraw-Hill, New York, 1977, p. 567.

find the conditions that will cause as many coefficients in the series expansion to vanish as possible. These conditions will then be used to determine the required form of the transfer function. One of the first to use this approach was W. E. Thomson* of the British Post Office Research Station. It turns out that the coefficients of the polynomials in $T(s)$ are closely related to Bessel polynomials. For this reason we will call the response that results the *Bessel-Thomson response*, although it will be called either Thomson or Bessel response in the literature.

We will illustrate the procedure just described in terms of an all-pole second-order transfer function:

$$T_2(s) = \frac{a_0}{s^2 + a_1 s + a_0} \tag{10.17}$$

For this function the phase is

$$\theta = -\tan^{-1}\left(\frac{a_1 \omega}{a_0 - \omega^2}\right) \tag{10.18}$$

Differentiating with respect to ω gives

$$D = -\frac{d\theta}{d\omega} = \frac{a_1}{a_0} \frac{1 + \omega^2/a_0}{1 + (a_1^2/a_0^2 - 2/a_0)\omega^2 + \omega^4} \tag{10.19}$$

Simple division of the numerator by the denominator gives a result in Taylor series form:

$$D = \frac{a_1}{a_0}\left[1 + \left(\frac{1}{a_0} - \frac{a_1^2}{a_0^2} + \frac{2}{a_0}\right)\omega^2 + \cdots\right] \tag{10.20}$$

For the second term to vanish we require that

$$\frac{a_1^2}{a_0^2} = \frac{3}{a_0} \tag{10.21}$$

or

$$a_1^2 = 3a_0 \tag{10.22}$$

In order to normalize D so that it will have the value of $D = 1$ which we require, we set $a_0 = a_1$, which, from Eq. (10.22), requires that $a_0 = a_1 = 3$. Then we have determined the form of the transfer function

$$T_2 = \frac{3}{s^2 + 3s + 3} \tag{10.23}$$

The corresponding delay is given by Eq. (10.19):

$$D_2(\omega) = \frac{3\omega^2 + 9}{\omega^4 + 3\omega^2 + 9} \tag{10.24}$$

---

* W. E. Thomson, "Delay Networks Having Maximally Flat Frequency Characteristics," *Proc. IEE*, pt. 3, vol. 96, pp. 487–490, 1949. Early papers in which the same problem was studied are by Z. Kiyasu, "On a Design Method of Delay Networks," *J. Inst. Elec. Comm. Eng. Japan*, vol. 26, pp. 598–610, 1943, and W. H. Huggins, "Network Approximation in the Time Domain," Rep E 5048A, Air Force Cambridge Research Labs, Cambridge, Mass., Oct. 1949.

Such calculations, though laborious, are routine. A similar analysis for a third-order all-pole function gives the transfer function

$$T_3 = \frac{15}{s^3 + 6s^2 + 15s + 15} \tag{10.25}$$

for which the delay is

$$D_3 = \frac{6\omega^4 + 45\omega^2 + 225}{\omega^6 + 6\omega^4 + 45\omega^2 + 225} \tag{10.26}$$

A clever method which is computationally simpler and more direct is due to Storch.* It starts from Eq. (10.10), the transfer function for an ideal time delay:

$$T(s) = e^{-s} = \frac{1}{e^s} \tag{10.27}$$

and makes use of the hyperbolic function identity

$$e^s = \sinh s + \cosh s \tag{10.28}$$

Then the transfer function becomes

$$T(s) = \frac{1}{e^s} = \frac{1}{\sinh s + \cosh s} = \frac{1/\sinh s}{1 + \coth s} \tag{10.29}$$

The series expansions of the hyperbolic functions of concern are

$$\cosh s = 1 + \frac{s^2}{2!} + \frac{s^4}{4!} + \frac{s^6}{6!} + \cdots \tag{10.30}$$

$$\sinh s = s + \frac{s^3}{3!} + \frac{s^5}{5!} + \frac{s^7}{7!} + \cdots \tag{10.31}$$

Dividing the first by the second, inverting, repeating the division, and continuing this process, we obtain the infinite continued fraction expansion of the coth s function:

$$\coth s = \frac{1}{s} + \cfrac{1}{\cfrac{3}{s} + \cfrac{1}{\cfrac{5}{s} + \cfrac{1}{\cfrac{7}{s} + \cdots}}} \tag{10.32}$$

In the Storch approach the continued fraction is simply truncated after n terms. The result is a quotient of polynomials where the numerator is identified with cosh s and the denominator with sinh s. The sum of the numerator and denominator polynomials is thus the approximation to e^s in Eq. (10.29). Storch showed that this truncation gave maximally flat delay at $\omega = 0$, no matter where the con-

* L. Storch, "Synthesis of Constant-Time-Delay Ladder Networks Using Bessel Polynomials," *Proc. IRE*, vol. 42, pp. 1666–1675, 1954.

tinued fraction is truncated. Further, he was first to associate the denominator polynomial thus obtained with a class of Bessel polynomials.

To show the simplicity of the Storch approach, we let $n = 2$ and then $n = 3$. For $n = 2$, Eq. (10.32) becomes

$$\coth s = \frac{1}{s} + \frac{1}{3/s} = \frac{3 + s^2}{3s} \tag{10.33}$$

Adding numerator and denominator, we obtain

$$s^2 + 3s + 3 \tag{10.34}$$

which is identical with the denominator of Eq. (10.23) which we obtained with great difficulty. For $n = 3$,

$$\coth s = \frac{1}{s} + \frac{1}{3/s + 1/(5/s)} = \frac{6s^2 + 15}{s^3 + 15s} \tag{10.35}$$

Again adding numerator and denominator gives

$$s^3 + 6s^2 + 15s + 15 \tag{10.36}$$

which is the same as the denominator of Eq. (10.25). We may use this procedure to generate all denominator polynomials. As an alternative, we may make use of the following recursion formula*:

$$\mathscr{B}_n = (2n - 1)\mathscr{B}_{n-1} + s^2 \,\mathscr{B}_{n-2} \tag{10.37}$$

and so generate the Bessel polynomial for any value of n. In forming the Bessel–Thomson function, the numerator of $T(s)$ is always chosen so that $T_n(j0) \cong 1$. Thus the general form of the function is

$$T_n(s) = \frac{\mathscr{B}_n(0)}{\mathscr{B}_n(s)} \tag{10.38}$$

We will be interested in properties of the Bessel polynomial $\mathscr{B}_n(s)$. But before we undertake this study, let us see how we have fared in making $D_n(\omega)$ maximally flat. For each $T_n(s)$ of Eq. (10.38) we can calculate the corresponding $D_n(\omega)$, just as was done in finding $D_2(\omega)$ in Eq. (10.24) and $D_3(\omega)$ in Eq. (10.26). The result of plotting each $D_n(\omega)$ as a function of ω is shown in Fig. 10.3. Sure enough, $D_n(\omega) \cong 1$ for small ω, and the delay is flat over a range of ω, with larger values of n giving a larger band of frequencies for flat response. Thus we have attained our objective of approximating the absolutely flat delay shown in Fig. 10.2c.

It was also our objective to achieve another flatness, namely, that of the magnitude function $|T_n(j\omega)|$. Here we have not done as well. The magnitude function is found from Eq. (10.38):

$$|T_n(j\omega)| = \left| \frac{\mathscr{B}_n(0)}{\mathscr{B}_n(j\omega)} \right| \tag{10.39}$$

A plot for $n = 4$ is shown in Fig. 10.4, and this response is characteristic for all n. We note the gradual transition from pass band to stop band. A corresponding

* We use the script $\mathscr{B}$ to avoid confusion with the Butterworth polynomial.

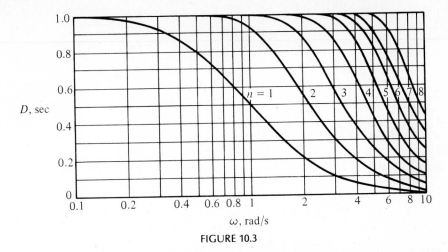

FIGURE 10.3

plot of the Butterworth response for $n = 4$ is shown for comparison. Hence we see that $|T_n(j\omega)|$ for the Bessel–Thomson case is neither constant over a large range of ω, nor does it have sharp cutoff in the transition from pass band to stop band, even compared with the Butterworth response (to say nothing of the Chebyshev response). In Section 10.4 we compare the Bessel–Thomson response with others in more detail.

10.3   BESSEL POLYNOMIALS

The procedure for generating Bessel polynomials described in the last section can be extended to any value of n. The result is the family of polynomials which begins

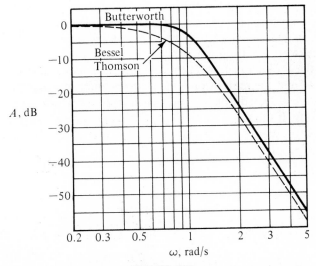

FIGURE 10.4

TABLE 10.1 Coefficients of $\mathscr{B}_n(s) = s^n + a_{n-1}s^{n-1} + \cdots + a_1 s + a_0$

n	a_0	a_1	a_2	a_3	a_4	a_5	a_6	a_7
1	1							
2	3	3						
3	15	15	6					
4	105	105	45	10				
5	945	945	420	105	15			
6	10,395	10,395	4,725	1,260	210	21		
7	135,135	135,135	62,370	17,325	3,150	378	28	
8	2,027,025	2,027,025	945,945	270,270	51,975	6,930	630	36

$$
\begin{aligned}
\mathscr{B}_0 &= 1 \\
\mathscr{B}_1 &= s + 1 \\
\mathscr{B}_2 &= s^2 + 3s + 3 \\
\mathscr{B}_3 &= s^3 + 6s^2 + 15s + 15 \\
\mathscr{B}_4 &= s^4 + 10s^3 + 45s^2 + 105s + 105 \\
&\vdots \quad \vdots
\end{aligned}
\tag{10.40}
$$

If we write the polynomials in the general form

$$
\mathscr{B}_n = s^n + a_{n-1}s^{n-1} + \cdots + a_1 s + a_0 = \sum_{j=0}^{n-1} a_j s^j \quad a_n = 1 \tag{10.41}
$$

then the coefficients can be tabulated as shown in Table 10.1, which gives values to $n = 8$, but could be routinely extended to higher n using the recursion relationship of Eq. (10.37).

Unlike for the Butterworth and Chebyshev responses, there is no simple rule to determine the roots of $\mathscr{B}_n(s) = 0$, which are the poles of the Bessel–Thomson response. But the roots can be routinely found by computer methods, and these are tabulated up to $n = 8$ in Table 10.2; the poles in the s plane are shown in Fig. 10.5 and tabulated in Table 10.3. It was shown by Henderson and Kautz* that the poles of the Bessel–Thomson response very nearly lie on eccentric circles. The centers for the circles for successive values of n are equally spaced along the positive real axis, and the radii of successive circles differ by equal amounts.

Filters are designed from ω_0 and Q specifications which may be found from pole locations. If the pole is located at $-\alpha \pm j\beta$, then, as we found in Eq. (8.91),

$$
\omega_0 = (\alpha^2 + \beta^2)^{1/2} \tag{10.42}
$$

and

$$
Q = \frac{1}{2\alpha} (\alpha^2 + \beta^2)^{1/2} \tag{10.43}
$$

---

* K. W. Henderson and W. H. Kautz, "Transient Responses of Conventional Filters," *IRE Trans. Circuit Theory*, vol. CT-5, pp. 333–347, 1958.

TABLE 10.2 Roots of $\mathscr{B}_n(s) = 0$ for $n = 1$ to $n = 8$ (poles of $T_n(s)$ for Bessel–Thomson response)

n							
1	-1.0000000						
2	-1.5000000	$\pm j0.8660254$					
3	$-2.3221854;$	-1.8389073	$\pm j1.7543810$				
4	-2.8962106	$\pm j0.8672341;$	-2.1037894	$\pm j2.6574180$			
5	$-3.6467386;$	-3.3519564	$\pm j1.7426614;$	-2.3246743	$\pm j3.5710229$		
6	-4.2483594	$\pm j0.8675097;$	-3.7357084	$\pm j2.6262723;$	-2.5159322	$\pm j4.4926730$	
7	$-4.9717869;$	-4.7582905	$\pm j1.7392861;$	-4.0701392	$\pm j3.5171740;$	-2.6856769	$\pm j5.4206941$
8	-5.5878860	$\pm j0.8676144;$	-5.2048408	$\pm j2.6161751$	-4.3682892	$\pm j4.414425;$	-2.8389840 $\pm j6.3539113;$

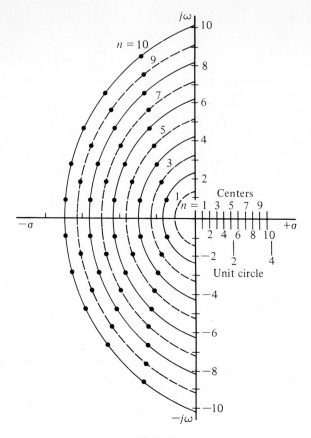

FIGURE 10.5

From these relationships the values of ω_0 and Q for $n = 2$ to $n = 8$ are tabulated in Table 10.4.

The relationship of the Bessel–Thomson poles to those previously found for the Butterworth and Chebyshev cases is of special interest. The general form of the relationship is illustrated by a specific example. Let us consider the case $n = 3$ for the Butterworth and Bessel–Thomson cases, and $n = 3$, $\alpha_{\text{max}} = 0.1$ dB for the Chebyshev response. For a meaningful comparison we must first determine the frequency at which the Bessel–Thomson response has a half-power value of

TABLE 10.3 $\mathcal{B}_n(s)$ **in factored form for $n = 1$ to $n = 6$**

n	$\mathcal{B}_n(s)$	$\mathcal{B}_n(0)$
1	$s + 1$	1
2	$s^2 + 3s + 3$	3
3	$(s^2 + 3.67782s + 6.45944)(s + 2.32219)$	15
4	$(s^2 + 5.79242s + 9.14013)(s^2 + 4.20758s + 11.4878)$	105
5	$(s^2 + 6.70391s + 14.2725)(s^2 + 4.64934s + 18.15631)(s + 3.64674)$	945
6	$(s^2 + 8.49672s + 18.80113)(s^2 + 7.47142s + 20.85282)(s^2 + 5.03186s + 26.51402)$	10,395

TABLE 10.4 Poles of $T_n(s)$ in terms of Q, ω_0 and σ

n	ω_0; Q are ordered pairs; $-\sigma$ is one entry				$\mathscr{B}_n(0)$
2	1.732; 0.577				3
3	2.542; 0.691	2.322			15
4	3.023; 0.522	3.389; 0.806			105
5	3.778; 0.564	4.261; 0.916	3.647		945
6	4.336; 0.510	4.566; 0.611	5.149; 1.023		10,395
7	5.066; 0.532	5.379; 0.661	6.050; 1.126	4.971	135,135
8	5.655; 0.506	5.825; 0560	6.210; 0.711	6.959; 1.226	2,027,025

$$\mathscr{B}_n(0) = \prod_{j=1}^{n-1 \text{ or } n} \omega_j^2 \times \begin{cases} 1, & n \text{ even} \\ \sigma_r, & n \text{ odd} \end{cases}$$

0.707. The transfer function is

$$T_3(s) = \frac{15}{s^3 + 6s^2 + 15s + 15} \tag{10.44}$$

Then the corresponding magnitude function with $s = j\omega$ is

$$|T_3(j\omega)| = \frac{15}{[(15 - 6\omega^2)^2 + (15\omega - \omega^3)^2]^{1/2}} \tag{10.45}$$

From this we determine that T_3 has a value of 0.707 when $\omega = 1.755666$. If we scale the $n = 3$ poles given in Table 10.2 by this amount, we obtain the scaled pole locations as

$$-0.32268 \quad \text{and} \quad -1.04741 \pm j0.99926 \tag{10.46}$$

The Butterworth poles lie on the unit circle shown in Fig. 10.6, and the Chebyshev poles are on an ellipse which is inside the unit circle. The Bessel–Thomson poles lie outside the unit circle. This is an interesting demonstration, which shows that moving the poles of $T(s)$ a small distance in the s plane changes the form of response achieved.

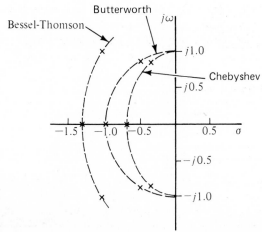

FIGURE 10.6

10.4 FURTHER COMPARISONS OF RESPONSES

We should remind ourselves that the Bessel–Thomson response resulted from imposing the requirements that

$$\left.\frac{dD_n(\omega)}{d\omega}\right|_{\omega=0} = 0 \quad \text{and} \quad D_n(0) = 1 \tag{10.47}$$

or that the delay be maximally flat at $\omega = 0$. From this we determined the required coefficients of $T_n(s)$ and the pole positions we have just discussed. We merely accepted the magnitude response that resulted and found that $|T_n(j\omega)| = 1$ only at low frequencies. In contrast, the Butterworth and Chebyshev responses were derived without reference to delay. But $D_n(\omega)$ can be found for any transfer function by simply determining the phase function $\theta(\omega)$ and then differentiating with respect to ω. We next inquire as to the delay responses associated with the Butterworth and Chebyshev transfer functions.

The Butterworth response was derived from the requirement that the magnitude function be maximally flat for small ω. If we determine the delay characteristics by the method just described, the result is as shown in Fig. 10.7 for $n = 1$ to $n = 10$. We see that the delay has a peak value, especially for larger values of n, at the normalized frequency $\omega = 1$. This may be contrasted to the Bessel–Thomson response shown in Fig. 10.3, which achieves the maximally flat response form. The reason for the peaking of the delay for the Butterworth response may be seen from the phase characteristic shown in Fig. 10.8, which compares the phase response for the Butterworth and Bessel–Thomson transfer functions for the specific case $n = 4$. While the Bessel–Thomson phase is nearly linear for small ω and then has ever decreasing slope as ω increases, the Butterworth response has a slope which becomes large near $\omega = 1$, and this results in the peak of delay.

The delay response for Chebyshev filters is more complicated, of course. It

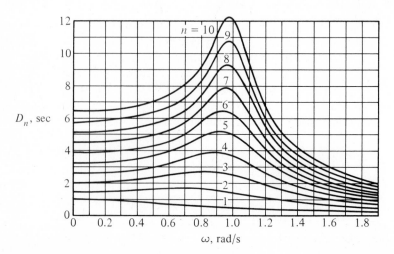

FIGURE 10.7 Reprinted by permission from H. J. Blinchikoff and A. I. Zverev, *Filtering in the Time and Frequency Domains,* John Wiley & Sons, Inc., New York, 1976.

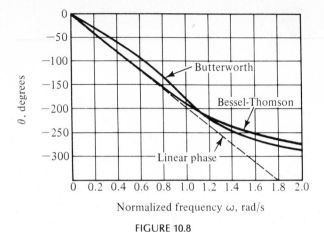

Normalized frequency ω, rad/s

FIGURE 10.8

was derived from the requirement of equal ripple in the pass band, and this resulted in a phase response that was much less linear than the Butterworth phase. In the Chebyshev case the response is determined by the ripple width α_{max} as well as by the order of the response n. Figure 10.9 shows the delay characteristics of the Chebyshev transfer function for $n = 2$ to $n = 5$ for four different values of ripple width.[*] In contrast with either the Bessel–Thomson or the Butterworth response, the Chebyshev delay response will cause considerable delay distortion.

Another distinguishing characteristic of the Bessel–Thomson filter is the time response that results from a step input, the so-called *step response*. When a step voltage is applied to a filter, as shown in Fig. 10.10, the resulting response is usually characterized by two quantities. The first is the *overshoot*, defined as

$$\% \text{ overshoot} = \frac{v_{2\,max} - v_{2\,final}}{v_{2\,final}} \times 100\% \tag{10.48}$$

The second quantity is the *rise time*, usually defined as the time for the response $v_2(t)$ to go from 10% to 90% of its final value. The two responses shown in Fig. 10.10 are for the case $n = 4$. The Butterworth response has considerable overshoot, while the Bessel–Thomson response does not. Also the Butterworth response has a smaller rise time than the Bessel–Thomson response. These same characteristics are shown in Fig. 10.11 for a pulse input and for $n = 8$ Butterworth and Bessel–Thomson filters. This absence of overshoot is a distinguishing characteristic of Bessel–Thomson filters and is sometimes the reason for their selection for a given application.

10.5   DESIGN OF BESSEL–THOMSON FILTERS

At this point in our development of a design procedure for the Butterworth and Chebyshev filters we showed that the specification of attenuations α_{min} and α_{max} together with related frequencies determined n, the order of the filter response,

[*] J. Vlach, *Computerized Approximation and Synthesis of Linear Networks,* Wiley, New York, 1969, pp. 266–267.

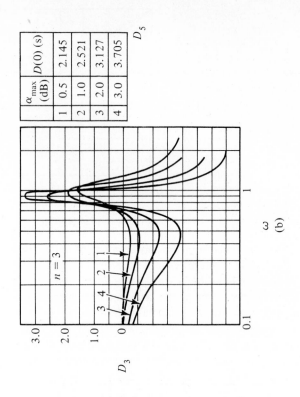

	α_{max} (dB)	$D(0)$ (s)
1	0.5	2.145
2	1.0	2.521
3	2.0	3.127
4	3.0	3.705

(b)

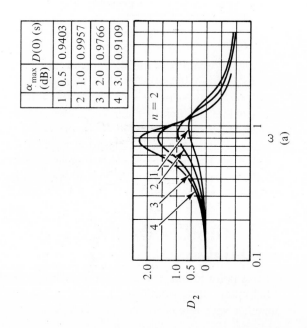

	α_{max} (dB)	$D(0)$ (s)
1	0.5	0.9403
2	1.0	0.9957
3	2.0	0.9766
4	3.0	0.9109

(a)

292

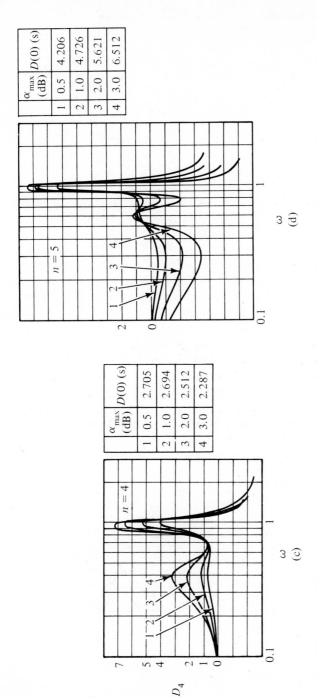

	α_{max} (dB)	$D(0)$ (s)
1	0.5	4.206
2	1.0	4.726
3	2.0	5.621
4	3.0	6.512

(d)

	α_{max} (dB)	$D(0)$ (s)
1	0.5	2.705
2	1.0	2.694
3	2.0	2.512
4	3.0	2.287

(c)

FIGURE 10.9 Reprinted by permission from J. Vlach, *Computerized Approximation and Synthesis of Linear Networks*, John Wiley & Sons, New York, 1969.

293

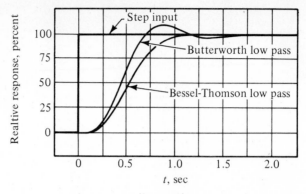

FIGURE 10.10

and specified half-power frequencies. There is no such direct analysis available for Bessel–Thomson filters; there is no formula for n. Instead, design is accomplished using guidelines often derived from prior experience.

First we see from the delay characteristics shown in Fig. 10.3 that the delay is constant over a larger band of frequencies for larger values of n. On the other hand, a large value of n implies a larger number of modules for its realization as a circuit. Hence we seek a compromise. Some of the guidelines we will use in selecting a suitable value for n are next enumerated:

1. As shown in Fig. 10.12a, we may specify the percentage deviation from the normalized delay of 1 second at a specific frequency ω_1.
2. Another specification that may be used in addition to delay deviation is the loss introduced by the filter at some frequency ω_1. This is shown in Fig. 10.12b.
3. In some cases the delay filter is serving a secondary function of acting as a lowpass filter. Then it is appropriate to specify the loss that is required at some specified frequency in the stop band. Such a loss would be calculated from the equation

$$\alpha(\omega_2) = 20 \log |T_n(j\omega_2)| = 20 \log \left| \frac{\mathscr{B}_n(0)}{\mathscr{B}_n(j\omega_2)} \right| \quad \text{dB} \qquad (10.49)$$

where ω_2 is the specified frequency in the stop band.

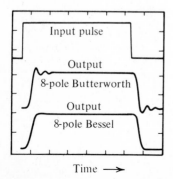

Time $\longrightarrow$ FIGURE 10.11

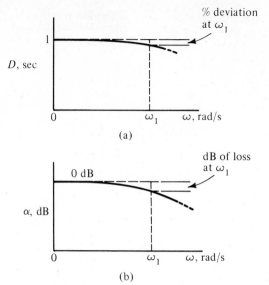

(a)

(b)   FIGURE 10.12

In connection with guidelines 1 and 2, we will make frequent reference to Fig. 10.3. It is also useful to refer to Table 10.5, which is due to Weinberg.* Its use will be illustrated by means of a design example.

Example 10.1   We are required to design a delay filter having at most a 10% deviation in delay at frequency 2.5, and at most 1 dB of loss at a frequency of 1.0. The filter is to provide 100 μs of delay. From Fig. 10.3 the 10% deviation requirement is not satisfied by $n = 3$, and so we have to use $n = 4$ (the equivalent of rounding up). From Table 10.5, we see that the 10% deviation occurs at the frequency $\omega = 2.84$ for $n = 4$, which is larger than the $\omega = 2.5$ that has been specified. In addition, we see that the 1 dB of loss for $n = 4$ occurs at $\omega = 1.25$, which is larger than the specified $\omega = 1$. Clearly, the specifications require a Bessel–Thomson response with $n = 4$.

The required form of the transfer function is obtained from Table 10.3:

$$T_4(s) = \frac{9.14013}{s^2 + 5.792425s + 9.14013} \frac{11.4878}{s^2 + 4.207585s + 11.4878} \tag{10.50}$$

As an alternative we may consult Table 10.4, from which the Q and ω_0 specifications for the two sections are seen to be

$$\omega_{01} = 3.023, \quad Q_1 = 0.522; \quad \omega_{02} = 3.389, \quad Q_2 = 0.806 \tag{10.51}$$

Thus the response required is fully specified. Should a detailed study of the delay characteristics be required, the delay response is made up of delays associated with the two parts of Eq. (10.50). Thus

$$D_4(\omega) = \frac{5.8\omega^2 + 53.01}{33.64\omega^2 + (9.14 - \omega^2)^2} + \frac{4.21\omega^2 + 48.37}{17.72\omega^2 + (11.49 - \omega^2)^2} \tag{10.52}$$

---

* L. Weinburg, *Network Analysis and Synthesis*, McGraw-Hill, New York, 1962; reprinted by Robert E. Krieger Publishing, Inc., Malabar, Fla., 1975, p. 502.

TABLE 10.5.   Time delay and loss characteristics of the Bessel–Thomson response*

Time delay table giving the frequency ω at which D deviates a specified amount from its zero-frequency value

n	ω for 1% deviation	ω for 10% deviation	ω for 20% deviation	ω for 50% deviation
1	0.10	0.34	0.50	1.00
2	0.56	1.09	1.39	2.20
3	1.21	1.94	2.29	3.40
4	1.93	2.84	3.31	4.60
5	2.71	3.76	4.20	5.78
6	3.52	4.69	5.95	6.97
7	4.36	5.64	6.30	8.15
8	5.22	6.59	7.30	9.33
9	6.08	7.55	8.31	10.50
10	6.96	8.52	9.33	11.67
11	7.85	9.49	10.34	12.84

Loss table giving the frequency ω at which the loss is a specified number of decibels from its zero-frequency value

n	ω for $\frac{1}{50}$ dB	ω for $\frac{1}{20}$ dB	ω for $\frac{1}{10}$ dB	ω for $\frac{1}{5}$ dB	ω for $\frac{1}{2}$ dB	ω for 1 dB	ω for 3 dB
1	0.07	0.11	0.14	0.21	0.35	0.51	1.00
2	0.11	0.18	0.26	0.36	0.57	0.80	1.36
3	0.14	0.23	0.34	0.48	0.75	1.05	1.75
4	0.17	0.28	0.40	0.56	0.89	1.25	2.13
5	0.20	0.32	0.45	0.64	1.01	1.43	2.42
6	0.22	0.36	0.50	0.71	1.12	1.58	2.70
7	0.24	0.39	0.54	0.77	1.22	1.72	2.95
8	0.26	0.41	0.59	0.83	1.31	1.85	3.17
9	0.28	0.44	0.62	0.88	1.40	1.97	3.39
10	0.30	0.47	0.66	0.93	1.48	2.08	3.58
11	0.31	0.49	0.69	0.98	1.55	2.19	3.77

* Reprinted by permission from L. Weinburg, *Network Analysis and Synthesis,* McGraw-Hill Book Company, New York, 1962; Robert E. Krieger Publishing Co., Inc., Malabar, Fla., 1975.

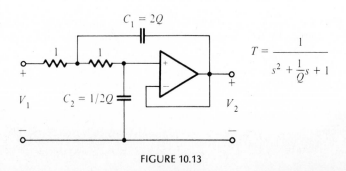

$$T = \frac{1}{s^2 + \frac{1}{Q}s + 1}$$

FIGURE 10.13

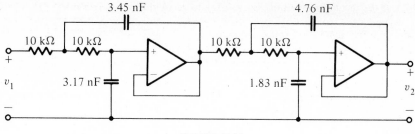

FIGURE 10.14

Similarly, the magnitude response is obtained from Eq. (10.50):

$$|T_4(j\omega)| = \frac{105}{[33.64\omega^2 + (9.14 - \omega^2)^2]^{1/2}[17.72\omega^2 + (11.49 - \omega^2)^2]^{1/2}} \qquad (10.53)$$

The lowpass filter that we will select for the realization of the delay filter is the $K = 1$ Sallen–Key circuit shown in Fig. 10.13. We only have to use the values of Q and ω_0 given by Eq. (10.51) and then scale to meet the 10^{-4} second delay specification.

We first determine the normalized element values for the circuit of Fig. 10.12. Using these values and those of Eq. (10.51), we obtain the following specifications for the stages:

Element	Stage 1	Stage 2
C_1	1.044	1.612
C_2	0.958	0.620
$R_1 = R_2$	1	1
k_f	3.023	3.389
k_m	10^4	10^4

Then we scale using the relationships from Appendix A:

$$C_{\text{new}} = \frac{D}{k_m k_f} C_{\text{old}} \quad \text{and} \quad R_{\text{new}} = k_m R_{\text{old}} \qquad (10.54)$$

Using these equations, we determine the following circuit values:

Element	Stage 1	Stage 2
C_1	3.45 nF	4.76 nF
C_2	3.17 nF	1.83 nF
$R_1 = R_2$	10 kΩ	10 kΩ

These values are shown on the circuit schematic of Fig. 10.14, and the design is complete.

PROBLEMS

10.1   For the circuit given in Fig. P10.1, determine the value of L and C such that the delay is maximally flat at $\omega = 0$ and the half-power frequency of $|T(j\omega)|$ occurs at $\omega = 1$.

10.2 Use the recursion equation, Eq. (10.37), to verify the entries in Table 10.1 for $n = 5$, 6, and 7.

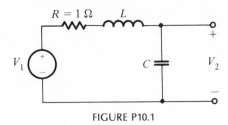

FIGURE P10.1

10.3 Accepting the factored polynomial given in Table 10.3 for $n = 6$, verify the entries in Table 10.4 for $n = 6$.

10.4 Design a Bessel-Thomson filter for $n = 3$ to provide 100 μs time delay. Do this with a cascade of two sections of RC op-amp circuits.

10.5 Design a fourth-order Bessel–Thomson filter to provide a time delay of 100 μs. Magnitude scale so that the elements in your filter are in a practical range.

10.6 A fifth-order Bessel–Thomson approximation function has a delay characteristic that is flat to within 5 percent of the *dc* delay up to 100 rad/s. Determine the loss at the frequencies of 250 rad/s and 600 rad/s.

10.7 A Bessel-Thomson approximation function is to be flat within 5 percent of the *dc* delay of 1 msec to the frequency of 1800 rad/s.
(a) Determine the appropriate Bessel–Thomson function.
(b) Determine the attenuation at a frequency of 1800 rad/s and 14,000 rad/s.

10.8 Design a delay filter having at most a 10 percent deviation in delay at the normalized frequency of 3.0 and at most a 1 dB deviation of loss at the normalized frequency of 1.35 rad/s. The filter is to provide 200 μs of delay. Complete the design to include magnitude scaling to give reasonable element values.

10.9 A delay filter is to be designed having at most a 10 percent deviation in delay at the normalized frequency of 4.0, and at most a 1 dB deviation of loss at the normalized frequency of 1.50 rad/s. The filter is to provide 100 μs of delay. Complete the design to include magnitude scaling to give practical element values.

10.10 The Gaussian magnitude response is given by the equation

$$|T(j\omega)| = e^{-\omega^2/2}$$

and a second-order approximation to this response is given by the equation

$$|T(j\omega)|^2 = \frac{1}{1 + \omega^2 + \frac{1}{2}\omega^4}$$

(a) Determine $T(s)$.
(b) Determine the pole locations for $T(s)$.
(c) Find the half-power frequency for $T(s)$.
(d) Compare the delay characteristics of the Gaussian and the Bessel-Thomson second-order functions.

Frequency
Transformations

Frequency transformation is one of the novel concepts of filter design, of equal importance to that of frequency and magnitude scaling and similar in some ways. It permits the design to be made in terms of the lowpass frequency characteristic, and then to be transformed to achieve some other form of response. In this chapter we extend the studies we initiated in Chapter 7 for the lowpass to bandpass transformation.

11.1 PROTOTYPE RESPONSE AND ITS TRANSFORMATION

Frequency transformation begins with a lowpass magnitude response. It may be any of those studied in previous chapters—Butterworth, Chebyshev, or Bessel–Thomson (shown in Fig. 11.1)—or one of the many other forms of lowpass response that may be found in the literature. For our description we represent all of these by the lowpass 'brick wall' introduced as our objective in Chapter 6. We will describe this model lowpass response as the *prototype response*, and indeed the circuits that we actually use in later studies for the lowpass case will be known as the *prototype circuit*. The prototype response of Fig. 11.1 is shown in Fig. 11.2a, where we have extended the frequency range to include both positive and negative values. The frequency for the prototype response is shown as Ω. We see that the pass band extends from $\Omega = -1$ to $\Omega = 1$, and that all other frequencies constitute the stop bands. In the lowpass to bandpass case studied in Section 7.1 this prototype response was transformed into the bandpass response shown in Fig. 11.2b, where we have used ω for the frequency in the transformed case.

The frequency Ω is the imaginary part of the complex frequency

$$S = \Sigma + j\Omega \qquad (11.1)$$

which is shown in Fig. 11.3. The part of the imaginary axis extending from -1 to $+1$ is transformed into the two segments of the imaginary axis shown in Fig. 11.3b, from ω_1 to ω_2 and from ω_3 to ω_4. The frequency ω is the imaginary part of

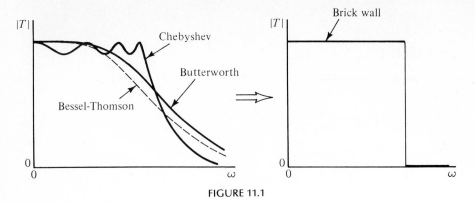

FIGURE 11.1

the complex frequency

$$s = \sigma + j\omega \qquad (11.2)$$

In Fig. 11.3 the function that accomplished this frequency transformation is designated as X_1; it satisfies the equation

$$\Omega = X_1(\omega) \qquad (11.3)$$

which we will restudy in the next section. Our objective is explained in terms of Fig. 11.4. Given the prototype response shown in Fig. 11.4a, we seek other frequency transformations, designed as $X_2(\omega)$, $X_3(\omega)$, and $X_4(\omega)$, to give the responses shown in the lower part of the figure, as well as any others we may wish to achieve. Note in terms of this figure that we do not transform $|T|$, shown as the brick wall, but *only* the frequency.

11.2 PROTOTYPE TO BANDPASS REVISITED

We seek the function $X_1(\omega)$ given by Eq. (11.3) that will transform the pass and stop bands of the prototype response to frequencies that will give the required bandpass response. We may visualize this requirement starting with the prototype response shown in Fig. 11.5a. Suppose that we rotate this response by 90° as illustrated in Fig. 11.5b. This figure also shows a plot of $X(\omega)$. In solving the equation $\Omega = X_1(\omega)$ suppose that we select five different values of Ω, denoted in the figure as Ω_a, Ω_1, 0, Ω_2, and Ω_b. By the graphical projections shown, we see that these five values of Ω correspond to five values of ω with the same subscript identifications. The prototype pass band from Ω_1 to Ω_2 is transformed into the band-

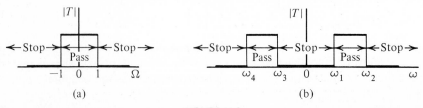

FIGURE 11.2

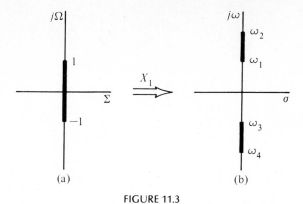

FIGURE 11.3

pass band from ω_1 to ω_2. The frequencies Ω_a and Ω_b are in the stop band, and they transform into stop band frequencies ω_a and ω_b. We seek the transformation $X_1(\omega)$ that accomplished our objective.

Let us examine the steps in Fig. 11.5 in more detail. Referring to Fig. 11.6a we see the prototype response and the identification of the pass and stop bands. In Fig. 11.6b we retain only the skeleton of Fig. 11.6a and rotate it through 90° to give Fig. 11.6c. We now start over with the objective $|T|$ as a function of ω, as shown in Fig. 11.6d. We seek a function X_1 in Fig. 11.6e which has stop and pass bands like those in Fig. 11.6d. The function that we will find must give these pre-scribed stop and pass bands. Since we must satisfy Eq. (11.3) by letting $\Omega = X_1$, we superimpose the requirements of Fig. 11.6c and e, as shown in Fig. 11.6f. Those regions that have the same requirements are cross-hatched, with double cross-hatching indicating the regions that specified stop bands and single cross-hatching for those in which Fig. 11.6c and e indicated pass-band requirements.

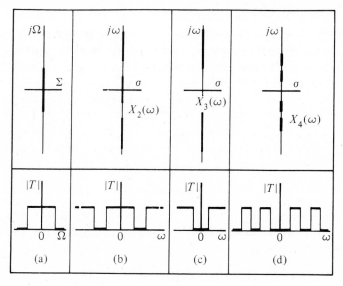

FIGURE 11.4

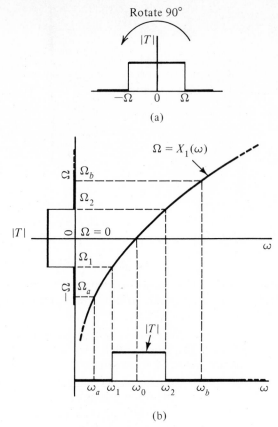

FIGURE 11.5

The remaining regions, without cross-hatching, had requirements with respect to stop and pass that were in conflict. The function that we seek, identified by the plot $X_1(\omega)$ in Fig. 11.5b, must pass through the cross-hatched regions to give the prescribed pass and stop bands of frequencies.

The function used in Chapter 7 was

$$X(\omega) = \Omega = \frac{1}{\text{bw}} \frac{\omega^2 - \omega_0^2}{\omega} \tag{11.4}$$

where

$$\omega_0^2 = \omega_1 \omega_2 \tag{11.5}$$

$$\text{bw} = \omega_2 - \omega_1 \tag{11.6}$$

The plot of this function Eq. (11.4), is shown in Fig. 11.7 with the layout patterned after Fig. 11.6b. Equation (11.4) was derived in Chapter 7 by imposing the requirement that

$$\Omega(\omega_1) = -1 \quad \text{and} \quad \Omega(\omega_2) = +1 \tag{11.7}$$

as shown by the circles on the $X(\omega)$ plot. The center frequency of the prototype

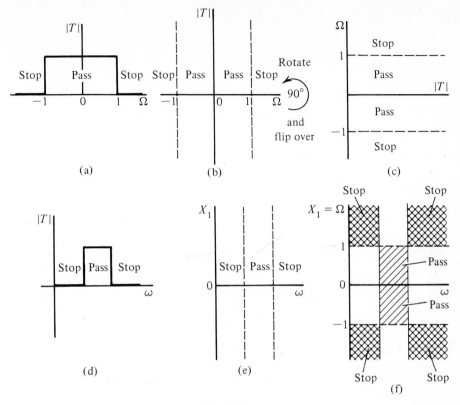

FIGURE 11.6

brick-wall response transformed to the frequency ω_0 in the bandpass response, where ω_0 is the geometrical mean of ω_1 and ω_2, given by Eq. (11.5). Comparing this frequency transformation function to the requirements imposed in Fig. 11.6f, we see that

$$X(0) = -\infty \qquad (11.8)$$

and that $X(\omega)$ remains negative up to the transition that occurs when $X = -1$, so identifying the prescribed stop band. Between the values of $X = -1$ and $X = +1$ we have the band of frequencies between ω_1 and ω_2 which is the pass band. As X exceeds the value of $+1$, we again require a stop band, which is the prescribed behavior for high frequencies. Drawing a cross-hatched region diagram like that of Fig. 11.6f is thus a good guide to the selection of the suitable function $X(\omega)$ to accomplish the desired objective.

We may generalize Eq. (11.4) from the ω axis to the s plane as

$$A(s) = j\,\Omega(\omega)\big|_{\omega=s/j} = j\,\frac{-s^2 - \omega_0{}^2}{\mathrm{bw}\,s/j} \qquad (11.9)$$

or

$$A(s) = \frac{1}{\mathrm{bw}}\,\frac{s^2 + \omega_0{}^2}{s} \qquad (11.10)$$

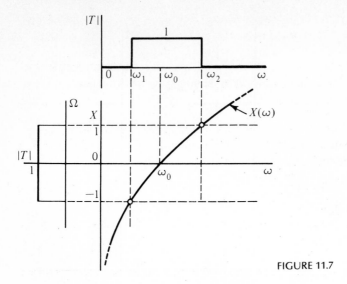

FIGURE 11.7

This function is the generalized frequency transformation

$$S = A(s) \tag{11.11}$$

It is interesting to observe from Eq. (11.10) that the poles and zeros of $A(s)$ are on the imaginary axis, with a pole at the origin, and complex imaginary zeros and $s = \pm j\omega_0$, as shown in Fig. 11.8. The function also has a pole at infinity. In this sense the poles and zeros alternate with increasing ω. This is one member of the family of frequency transformations that we describe next.

11.3 FOSTER REACTANCE FUNCTIONS

Following the lead of the last section where $A(s)$ was found to have simple poles and zeros interlaced on the imaginary axis, let us construct a class of functions

$$A_n(s) = H \frac{p(s)}{q(s)} \tag{11.12}$$

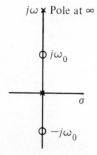

FIGURE 11.8

where n is used to indicate different functions, H is a real and positive constant, and the poles and zeros have these properties:

1. All poles and zeros of $A(s)$ are simple and are located on the imaginary axis. Since they must occur as conjugate imaginary terms, they will have the form

$$(s + j\omega_0)(s - j\omega_0) = s^2 + \omega_0^2 \tag{11.13}$$

2. Poles and zeros are interlaced, meaning that they alternate with increasing ω.
3. A simple pole or zero is located at the origin, $s = 0$. Thus either $p(s)$ or $q(s)$, but not both, will have an s multiplier:

$$p(s) = Ks(s^2 + \omega_1^2)(s^2 + \omega_2^2) \cdots \tag{11.14}$$

This implies that either $p(s)$ is even and $q(s)$ is odd or vice versa.
4. A simple pole or a simple zero is located at infinity. This means that the degree of difference of $p(s)$ and $q(s)$ is 1.

It is simple to construct such functions. Several that satisfy the requirements are the following:

$$A_1(s) = Hs \tag{11.15}$$

$$A_2(s) = \frac{H}{s} \tag{11.16}$$

$$A_3(s) = H \frac{s^2 + \omega_0^2}{s} \tag{11.17}$$

$$A_4(s) = H \frac{s}{s^2 + \omega_0^2} \tag{11.18}$$

$$A_5(s) = \frac{H(s^2 + \omega_1^2)}{s(s^2 + \omega_2^2)} \tag{11.19}$$

In general,

$$A(s) = H \frac{(s^2 + \omega_1^2)(s^2 + \omega_3^2)(s^2 + \omega_5^2) \cdots}{s(s^2 + \omega_2^2)(s^2 + \omega_4^2)(s^2 + \omega_6^2) \cdots} \tag{11.20}$$

where

$$0 \le \omega_1 < \omega_2 < \omega_3 < \omega_4 < \omega_5 < \omega_6 < \cdots \tag{11.21}$$

To see whether these functions are useful as transformations, we must compute the corresponding $X(\omega)$ functions. When $s = j\omega$, all of the factors in Eq. (11.20) are real, except for the s in the denominator (or the s in the numerator if $\omega_1 = 0$). Hence we conclude that $A(j\omega)$ is purely imaginary. The imaginary part is simply found from

$$X(\omega) = \frac{1}{j} A(j\omega) \tag{11.22}$$

Before we apply this equation to the family of $A(s)$ functions just given, we observe the behavior of $X(\omega)$ near a pole of $A(s)$ and also near a zero of $A(s)$. If $A(s)$ has a zero at the frequency ω_1, then the $X(\omega)$ corresponding to this $A(s)$ will have the form

$$X(\omega) = X_1(\omega)(-\omega^2 + \omega_1^2) \tag{11.23}$$

For ω slightly less than ω_1, the term in Eq. (11.23) will be slightly negative; when $\omega = \omega_1$, $X(\omega)$ will be zero; and when ω is slightly larger than ω_1, then the term will be slightly positive. This behavior is shown in Fig. 11.9a and assumes that $X_1(\omega)$ remains positive near ω_1. On the other hand, if $A(s)$ has a pole at ω_2, then the form of $X(\omega)$ will be

$$X(\omega) = X_2(\omega) \frac{1}{-\omega^2 + \omega_2^2} \tag{11.24}$$

Assume that $X_2(\omega)$ remains negative near ω_2. Then examining the behavior of $X(\omega)$ near ω_2 shows that it becomes infinity at ω_2 and there changes from $+\infty$ to $-\infty$. This is shown in Fig. 11.9b.

Next we apply Eq. (11.22) to the calculation of $X(\omega)$ for the functions of Eqs. (11.15)–(11.19):

$$X_1 = \omega H \tag{11.25}$$

$$X_2 = \frac{-H}{\omega} \tag{11.26}$$

$$X_3 = -H \frac{\omega_0^2 - \omega^2}{\omega} \tag{11.27}$$

$$X_4 = H \frac{\omega}{-\omega^2 + \omega_0^2} \tag{11.28}$$

$$X_5 = \frac{-H}{\omega} \frac{-\omega^2 + \omega_1^2}{-\omega^2 + \omega_2^2} \tag{11.29}$$

These functions are shown in Fig. 11.10 together with the pole–zero plots for the given $A(s)$. From these plots we observe two important facts:

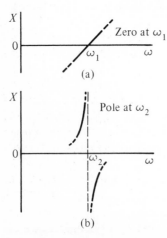

(a)

(b)   FIGURE 11.9

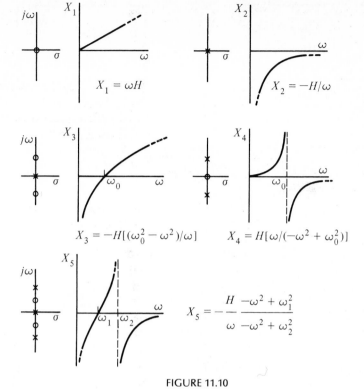

FIGURE 11.10

1. $X(0)$ has either the value 0 or the value $-\infty$.
2. The slope of X with respect to ω is always positive.

$$\frac{dX(\omega)}{d\omega} > 0 \tag{11.30}$$

Clearly this positive slope property is directly related to the interlacing of poles and zeros, since the condition $X = 0$ identifies a zero of $A(s)$ while $X = \pm\infty$ identifies a pole of $A(s)$.

The functions we have described are known as reactance functions, and these properties were first described by Foster in 1924. In recognition of Foster's work,* we will refer to them as *Foster functions.*

We have thinly disguised the fact that $A(s)$ is the driving-point impedance or admittance of an LC circuit. The choice of $A(s)$ as our frequency transformation relates directly to that fact. In the prototype circuit shown in Fig. 11.11a the realization of the lowpass characteristic was originally made by the use of inductors and capacitors only. The theory of such realizations was developed in the 1920s and 1930s, and the circuits were known as lossless transmission filters. Now the filter shown in Fig. 11.11b would have a different frequency characteristic such as bandpass or highpass, but it too would be made up of inductors and capacitors

---

* R. M. Foster, "A Reactance Theorem," *Bell Sys. Tech. J.,* vol. 3, pp. 259–267, 1924.

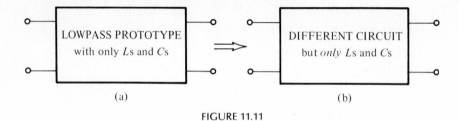

FIGURE 11.11

since op amps were not available for such realizations in those days. Now the impedance and admittance of the inductors and capacitors in the prototype would be

$$Z_L = LS \quad \text{and} \quad Y_C = CS \qquad (11.31)$$

When the frequency transformation $S = A(s)$ is applied, then

$$Z_L = L\,A(s) \quad \text{and} \quad Y_C = C\,A(s) \qquad (11.32)$$

must represent an LC circuit. Thus we see that L and C are scaling factors, and $A(s)$ must represent the input impedance or admittance of an LC circuit. Since LC circuits are lossless,

$$A(s)|_{s=j\omega} = 0 + jX(\omega) \qquad (11.33)$$

where $X(\omega)$ is known as a reactance function. In Section 11.5 we consider the problem of finding circuits that have the impedances specified by Eq. (11.32) for several useful frequency transformations. Actually the number of frequency transformations that we will use is so limited that it is not worthwhile considering the general properties of $A(s)$ as enumerated by Foster. They are, however, available in elementary textbooks.[†]

11.4 HIGHPASS AND BANDSTOP CASES

We seek a frequency transformation from the lowpass prototype to the highpass case shown at the top of Fig. 11.12. The figure shows a stop band from 0 to ω_0 and a pass band beginning at ω_0 and extending to all higher frequencies. From the discussions in the last section, we require a reactance transformation $X(\omega)$ that is confined to the cross-hatched areas shown in Fig. 11.12. When $\omega = 0$, $X(0)$ must be either 0 or $-\infty$. But it cannot be zero because that is not in the cross-hatched region. At frequency ω_0, X must be either $+1$ or -1, corresponding to points Q and P in the figure. But if $X(\omega)$ begins at $-\infty$ and is required to have a positive slope, then P is the only possibility. Thus we see that the required form of the reactance function is that shown in Fig. 11.13, which is described by Eq. (11.26):

$$X(\omega) = \frac{-H}{\omega} \qquad (11.34)$$

[†] M. E. Van Valkenburg, *Introduction to Modern Network Synthesis*, Wiley, New York, 1960, chap. 5.

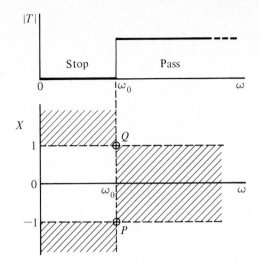

FIGURE 11.12

To evaluate H, we observe that

$$X(\omega_0) = -1 = \frac{-H}{\omega_0} \tag{11.35}$$

so that $H = \omega_0$. Then

$$jX(\omega) = \frac{\omega_0}{j\omega} \tag{11.36}$$

or

$$S(s) = \left.\frac{\omega_0}{j\omega}\right|_{j\omega=s} = \frac{\omega_0}{s} \tag{11.37}$$

which is then the lowpass to highpass frequency transformation. Some use of this

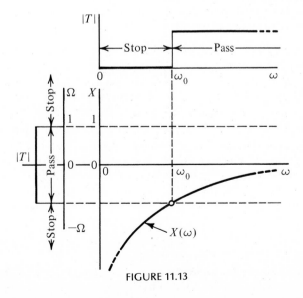

FIGURE 11.13

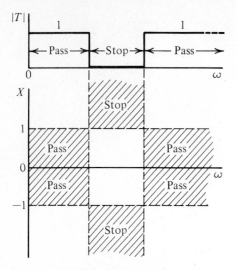

FIGURE 11.14

fact was made in Section 6.7, which might now be reviewed for a more complete understanding of *RC–CR* transformation.

The same procedure may be applied to obtain the frequency response shown at the top of Fig. 11.14. This is known as a *bandstop* or *band-elimination* frequency response. Both low and high frequencies are passed through this filter, but frequencies between ω_1 and ω_2 are stopped. Figure 11.14 also shows the regions through which the $X(\omega)$ curve must pass. Since it is clear that the response must begin with $X(0) = 0$, then the positive slope requirement implies that the only solution is that shown in Fig. 11.15, which has the form of Eq. (11.28):

$$X(\omega) = H \frac{\omega}{-\omega^2 + \omega_0^2} \tag{11.38}$$

We come to this conclusion by observing that $X(\omega)$ has a zero at $\omega = 0$ and a pole at ω_0. This pole–zero configuration is shown in connection with X_4 in Fig. 11.10.

To determine the value of H in Eq. (11.38) we substitute the known values of H at ω_1 and ω_2. Thus

$$X(\omega_1) = +1 = \frac{H\omega_1}{-\omega_1^2 + \omega_0^2} \tag{11.39}$$

$$X(\omega_2) = -1 = \frac{H\omega_2}{-\omega_2^2 + \omega_0^2} \tag{11.40}$$

These two equations may be solved to determine that

$$H = \omega_2 - \omega_1 = \text{bw} \tag{11.41}$$

and

$$\omega_0^2 = \omega_1 \omega_2 \tag{11.42}$$

Then the required frequency transformation is

$$S = A(s) = \text{bw} \frac{s}{s^2 + \omega_0^2} \tag{11.43}$$

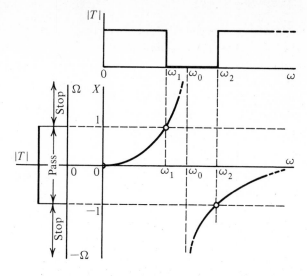

FIGURE 11.15

It is interesting to observe that this is the reciprocal of the frequency transformation from the lowpass to the bandpass case.

11.5 TRANSFORMATIONS OF *LC* CIRCUITS

We now turn to the use of the frequency transformation to convert the lowpass prototype circuit into some other kind of circuit, a bandpass circuit, for example. We assume that the prototype circuit is made up of R's, L's, and C's, but only L's and C's have impedances (or admittances) that change with frequency. Let the L's and C's in the prototype circuit be distinguished by the subscript p, standing for prototype. We begin with Eq. (11.32) which we now write as

$$Z = L_p A(s) \qquad (11.44)$$

for every inductor in the prototype, and

$$Y = C_p A(s) \qquad (11.45)$$

for every capacitor in the prototype circuit. We will consider the frequency transformations in the order in which they were introduced.

Case I: Lowpass to Bandpass

This transformation is

$$A(s) = \frac{1}{\text{bw}} \frac{s^2 + \omega_0^2}{s} \qquad (11.46)$$

Substituting this $A(s)$ into Eq. (11.44), we obtain

$$Z = \frac{L_p}{\text{bw}} s + \frac{L_p \omega_0^2}{\text{bw } s} = L_1 s + \frac{1}{C_1 s} \qquad (11.47)$$

From this

$$L_1 = \frac{L_p}{\text{bw}} \quad \text{and} \quad C_1 = \frac{\text{bw}}{L_p \omega_0^2} \tag{11.48}$$

The fact that the impedances are added in Eq. (11.47) implies a series connection. Thus we see that each inductor transforms into a series LC circuit, as shown in Fig. 11.16. Similarly, for each capacitor in the prototype circuit,

$$Y = \frac{C_p}{\text{bw}} s + \frac{C_p \omega_0^2}{\text{bw}} \frac{1}{s} = C_2 s + \frac{1}{L_2 s} \tag{11.49}$$

The addition of admittances implies a parallel connection with element values identified in this equation as

$$L_2 = \frac{\text{bw}}{C_p \omega_0^2} \quad \text{and} \quad C_2 = \frac{C_p}{\text{bw}} \tag{11.50}$$

as shown in the lower part of Fig. 11.16. Thus if all capacitors and inductors in the prototype are converted as prescribed in Fig. 11.16, the circuit will be changed from a lowpass to a bandpass circuit.

Case II: Lowpass to Highpass

The appropriate equation for $A(s)$ is given by Eq. (11.37), which is $A(s) = \omega_0/s$. Hence

$$Z = L_p \frac{\omega_0}{s} = \frac{1}{C_1 s} \tag{11.51}$$

or

$$C_1 = \frac{1}{L_p \omega_0} \tag{11.52}$$

Similarly,

$$Y = C_p \frac{\omega_0}{s} = \frac{1}{L_2 s} \tag{11.53}$$

or

$$L_2 = \frac{1}{C_p \omega_0} \tag{11.54}$$

This result is shown in Fig. 11.17. To convert from a lowpass to a highpass circuit, all inductors are changed to capacitors, and all capacitors are changed to inductors, with numerical values as given in Fig. 11.17.

Case III: Lowpass to Bandstop

The frequency transformation for this case was given by Eq. (11.43):

$$A(s) = \text{bw} \frac{s}{s^2 + \omega_0^2} \tag{11.55}$$

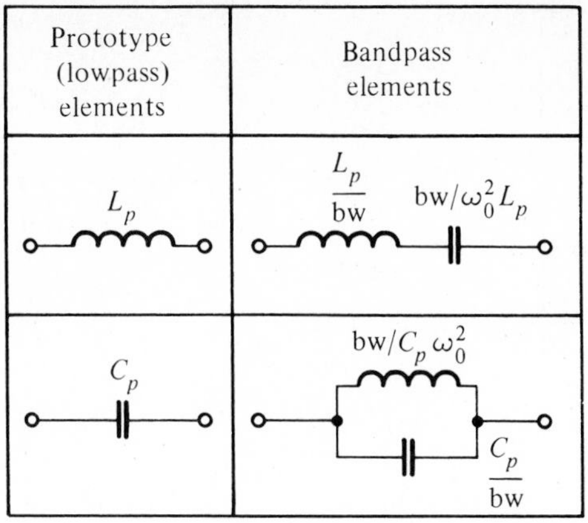

Prototype (lowpass) elements	Bandpass elements
L_p	$\dfrac{L_p}{\text{bw}}$ $\quad$ $\text{bw}/\omega_0^2 L_p$
C_p	$\text{bw}/C_p \omega_0^2$ $\quad$ $\dfrac{C_p}{\text{bw}}$

FIGURE 11.16

For each inductor in the prototype circuit the impedance is

$$L_pA(s) = \frac{1}{(1/\text{bw }L_p)s + [1/(\text{bw }L_p/\omega_0^2)]\,(1/s)} = \frac{1}{C_1 s + (1/L_1 s)} \quad (11.56)$$

We have obtained this equation by dividing numerator and denominator of Eq. (11.55) by s and then manipulating the constants so that an identification can be made for C_1 and L_1 in the last equation. It is

$$C_1 = \frac{1}{\text{bw }L_p} \quad \text{and} \quad L_1 = \frac{\text{bw }L_p}{\omega_0^2} \quad (11.57)$$

as shown by the first row in Fig. 11.18. The admittance function is similarly treated:

$$C_pA(s) = \text{bw }C_p \frac{1}{s + \omega_0^2(1/s)} \quad (11.58)$$

or the impedance is the reciprocal of this quantity:

$$Z = \frac{1}{\text{bw }C_p}s + \frac{1}{(\text{bw }C_p/\omega_0^2)s} = L_2 s + \frac{1}{C_2 s} \quad (11.59)$$

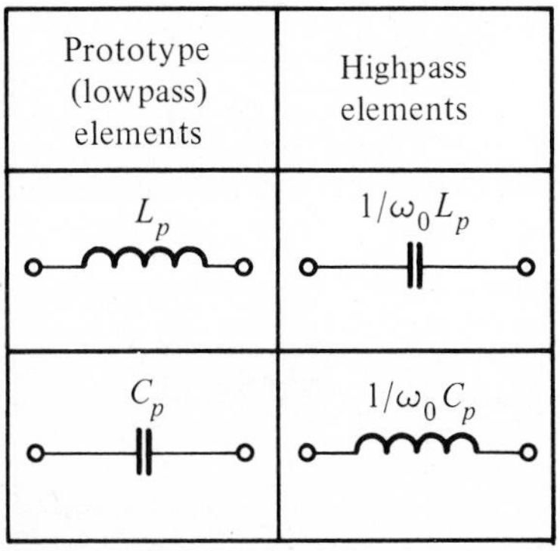

Prototype (lowpass) elements	Highpass elements
L_p	$1/\omega_0 L_p$
C_p	$1/\omega_0 C_p$

FIGURE 11.17

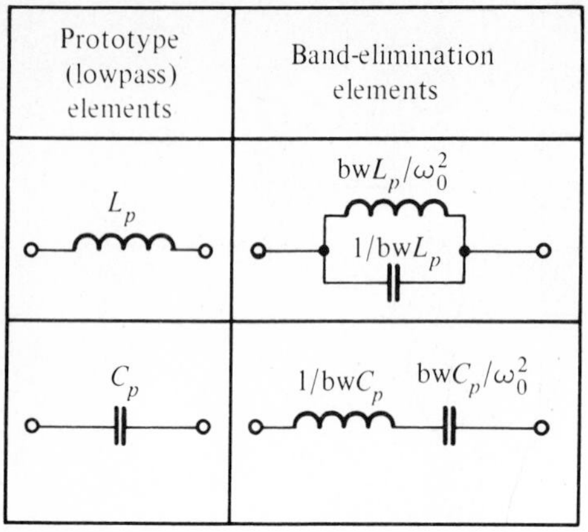

Prototype (lowpass) elements	Band-elimination elements
L_p	bwL_p/ω_0^2 $1/bwL_p$
C_p	$1/bwC_p$ bwC_p/ω_0^2

FIGURE 11.18

Hence we have the element values

$$L_2 = \frac{1}{bw\ C_p} \quad \text{and} \quad C_2 = \frac{bw\ C_p}{\omega_0^2} \tag{11.60}$$

as shown by the second row in Fig. 11.18. The three frequency transformations we have found are combined in Table 11.1.

The simplest passive lowpass circuit is the RC circuit shown in Fig. 11.19a. Not only is it the simplest, but it is probably the one most frequently used in applications without severe specifications. This circuit will serve as an example to illustrate the transformation of LC circuits. If we apply the lowpass to bandpass transformation to Fig. 11.19a, we obtain the circuit shown in Fig. 11.19b, where element values can be found using the formulas in Table 11.1. That this is, indeed, a bandpass circuit may be seen by noting that at both low and high frequencies the inductor or the capacitors short the output. At some intermediate frequency the L_1C_1 circuit is in parallel resonance, represents an open circuit, and so there is direct transmission through the circuit. Applying the lowpass to highpass transformation gives us the circuit shown in Fig. 11.19c. Since the inductor is

TABLE 11.1

Prototype (lowpass) elements	Highpass elements	Bandpass elements	Bandstop elements
L_p	$1/\omega_0 L_p$	L_p/bw $bw/\omega_0^2 L_p$	bwL_p/ω_0^2 $1/bwL_p$
C_p	$1/\omega_0 C_p$	$bw/C_p\omega_0^2$ C_p/bw	$1/bwC_p$ $bw/C_p\omega_0^2$

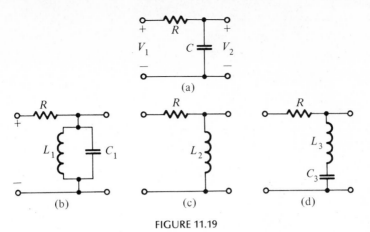

FIGURE 11.19

a short circuit at low frequencies and an open circuit at high frequencies (as an approximation), we see the highpass nature of this circuit. Finally the lowpass to bandstop transformation gives the circuit shown in Fig. 11.19d. For this circuit there is transmission at both low and high frequencies, but when the L_3C_3 series circuit is in resonance, it becomes a short circuit, and thus intermediate frequencies are eliminated.

The manner in which we most frequently use the transformation of LC circuits is illustrated by Fig. 11.20. The circuit shown of Fig. 11.20a is a lowpass filter of a type we will study in Chapter 14. Depending upon the element values, such a circuit might be a Butterworth circuit or a Chebyshev circuit. To determine the form of a bandpass filter, we use the transformations for the lowpass to bandpass case shown in Table 11.1 and obtain a circuit of the form shown in Fig. 11.20b. The element values in this circuit will depend on the values from which they are found in the prototype, and also on the values of ω_0 and the desired bandwidth. This is a powerful idea, and it forms the basis of several of the chapters to appear later in this book. It is of great importance to a filter designer, because it indicates that tables need be prepared for the lowpass case only. If an-

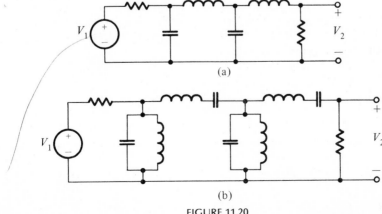

FIGURE 11.20

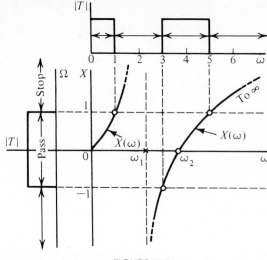

FIGURE 11.21

other kind of frequency response is desired, then the appropriate structure of the filter may be obtained by using a transformation of the prototype circuit, as we have indicated by examples.

11.6 MULTIPLE PASS-BAND TRANSFORMATIONS

Although the frequency responses that we will most often need for the solution of filter design problems are those that have been considered thus far (lowpass, bandpass, highpass, bandstop), the concept of the frequency transformation is general and can be applied to specific problems that do not meet the requirements of the cases considered. We will show this by means of two examples.

Example 11.1 The frequency characteristic shown on the top of Fig. 11.21 indicates that we require a pass band from 0 to 1 and another pass band from 3 to 5. We note that these numbers are small integers, and that they would likely result from frequency scaling in any actual application. Using the method already developed, we see that the required form for $X(\omega)$ is as shown in Fig. 11.21. The form of $X(\omega)$ indicates that $A(s)$ has the pole-zero locations given in Fig. 11.22. We seek to determine these locations so that we may determine $A(s)$. The required form of $X(\omega)$ is

$$X(\omega) = H\omega \, \frac{\omega_2{}^2 - \omega^2}{\omega_1{}^2 - \omega^2} \qquad (11.61)$$

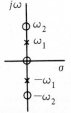

FIGURE 11.22

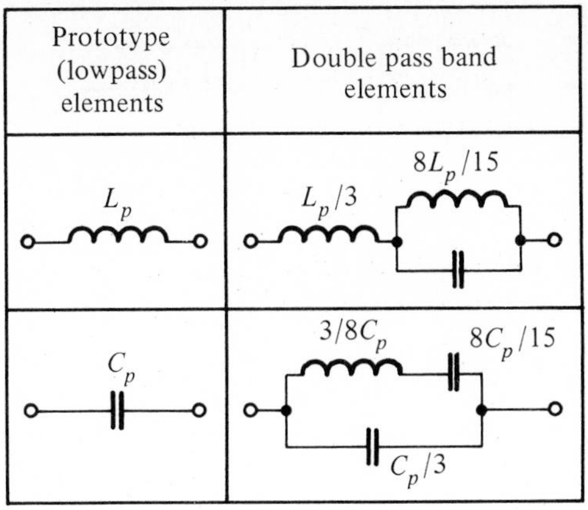

Prototype (lowpass) elements	Double pass band elements
L_p	$L_p/3$ ⌇ $8L_p/15$
C_p	$3/8C_p$ ⌇ $8C_p/15$ ⌇ $C_p/3$

FIGURE 11.23

where $\pm j\omega_1$ is the pole position, and $\pm j\omega_2$ is the zero position. In this equation we have three unknown quantities which we must determine. We shall do so by making use of three requirements that are imposed on $X(\omega)$, namely,

$$X(1) = +1, \qquad X(3) = -1, \qquad X(5) = +1 \qquad (11.62)$$

which are the requirements indicated by $X(\omega)$ crossing the $X = \pm 1$ lines. Substituting these requirements into Eq. (11.61) gives three equations:

$$X(1) = \frac{H(\omega_2{}^2 - 1)}{\omega_1{}^2 - 1} = 1 \qquad (11.63)$$

$$X(3) = \frac{3H(\omega_2{}^2 - 9)}{\omega_1{}^2 - 9} = -1 \qquad (11.64)$$

$$X(5) = \frac{5H(\omega_2{}^2 - 25)}{\omega_1{}^2 - 25} = 1 \qquad (11.65)$$

These equations are actually linear in three unknowns: $\omega_1{}^2$, $\omega_2{}^2$, and H. The three simultaneous equations may be routinely solved to give

$$H = \frac{1}{3}, \qquad \omega_1{}^2 = 5, \qquad \omega_2{}^2 = 13 \qquad (11.66)$$

Then we have determined the unknown quantities in Eq. (11.62), and we may write the result:

$$A(s) = \frac{1}{3} s \frac{s^2 + 13}{s^2 + 5} \qquad (11.67)$$

Using the methods developed in the last section, we see that the table of LC element transformations is as given in Fig. 11.23.

Example 11.2 As a second example we consider the specifications indicated by Fig. 11.24, which shows that we require a pass band between 1 and 2 as well as for all frequencies in excess of 3. (Again, these are scaled values.) The form of $X(\omega)$ which satisfies these requirements is shown in Fig. 11.24:

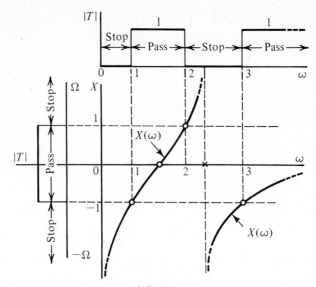

FIGURE 11.24

$$X(\omega) = \frac{-H(\omega_1^2 - \omega^2)}{\omega(\omega_2^2 - \omega^2)} \tag{11.68}$$

from which

$$A(s) = \frac{H(s^2 + \omega_1^2)}{s(s^2 + \omega_2^2)} \tag{11.69}$$

Here ω_1 is the zero frequency and ω_2 the pole frequency identified in the figure. The requirements imposed on Eq. (11.68) are

$$X(1) = -1, \qquad X(2) = +1, \qquad X(3) = -1 \tag{11.70}$$

As in the last example, imposing these conditions on Eq. (11.68) results in three simultaneous equations in ω_1^2, ω_2^2, and H. The routine solution of these equations gives

$$\omega_1^2 = 3, \qquad \omega_2^2 = 5, \qquad H = 2 \tag{11.71}$$

so that the required frequency transformation is

$$A(s) = \frac{2(s^2 + 3)}{s(s^2 + 5)} \tag{11.72}$$

which is represented by the poles and zeros shown in Fig. 11.25. This may also be written

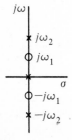

FIGURE 11.25

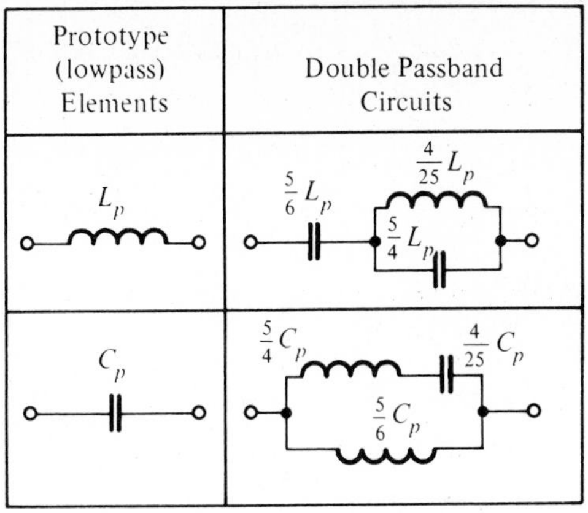

Prototype (lowpass) Elements	Double Passband Circuits
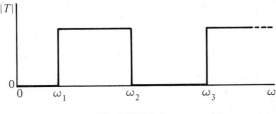 L_p	$\frac{5}{6}L_p$ $\frac{4}{25}L_p$ $\frac{5}{4}L_p$
C_p	$\frac{5}{4}C_p$ $\frac{4}{25}C_p$ $\frac{5}{6}C_p$

FIGURE 11.26

as a partial fraction expansion:

$$A(s) = \frac{6/5}{s} + \frac{(4/5)s}{s^2 + 5} \qquad (11.73)$$

and from this form the *LC* circuit transformations shown in Fig. 11.26 are easily found.

PROBLEMS

11.1 For this problem, we are studying the frequency transformation $X(\omega)$ which transforms the lowpass prototype characteristic into the frequency characteristic shown in Fig. P11.1.
(a) Sketch $X(\omega)$ and identify pole and zero frequencies.
(b) Write $A(s)$ in its general form.

FIGURE P11.1

11.2 Repeat Problem 11.1 for the frequency characteristic shown in Fig. P11.2.
11.3 Repeat Problem 11.1 for the frequency characteristic shown in Fig. P11.3.
11.4 Repeat Problem 11.1 for the frequency characteristic shown in Fig. P11.4.
11.5 Repeat Problem 11.1 for the frequency characteristic shown in Fig. P11.5.
11.6 If we are given the filter characteristic having two passbands shown in Fig. P11.6, and we wish to study the frequency transformation of the lowpass prototype into this characteristic, then
(a) Determine an expression for $A(s)$ similar to Eq. (11.72).
(b) Construct a chart like that given in Fig. 11.26.
11.7 Given the magnitude characteristic shown in Fig. 11.24 except that the cutoff frequency at $\omega = 3$ is moved to $\omega = 4$.

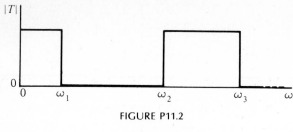

FIGURE P11.2

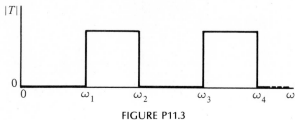

FIGURE P11.3

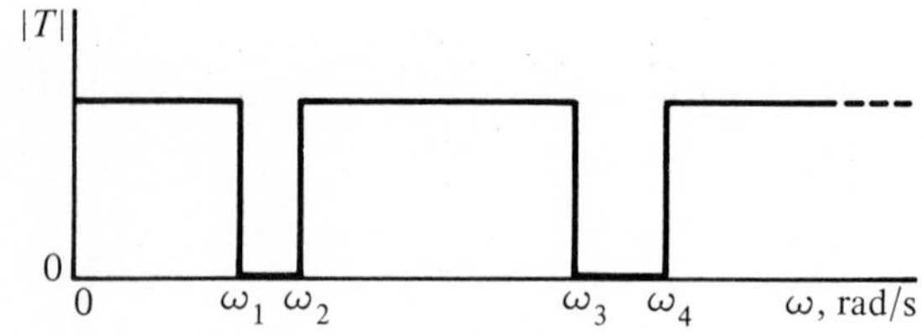

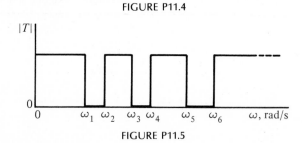

FIGURE P11.4

FIGURE P11.5

(a) Determine an expresion for $A(s)$ similar to Eq. (11.72).

(b) Construct a chart like that given in Fig. 11.26.

11.8 Given the magnitude characteristic shown in Fig. 11.24 except that the cutoff frequency at $\omega = 3$ is moved to $\omega = 5$.

(a) Determine an expression for $A(s)$ similar to Eq. (11.72).

(b) Construct a chart like that given in Fig. 11.26.

11.9 Given the magnitude characteristic shown in Fig. P11.9 with the cutoff frequencies for the two passbands indicated.

(a) Determine the expression for $A(s)$.

(b) Construct a chart showing new element sizes in terms of those of the prototype, like Fig. 11.26.

11.10 Figure P11.10 shows a double passband frequency characteristic which we wish to relate to the prototype. For the numerical values given:

(a) Determine $A(s)$.

(b) Construct a column that might fit into Table 11.1.

11.11 Repeat Problem 11.10 for the filter characteristic shown in Fig. P11.11.

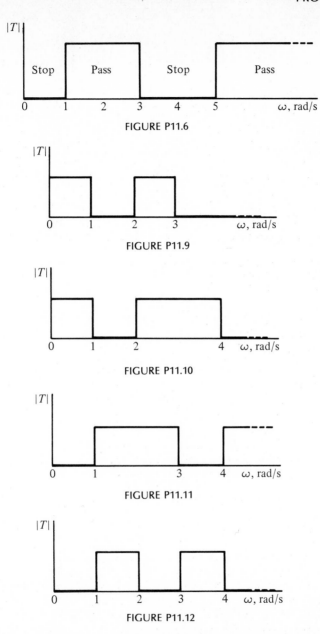

FIGURE P11.6

FIGURE P11.9

FIGURE P11.10

FIGURE P11.11

FIGURE P11.12

11.12 Figure P11.12 shows a double passband frequency characteristic which we wish to relate to the prototype. For the numerical values given, show that

$$A(s) = 0.5 \frac{(s^2 + 2.23)(s^2 + 10.77)}{s(s^2 + 7)}$$

Construct an entry that might fit into Table 11.1.

11.13 Consider the double-notch filter characteristic shown in Fig. P11.4 with $\omega_1 = 1$, $\omega_2 = 2$, $\omega_3 = 3$, and $\omega_4 = 4$. For these numerical values,
   (a) Determine $A(s)$.
   (b) Construct entries that might fit into Table 11.1.

CHAPTER 12

Highpass
and
Band-
Elimination
Filters

This chapter and the next are concerned with the remaining general classes of filters. These filters have several design procedures in common. One such procedure is pole reciprocation; another is the introduction of zeros on different parts of the imaginary axis (including the origin). We also describe op-amp circuits for the realization of the required poles and zeros that come from the filter specifications.

12.1 LOWPASS TO HIGHPASS TRANSFORMATION

There are signal-processing applications in which low-frequency components of the input must be eliminated. This is illustrated in Fig. 12.1 which shows a high-frequency signal distorted by a low-frequency oscillation. Our philosophy of design for highpass filters is similar to that introduced in earlier chapters and is illustrated by Fig. 12.2. We begin with highpass filter specifications and then employ a frequency transformation to obtain the equivalent lowpass specifications. From these specifications the order of the filter is determined for a given form of magnitude response, and also such quantities as the half-power frequency. The frequency transformation is then used once more to obtain the required highpass transfer function from which the filter may be designed. In some cases some of these steps may be skipped by making use of the RC–CR transformation discussed in Section 6.7.

We let ω_P be the frequency at the edge of the pass-band over which the attenuation is less than $\alpha_{\max}$. The transformation from lowpass to highpass was

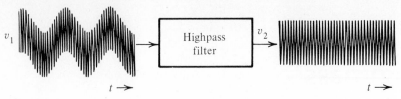

FIGURE 12.1

found in Chapter 11 to be

$$S = \frac{\omega_P}{s} \tag{12.1}$$

With imaginary values for both S and s, this equation becomes

$$\Omega = \frac{-\omega_P}{\omega} \tag{12.2}$$

An interpretation of the negative sign in this equation may be made with the help of Fig. 12.3, which shows how the lowpass prototype brick-wall response transforms into the highpass brick-wall response. From the figure we see that the use of Eq. (12.2) transforms the frequency range of $\Omega = 0$ to $\Omega = 1$ to the range of $\omega = -\omega_P$ to $\omega = -\infty$, as indicated by the cross-hatched parts of the figure. On the other hand the frequency range between $\Omega = 0$ and $\Omega = -1$ transforms into the range between $\omega = \omega_P$ and $\omega = +\infty$, shown by the dotted parts of the figure. Because there is symmetry of the transfer functions $|T|$ with respect to both $\Omega = 0$ and $\omega = 0$, Eq. (12.2) may be used without the negative sign. As a consequence, we may use the simplification of Fig. 12.3 shown as Fig. 12.4.

Of course we have no brick-wall response filters, but responses such as Butterworth or Chebyshev, as studied in previous chapters. A more realistic $|T|$ characteristic than that of Fig. 12.4 is shown in Fig. 12.5a. The corresponding $\alpha = -20 \log|T|$ dB response is shown in Fig. 12.5b. We must find an appropriate

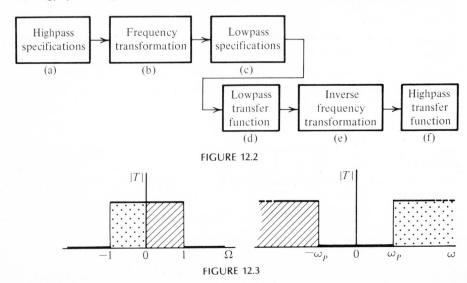

FIGURE 12.2

FIGURE 12.3

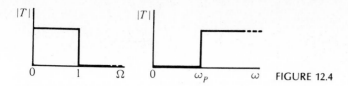

FIGURE 12.4

form for $T(s)$ from specifications given in terms of $\alpha\,(\omega)$. These specifications will take the form shown in Fig. 12.5b. At the frequency ω_p it is specified that the attenuation be $\alpha_{\max}$ at most, and at the frequency ω_S the attenuation must be at least $\alpha_{\min}$. We normally represent these specifications in the form shown in Fig. 12.5c. Given ω_p, ω_S, $\alpha_{\max}$, and $\alpha_{\min}$, we seek $\alpha(\omega)$ having the form shown in the figure. As was the case in our previous studies, we can seldom meet the specifications exactly because the order of the filter n must be an integer. But we can always meet the required upper value of attenuation in the pass band and the required minimum value in the stop band.

The highpass filter specifications of Fig. 12.5 are repeated in Fig. 12.6a. To this we will apply the frequency transformation. From Eq. (12.2), without the negative sign,

$$\Omega = \frac{\omega_P}{\omega} \tag{12.3}$$

we may find the corresponding value of Ω. We see that

$$\omega = \omega_P \text{ corresponds to } \Omega = 1 \tag{12.4}$$

$$\omega = \omega_S \text{ corresponds to } \Omega = \frac{\omega_P}{\omega_S} \tag{12.5}$$

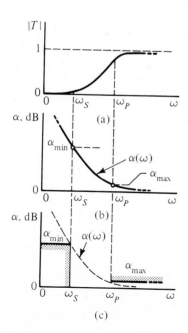

FIGURE 12.5

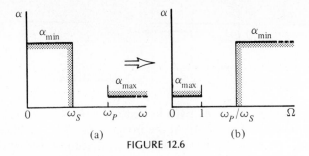

FIGURE 12.6

as shown in Fig. 12.6. This is then the prototype lowpass filter for which n and Ω_0 may be determined as in Chapter 6 or Chapter 8. If we assume that the response desired is Butterworth, then n is determined from Eq. (6.46) and Ω_0 from Eq. (6.41). In Eq. (6.46) ω_s/ω_p is replaced by ω_p/ω_S, of course, since we are considering a highpass rather than a lowpass response.

Example 12.1 The specifications for a highpass filter are shown in Fig. 12.7a, indicating that 25 dB of attenuation is required in the stop band and 1 dB in the pass band. From Eq. (6.46) we find n as

$$n = \frac{\log\left[(10^{25/10} - 1)/(10^{1/10} - 1)\right]}{2 \log 3.5} = 2.836 \tag{12.6}$$

which requires that we round up to $n = 3$. The half-power frequency for the lowpass prototype is given by Eq. (6.41) as

$$\Omega_0 = \frac{1}{(10^{1/10} - 1)^{1/6}} = 1.2526 \text{ rad/s} \tag{12.7}$$

For the highpass filter the corresponding number is found from Eq. (12.2), which becomes

$$\omega_0 = \frac{\omega_P}{\Omega_0} = 3492.8 \text{ rad/s} \tag{12.8}$$

We may now take two approaches. Since we have found that an $n = 3$ Butterworth response is required, the lowpass transfer function is known from Chapter 6 as

$$T(S) = \frac{1}{(S^2 + S + 1)(S + 1)} \tag{12.9}$$

This lowpass prototype response has its half-power frequency at $\Omega = 1$. We know that the frequency transformation

$$S = \frac{1}{s} \tag{12.10}$$

gives the same half-power frequency for the highpass Butterworth response, as shown in Fig. 12.8. Substituting Eq. (12.10) into Eq. (12.9) gives us

$$T(s) = \frac{s^3}{(s^2 + s + 1)(s + 1)} \tag{12.11}$$

Since this transfer function has its half-power frequency at $\omega = 1$, as shown in Fig. 12.9,

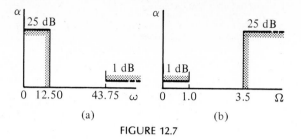

(a)   (b)

FIGURE 12.7

we may determine the required transfer function by frequency scaling with $k_f = 3492.8$. We must then find a circuit realization for this transfer function.

Another approach makes use of the *RC–CR* transformation given in Chapter 6. The Geffe circuit given in Fig. 6.32 was determined for $n = 3$ with a half-power frequency at $\omega = 1$. Hence it may be used as a realization provided that frequency is scaled with $k_f = 3492.8$. If we specify that the capacitors in the realization have the values of $0.1\ \mu F$, then from scaling equations for C,

$$k_m = \frac{C_{\text{old}}}{k_f C_{\text{new}}} = 2863 \qquad (12.12)$$

The scaled circuit that meets the specifications is shown in Fig. 12.10.

12.2 POLE RECIPROCATION

We next examine in greater detail the steps taken in going from Eq. (12.9) to Eq. (12.11). The lowpass to highpass transformation of Eq. (12.1) can be written with

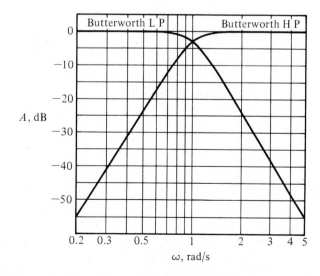

FIGURE 12.8

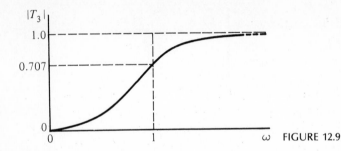

FIGURE 12.9

$\Omega_p = 1$ without loss of generality, since we can always frequency scale to restore this frequency. Let us apply this frequency transformation to a general lowpass transfer function:

$$T(S) = \frac{p_1 p_2 \cdots p_n}{(S + p_1)(S + p_2) \cdots (S + p_n)} \qquad (12.13)$$

which has been normalized such that $T(0) = 1$. Substituting $S = 1/s$ gives us

$$T(s) = \frac{p_1 p_2 \cdots p_n}{(1/s + p_1)(1/s + p_2) \cdots (1/s + p_n)} \qquad (12.14)$$

or

$$T(s) = \frac{s^n}{(s + 1/p_1)(s + 1/p_2) \cdots (s + 1/p_n)} \qquad (12.15)$$

The formation of $1/p_i$ from p_i is known as pole reciprocation. Thus we see that the application of the frequency transformation to a lowpass transfer function of order n is accomplished by reciprocation of its n poles and the introduction of n zeros at the origin.

If the pole p_i is real, then $1/p_i$ is real. However, for a complex pole, written in the polar form

$$p_i = r_i e^{j\alpha_i} \qquad (12.16)$$

then

$$\frac{1}{p_i} = \frac{1}{r_i} e^{-j\alpha_i} \qquad (12.17)$$

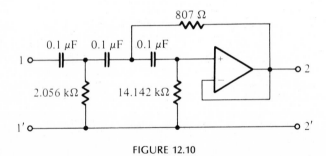

FIGURE 12.10

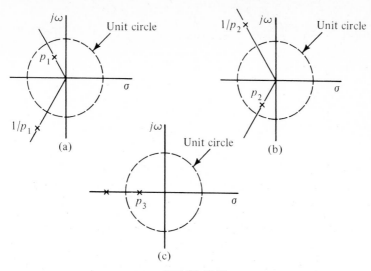

FIGURE 12.11

Such complex poles must occur in conjugate pairs, of course. Reciprocation is illustrated in Fig. 12.11. If the pole p_1 is in the second quadrant, then $1/p_1$ is in the third quadrant. The conjugate of p_1, which is p_2, is in the third quadrant, and $1/p_2$ is in the second quadrant. Thus conjugate poles inside of the unit circle become conjugate poles outside of the unit circle, and ω_0 is changed, but not Q. In Fig. 12.11c we show that reciprocation of a simple real pole on the negative real axis produces a pole that is also on the negative real axis.

The application of pole reciprocation for the Butterworth response is very direct. Since all poles lie on the unit circle, reciprocation does not change positions at all—neither Q nor ω_0 changes. Figure 12.12 illustrates this for a third-order Butterworth response, described by Eq. (12.9). The poles remain fixed in position, but three zeros are introduced at the origin, as in Eq. (12.11). The illustration for a third-order Butterworth response is general: The poles for a lowpass Butterworth response become the poles for a highpass Butterworth response, and n zeros are placed at the origin.

This simplicity of pole reciprocation for the Butterworth response does not extend to other forms of response, such as the Chebyshev response. For the Chebyshev case the poles are located on an ellipse, and the ellipse may have a

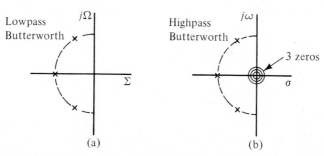

FIGURE 12.12

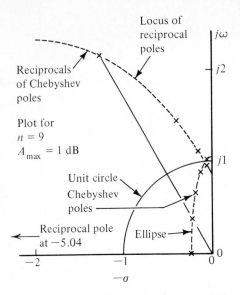

FIGURE 12.13

major semiaxis which is larger or smaller than the unit circle. In either case the computation of the new pole locations is accomplished by using Eq. (12.17). The graphical construction of the new pole positions is illustrated in Fig. 12.13. The locus of the reciprocal poles is ellipselike, but not an ellipse. A typical locus is given in Fig. 12.14 which shows that pole reciprocation produces an ellipselike figure in which positions of the major and minor semiaxes are reversed. While it is difficult to describe the locus, it is easy to determine the pole positions using Eq. (12.17). As in the Butterworth case, n zeros are introduced at the origin of the s plane by the reciprocation of a lowpass function with n poles.

The relationship between such features as the maximally flat part of the response for the lowpass and highpass cases is illustrated by Fig. 12.15. Note that the highpass response is maximally flat near infinity rather than at $\omega = 0$ as in

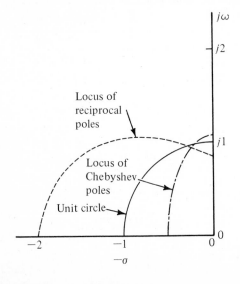

FIGURE 12.14

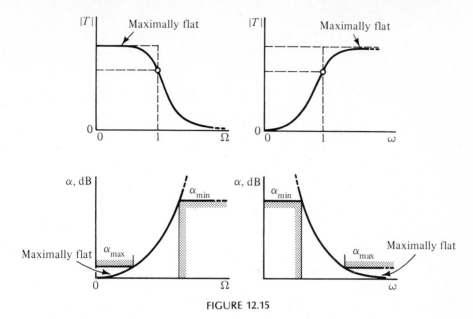

FIGURE 12.15

the lowpass case. A similar comparison is made for the Chebyshev case in Fig. 12.16. Compare the frequency range over which the magnitude response is equal ripple for the lowpass and highpass cases.

Example 12.2 Figure 12.17a shows the specifications for a Chebyshev highpass filter. From the figure we see that 25 dB of attenuation is required in the stop band extending from 0 to 3000 rad/s, while in the pass band, which is all frequencies above 6990 rad/s, we require that the attenuation be less than or equal to 1 dB. We first determine the equiva-

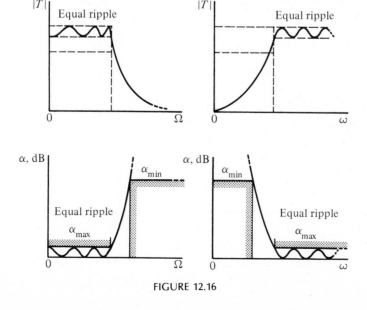

FIGURE 12.16

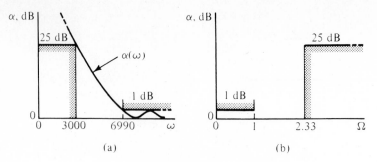

(a)   (b)

FIGURE 12.17

lent lowpass prototype, using the method outlined in the last section. This is shown in Fig. 12.17b. Making use of Eq. (8.39) for n and Eq. (8.35) for half-power frequency for the Chebyshev case, we find

$$n = 2.85 \quad \text{and} \quad \Omega_{hp} = 1.095 \tag{12.18}$$

We must round up the value of n to the next integer, and we may use the equation for the frequency transformation, Eq. (12.2), to find the values

$$n = 3 \quad \text{and} \quad \omega_{hp} = 6384 \text{ rad/s} \tag{12.19}$$

For the Chebyshev case with $n = 3$ we determine the position of the poles to be

$$s_1 = -0.494 \tag{12.20}$$

$$s_2, s_3 = -0.247 \pm j0.966$$

The complex conjugate pole location corresponds to $\omega_0 = 0.999$ and $Q = 2.02$. Then $1/s_1 = -2.02$; with ω_0 so close to 1, the reciprocation of the complex poles leaves them unchanged. Hence the transfer function is

$$T(s) = \frac{s}{s + 2.024} \times \frac{s^2}{s^2 + 0.494s + 0.997} \tag{12.21}$$

The pole and zero locations for this transfer function are shown in Fig. 12.18. These poles and zeros correspond to the plot of $\alpha(\omega)$ in Fig. 12.17a after suitable frequency scaling.

To complete the design specified by Eq. (12.21) requires that we have a highpass circuit for each of the transfer functions. There is no need to construct a new catalog since that produced for the lowpass case in earlier chapters provides a source after the *RC–CR* transformation is applied. For example, the Sallen–Key circuit of Fig. 6.19b has a high-pass equivalent, which is shown in Fig. 12.19, found from the *RC–CR* transformation of Section 6.7. With the element values as shown, the circuit has a half-power frequency of

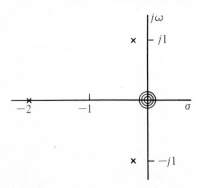

FIGURE 12.18

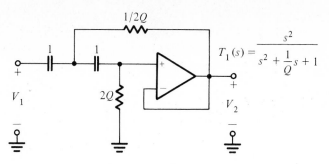

FIGURE 12.19

$\omega_0 = 1$. If it is scaled in frequency by the value given in Eq. (12.19), namely, $k_f = 6384$, then it will match the highpass specifications. The first-order section shown in Fig. 12.20 is required to produce a pole at $s = -2.024$, and hence for this section we frequency scale by $k_f = 12,920$. The two sections are isolated, of course, and so they may be magnitude scaled separately to produce the final design shown in Fig. 12.21. The choice was made that all capacitors in the final design should have the value of 0.01 μF, and this required that

$$k_m = \frac{10^8}{k_f} \qquad (12.22)$$

Thus we find that $k_m = 7740$ for the first stage and $k_m = 15,664$ for the second stage, and with these values the resistor values are determined as shown.

12.3 BAND-ELIMINATION (NOTCH) FILTERS

In Chapter 11 we found that the lowpass to bandpass and lowpass to band-elimination (bandstop) frequency transformations were inverses of each other. Just as in the bandpass case, there are signal frequencies to be preserved, so in the band-elimination case there are signal frequencies, or even a single signal frequency, to be eliminated (rejected). This is illustrated in Fig. 12.22. In Fig. 12.22a an intermediate frequency is eliminated, leaving intact the low- and high-frequency components. In Fig. 12.22b 60-Hz noise is eliminated from an electrocardiogram (EKG), leaving an undistorted signal.

The brick-wall characterization of the band-elimination filter characteristic is shown in Fig. 12.23a, a typical realistic response in Fig. 12.22b, and the attenuation characteristic of Fig. 12.23b is shown in Fig. 12.23c. To match a specified frequency response such as that of Fig. 12.23c, we begin with specifications shown in Fig. 12.24, which are very similar to those used for the bandpass and highpass cases already considered. In terms of Fig. 12.24, we specify that over a band of frequencies from ω_3 to ω_4 the attenuation be equal to or greater than a

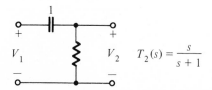

FIGURE 12.20

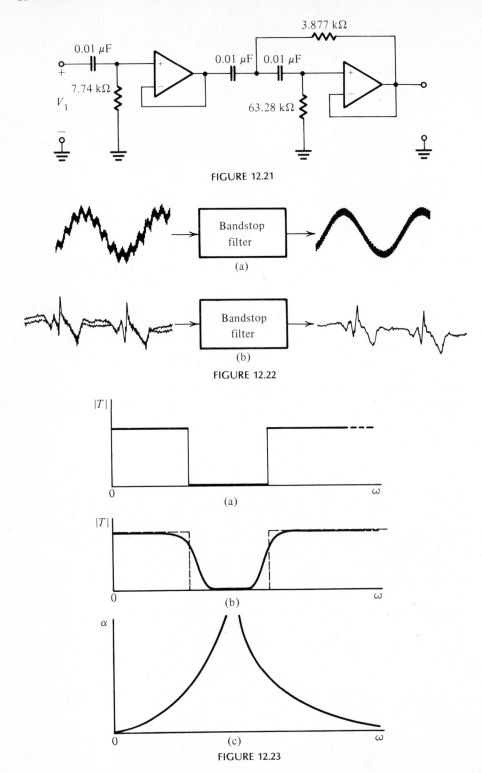

FIGURE 12.21

FIGURE 12.22

FIGURE 12.23

minimum attenuation α_{min}. At a specified low frequency ω_1 we specify that the attenuation be less than a certain maximum value α_{max}, and that this also apply for frequencies in excess of ω_2. The bandwidth of the filter is defined as bw = $\omega_2 - \omega_1$, as shown in the figure. The frequency ω_0 is the geometrical mean given by

$$\omega_0^2 = \omega_1\omega_2 = \omega_3\omega_4 \tag{12.23}$$

The philosophy of design developed thus far requires that we find the corresponding lowpass characteristic, the prototype. We do this by making use of the frequency transformation

$$S = \text{bw}\ \frac{s}{s^2 + \omega_0^2} \tag{12.24}$$

$$\Omega = \text{bw}\ \frac{\omega}{\omega_0^2 - \omega^2} \tag{12.25}$$

We begin with the band-elimination characteristic of Fig. 12.24 and determine Ω for values of ω between $\omega = 0$ and $\omega = \omega_0$. First at $\omega = 0$, $\Omega = 0$. At $\omega = \omega_1$,

$$\Omega_P = \frac{(\omega_2 - \omega_1)\,\omega_1}{\omega_1\omega_2 - \omega_1^2} = 1 \tag{12.26}$$

This is shown on Fig. 12.25. At the frequency ω_3 we have

$$\Omega_S = \frac{(\omega_2 - \omega_1)\,\omega_3}{\omega_3\omega_4 - \omega_3^2} = \frac{\omega_2 - \omega_1}{\omega_4 - \omega_3} \tag{12.27}$$

which is again shown in Fig. 12.25. Finally let $\omega = \omega_0$, for which we see from Eq. (12.25) that $\Omega = \infty$. The identification of frequencies for the lowpass prototype is now complete. The lowpass specifications are

$$\alpha_{max}, \qquad \alpha_{min}, \qquad \Omega_P = 1, \qquad \Omega_S = \frac{\omega_2 - \omega_1}{\omega_4 - \omega_3} \tag{12.28}$$

From the lowpass prototype we can determine the required order of the frequency response and so of the filter, and also the half-power frequency Ω_0. In the case of the Butterworth response this is accomplished by the use of Eqs. (6.46) and (6.41), and for the Chebyshev response we use Eqs. (8.39) and (8.35).

Once the required values of n and Ω_0 are determined, we may use the frequency transformation of Eq. (12.24) in reverse and find the location of the poles and zeros of $T(s)$. As a simple example, suppose that $\alpha_{max} = 3$ dB and that n was determined to be 3. The prototype Butterworth response would then be the familiar

$$T(S) = \frac{1}{(S^2 + S + 1)(S + 1)} \tag{12.29}$$

from Chapter 6. If we select bw = 1 and $\omega_0 = 1$ for simplification, then substituting

$$S = \frac{s}{s^2 + 1} \tag{12.30}$$

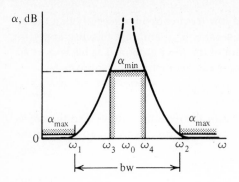

FIGURE 12.24

into Eq. (12.29) gives

$$T(s) = \frac{(s^2 + 1)^3}{(s^4 + s^3 + 3s^2 + s + 1)(s^2 + s + 1)} \tag{12.31}$$

which suggests the form that will result. As usual, we will seek an alternative to factoring the fourth-order polynomial.

Before describing an algorithm to determine this $T(s)$, consider the first-order Butterworth response for the prototype,

$$T_1(S) = \frac{1}{S + 1} \tag{12.32}$$

If we substitute Eq. (12.24) into this equation, we obtain

$$T_1(s) = \frac{s^2 + \omega_0^2}{s^2 + (\omega_0/Q)s + \omega_0^2} \tag{12.33}$$

since for the Butterworth response bw = BW = ω_0/Q. This is the standard form of the transfer function for a notch filter, given by Eq. (5.66), which was one of the realizations of the biquad circuit.

If we are allowed only one section of a filter that achieves the transfer function of Eq. (12.33), the conventional notch filter, then we may consider a different design problem. As illustrated in Fig. 12.26, a band of frequencies is identified from ω_3 to ω_4, over which the attenuation is required to be at least α_1. Assuming that ω_0 is specified, what value of Q will give the desired attenuation over this band of frequencies? The figure reminds us that for a specified ω_0 and Q, the val-

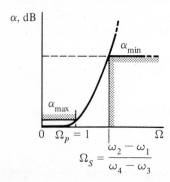

FIGURE 12.25

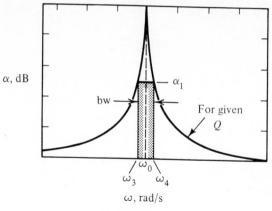

FIGURE 12.26

ues of α_1 and bandwidth bw cannot be specified independently. We begin by letting $s = j\omega$ in Eq. (12.33), which gives

$$T_1(j\omega) = \frac{\omega_0{}^2 - \omega^2}{\omega_0{}^2 - \omega^2 + j\,\omega\omega_0/Q} \qquad (12.34)$$

If we divide the denominator by the numerator, then

$$T_1(j\omega) = \frac{1}{1 + j\,(1/Q)\,\omega\omega_0/(\omega_0{}^2 - \omega^2)} \qquad (12.35)$$

The logarithm of magnitude of this function may be found and then multiplied by 20 to give the attenuation

$$\alpha = 10 \log \left[1 + \frac{1}{Q^2} \left| \frac{\omega\omega_0}{\omega_0{}^2 - \omega^2} \right|^2 \right] \quad \text{dB} \qquad (12.36)$$

This is the equation that is graphed in Fig. 12.26. From the figure we see that $\alpha = \alpha_1$ at the two frequencies ω_3 and ω_4. Considering only the squared function in Eq. (12.36) and noting that $\omega_0{}^2 = \omega_3\omega_4$,

$$\frac{\omega\omega_0}{\omega_0{}^2 - \omega^2} = \frac{\omega_3\omega_0}{\omega_3\omega_4 - \omega_3{}^2} = \frac{\omega_0}{\omega_4 - \omega_3} = \frac{\omega_0}{\text{bw}} \qquad (12.37)$$

$$= \frac{\omega_4\omega_0}{\omega_3\omega_4 - \omega_4{}^2} = \frac{\omega_0}{\omega_3 - \omega_4} = \frac{-\omega_0}{\text{bw}} \qquad (12.38)$$

So at both frequencies Eq. (12.36) becomes

$$\alpha_1 = 10 \log \left[1 + \left| \frac{\omega_0}{\text{bw}\,Q} \right|^2 \right] \quad \text{dB} \qquad (12.39)$$

This is the desired relationship, which may also be written in the form

$$\frac{\omega_0}{\text{bw}\,Q} = \sqrt{10^{\alpha_1/10} - 1} \qquad (12.40)$$

From these two equations a number of interesting observations may be made.

First, since the 3-dB bandwidth is

$$\text{BW} = \frac{\omega_0}{Q} \tag{12.41}$$

then Eq. (12.40) may be written

$$\frac{\text{bw}}{\text{BW}} = (10^{\alpha_1/10} - 1)^{-1/2} \tag{12.42}$$

which tells us that the ratio of the bandwidth to the 3-dB bandwidth is a constant which is determined by the attenuation α_1. This equation may be written in a slightly different form:

$$\text{bw} \, (10^{\alpha_1/10} - 1)^{1/2} = \text{BW} = \frac{\omega_0}{Q} \tag{12.43}$$

Thus if ω_0 and Q are specified, then the product of bw and a factor involving only the attenuation α_1 is a constant, which is BW. These factors are given in Fig. 12.27, which shows the form of the attenuation for two values of Q.

One of the most common uses of a notch filter is for the removal of power-line pickup. The ac power frequency is 60 Hz (50 Hz in some countries) and 400 Hz in airborne electronic systems. This becomes especially important in medical electronics where signal levels are often very low. Suppose that the specifications for a notch filter require that $\alpha_1 = 60$ dB centered at the frequency of 400 Hz, with a bandwidth of 0.01 Hz. Then Eq. (12.40) may be used to find that $Q = 40$ is required. With this value of Q, and with a bandwidth of 1 Hz, the value of α_1 is 20 dB.

Using Eq. (12.43) we may study the design of a notch filter for use at 60 Hz, having a Q of 6. For this filter the 3-dB bandwidth is 75.4 rad/s. For an attenuation $\alpha_1 = 20$ dB, then bw $= 7.58$ rad/s. For an attenuation of 40 dB, this same filter would have the value bw $= 0.754$ rad/s. The choice of a value of bw for design may depend on the drift in the power-system frequency; for commercial power systems this drift is small, while for airborne systems it may be large.

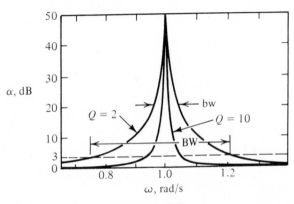

FIGURE 12.27

12.4 POLE AND ZERO LOCATIONS FOR THE BAND-ELIMINATION CASE

In the last section we described the procedure for going from the band-elimination specifications to the lowpass prototype values of n and Ω_0. Our next task is to go from the lowpass pole locations back to the band-elimination pole and zero locations through the lowpass to band-elimination frequency transformation. This we will do through a somewhat circuitous route. In Fig. 12.28a we show once again the lowpass to bandpass transformation. If the same transformation is applied to the highpass characteristic shown in Fig. 12.28b, the result is seen to be the band-elimination response. This suggests a two-step frequency transformation, illustrated in Fig. 12.29: lowpass to highpass, followed by highpass to bandpass. As we see from the figure, the net result is equivalent to lowpass to band-elimination. The steps shown in Fig. 12.29 are equivalent to these algebraic steps:

$$S = \frac{1}{\hat{S}} \qquad \text{(lowpass to highpass)} \qquad (12.44)$$

and

$$\hat{S} = \frac{s^2 + \omega_0{}^2}{\text{bw } s} \qquad (12.45)$$

These two steps are equivalent to

$$S = \frac{1}{\hat{S}} = \frac{\text{bw } s}{s^2 + \omega_0{}^2} \qquad (12.46)$$

which is the lowpass to band-elimination transformation.

An example will illustrate this procedure. We begin with the third-order Butterworth response

$$T_{\text{LP}}(S) = \frac{1}{(S^2 + S + 1)(S + 1)} \qquad (12.47)$$

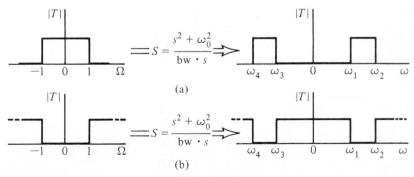

(a)

(b)

FIGURE 12.28

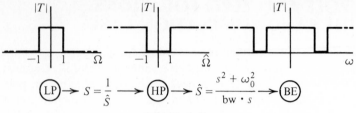

FIGURE 12.29

Then from Eq. (12.44)

$$T_{\text{HP}}(\hat{S}) = \frac{1}{(1/\hat{S}^2 + 1/\hat{S} + 1)(1/\hat{S} + 1)} \tag{12.48}$$

$$= \frac{\hat{S}^3}{(\hat{S}^2 + \hat{S} + 1)(\hat{S} + 1)} \tag{12.49}$$

Then applying the transformation of Eq. (12.45)

$$T_{\text{BE}}(s) = \frac{(1/bw^3)(1/s^3)(s^3)(s^2 + \omega_0{}^2)^3}{\{[(s^2 + \omega_0{}^2)/bw\ s]^2 + (s^2 + \omega_0{}^2)/bw\ s + 1\}[s^2 + \omega_0{}^2)bw\ s + 1]} \tag{12.50}$$

After cancellation of common factors, there remains

$$T_{\text{BE}}(s) = \frac{(s^2 + \omega_0{}^2)^3}{\text{sixth-order polynomial in } s} \tag{12.51}$$

This transfer function has three zeros at $j\omega_0$ (and another three at $-j\omega_0$). We are especially interested in pole locations. In going from the lowpass to the highpass functions using the frequency transformation $S = 1/\hat{S}$, it was necessary to find the reciprocals of the poles. But we recall that reciprocation of Butterworth poles leaves them unchanged in position. Hence the poles of $T_{\text{BE}}(s)$ of Eq. (12.51) are identical to the bandpass poles. These bandpass poles were found in Chapter 6 through the use of the Geffe algorithm, and thus we may use the Geffe algorithm to find the $T_{\text{BE}}(s)$ poles. The pole and zero locations through the various steps of this example are shown in Fig. 12.30.

The preceding example involved a Butterworth response for the prototype. This is not necessary, of course, since the prototype response may be Chebyshev, Bessel–Thomson, or any other such response. However, the example allows us to identify the steps involved in finding the band-elimination function poles and zeros. We begin with a lowpass transfer function which has been normalized so that $T_{\text{LP}}(0) = 1$:

$$T_{\text{LP}} = \frac{p_1 p_2 p_3 \cdots p_n}{(S + p_1)(S + p_2)(S + p_3) \cdots (S + p_n)} \tag{12.52}$$

We next perform the step of pole reciprocation, giving

$$T_{\text{HP}} = \frac{\hat{S}^n}{(\hat{S} + 1/p_1)(\hat{S} + 1/p_2)(\hat{S} + 1/p_3) \cdots (\hat{S} + 1/p_n)} \tag{12.53}$$

Finally, to this function we apply the lowpass to bandpass frequency transforma-

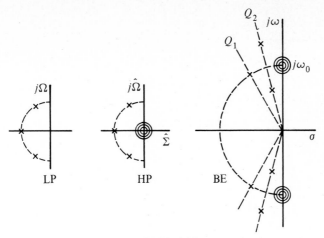

FIGURE 12.30

tion and obtain

$$T_{BE} = \frac{(s^2 + \omega_0^2)^n}{2n \text{ poles found by Geffe method}} \qquad (12.54)$$

where the Geffe method is applied to the reciprocated poles.

From this discussion we see that the determination of pole and zero locations for the band-elimination function may be carried out using a step-by-step procedure:

1. From the band-elimination specifications determine the lowpass prototype which is specified by the values of n and of Ω_0.
2. Locate the prototype poles (Butterworth, Chebyshev, etc.).
3. Reciprocate the prototype poles. Apply the Geffe algorithm to the reciprocated poles.
4. Locate n zeros at $j\omega_0$ (and n zeros at $-j\omega_0$).

With all of the poles and zeros located, we may then group the pole pairs and zero pairs with the object of finding a circuit realization for each.

Example 12.3   The specifications for a band-elimination filter are shown in Fig. 12.31a, from which we see that bw = 2000 rad/s and $\omega_0 = 1732$ rad/s. The equivalent lowpass prototype specifications are shown in Fig. 12.31b. Using Eq. (6.46), we find that $n = 2.98$, which is rounded up to $n = 3$. It is required that the response be maximally flat in the pass bands, as shown in Fig. 12.31a. The lowpass prototype poles are shown in Fig. 12.30, where the poles are located at $-1 + j0$ and $-0.5, j0.866$. The routine application of the Geffe algorithm of Chapter 7 gives

$$Q = 0.866 \quad \text{with} \quad \omega = 1732 \text{ rad/s}$$

$$Q = 1.985 \quad \text{with} \quad \omega = 1054 \text{ and } 2846 \text{ rad/s} \qquad (12.55)$$

There are three zeros at $j1732$. All of the poles and zeros are shown in Fig. 12.32.

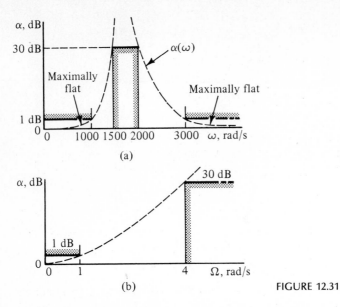

(a)

(b)

FIGURE 12.31

Example 12.4 Calculations for the last example were made simple by the specification that the response be Butterworth, thereby avoiding the need for pole reciprocation. In this example we retain the same specifications as in the last example, but specify that the response be Chebyshev, which is to say that we require that the response be equal ripple in both pass bands. The desired form of the response then is as shown in Fig. 12.33.

We first calculate n using Eq. (8.39) and find that $n = 2.34$, which we round up to 3. The poles for the lowpass prototype are then found as outlined in Chapter 8. The constant a is calculated from Eq. (8.75). Using $\alpha_{max} = 1$ dB as specified and $n = 3$, it is found that $a = 0.47599$, from which we find the major and minor axes of the Chebyshev ellipse. Thus $\cosh a = 1.11544$ and $\sinh a = 0.49417$. Then the prototype pole locations are found to be

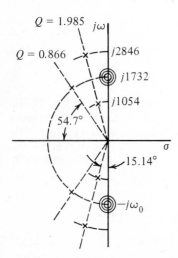

FIGURE 12.32

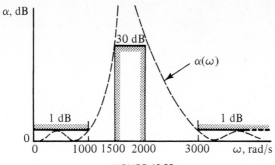

FIGURE 12.33

$$p_1 = -0.49417 \qquad p_2, p_3 = -0.24709 \pm j0.96600 \qquad (12.56)$$

from Eqs. (8.65) and (8.66). These are shown in Fig. 12.34a. The reciprocation of these poles introduces three zeros at the origin, and the new pole locations are

$$p_{1h} = -2.02360 \qquad p_{2h}, p_{3h} = -0.24852 \pm j0.97162 \qquad (12.57)$$

as shown in Fig. 12.34b. These values are used to find the band-elimination poles using the Geffe algorithm. As in the last example, bw = 2000 rad/s and ω_0 = 1732 rad/s, so that $q_c = \omega_0/\text{bw} = 0.866$.

We first consider the first-order factor $(S + 2.02360)$. From Eq. (7.23) we find that $Q = 0.42797$. Since this is less than $Q = 0.5$, the poles are both on the negative real axis. Then

$$S + 2.02360 = \frac{s^2 + 3 \times 10^6}{2000s} + 2.02360 \qquad (12.58)$$

results in the polynomial

$$s^2 + 4047s + 3 \times 10^6 = (s + 3070)(s + 977) \qquad (12.59)$$

and these poles are shown in Fig. 12.35.

The complex conjugate poles of Eq. (12.57) are transformed using the Geffe al-

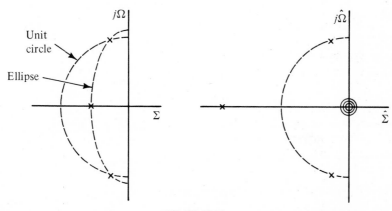

FIGURE 12.34

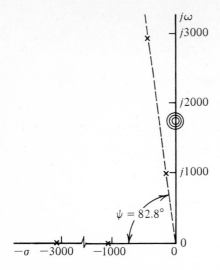

FIGURE 12.35

gorithm. Following the steps from Eqs. (7.39) to (7.47), we have

$$C = 1.000$$
$$D = 0.5739$$
$$E = 5.333$$
$$G = 5.20866$$
$$Q = 4.0004$$
$$K = 1.1480$$
$$W = 1.71186$$
$$\omega_{T1} = 2965 \text{ rad/s}$$
$$\omega_{T2} = 1011 \text{ rad/s}$$

(12.60)

From the value of Q we find the angle of the complex conjugate poles

$$\psi = \cos^{-1} \frac{1}{2Q} = 82.8°$$

(12.61)

and from this we determine the pole locations

$$-370.6 \pm j2918.7$$
$$-126.4 \pm j1003.1$$

(12.62)

These poles, those of Eq. (12.59), and the six zeros at $\pm j1732$ are shown in Fig. 12.35. This is the pole–zero configuration that results in the attenuation as a function of frequency, shown as the dashed line in Fig. 12.33.

12.5 THE NOTCH RESPONSE GENERALIZED

To accommodate the possibilities of relative positions of poles and zeros for the band-elimination case, as shown by Figs. 12.32 and 12.35, it is clear that the second-order notch characteristic given in Eq. (5.66) should be generalized as

$$T(s) = \frac{s^2 + \omega_z^2}{s^2 + (\omega_0/Q)s + \omega_0^2} \tag{12.63}$$

At the notch frequency ω_z the transfer function has the value of zero, or

$$T(j\omega_z) = 0 \tag{12.64}$$

From Eq. (12.63) we see that at high and low frequencies the asymptotic values are

$$T(0) = \frac{\omega_z^2}{\omega_0^2} \quad \text{and} \quad T(j\infty) = 1 \tag{12.65}$$

The relative magnitudes of ω_z and ω_0 give rise to the names of the corresponding filter characteristics:

$\dfrac{\omega_z}{\omega_0} < 1$ defines a highpass notch characteristic

$\dfrac{\omega_z}{\omega_0} = 1$ defines a regular notch characteristic   (12.66)

$\dfrac{\omega_z}{\omega_0} > 1$ defines a lowpass notch characteristic

Using the zero as the basis for comparison, we see that the three cases are as shown in Fig. 12.36, Fig. 12.36a being a lowpass notch, Fig. 12.36b a regular notch, and Fig. 12.36c a highpass notch. With frequency normalized so that $\omega_z = 1$, then the appropriate plots of the magnitude function $|T(j\omega)|$ are as shown in Fig. 12.37. With the pole chosen as the basis for comparison, as shown in Fig. 12.38, then the magnitude responses for the three cases are as shown in Fig. 12.39.

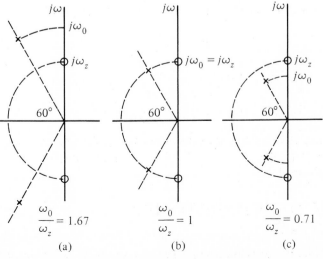

FIGURE 12.36

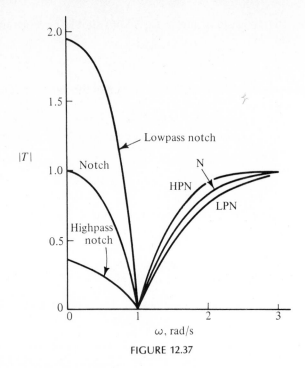

FIGURE 12.37

Returning to the band-elimination pole–zero configurations shown in Figs. 12.32 and 12.35, but especially that of Fig. 12.32, we see that a grouping of the poles and zeros that would seem to be reasonable is as shown in Fig. 12.40. If we can find a circuit that corresponds to each of the kinds of filter characteristics, then we can realize a band-elimination filter by the cascade connection of these three kinds of notch filters, again as shown in Fig. 12.40. Circuits that provide the three kinds of frequency characteristics will be studied next.

12.6 NOTCH CIRCUITS

A circuit that provided the notch filter characteristic was studied in Chapter 5. One of the modes of operation of the biquad circuit was that shown in Fig. 5.36 for which $T(s)$ was given by Eq. (5.65). This is the same as Eq. (12.63), except that $\omega_z = \omega_0$. Other circuits that provide this same notch characteristic are given in Section 12.8.

The circuit shown in Fig. 12.41 is due to Bainter* and is capable of the three modes of operation given by Eqs. (12.66). The Bainter circuit is simple to analyze since it is composed of simple units with two feedback paths. The transfer function is

$$T(s) = \frac{s^2 + R_2/R_1R_3R_5C_1C_2}{s^2 + [(R_5 + R_6)/R_5R_6C_2]s + 1/R_4R_5C_1C_2}$$ (12.67)

---

* J. R. Bainter, "Active Filter Has Stable Notch, and Response Can be Regulated," *Electronics,* pp. 115–117, Oct. 2, 1975.

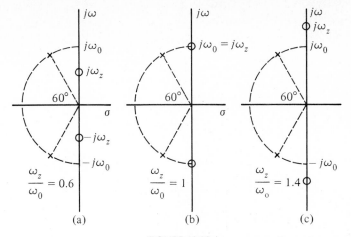

FIGURE 12.38

Comparing this to the standard form of the notch transfer function given by Eq. (12.63), we find that

$$\omega_0^2 = \frac{1}{R_4 R_5 C_1 C_2} \tag{12.68}$$

$$\omega_z^2 = \frac{R_2}{R_1 R_3 R_5 C_1 C_2} \tag{12.69}$$

$$Q = \left(\frac{C_2}{R_4 R_5 C_1}\right)^{1/2} \frac{R_5 R_6}{R_5 + R_6} \tag{12.70}$$

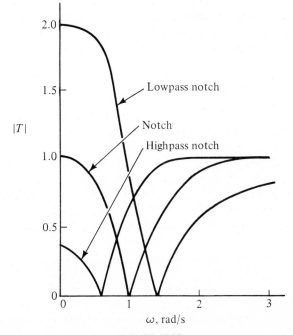

FIGURE 12.39

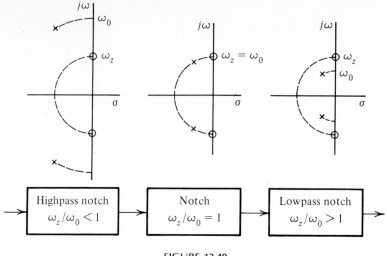

FIGURE 12.40

We first note that this circuit is capable of orthogonal tuning. The tuning sequence that accomplishes this is the following:

1. R_4 tunes ω_0. Both ω_z and Q are not functions of R_4.
2. With R_4 fixed, then R_6 tunes Q. This does not affect either ω_0 or ω_z.
3. Finally, R_1 tunes ω_z. This does not affect either ω_0 or Q.

Design equations are needed since there are eight circuit elements in the Bainter circuit of Fig. 12.41 and only two specifications, ω_z and Q, since we may normalize frequency such that $\omega_0 = 1$. First let $R_2/R_1 = K$, which is the gain of inverting stage. We make the choices $C_1 = C_2 = 1$ and $R_5 = R_6$. The consequence

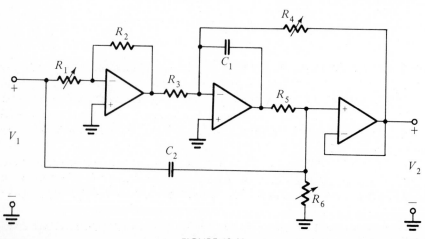

FIGURE 12.41

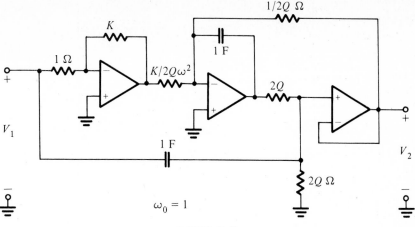

FIGURE 12.42

of this capacitor choice is that, from Eq. (12.68), $R_4R_5 = 1$ so that $\omega_0 = 1$. Then our design equations become the following:

$$C_1 = C_2 = 1$$

$$R_1 = 1$$

$$R_2 = K$$

$$R_3 = \frac{K}{2\,\omega_z^2 Q} \tag{12.71}$$

$$R_4 = \frac{1}{2Q}$$

$$R_5 = 2Q$$

$$R_6 = 2Q$$

These values are shown on the circuit of Fig. 12.42. Other choices are possible, of course, but these values will give a circuit which may be frequency scaled to the required value of ω_0, and then magnitude scaled to give appropriate element values. In this realization the resistor R_3 is the single element that is adjusted to give the three forms of response described by Eq. (12.66): highpass notch, notch, or lowpass notch.

Another notch circuit design which finds use in practical realizations is due to Boctor,* and two forms of the circuit are shown in Fig. 12.43. That shown in Fig. 12.43a is the lowpass notch (LPN) realization. Analysis of the circuit to de-

*S. A. Boctor, "Single Amplifier Functionally Tunable Low-Pass Notch Filter," *IEEE Trans. Circuits and Syst.*, vol. CAS-22, pp. 875–881, 1975.

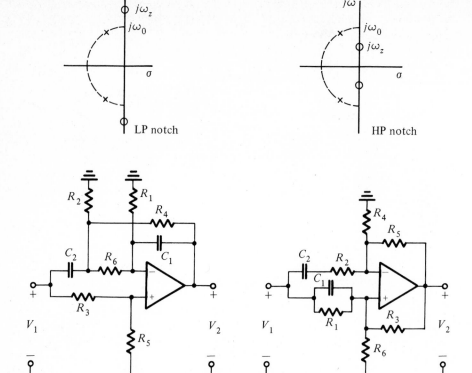

FIGURE 12.43

termine the transfer function is tedious, but routine analysis leads to the equation

$$T(s) = k \frac{s^2 + s\left(\dfrac{1}{R_6 C_2} + \dfrac{1}{R_{24} C_2} + \dfrac{1}{R_1 C_1} - \dfrac{1}{R_6 C_1} \cdot \dfrac{R_3}{R_5}\right) + \dfrac{R_1 + R_{24} + R_6}{R_1 R_{24} R_6 C_1 C_2}}{s^2 + s\left(\dfrac{1}{R_6 C_2} + \dfrac{1}{R_{24} C_2}\right) + \dfrac{1}{R_4 R_6 C_1 C_2}} \qquad (12.72)$$

where

$$R_{24} = \frac{R_2 R_4}{R_2 + R_4} \quad \text{and} \quad k = \frac{R_5}{R_3 + R_5} \qquad (12.73)$$

We require design equations that will give one realization. The usual specifications will be the value of ω_0 which will be scaled to 1, the value of ω_z which has been scaled just as ω_0, and the circuit Q required. Since there are eight elements in the Boctor circuit, there is considerable latitude in the choice of element sizes. We first define a new constant:

$$\frac{\omega_0^2}{\omega_z^2} < k_1 < 1 \qquad (12.74)$$

which will be arbitrarily chosen to ensure that all element values are positive. With $\omega_0 = 1$ and thus $\omega_z/\omega_0 = \omega_z$, element values are found from the following equations*:

$$R_1 = \frac{2}{k_1\omega_z^2 - 1} \tag{12.75}$$

$$R_2 = \frac{1}{1 - k_1} \tag{12.76}$$

$$R_3 = \frac{1}{2}\left|\frac{k_1}{Q^2} + k_1\omega_z^2 - 1\right| \tag{12.77}$$

$$R_4 = \frac{1}{k_1} \tag{12.78}$$

$$R_5 = R_6 = 1 \tag{12.79}$$

$$C_1 = \frac{k_1}{2Q} \tag{12.80}$$

$$C_2 = 2Q \tag{12.81}$$

Once these circuit element values are determined, we frequency scale by the factor $k_f = \omega_0$, and then magnitude scale to obtain convenient (practical) element values. The design found by the Cioffi equations above is not unique, of course, and other design values may be found.

To complete the design we must find k which is the gain of the circuit given by Eq. (12.72). Since we have selected $R_5 = 1$, Eq. (12.73) becomes

$$k = \frac{1}{R_3 + 1} \tag{12.82}$$

If we now substitute (12.73) R_3 in Eq. (12.77), we have

$$k = \frac{1}{1/2(k_1/Q^2 + k_1\Omega_z^2 - 1)} \tag{12.83}$$

which is the desired result.

A similar procedure may be given for the Boctor highpass notch (HPN) circuit of Fig. 12.43b. This circuit provides a high-frequency gain, $G > 1$ and applies only when Q satisfies the relationship

$$Q < \frac{1}{1 - \omega_z^2/\omega_0^2} \tag{12.84}$$

As before, we normalize frequency such that $\omega_0 = 1$, so that ω_z, which appears in

---

*The algorithmic forms of these design equations were found by John M. Cioffi while he was a junior at the University of Illinois in 1977.

the equations to follow, is really ω_z/ω_0. Two new quantities are first defined:

$$\gamma = \frac{1}{Q(1 - \omega_z^2)} \tag{12.85}$$

$$\alpha = \frac{\gamma^2 - 1}{1 + \gamma^2 \omega_z^2} \tag{12.86}$$

Then the following element values may be used:

$$R_1 = \left|\frac{1 + \alpha}{\alpha \gamma \omega_z}\right|^2 \tag{12.87}$$

$$R_2 = 1 \tag{12.88}$$

$$R_3 = G \frac{(1 + \alpha)^2}{\alpha \gamma^2} \tag{12.89}$$

$$R_4 = \frac{G}{G - 1} \frac{1}{\alpha} \tag{12.90}$$

$$R_5 = \frac{G}{\alpha} \tag{12.91}$$

$$R_6 = \frac{G(1 + \alpha)^2}{\alpha \gamma^2 [G(1 + \alpha - \alpha \omega_z^2) - 1]} \tag{12.92}$$

$$C_1 = C_2 = \frac{\alpha \gamma}{1 + \alpha} \tag{12.93}$$

Once these element values are determined, then the usual frequency and magnitude scaling completes the design.

Example 12.5 Suppose that a design requires a highpass notch filter for which $Q = 1$ and $\omega_0 = \sqrt{2}\,\omega_z$. At high frequencies the filter is to provide 12 dB of gain. From these specifications we see that since

$$T(s) = G \frac{s^2 + \omega_z^2}{s^2 + (\omega_0/Q)s + \omega_0^2} \tag{12.94}$$

then $G = 4$, and the low-frequency gain is

$$T(0) = G \frac{\omega_z^2}{\omega_0^2} = 2 \text{ or } 6 \text{ dB} \tag{12.95}$$

Suppose further that the notch frequency corresponds to 60 Hz or 377 rad/s. Then the desired response is as shown in Fig. 12.44.

To determine the circuit element values we note that the specifications and Eqs. (12.85) and (12.86) give us these values:

$$\omega_z^2 = \frac{1}{2}, \qquad \gamma = 2, \qquad \alpha = 1, \qquad G = 4 \tag{12.96}$$

From these values all element values are determined by the use of Eqs. (12.87)–(12.93).

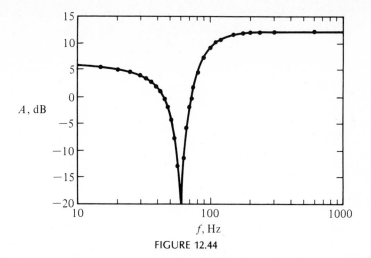

FIGURE 12.44

These element values must be scaled in frequency using $k_f = 2 (2\pi \times 60)$, and we select the value of $k_m = 12,500$. The results are shown in the following table:

Element	Normalized values	Scaled values
R_1	2 Ω	25 kΩ
R_2	1 Ω	12.5 kΩ
R_3	4 Ω	50 kΩ
R_4	$\frac{4}{3}$ Ω	16.67 kΩ
R_5	4 Ω	50 kΩ
R_6	0.8 Ω	10 kΩ
$C_1 = C_2$	1 F	0.15005 μF

The circuit with these element values is shown in Fig. 12.45, and the design is complete.

12.7 BAND-ELIMINATION FILTER DESIGN

With circuits that realize the three kinds of notch characteristics available, we may now put everything together and illustrate the design of band-elimination filters. We do this with two examples.

Example 12.6 We return to Example 12.3 in which the specifications for a band-elimination filter were determined and summarized in Fig. 12.32. The filter specifications are given in Fig. 12.46 as well as the design strategy to be used, which is to realize the specifications as a cascade of three notch filters. We have scaled to the pole frequency $\omega_0 = 1$ for each stage, and as the final step we will frequency scale to restore the zero position to the required frequency. We make the decision to use the Bainter circuit with $K = \omega_z^2$, and we will magnitude scale such that all capacitors in the circuit realization have the same value of 0.1 μF. The resulting circuit design is given in Fig. 12.47 (p. 355).

Example 12.7 The filter specifications for this example are given in Fig. 12.48 along with the requirement that the magnitude response be maximally flat in the pass band. The 3-dB

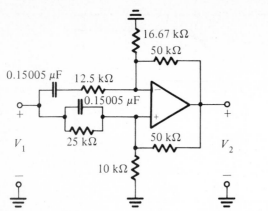

FIGURE 12.45

bandwidth of the filter is 450 rad/s centered on 1000 rad/s, and it is required that there be 32 dB of attenuation in the 60-rad/s band centered on 1000 rad/s. The corresponding lowpass prototype filter is found from Eq. (12.27) which gives $\Omega_P = 1$ and $\Omega_S = 7.5$, as shown in Fig. 12.49. From the lowpass prototype n is found to be 1.83 using Eq. (6.46), and is rounded up to $n = 2$. The half-power frequency is given as part of the specifications so that for the prototype $\omega_0 = 1$.

To apply the Case II Geffe algorithm, we use the value $\Sigma_2 = 0.707$ and $q_c = \omega_0/\mathrm{BW} = 1000/450 = 2.222$. The result is

$$Q = 3.160 \quad \text{and} \quad \omega_0 = 900.62 \text{ and } 1110.3 \text{ rad/s} \quad (12.97)$$

Thus the required pole and zero locations are those shown in Fig. 12.50. These will be realized by a cascade connection of a highpass notch circuit and a lowpass notch circuit for which the specifications are summarized in Fig. 12.51. We make the decision to use the two Boctor circuits given in Fig. 12.43 to realize the specifications. For the highpass notch we make the choice of $G = 4$. This gives gain which is not really needed and can be removed if desired by a resistive voltage-divider circuit. Then we use Eqs. (12.85) to (12.93) to determine the element values which are summarized in Table 12.1. For the lowpass notch, we make the arbitrary choice $k_1 = 0.9$ and then use Eqs. (12.75) to (12.81) to obtain the values given in the table. The values of k_f are specified by the required position of the imaginary-axis zeros, and k_m is chosen to make as many capacitors be of value 1 μF as possible. The final design is shown in Fig. 12.52 without an auxiliary circuit for gain adjustment. The circuit meets the specifications given in Fig. 12.48.

TABLE 12.1 Boctor circuit values

| Element | Highpass notch circuit | | Lowpass notch circuit | |
	Normalized	*Scaled*	*Normalized*	*Scaled*
R_1	3.830	19.57 kΩ	18.27	128.2 kΩ
R_2	1.000	5.109 kΩ	10.0	70.17 kΩ
R_3	6.365	32.52 kΩ	10.83	75.99 kΩ
R_4	2.603	13.30 kΩ	1.11	7.789 kΩ
R_5	7.810	39.90 kΩ	1.00	7.017 kΩ
R_6	9.595	49.02 kΩ	1.00	7.017 kΩ
C_1	0.5673	1 μF	0.1424	0.0225 μF
C_2	0.5673	1 μF	6.32	1 μF
	$k_f = 1110.3$		$k_f = 900.62$	
	$k_m = 5109.4$		$k_m = 70.17$	

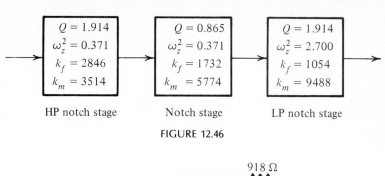

FIGURE 12.46

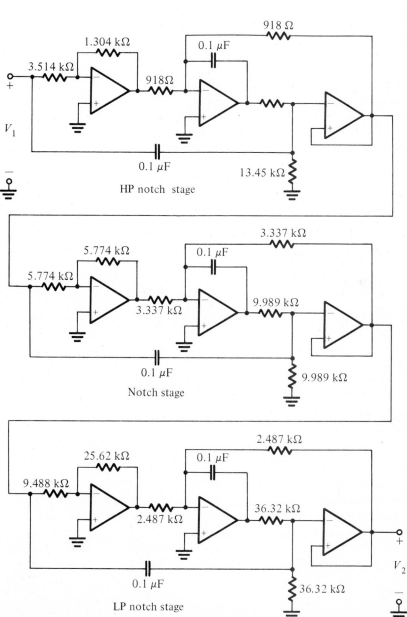

FIGURE 12.47

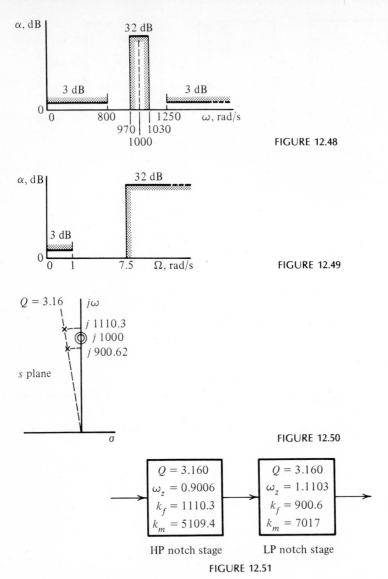

FIGURE 12.48

FIGURE 12.49

FIGURE 12.50

FIGURE 12.51

12.8 OTHER USEFUL CIRCUITS

The designer always has several alternatives in selecting a circuit to use in solving a particular problem. The choices given thus far in this chapter are the Boctor circuits and the Bainter circuit. We may use the biquad circuits from Chapter 5. Additional circuits based on the Friend circuit of Chapter 7 are presented next.

The circuit given in Fig. 12.53 will be shown to provide notch and highpass notch characteristics. To simplify analysis, we first let

$$C_1 = k_1 C \qquad (12.98)$$

and for the voltage-divider circuit connected to the + terminal of the op amp, we let

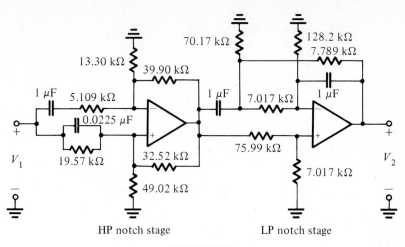

HP notch stage   LP notch stage

FIGURE 12.52

$$k_2 = \frac{R_4}{R_3 + R_4} \tag{12.99}$$

Further, we normalize by letting $R_1 = 1$. Then routine analysis of the circuit gives the transfer function

$$T = k_3 \frac{s^2 + 1/R_2C^2(k_1 + 1)}{s^2 + [(2 + k_1)/R_2C]s + 1/R_2C^2} \tag{12.100}$$

where $k_3 = k_2(k_1 + 1)$. This form of the equation requires that the coefficient of s in the numerator of this equation vanish:

$$k_2\left(\frac{2 + k_1}{R_2}\right) + k_2 = 1 \tag{12.101}$$

Equation (12.100) may be compared to the standard form of a notch characteristic which is

$$T = K\frac{s^2 + \omega_z^2}{s^2 + (\omega_0/Q)s + \omega_0^2} \tag{12.102}$$

From this comparison we see that

$$k_1 = \frac{\omega_0^2}{\omega_z^2} - 1 \tag{12.103}$$

Since k_1 must be positive, we see that it is necessary that

$$\omega_z \leq \omega_0 \tag{12.104}$$

meaning that the circuit may be used as a notch circuit with the equality sign or as a highpass notch. Using the relationships given, the following results are derived:

$$k_2 = \frac{(2 + k_1)Q^2}{(2 + k_1)Q^2 + 1} \tag{12.105}$$

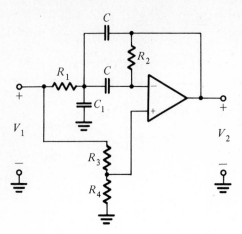

FIGURE 12.53

$$k_3 = k_2 \left(\frac{\omega_0^2}{\omega_z^2} \right) \tag{12.106}$$

and the following element values are found:

$$R_2 = Q^2(k_1 + 2)^2 \tag{12.107}$$

If

$$R_3 = 1 \tag{12.108}$$

then

$$R_4 = Q^2(k_1 + 2) \tag{12.109}$$

Finally,

$$C = \frac{1}{Q(2 + k_1)}, \qquad \omega_0 = 1 \tag{12.110}$$

The complete design of the circuit is summarized in Fig. 12.54. The systematic

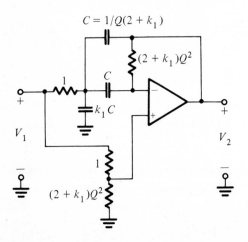

FIGURE 12.54

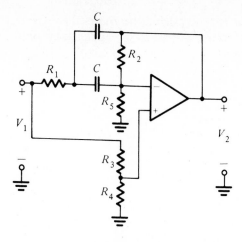

FIGURE 12.55

design may be accomplished by the following steps:

1. Find k_1 from Eq. (12.103).
2. Find k_2 from Eq. (12.105) and k_3 (the dc gain) from Eq. (12.106).
3. Determine all element values from the equations given.
4. Frequency scale by letting $k_f = \omega_0$.
5. Magnitude scale to given convenient element sizes.

   The circuit that may be used for lowpass notch applications is shown in Fig. 12.55. As before, we normalize frequency by letting $\omega_0 = 1$. Assume that we are given Q and ω_z, and that we will be interested in knowing the dc gain. The same procedure as that outlined for the circuit of Fig. 12.53 may be used. The resulting element sizes are shown in Fig. 12.56 from which a design may be made.
   In solving for R_s it is found that

$$R_s = \frac{4Q^2}{\omega_z^2 - 1} \tag{12.111}$$

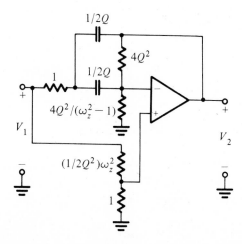

FIGURE 12.56

Since R_s must be positive, we see that it is necessary that ω_z be greater than $\omega_0 = 1$, which implies that the circuit may be used for lowpass notch applications only. The low frequency (dc) gain of the circuit is

$$k_3 = \frac{1}{1 + \omega_z^2/2Q^2} \tag{12.112}$$

which indicates that the circuit introduces loss.

PROBLEMS

12.1 Design a highpass filter with a Butterworth response to meet the attenuation specifications given in Fig. P12.1. In addition, determine the attenuation your filter realizes at the two frequencies, $\omega = 1000$ rad/s and $\omega = 800$ rad/s.

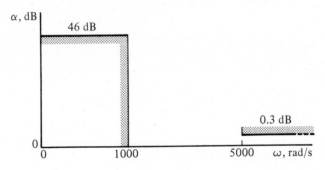

FIGURE P12.1

12.2 A highpass filter with a Butterworth response must have at least 55 dB of attenuation below 300 Hz, and the attenuation must be no more than 0.5 dB above 1500 Hz. (See Fig. P12.2.)
  (a) Design the filter using a Sallen–Key circuit.
  (b) Modify the circuit to provide a gain enhancement of 10 dB (flat for all frequencies) without the addition of an op amp.

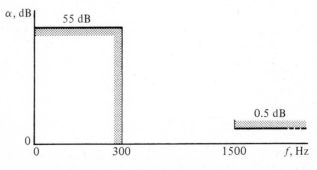

FIGURE P12.2

12.3 A highpass filter is required and the decision is made that it should have a Chebyshev response in the passband. The specifications are those given in Fig. P12.3.
(a) Design a circuit based on Sallen–Key circuits.
(b) Determine the loss provided a frequency of 800 rad/s.

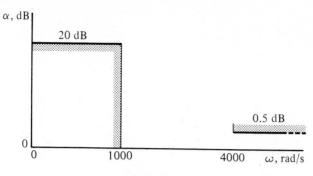

FIGURE P12.3

12.4 A highpass filter with a Chebyshev response must have at least 46 dB of attenuation below 1000 rad/s, and no more than 0.3 dB of attenuation above 4500 rad/s. (See Fig. P12.4.)
(a) Design a circuit to meet the specifications.
(b) Determine the loss actually realized by your circuit at $\omega = 1000$ and $\omega = 800$ rad/s.

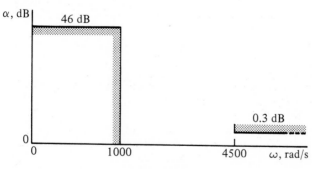

FIGURE P12.4

12.5 The circuit shown in Fig. P12.5 is known as a twin-T (see Problem 9.11) and has notch filter characteristics. Show that the voltage-ratio transfer function is

$$\frac{V_2}{V_1} = \frac{s^2 + \omega_0^2}{s^2 + \omega_0/Q\,s + \omega_0^2}$$

where $\omega_0 = 1/RC$ and $Q = \frac{1}{4}[1/(1-k)]$.

12.6 Based on the results of Problem 12.5, design a notch filter centered on $f_0 = 1000$ Hz which provides at least 20 dB of attenuation over a bandwidth of 2 Hz. Repeat for a bandwidth of 10 Hz.

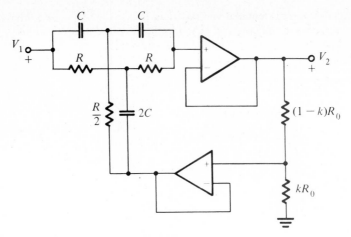

FIGURE P12.5

12.7 Given that the prototype from which a bandstop filter is to be designed is third-order Butterworth. The bandstop filter is to be designed with $\omega_0 = 1000$ rad/s, and the half-power bandwidth is to be 100 rad/s.
(a) Determine the pole and zero locations for the backstop filter.
(b) Prepare a block diagram showing the ω_0 and Q and ω_z for each of the cascade-connected stages.

12.8 Design a bandstop filter with a Butterworth response having the frequency characteristic shown in Fig. P12.8. Magnitude scale so that element values are in a practical range. Calculate the attenuation your design actually achieves at $f = 800$ Hz and $f = 850$ Hz.

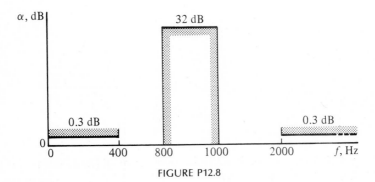

FIGURE P12.8

12.9 Reconsider the design of the bandstop filter carried out as Example 12.6. Since the Boctor circuits are less sensitive and easier to tune than the Bainter circuits, redesign the filter using Boctor circuits.

12.10 Design a bandstop filter with a Chebyshev response having the frequency response shown in Fig. P12.10. Magnitude scale so that element values are in a practical range.

12.11 Design a bandstop filter with a Chebyshev response with the frequency response shown in Fig. P12.11. Magnitude scale so that the element values in the filter are in a practical range.

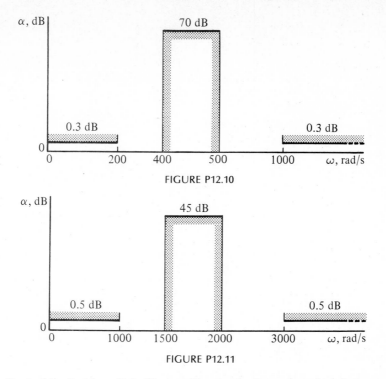

FIGURE P12.10

FIGURE P12.11

12.12 Reconsider Problem 12.8. The specification is changed so that the response is Chebyshev rather than Butterworth. Design a filter to meet the new specification. Magnitude scale your element sizes so that they are in a practical range.

CHAPTER 13

Inverse

Chebyshev

and

Cauer

Filters

The magnitude responses to be considered in this chapter are shown in Fig. 13.1. Both are lowpass responses. That shown in Fig. 13.1a is maximally flat in the pass band and equal ripple in the stop band, while that of Fig. 13.1b is equal ripple in both the pass band and the stop band. The first of these is known as the inverse Chebyshev response and the second as the Cauer response.

13.1 THE INVERSE CHEBYSHEV RESPONSE

All forms of lowpass responses are competitive in the sense that the designer may choose one or another depending on its advantages or disadvantages. Many designers prefer the inverse Chebyshev response for filter design. One of the objectives of our study will be to discover why this might be. We do this by comparing this response to the Butterworth and Chebyshev responses studied in Chapters 6 and 8.

We first show that the inverse Chebyshev response function may be generated in three steps. As shown in Fig. 13.2, we begin with the lowpass Chebyshev response of Chapter 8, which is

$$|T_C(j\omega)|^2 = \frac{1}{1 + \varepsilon^2 C_n^2(\omega)} \tag{13.1}$$

which is shown in Fig. 13.2a. We next subtract this function from 1, giving

$$1 - |T_C(j\omega)|^2 = \frac{\varepsilon^2 C_n^2(\omega)}{1 + \varepsilon^2 C_n^2(\omega)} \tag{13.2}$$

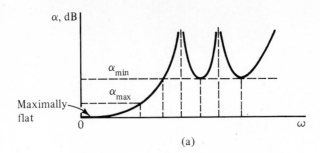

(a)

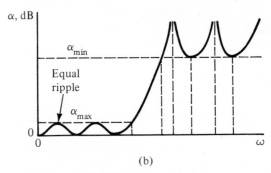

(b)

FIGURE 13.1

which is the response shown in Fig. 13.2b. Finally we invert frequency by replacing ω with $1/\omega$, giving the inverse Chebyshev response

$$|T_{IC}(j\omega)|^2 = \frac{\varepsilon^2 C_n^2(1/\omega)}{1 + \varepsilon^2 C_n^2(1/\omega)} \tag{13.3}$$

Given this response, we have a number of questions to answer: (1) What are the characteristics of the response? (2) Where are the poles and zeros of $T_{IC}(s)$ located in the s plane? (3) How does the form of the response relate to specifications? We first drop the subscript IC and then proceed.

We direct special attention to the frequency $\omega = 1$. Since $C_n(1) = 1$ for all n, then from Eq. (13.3),

$$|T(j1)|^2 = \frac{\varepsilon^2}{1 + \varepsilon^2} \tag{13.4}$$

Referring to Fig. 13.2a, we see that the ripple band which extends from 0 to 1 for the Chebyshev response corresponds to the range of 1 to ∞ for the inverse Chebyshev response of Fig. 13.2c. Thus the upper limit of $|T(j\omega)|^2$ in the stop band of the inverse Chebyshev response is the same as that for $\omega = 1$, that is, the value given by Eq. (13.4). In terms of the attenuation given by

$$\alpha = -10 \log |T(j\omega)|^2 \tag{13.5}$$

the minimum value of the attenuation in the stop band is, from Eq. (13.4),

$$\alpha_{\min} = -10 \log |T(j1)|^2 = 10 \log \left(1 + \frac{1}{\varepsilon^2}\right) \quad \text{dB} \tag{13.6}$$

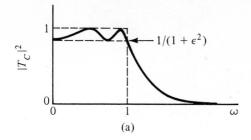

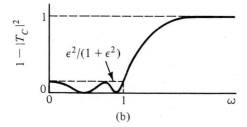

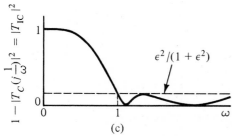

(a)

(b)

(c)

FIGURE 13.2

Solving this equation for ε, gives

$$\varepsilon = (10^{\alpha_{\min}/10} - 1)^{-1/2} \tag{13.7}$$

If Eq. (13.3) is written in the form found by dividing the denominator by the numerator,

$$|T(j\omega)|^2 = \frac{1}{1 + 1/\varepsilon^2 C_n^2(1/\omega)} \tag{13.8}$$

then we may make use of Eq. (13.5) to determine an expression for the attenuation

$$\alpha = 10 \log \left[1 + \frac{1}{\varepsilon^2 C_n^2(1/\omega)} \right] \quad \text{dB} \tag{13.9}$$

This is the general form of the inverse Chebyshev response, and a typical plot of this response is shown in Fig. 13.3 for both $|T|^2$ and α. We are especially interested in the behavior of $|T|^2$ and α in the stop band, which is for frequencies in excess of 1. At the frequency $\omega = 1$ the attenuation is $\alpha_{\min}$, and this corresponds to the value $\varepsilon^2/(1 + \varepsilon^2)$ for $|T|^2$. The equal-ripple nature of $|T|^2$ as shown in Fig. 13.3a indicates oscillations between this value and 0. From Eq. (13.3) the limits of

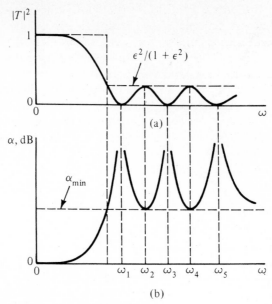

FIGURE 13.3

the ripples correspond to

$$C_n^2\left(\frac{1}{\omega_k}\right) = 0 \quad \text{and} \quad C_n^2\left(\frac{1}{\omega_k}\right) = 1 \tag{13.10}$$

We wish to solve for these frequencies. Evidently they correspond to

$$\cos n \cos^{-1}\left(\frac{1}{\omega_k}\right) = \begin{cases} 0 \\ \pm 1 \end{cases} \tag{13.11}$$

since the Chebyshev function is expressed in terms of trigonometric functions in the range of ω from 1 to ∞. Let $\theta_k = \cos^{-1}(1/\omega_k)$ and observe that

$$\cos n\theta_k = 0 \quad \text{when} \quad n\theta_k = k\frac{\pi}{2} \text{ for } k \text{ odd} \tag{13.12}$$

$$\cos n\theta_k = 1 \quad \text{when} \quad n\theta_k = k\frac{\pi}{2} \text{ for } k \text{ even} \tag{13.13}$$

Thus all frequencies for maxima and minima are given by one equation:

$$\cos^{-1}\left(\frac{1}{\omega_k}\right) = \theta_k = \frac{k\pi}{2n} \tag{13.14}$$

or

$$\omega_k = \sec\left|\frac{k\pi}{2n}\right| \tag{13.15}$$

where odd values of k correspond to $|T|^2 = 0$ and even values of k to the value $|T|^2 = \varepsilon^2/(1 + \varepsilon^2)$, as shown in Fig. 13.3a. In terms of the attenuation plot, $\alpha = \infty$

for k odd, and $\alpha = \alpha_{\min}$ for $\omega = 1$ and k even. The oscillations in Fig. 13.3 have been shown as almost sinusoidal for clarity. In reality the oscillations are bunched near $\omega = 1$, as an example will illustrate. If we let $n = 9$, then Eq. (13.15) may be used to compute the values given in Table 13.1, and the actual variation of $|T|^2$ in the stop band is as shown in Fig. 13.4.

From Figs. 13.3 and 13.4 it is seen that the stop band begins at the frequency $\omega = 1$. It is crucial that frequency scaling be done such that this is the case. This is shown in Fig. 13.5. Another frequency is also defined as being the end of the pass band, ω_p. This frequency must be a value that results from the scaling of the stop band frequency to $\omega_s = 1$.

TABLE 13.1

k	$k\theta$	ω_k	α
0	0	1.000	$\alpha_{\min}$
1	10	1.015	∞
2	20	1.064	$\alpha_{\min}$
3	30	1.155	∞
4	40	1.305	$\alpha_{\min}$
5	50	1.555	∞
6	60	2.000	$\alpha_{\min}$
7	70	2.924	∞
8	80	5.760	$\alpha_{\min}$
9	90	∞	∞

So much for the features of the stop band. To observe the interesting characteristic of the pass band, we note that the asymptotic form of the Chebyshev function for large ω is*

$$C_n(\omega) \cong 2^{n-1}\,\omega^n, \quad \omega \gg 1 \tag{13.16}$$

Hence we know that near $\omega = 0$

$$C_n\left(\frac{1}{\omega}\right) \cong 2^{n-1}\left(\frac{1}{\omega}\right)^n, \quad \omega \ll 1 \tag{13.17}$$

If we substitute this approximate form into Eq. (13.8), there results

$$|T|^2 \cong \frac{1}{1 + 1/\varepsilon^2[2^{n-1}\,(1/\omega^n)]^2} \tag{13.18}$$

$$\cong \frac{1}{1 + \omega^{2n}/\varepsilon^2 2^{2n-2}} = \frac{1}{1 + (\omega/\omega_k)^{2n}} \tag{13.19}$$

where

$$\omega_k = (\varepsilon^2 2^{2n-2})^{1/2n} = (\varepsilon 2^{n-1})^{1/n} \tag{13.20}$$

Since Eq. 13.19 has the form of the equation for the Butterworth response, Eq.

* For example, see G. Daryanani, *Principles of Active Network Synthesis and Design*, Wiley, New York, 1976, p. 111.

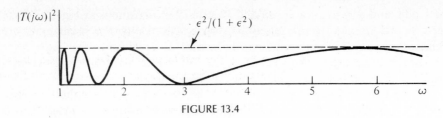

FIGURE 13.4

(6.10), we may conclude that for small ω the inverse Chebyshev response is maximally flat. We will show later that if ω_k given by Eq. (13.20) is larger than 1, then this response is "flatter" than the Butterworth response.

13.2 FROM SPECIFICATIONS TO POLE AND ZERO LOCATIONS

As was done for the Butterworth and Chebyshev lowpass responses, we determine the value of n that will satisfy specified α_{min} and α_{max}. We begin with Eq. (13.9) and apply the specifications shown in Fig. 13.5. In that figure we see that the end of the pass band, where the attenuation is less than α_{max}, is ω_p. Then

$$\alpha_{max} = 10 \log \left[1 + \frac{1}{\epsilon^2 \, C_n^{\,2}(1/\omega_p)} \right] \tag{13.21}$$

If we substitute the value of ϵ^2 found in Eq. (13.7) into this equation, we may then solve for $C_n^{\,2}(1/\omega_p)$:

$$C_n^{\,2}\left(\frac{1}{\omega_p}\right) = \frac{10^{\alpha_{min}/10} - 1}{10^{\alpha_{max}/10} - 1} \tag{13.22}$$

This equation applies to the pass band for which $\omega < 1$, where the Chebyshev function is expressed in terms of hyperbolic functions. If we extract the square root of Eq. (13.22), then it may be written

$$\cosh n \cosh^{-1}\left(\frac{1}{\omega_p}\right) = \left[\frac{10^{\alpha_{min}/10} - 1}{10^{\alpha_{max}/10} - 1} \right]^{1/2} \tag{13.23}$$

Solving this equation for n, we obtain finally

$$n = \frac{\cosh^{-1}\left[(10^{\alpha_{min}/10} - 1)/(10^{\alpha_{max}/10} - 1)\right]^{1/2}}{\cosh^{-1}(1/\omega_p)} \tag{13.24}$$

This equation may be compared to that found by the Chebyshev response, Eq.

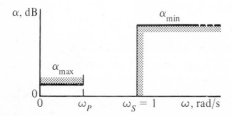

FIGURE 13.5

(8.39). They are seen to be identical except that Eq. (13.24) is scaled such that $\omega_s = 1$, while Eq. (8.39) is scaled such that $\omega_p = 1$. Thus if written in terms of ω_s/ω_p, the two equations are the same. We conclude that the Chebyshev and the inverse Chebyshev responses require the same value of n to satisfy the general specifications

$$\alpha_{max}, \qquad \alpha_{min}, \qquad \omega_p, \qquad \omega_s$$

If we next return to Eq. (13.3), we see that the requirement that $|T|^2$ have the value $\frac{1}{2}$ defines the half-power frequency, and that this occurs when

$$\varepsilon^2 C_n{}^2\left(\frac{1}{\omega}\right) = 1 \tag{13.25}$$

This is the same equation as Eq. (8.30), except that $1/\omega$ has replaced ω. Since Eq. (8.30) was solved to give Eq. (8.34), we see that the half-power frequency for the inverse Chebyshev response is

$$\omega_{hp} = \frac{1}{\cosh\left[\dfrac{1}{n}\cosh^{-1}\left(\dfrac{1}{\varepsilon}\right)\right]} < 1 \tag{13.26}$$

This reminds us that ω_{hp} will be less than 1 for the inverse Chebyshev response, just as the Chebyshev response was greater than 1, and that the two quantities are reciprocals.

Having found the value of n that will satisfy the specifications of Eq. (13.24), we may now determine the location of the poles and zeros. For the magnitude-squared function we may identify numerator and denominator as

$$|T|^2 = \left.\frac{p(s)p(-s)}{q(s)q(-s)}\right|_{s=j\omega} \tag{13.27}$$

Equating this result to Eq. (13.3), we make the identifications

$$p(s)p(-s)|_{s=j\omega} = \varepsilon^2 C_n{}^2\left(\frac{1}{\omega}\right) \tag{13.28}$$

from which the zero locations for $T(s)$ will be found, and

$$q(s)q(-s)|_{s=j\omega} = 1 + \varepsilon^2 C_n{}^2\left(\frac{1}{\omega}\right) \tag{13.29}$$

Now the only values of ω in Eq. (13.28) that will cause the function to become zero are those for which $\omega > 1$ when the Chebyshev function is expressed as trigonometric functions. Thus zeros are defined when

$$C_n{}^2\left(\frac{1}{\omega_k}\right) = 0 \tag{13.30}$$

This is identical to Eq. (13.10), and the solution to this requirement was given by Eq. (13.15) for k odd,

$$\omega_k = \sec\left|\frac{k\pi}{2n}\right|, \qquad k = 1, 3, 5, \ldots, n \tag{13.31}$$

Equation (13.29), which defines the poles of the inverse Chebyshev response, has the same form as the denominator of Eq. (8.10), except that $1/\omega$ has replaced ω. Since we have already determined the location of the Chebyshev poles in Section 8.3, the poles for the inverse Chebyshev response may be determined by a simple procedure: find the Chebyshev poles by the methods outlined in Section 8.3, and then find the inverse Chebyshev poles by pole reciprocation as described in Section 12.2. Thus a characteristic pole and zero location pattern for the inverse Chebyshev response is that shown in Fig. 13.6 for the case $n = 9$. We will now illustrate this general procedure in terms of an example.

Example 13.1 The specifications for an inverse Chebyshev filter are given in terms of the notation of Fig. 13.5 as

$$\alpha_{\min} = 18 \text{ dB}, \quad \alpha_{\max} = 0.25 \text{ dB}, \quad \omega_s = 1400, \quad \omega_p = 1000 \qquad (13.32)$$

The required value of n is found using Eq. (13.24):

$$n = \frac{\cosh^{-1} \sqrt{(10^{1.8} - 1)/(10^{0.025} - 1)}}{\cosh^{-1} 1.4} = 4.81 \qquad (13.33)$$

which is rounded up to $n = 5$. With n determined, we next determine the location of the inverse Chebyshev poles. We first find ε from

$$\varepsilon = \frac{1}{\sqrt{10^{0.1\alpha_{\min}} - 1}} = \frac{1}{\sqrt{10^{1.8} - 1}} = 0.1269 \qquad (13.34)$$

Then from Eq. (8.55),

$$a = \frac{1}{5} \sinh^{-1}\left(\frac{1}{0.1269}\right) = 0.5523 \qquad (13.35)$$

The Butterworth angles are determined as in Chapter 6, and are

$$\psi_k = 0, \pm 36°, \pm 72° \qquad (13.36)$$

Then the Chebyshev pole positions are calculated using Eqs. (8.73) and (8.74), repeated here:

$$\sigma_k = -\cos \psi_k \sinh a$$
$$\omega_k = \pm \sin \psi_k \cosh a \qquad (13.37)$$

We will also make use of the equations

$$\omega_{0k} = \sqrt{\sigma_k^2 + \omega_k^2} \qquad (13.38)$$

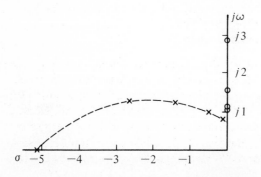

FIGURE 13.6

$$Q_k = \frac{1}{2 \cos \tan^{-1}(\omega_k/\sigma_k)} \qquad (13.39)$$

to determine the pole positions. Then we determine the following values:

k	σ_k	$j\omega_k$	$\tan^{-1}(\omega_k/\sigma_k)$	ω_{0k}	Q_k
0	−0.5808	0	0	0.5808	0.5
1	−0.4699	±j0.6797	55.3°	0.8263	0.8792
2	−0.1795	±j1.0998	80.7°	1.1144	3.104

To determine the positions of the inverse Chebyshev poles, we find the reciprocal of ω_{0k}, but keep the same values of Q_k. The required values are:

w_{0k}	Q_k	$\alpha \pm j\beta$
1.7218	0.5	−1.7218 ± j0
1.2102	0.8792	−0.6882 ± j0.9954
0.8973	3.104	−0.1445 ± j0.8856

Finally the zeros are determined from Eq. (13.31), and the values are as follows:

k	ω_{zk}
1	1.051
3	1.701
5	∞

The poles and zeros that have been calculated are shown in Fig. 13.7 for the upper half-plane. These locations are based on assuming that $\omega_s = 1$, but since the specified ω_s was 1400 rad/s, frequency scaling using $k_f = 1400$ will be required. To find a circuit that will realize the poles and zeros of Fig. 13.7, we pair them off as shown in Fig. 13.8, (a) such that the circuit realization will consist of two lowpass notch circuits connected in cascade with a first-order circuit to realize the pole on the negative real axis. Circuits suitable for this realization were studied in the last chapter. A realization based on the circuit of Fig. 12.56 is shown in Fig. 13.8(b).

13.3 COMPARISONS OF INVERSE CHEBYSHEV RESPONSE WITH OTHER RESPONSES

The example of the last section has shown that the pole and zero locations for the inverse Chebyshev response can be determined by a simple step-by-step procedure, and that each step can be accomplished by an inexpensive calculator. We next address this question: how does this new response compare with those found previously for the lowpass filter in Chapters 6 and 8. Some of the quantities we will compare are: (1) the pass-band response, (2) the stop-band response, (3) the transition from pass band to stop band, (4) the values of Q required, (5) circuits

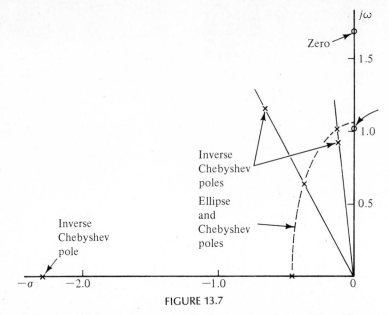

FIGURE 13.7

for the realization of the responses, and (6) the time delay. We will make comparisons using the specifications given in Example 13.1 and the Butterworth and Chebyshev responses that meet the same specifications.

We must first decide on how to frequency scale for comparison. We recall that $\omega = 1$ is the half-power frequency for the Butterworth response, the end of the ripple band for the Chebyshev response, and the beginning of the stop band for the inverse Chebyshev response. Hence this decision concerning frequency scaling is not a trivial one. We will frequency scale so that $\omega = 1$ is the half-power frequency for all three responses; this is an arbitrary choice since we might have decided to let $\omega = 1$ be the beginning of the stop band for all three responses.

For the Butterworth response we know that $\omega = 1$ is the half-power frequency for all values of n. For the Chebyshev response we satisfy the specification that $\alpha_{max} = 0.25$ dB such that

$$\varepsilon = [10^{\alpha_{max}/10} - 1]^{1/2} = 0.2434 \qquad (13.40)$$

Since we know that $n = 5$ from Example 13.1, we determine the half-power frequency to be

$$\omega_{hp} = \cosh\left[\frac{1}{n}\cosh^{-1}\left(\frac{1}{\varepsilon}\right)\right] = 1.0887 \qquad (13.41)$$

For the inverse Chebyshev response we meet the specification $\alpha_{max} = 0.25$ dB by requiring that

$$0.25 = 10\log\left[1 + \frac{1}{\varepsilon^2 C_5^2(1.4)}\right] \qquad (13.42)$$

from which $\varepsilon = 0.1076$, and then

$$\omega_{hp} = \frac{1}{\cosh\left[\frac{1}{n}\cosh^{-1}\left(\frac{1}{\varepsilon}\right)\right]} = 0.8508 \qquad (13.43)$$

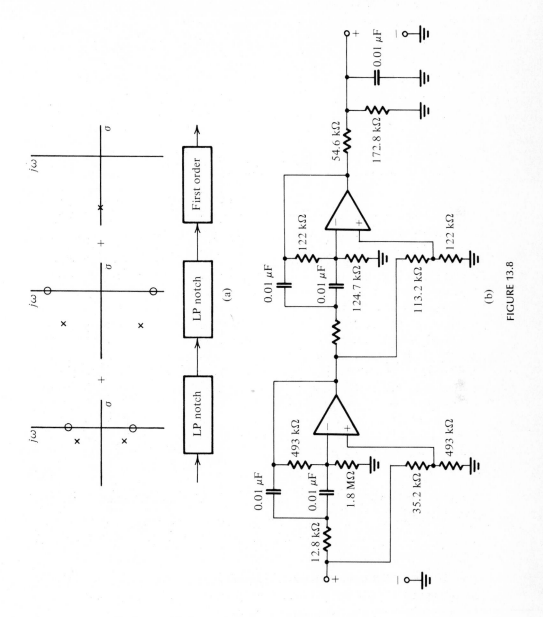

FIGURE 13.8

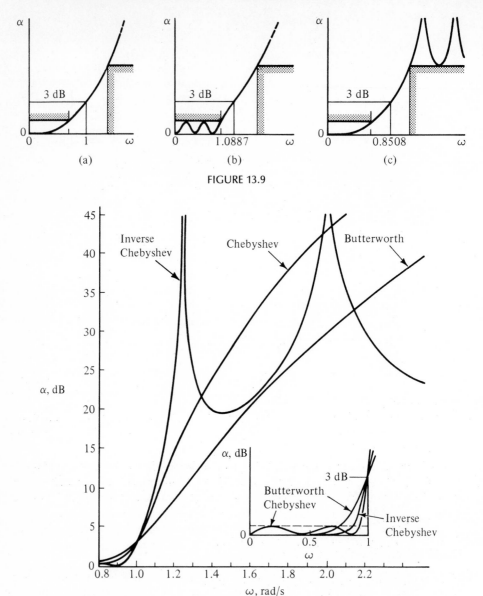

FIGURE 13.9

FIGURE 13.10

These values with the characteristic shapes of the three responses are shown in Fig. 13.9. We must frequency scale such that $\omega = 1$ is common.

Next we tabulate the attenuation formulas from which we may make plots for comparison. The Butterworth response is

$$\alpha(\omega) = 10 \log (1 + \omega^{10}) \quad \text{dB} \tag{13.44}$$

The Chebyshev response is, for the pass band,

$$\alpha(\omega) = 10 \log[1 + (0.2434)^2(\cos 5 \cosh^{-1} 1.0887\omega)^2] \quad \text{dB} \tag{13.45}$$

and for the stop band cos is replaced by cosh. The inverse Chebyshev response is,

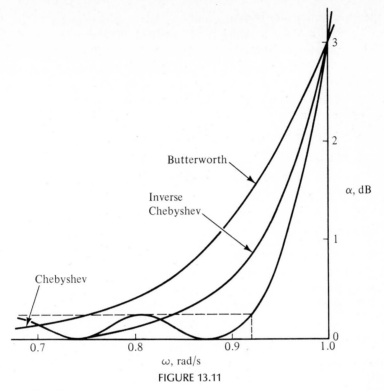

FIGURE 13.11

for the pass band,

$$\alpha(\omega) = 10 \log \left[1 + \frac{1}{(0.1076)^2 \cosh 5[\cosh^{-1}(1.1754/\omega)]^2} \right] \quad (13.46)$$

and for the stop band cosh is replaced by cos. These three equations are plotted in Fig. 13.10, and also in Fig. 13.11 for frequencies approaching the half-power frequency. From these plots we may make comparisons for this special case, which turn out to be fairly general comparisons for other specifications.

1.   Passband Response

The inverse Chebyshev is always better than the Butterworth in the pass band, and it is better than the Chebyshev except near $\omega = 1$. For low frequencies the inverse Chebyshev is always the best of the three. A heuristic suggestion for the reason for this comes from the approximation given by Eq. (13.20), where ω_k is a fictional half-power frequency under the assumption that $\omega < 1$. When true, the response behaves as if this were the actual response. For this particular example, $\omega_k = 1.115$.

2.   The Stop-Band Response

We see that the inverse Chebyshev response meets the $\alpha_{\min}$ specification at the lowest frequency. For this particular example we matched the specification at the frequency at which $\alpha_{\max}$ was specified. Due to the rounding up of n we cannot meet the specifications for both α_{max} and α_{min}, and thus the value actually realized was $\alpha_{min} = 19.4$ dB rather than the 18 dB specified. In other words, we more than meet the specifications. From Fig. 13.10 we see that the Butterworth and Chebyshev responses have greater attenuations than the inverse Chebyshev response for higher values of frequency. This is generally not significant if indeed α_{min} is the minimum value of attenuation required for the entire stop band.

3. The Transition Band

The inverse Chebyshev response has the most rapid change from α_{max} to α_{min}, or the sharpest cutoff. The Butterworth response is the worst, with the Chebyshev in between.

4. The Q Required

As shown in Example 13.1, the Q for the inverse Chebyshev response is the same as the Q required for the Chebyshev response for the same ε, and both are greater than the Q required for the Butterworth response. For this particular example, none of the Q values are excessive.

5. Circuit Realizations

The Butterworth and Chebyshev responses can be realized with various forms of the Sallen–Key circuit of Chapter 6, while the Boctor circuit or equivalent is required for the inverse Chebyshev response. The Boctor circuit is more difficult to design than the Sallen–Key circuit.

6. Time Delay

In Chapter 18 we show that the Butterworth response has a superior time delay characteristic, the Chebyshev response being much inferior, with the inverse Chebyshev response being intermediate between the two extremes.

13.4 CAUER MAGNITUDE RESPONSE

A comparison of the three responses shown in Fig. 13.10 provides us with a rationale for introducing yet another response. We recall from Chapter 6 that the Butterworth response was derived with the objective of reducing the error between the response and the brick-wall response as much as possible for frequencies near $\omega = 0$, and this resulted in a maximally flat magnitude characteristic. The Chebyshev response of Chapter 8 distributed this error over the entire pass band, and this resulted in an equal-ripple response. The result was that the size of the frequency band between the end of the pass band and the beginning of the stop band, $\omega_s - \omega_p$, was reduced for a given order n. We will call this the *transition band*, and speak of a small transition band as a *sharp cutoff* characteristic. In both the Butterworth and the Chebyshev cases the way we meet the attenuation specifications α_{max} and α_{min} in the transition band is to control the size of n. We found that for such specifications the Chebyshev required a smaller value of n than the Butterworth. The inverse Chebyshev response provided a fresh approach. By introducing zeros of $T(s)$ on the imaginary axis beyond ω_s, we were able to decrease the transition band. The price paid for this was that the asymptotic increase in attenuation at higher frequencies was reduced. However, this is seldom important since our objective is to attain a specified α_{min} at the frequency ω_s and at all higher frequencies. Thus an asymptotic increase of 0 dB per octave would be satisfactory, even though this cannot be achieved. Again referring to Fig. 13.10, we see that the Butterworth and Chebyshev responses can be regarded as inefficient in the stop band in the sense that they provide more attenuation than is needed. With the introduction of zeros in the inverse Chebyshev response, the attenuation is more evenly distributed over the pass band.

Now the inverse Chebyshev response is maximally flat in the pass band. We have observed that the Chebyshev response has a sharper cutoff rate than the Butterworth response, and this suggests that a new response with equal ripple in both the pass band and the stop band would be superior to even the inverse Chebyshev response. We shall show that indeed this is the case, as was first demonstrated by Wilhelm Cauer. If we take as our specification the requirement that $\omega_s/\omega_p = 1.5$, $\alpha_{min} = 50$ dB, and $\alpha_{max} = 0.5$ dB, then for a Butterworth response it is necessary that $n = 17$, and for a Chebyshev response that $n = 8$. For a Cauer response it is necessary only that $n = 5$. This is a significant improvement.

Cauer first used his new theory in solving a filter problem for the German telephone industry. His new design achieved specifications with one less inductor than had ever been done before. The world first learned of the Cauer method not through scholarly publication but through a patent disclosure, which eventually reached the Bell Laboratories. Legend has it that the entire Mathematics Department of Bell Laboratories spent the next two weeks at the New York Public Library studying elliptic functions. Cauer had studied mathematics under Hilbert at Goettingen, and so elliptic functions and their applications were familiar to him.

Given the objective of attaining a response which is equal ripple in both the pass band and the stop band, what are the problems to be considered in determining $T(s)$? As shown in Fig. 13.12, the poles of $T(s)$ must be placed in such a way that we have equal ripple in the pass band. In our study of the Chebyshev response these poles were located on an ellipse. But with zeros on the imaginary axis, what will be their positions? Given that we must have zeros of $T(s)$ on the imaginary axis, where should they be located? We wish to locate them such that equal ripple is achieved in the stop band, which requires that $\alpha_1 = \alpha_2 = \alpha_{min}$.

Let us review the Chebyshev response of Chapter 8 with the objective of seeing why it does not meet our needs. It is

$$|T(j\omega)|^2 = \frac{1}{1 + \epsilon^2 C_n^2(\omega)} \tag{13.47}$$

Now $C_n(\omega)$ is a polynomial in ω and so may be characterized by its zeros alone, since all poles are at infinity. We require a function to replace $C_n(\omega)$ that has

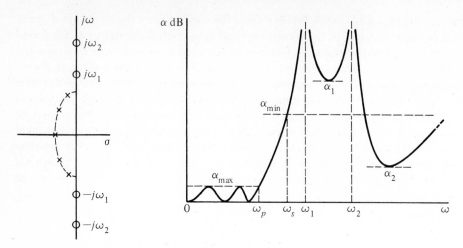

FIGURE 13.12

poles at finite frequencies to produce the required zero values for $|T(j\omega)|$. Furthermore these poles must be in the pass band, as we have just seen. Since $C_n(\omega)$ does not meet our needs, suppose that we replace it by a new function which is a quotient of polynomials:

$$R_n(s) = \frac{P(s)}{Q(s)} \quad \text{or} \quad R_n^2(\omega) = \frac{P(j\omega)P(-j\omega)}{Q(j\omega)Q(-j\omega)} \tag{13.48}$$

Then the new function replacing Eq. (13.47) will be

$$|T(j\omega)|^2 = \frac{1}{1 + {}^2R_n^2(\omega)} \tag{13.49}$$

or

$$|T(j\omega)|^2 = \frac{Q(j\omega)Q(-j\omega)}{Q(j\omega)Q(-j\omega) + \varepsilon^2 P(j\omega)P(-j\omega)} \tag{13.50}$$

From this we see that the poles of R_n will be the required zeros of T. The attenuation corresponding to Eq. (13.47) is

$$\alpha = 10 \log [1 + \varepsilon^2 R_n^2(\omega)] \tag{13.51}$$

Before studying this equation further, let us make the decision that the end of the ripple band will be the normalized frequency $\omega_p = 1$, and further that we will later find that $R_n(1) = 1$. Then from the last equation,

$$\alpha(1) = \alpha_{max} = 10 \log (1 + \varepsilon^2) \tag{13.52}$$

or

$$\varepsilon = (10^{\alpha_{max}/10} - 1)^{1/2} \tag{13.53}$$

Now referring to Fig. 13.12 and the requirement that $\alpha_1 = \alpha_2 = \alpha_{min}$, we shall require that $R_n = \pm L$ at the frequencies at which α has a minimum value in the stop band, as well as at ω_s. This will require that

$$\alpha_{min} = 10 \log (1 + \varepsilon^2 L^2) \tag{13.54}$$

or that

$$L^2 = \frac{10^{\alpha_{min}/10} - 1}{10^{\alpha_{max}/10} - 1} \tag{13.55}$$

This quotient has appeared before in the expression for n for both the Butterworth and the Chebyshev responses. From this we see that from the usual specifications we may determine ε and L. To go further, we must know more about our new function R_n. This we do in the next section.

13.5   CHEBYSHEV RATIONAL FUNCTIONS

It should be no surprise that $R_n(\omega, L)$ the *Chebyshev rational functions* exactly meet the requirements outlined in the last section. The behavior of such functions is illustrated for the case $n = 4$ in Fig. 13.13. From that figure we see the following properties of $R_4(\omega, L)$:

1. In the frequency range $-1 < \omega < 1$ there are four zeros and an equal-ripple response with

$$|R_4(\omega, L)| \leq 1 \tag{13.56}$$

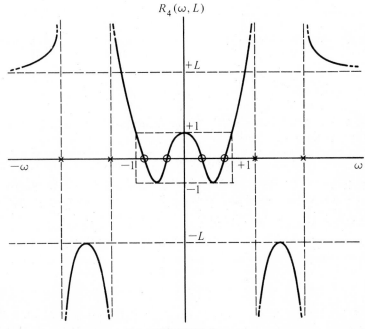

FIGURE 13.13

2.   At the frequency that we have chosen to be the end of the pass band, $\omega = \pm 1$, we see that $R_4(\pm 1) = 1$, but that in general

$$R_n(1) = 1 \quad \text{and} \quad R_n(-1) = (-1)^n \tag{13.57}$$

as was the case for the Chebyshev functions C_n.

3.   For $|\omega| > 1$ there are four poles and the lower limit of the function,

$$|R_4(\omega, L)| \geq L \tag{13.58}$$

This special case $n = 4$ generalizes as shown in Fig. 13.14, which tabulates $R_n(\omega, L)$ for $n = 2$ to $n = 5$, and also compares these functions to the corresponding Chebyshev functions $C_n(\omega)$.

The Chebyshev rational functions may be constructed from the poles and

Chebyshev polynomial   Chebyshev rational function

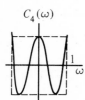 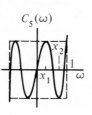 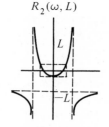 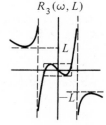

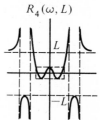 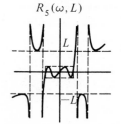

CHARACTERISTICS

$C_n(\omega)$	$R_n(\omega, L)$
1. $C_n(\omega)$ is even (odd) if n is even (odd)	1. $R_n(\omega, L)$ is even (odd) if n is even (odd)
2. $C_n(\omega)$ has all its zeros in $|\omega| < 1$	2. $R_n(\omega, L)$ has all its zeros in the range $|\omega| < 1$ and all its poles in the range $|\omega| > 1$
3. $C_n(\omega)$ oscillates between ± 1 in the pass band	3. $R_n(\omega, L)$ oscillates between ± 1 in the pass band
4. $C_n(\omega) = 1$ at $\omega = 1$	4. $R_n(\omega, L) = 1$ at $\omega = 1$
5. $C_n(\omega)$ increases monotonically for $|\omega| > 1$ (in the stop band)	5. $R_n(\omega, L)$ oscillates between $\pm L$ and ∞ in the stop band.

FIGURE 13.14

zeros illustrated in Fig. 13.14. For $n = 4$,

$$R_4(s, L) = C_4 \frac{(s^2 + \omega_{z1}^2)(s^2 + \omega_{z2})^2}{(s^2 + \omega_{p1}^2)(s^2 + \omega_{p2})^2} \qquad (13.59)$$

and for $n = 5$,

$$R_5(s, L) = C_5 s \frac{(s^2 + \omega_{z1}^2)(s^2 + \omega_{z2}^2)}{(s^2 + \omega_{p1}^2)(s^2 + \omega_{p2}^2)} \qquad (13.60)$$

The corresponding pole–zero locations for these two cases are shown in Fig. 13.15. Clearly, C_4 and C_5 are constants that depend on L.

With this knowledge of $R_n(\omega, L)$ we may turn to the Cauer magnitude function with special attention to the frequencies at which the response is either a maximum or a minimum. That function is

$$|T(j\omega)|^2 = \frac{1}{1 + \epsilon^2 R_n^2(\omega, L)} \qquad (13.61)$$

A typical response is shown in Fig. 13.16. In terms of this response, we make the following observations:

1. The zeros of $R(\omega, L)$, designated ω_{zr}, correspond to the equation $|T(j\omega)|^2 = 1$. These are seen to be located in the pass band as required.
2. The poles of $R_n(\omega, L)$, designated ω_{pr}, correspond to the values of frequency at which $|T(j\omega)|^2 = 0$. These are shown in Fig. 13.16 in the stop band.
3. When $R_n(\omega, L) = 1$, we define the attenuation α_{max} which occurs at $\omega = 1$ and various other frequencies in the pass band. Under this condition,

$$|T(j\omega)|^2 = \frac{1}{1 + \epsilon^2} \qquad (13.62)$$

4. Similarly, we are interested in frequencies at which $R_n(\omega, L) = \pm L$, where

$$|T(j\omega)|^2 = \frac{1}{1 + \epsilon^2 L^2} \qquad (13.63)$$

These relate to the attenuation having the value α_{min}.

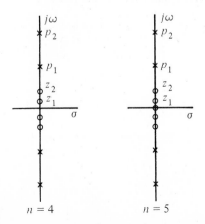

$n = 4$ $n = 5$ FIGURE 13.15

TABLE 13.2 Cauer magnitude response characteristics

Band	$\|T(j\omega)\|^2$	Attenuation α	$R_n(\omega, L)$	Frequency notation	Frequency name
Pass	1	$\alpha = 0$	0	$\omega_{z\nu}$	Attenuation zeros, zeros of R_n
	$\dfrac{1}{1 + \epsilon^2}$	$\alpha = \alpha_{max}$	± 1	$\omega_{y\nu}$	
Stop		$\alpha = \infty$	∞	$\omega_{p\nu}$	Attenuation poles, transmission zeros
	$\dfrac{1}{1 + \epsilon^2 L^2}$	$\alpha = \alpha_{min}$	$\pm L$	$\omega_{x\nu}$	

This information is summarized in Table 13.2. An interesting relationship between these frequencies is

$$\omega_{p\nu} = \frac{1}{\omega_{z\nu}} \tag{13.64}$$

From this equation we see that the magnitude response of Fig. 13.16 is somewhat distorted in the interest of clarity of the definitions of frequency. An actual response for an $n = 5$ Cauer filter is given in Fig. 13.17, which shows the character-

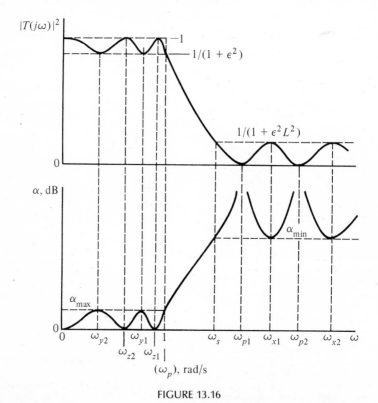

FIGURE 13.16

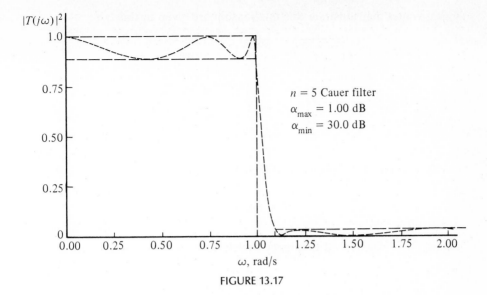

FIGURE 13.17

istic bunching of the ripples near $\omega = 1$ in the pass band, and even more so in the stop band.

13.6 CAUER FILTER DESIGN

There remains the problem of the determination of the pole and zero locations for $T(s)$ from Eq. (13.50). Unfortunately this is not as simple as it was for the Butterworth and Chebyshev cases. Cauer found a systematic method for calculating this information, later published in his textbook.* It is rather involved, making use of elliptic functions first studied by Jacobi in 1826. Those who design Cauer (or elliptic) filters generally do so by the use of the extensive tables that are available, and we will follow this procedure.†

Some tables are given in terms of parameters that have their origins in transmission-line theory. One of these is the reflection coefficient, usually expressed as a percentage. If the reflection coefficient has the magnitude $|\rho|$, then it is related to the pass-band attenuation limit by the equation

$$\alpha_{max} = -10 \log (1 - |\rho|^2) \qquad (13.65)$$

or the inverse relationship

$$|\rho| = (1 - 10^{-\alpha_{max}/10})^{1/2} \qquad (13.66)$$

*W. Cauer, *Synthesis of Linear Communication Networks*, McGraw-Hill, New York, 1958, App. 3. A more recent treatment may be found in R. W. Daniels, *Approximation Methods for Electronic Filter Design*, McGraw-Hill, New York, 1974, chap. 5.
†As an example of an alternative, we note that in 1975 Sidney Darlington programmed a hand-held calculator, the HP 65, to compute the elliptic filter parameters.

Typical values that illustrate this relationship are given in this table:

$|\rho|$ (%)	α_{max} (dB)
5	0.0109
10	0.0439
15	0.0988
25	0.2800
50	1.250

The other parameter is known as the modular angle θ, which is defined as

$$\theta = \sin^{-1}\left(\frac{\omega_p}{\omega_s}\right) \tag{13.67}$$

Since we have normalized frequency by letting $\omega_p = 1$, this equation becomes

$$\theta = \sin^{-1}\left(\frac{1}{\omega_s}\right) \tag{13.68}$$

We will illustrate the use of these parameters by two examples.

Example 13.2 For this example we make use of the tables due to Christian and Eisenmann.* A sample page from these tables is shown as Table 13.3. In design, the table is used in an iterative manner as follows:

1. Select a trial value for n.
2. Find the appropriate modular angle from Eq. (13.68). If this value is not in the table, take the next larger value (round up).
3. Find the appropriate reflection coefficient from Eq. (13.66). If this value is not in the table, take the next lower value (round down).
4. Check α_{min} to see if it meets specifications. If not, take a higher value of n.

For this example suppose that the specifications are as shown in Fig. 13.18, with $\alpha_{max} = 0.30$ dB, $\alpha_{min} = 35$ dB, and $\omega_s/\omega_p = 2.5667$. Suppose that we select $n = 3$, which brings us to the sample page shown in Table 13.3. From Eq. (13.67) we compute

$$\theta = \sin^{-1}\left(\frac{1}{2.5667}\right) = 22.93° \text{ (round up to } 23°) \tag{13.69}$$

Then from Eq. (13.66) we compute the reflection coefficient as

$$|\rho| = (1 - 10^{0.03})^{1/2} = 0.258 \tag{13.70}$$

which we round down to 25 percent. Now from Table 13.3 we see that α_{min} has the value of 35.75 dB, which is only slightly larger than the specified 35 dB. Hence we have located the proper entry in the table, and the transfer function $T(s)$ may be constructed. In the

---

*E. Christian and E. Eisenmann, *Filter Design Tables and Graphs,* Transmission Networks International, Inc., Knightdale, N. C., 1977. Other recommended tables are given in A. I. Zverev, *Handbook of Filter Synthesis,* Wiley, New York, 1967; R. Saal and E. Ulbrich, "On the Design of Filters by Synthesis," *IRE Trans. Circuit Theory,* pp. 284–327, Dec. 1958; and R. Saal, *Handbook of Filter Design,* AEG-Telefunken, Germany, 1979.

TABLE 13.3

ATTEN.ZEROS 0.873337				-	-	-	C63 θ=21°
ATTEN.POLES 3.1951359428				-	-	-	Ω$_s$=2.79043
p[%]	A$_{min}$[db]	C	i	-a$_i$	±b$_i$	-a$_{i+1}$	±b$_{i+1}$
5	23.95	1.9429170	0	1.7665932282	-	0.6259516014	1.6070176815
10	29.99	3.9005253	0	1.2727795117	-	0.5082018936	1.3409348481
15	33.56	5.8880782	0	1.0398432873	-	0.4350042932	1.2157965985
25	38.18	10.020631	0	0.7841084387	-	0.3421571620	1.0873006154
50	45.17	22.406812	0	0.4722896604	-	0.2138298877	0.9586309656

ATTEN.ZEROS 0.874055				-	-	-	C63 θ=22°
ATTEN.POLES 3.0541207738				-	-	-	Ω$_s$=2.66947
p[%]	A$_{min}$[db]	C	i	-a$_i$	±b$_i$	-a$_{i+1}$	±b$_{i+1}$
5	22.71	1.7663230	0	1.7890385081	-	0.6114453202	1.6055856285
10	28.75	3.5460020	0	1.2835163993	-	0.5007543311	1.3411450462
15	32.32	5.3529039	0	1.0469232254	-	0.4300543688	1.2163441933
25	36.94	9.1098441	0	0.7883084966	-	0.3392685670	1.0880099129
50	43.93	20.370231	0	0.4742274346	-	0.2125680948	0.9593738545

ATTEN.ZEROS 0.874806				-	-	-	C63 θ=23°
ATTEN.POLES 2.9255678744				-	-	-	Ω$_s$=2.55930
p[%]	A$_{min}$[db]	C	i	-a$_i$	±b$_i$	-a$_{i+1}$	±b$_{i+1}$
5	21.53	1.6122542	0	1.8130820356	-	0.5964162248	1.6038317587
10	27.56	3.2366994	0	1.2949359905	-	0.4929896399	1.3412780502
15	31.13	4.8859929	0	1.0544314553	-	0.4248823821	1.2168727115
25	35.75	8.3152312	0	0.7927509004	-	0.3362448311	1.0887340063
50	42.74	18.593422	0	0.4762732630	-	0.2112454008	0.9601468591

ATTEN.ZEROS 0.875591				-	-	-	C63 θ=24°
ATTEN.POLES 2.8079253038				-	-	-	Ω$_s$=2.45859
p[%]	A$_{min}$[db]	C	i	-a$_i$	±b$_i$	-a$_{i+1}$	±b$_{i+1}$
5	20.40	1.4770417	0	1.8388119027	-	0.5808914734	1.6017256339
10	26.42	2.9652520	0	1.3070657262	-	0.4849131303	1.3413223031
15	29.99	4.4762268	0	1.0623019725	-	0.4194897651	1.2173761393
25	34.61	7.6178703	0	0.7974415777	-	0.3330856467	1.0894706230
50	41.59	17.034076	0	0.4784286223	-	0.2098610835	0.9609499317

* Reprinted by permission from Erich Christian and Egon Eisenmann, *Filter Design Tables and Graphs*, Transmission Networks International, Inc., Knightdale, N.C., 1977.

chart of Table 13.3 the pole locations are indicated as

$$-a_0 = -0.7927509004$$

$$-a_1 \pm jb_1 = -0.3362448311 \pm j1.0887340063 \quad (13.71)$$

The zero of $T(s)$ is labeled in the chart as "atten. pole," and so is $\omega_z = 2.9255678744$. The

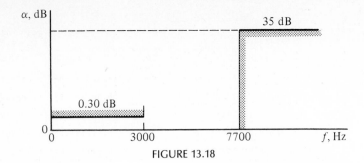

FIGURE 13.18

constant C in the chart is actually the reciprocal of the constant in $T(s)$. Hence we have determined $T(s)$ which is, before roundoff,

$$T(s) = \frac{0.1202612382(s^2 + 8.558947385)}{(s + 0.7927509004)(s^2 + 0.672489662s + 1.298402322)} \qquad (13.72)$$

The attenuation plot corresponding to this $T(s)$ is shown in Fig. 13.19 with frequency scaled by $3000 \times 2\pi$. Notice that the chart gives the frequency at which the attenuation is zero in the pass band.

The transfer function $T(s)$ given in the last equation has a degree difference of 1 between the numerator and the denominator and hence represents a system with a rolloff rate of -6 dB per octave (with attenuation having a corresponding increase of 6 dB per octave). For the corresponding all-pole transfer functions, such as those previously described for the Butterworth and Chebyshev responses, the rolloff rate would be -18 dB per octave. As we have noted earlier, this large attenuation for high frequencies is essentially wasted simply to achieve a rapid increase in attenuation between ω_p and ω_s. This characteristic of Cauer filters is illustrated by the attenuation plots with a logarithmic frequency scale in Fig. 13.20.

Example 13.3   For this example suppose that all of the specifications of the last example are the same, except that an $\alpha_{\min}$ of 55 dB is required. We proceed by consulting the Christian and Eisenmann charts once more, starting with the same θ and $|\rho|$, first trying $n = 4$ (that number being indicated by the number of poles given for $T(s)$). The search ends with

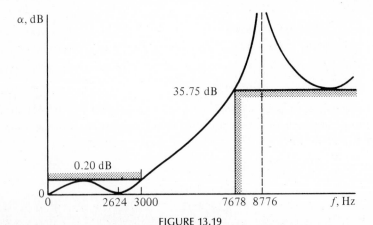

FIGURE 13.19

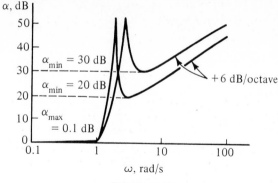

FIGURE 13.20

the $n = 4$ entry which gives the following specifications for $T(s)$, rounded off this time (see Table 13.4):

$$\text{poles:} \quad -0.1576 \pm j1.050 \quad \text{and} \quad -0.5117 \pm j0.4633$$

$$\text{zeros:} \quad \pm j2.793 \quad \text{and} \quad \pm j6.450 \tag{13.73}$$

The chart gives the value $C = 602.26991$ which is the reciprocal of the value needed. The chart also indicates that there are attenuation zeros at the pass band frequencies of 0.396244 and 0.929342 (before rounding). Hence we have determined the required transfer function to be

$$T(s) = \frac{1}{602.3} \frac{s^2 + 7.8008}{s^2 + 1.0234s + 0.4765} \frac{s^2 + 41.6025}{s^2 + 0.3152s + 1.1273} \tag{13.74}$$

where the poles and zeros have been grouped as shown in Fig. 13.21. A plot of $|T(j\omega)|$ is shown in Fig. 13.22, and a circuit realization using the Bainter circuit of Chapter 12 is shown in Fig. 13.23. For this particular example the equations for n derived in Chapters 6 and 8 may be used to show that a Butterworth response with $n = 9$ or a Chebyshev response with $n = 6$ would be required to meet the specifications that have beeen met with an $n = 4$ Cauer response.

All tables used for the design of Cauer filters are compiled in terms of the four parameters

$$\alpha_{\max}, \quad \alpha_{\min}, \quad \omega_s/\omega_p, \quad n \tag{13.75}$$

although these are not independent. There are many ways in which such tables can be arranged. An arrangement that is completely different from the Christian–Eisenmann tables is shown in Table 13.5. Some of the features of this arrangement are: (1) It is tabulated for only one value of $\alpha_{\max}$. (2) It is tabulated for only three different values of ω_s/ω_p. (3) It is tabulated for only three different values of n for which $\alpha_{\min}$ is then specified. For these choices, factors in the numerator and denominator polynomials of $T(s)$ are given, along with the values of C. This value of C is that for which

$$|T(j\omega)|_{\max} = 1 \tag{13.76}$$

This occurs at $\omega = 0$ and other frequencies for n odd. For n even, the value of $|T(j0)|$ is given by Eq. (13.62) with ε determined by Eq. (13.53). In other words, C

TABLE 13.4

ATTEN.ZEROS 0.393946		0.928433		-		-	C04 A θ=21°
ATTEN.POLES 3.0055301103		7.0832867932		-		-	Ω_s=2.79043
p[%]	A_{min}[db]	C	i	$-a_i$	$\pm b_i$	$-a_{i+1}$	$\pm b_{i+1}$
5	44.59	169.60913	1	0.3489946588	1.3292297324	1.0105569740	0.6287884817
10	50.64	340.50075	1	0.2809015449	1.1887919674	0.7788363122	0.5384686667
15	54.22	514.60643	1	0.2409491236	1.1211296653	0.6556220559	0.4983638864
25	58.84	874.76227	1	0.1906792960	1.0502451812	0.5094329854	0.4583916986
50	65.83	1956.0279	1	0.1203692610	0.9776078228	0.3160684197	0.4193655265

ATTEN.ZEROS 0.395077		0.928877		-		-	C04 A θ=22°
ATTEN.POLES 2.8738668365		6.7568381969		-		-	Ω_s=2.66947
p[%]	A_{min}[db]	C	i	$-a_i$	$\pm b_i$	$-a_{i+1}$	$\pm b_{i+1}$
5	42.93	140.16988	1	0.3437464121	1.3274986803	1.0136162089	0.6363309197
10	48.99	281.39964	1	0.2777570950	1.1881222160	0.7868632500	0.5429809525
15	52.56	424.78973	1	0.2386501509	1.1208443358	0.6571938391	0.5017539853
25	57.18	722.92876	1	0.1891653697	1.0502863330	0.5105467413	0.4608032745
50	64.17	1616.5179	1	0.1195952888	0.9779113898	0.3166930721	0.4209577269

ATTEN.ZEROS 0.396266		0.929342		-		-	C04 A θ=23°
ATTEN.POLES 2.7538668301		6.4585483632		-		-	Ω_s=2.55930
p[%]	A_{min}[db]	C	i	$-a_i$	$\pm b_i$	$-a_{i+1}$	$\pm b_{i+1}$
5	41.35	116.77513	1	0.3382960879	1.3256582733	1.0167865129	0.6443387771
10	47.40	234.43324	1	0.2744801722	1.1874075466	0.7829772329	0.5477569826
15	50.98	353.89112	1	0.2362506919	1.1205372012	0.6588368120	0.5053369955
25	55.60	602.26991	1	0.1875829230	1.0503250899	0.5117132703	0.4633480507
50	62.59	1346.7165	1	0.1187851388	0.9782279100	0.3173485950	0.4226347239

ATTEN.ZEROS 0.397515		0.929829		-		-	C04 A θ=24°
ATTEN.POLES 2.6441347513		6.1849029716		-		-	Ω_s=2.45859
p[%]	A_{min}[db]	C	i	$-a_i$	$\pm b_i$	$-a_{i+1}$	$\pm b_{i+1}$
5	39.83	98,001287	1	0.3326493940	1.3237053902	1.0200600266	0.6528293712
10	45.88	196.74361	1	0.2710725253	1.1866465282	0.7851754302	0.5528049825
15	49.46	296.99634	1	0.2337514336	1.1202075485	0.6605492467	0.5091184245
25	54.07	505.44348	1	0.1859319656	1.0503613909	0.5129314316	0.4660295479
50	61.06	1130.2060	1	0.1179385190	0.9785579244	0.3180341043	0.4243987964

* Reprinted by permission from Erich Christian and Egon Eisenmann, *Filter Design Tables and Graphs*, Transmission Networks International, Inc., Knightdale, N.C., 1977.

can be determined from other information that is given, but it is compiled for the user's convenience.

Example 13.4 The specifications for this example will be matched exactly to the values

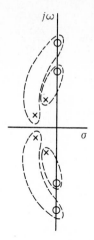

FIGURE 13.21

given in Table 13.5. We require a lowpass filter for which

$$\alpha_{\min} = 66.1 \text{ dB}, \qquad \alpha_{\max} = 0.5 \text{ dB}, \qquad \omega_p = 2000 \text{ rad/s} \qquad \omega_s = 4000 \text{ rad/s} \quad (13.77)$$

The values in Table 13.5 indicate the need for a filter with $n = 5$. The poles and zeros of $T(s)$ may be arranged in the following form for design, with the constants adjusted so that Eq. (13.76) is satisfied for each factor:

$$T(s) = \frac{0.392612}{s + 0.392612} \times \frac{0.12028(s^2 + 4.36495)}{s^2 + 0.58054s + 0.52500} \times \frac{0.097847(s^2 + 10.56773)}{s^2 + 0.19255s + 1.03402} \quad (13.78)$$

We will identify these factors by the transfer functions with the subscript indicating the module numbers

$$T(s) = T_1(s)T_2(s)T_3(s) \quad (13.79)$$

Each of the three modules may now be designed. It should be recognized that the values given in Eq. (13.78) are on the basis that $\omega_p = 1$, so that frequency scaling using $k_f = 2000$ will be required. Magnitude scaling may be done separately for each module by selecting a value of k_m that gives reasonable element sizes.

Module 1

This module will be realized by the circuit shown in Fig. 13.21 and adjusted so that $R =$

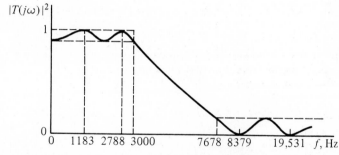

FIGURE 13.22

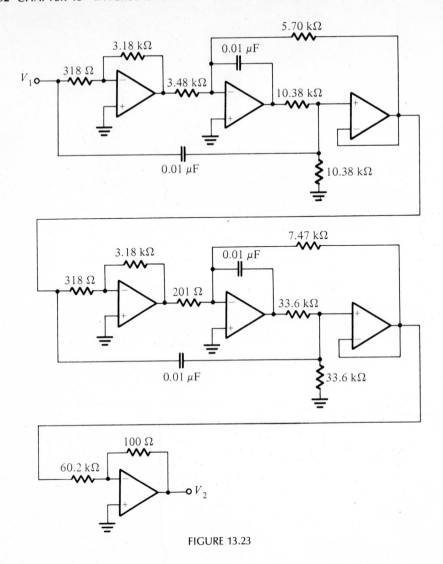

FIGURE 13.23

$R_1 = R_2$. Then the design equation becomes

$$\frac{1}{RC_2} = 0.392612 \qquad (13.80)$$

which is satisfied if $R = 1$ and $C_2 = 2.547$. If $k_f = 2000$ and $k_m = 10^5$, then the element values are $R = R_1 = R_2 = 10^5 \, \Omega$ and $C_2 = 12.74$ nF.

Module 2

For $T_2(s)$ we see that $\omega_0 = (0.52500)^{1/2} = 0.72467$, $\omega_z = (4.36495)^{1/2} = 2.089$, and $Q = 0.72467/0.58054 = 1.284$. The ratio $\omega_z/\omega_0 = 2.883$ indicates the need for a lowpass notch

TABLE 13.5 Cauer parameters for $\alpha_{max} = 0.5$ dB

n	α_{min}	Constant C	Numerator of $T(s)$	Denominator of $T(s)$
				(A) $\omega_s/\omega_p = 1.5$
2	8.3	0.38540	$s^2 + 3.92705$	$s^2 + 1.03153s + 1.60319$
3	21.9	0.31410	$s^2 + 2.80601$	$(s^2 + 0.45286s + 1.14917)(s + 0.766952)$
4	36.3	0.015397	$(s^2 + 2.53555)(s^2 + 12.09931)$	$(s^2 + 0.25496s + 1.06044)(s^2 + 0.92001s + 0.47183)$
5	50.6	0.019197	$(s^2 + 2.42551)(s^2 + 5.43764)$	$(s^2 + 0.16346s + 1.03189)(s^2 + 0.57023s + 0.57601)(s + 0.42597)$
				(B) $\omega_s/\omega_p = 2.0$
2	13.9	0.20133	$s^2 + 7.4641$	$s^2 + 1.24504s + 1.59179$
3	31.2	0.15424	$s^2 + 5.15321$	$(s^2 + 0.53787s + 1.14849)(s + 0.69212)$
4	48.6	0.0036987	$(s^2 + 4.59326)(s^2 + 24.22720)$	$(s^2 + 0.30116s + 1.06258)(s^2 + 0.88456s + 0.41032)$
5	66.1	0.0046205	$(s^2 + 4.36495)(s^2 + 10.56773)$	$(s^2 + 0.19255s + 1.03402)(s^2 + 0.58054s + 0.52500)(s + 0.392612)$
				(C) $\omega_s/\omega_p = 3.0$
2	21.5	0.083974	$s^2 + 17.48528$	$s^2 + 1.35715s + 1.55532$
3	42.8	0.063211	$s^2 + 11.82781$	$(s^2 + 0.58942s + 1.14559)(s + 0.65263)$
4	64.1	0.00062046	$(s^2 + 10.4554)(s^2 + 58.471)$	$(s^2 + 0.32979s + 1.06328)(s^2 + 0.86258s + 0.37787)$
5	85.5	0.00077547	$(s^2 + 9.8955)(s^2 + 25.0769)$	$(s^2 + 0.21066s + 1.0351)(s^2 + 0.58441s + 0.496388)(s + 0.37452)$

circuit, and for this purpose we select the Boctor circuit of Fig. 12.43. Design will be accomplished using the Cioffi algorithm which begins with Eq. (12.74). To avoid confusion, let us designate $\omega_z/\omega_0 = \Omega_z$ to use in Eq. (12.75). For the Boctor lowpass notch circuit the constant multiplier is

$$k = \frac{R_5}{R_3 + R_5} \tag{13.81}$$

In the Cioffi algorithm $R_5 = 1$ and R_3 is given by Eq. (12.77). Combining these equations, we have

$$2k = \frac{1}{k_1/Q^2 + k_1\Omega_z^2 + 1} \tag{13.82}$$

and this equation may now be used to determine the value of k_1 that is required. Substituting values into Eq. (13.82) gives

$$2 \times 0.12028 = \frac{1}{0.8013k_1 + 8.31k_1 + 1} \tag{13.83}$$

from this we find that $k_1 = 0.3465$, which satisfies Eq. (12.74) since $0.1203 < 0.3465 < 1$. We will scale using $k_f = 2000$ as required and make the choice that $k_m = 10^5$. The element values are seen to be the following:

Design values	Scaled values
$R_1 = \dfrac{2}{k_1\Omega_z^2 - 1} = 1.064$	$R_1 = 1.064 \times 10^5\ \Omega$
$R_2 = \dfrac{1}{1 - k_1} = 1.530$	$R_2 = 1.53 \times 10^5\ \Omega$
$R_3 = \dfrac{1}{2}\left(\dfrac{k_1}{Q^2} + k_1\Omega_z^2 - 1\right) = 1.051$	$R_3 = 1.051 \times 10^5\ \Omega$
$R_4 = \dfrac{1}{k_1} = 2.886$	$R_4 = 2.886 \times 10^5\ \Omega$
$R_5 = R_6 = 1$	$R_5 = R_6 = 10^5\ \Omega$
$C_1 = \dfrac{k_1}{2Q} = 0.13883$	$C_1 = 694.1\ \text{pF}$
$C_2 = 2Q = 2.496$	$C_2 = 12.48\ \text{nF}$

Module 3

For this module we see from the equation for $T_3(s)$ that $\omega_0 = 1.0169$, $\omega_z = 3.251$, $\Omega_z = 3.197$, and $Q = 5.281$. From Eq. (13.82) we determine that the value $k_1 = 0.4006$ is required which satisfies Eq. (12.74), since $0.978 < 0.4006 < 1$. The second lowpass notch Boctor circuit is designed as follows:

Design values	Scaled values
$R_1 = 0.6464$	$R_1 = 6.464 \times 10^4 \ \Omega$
$R_2 = 1.668$	$R_2 = 1.668 \times 10^5 \ \Omega$
$R_3 = 1.554$	$R_3 = 1.554 \times 10^5 \ \Omega$
$R_4 = 2.496$	$R_4 = 2.496 \times 10^5 \ \Omega$
$R_5 = R_6 = 1$	$R_5 = R_6 = 10^5 \ \Omega$
$C_1 = 0.03793$	$C_1 = 189.65 \ \text{pF}$
$C_2 = 10.562$	$C_2 = 52.81 \ \text{nF}$

The design is now complete and is shown in Fig. 13.24. To realize the specifications of this example using a Butterworth response would require that $n = 13$ and for a Chebyshev response that $n = 8$. With the Cauer parameters the specifications have been realized with $n = 5$.

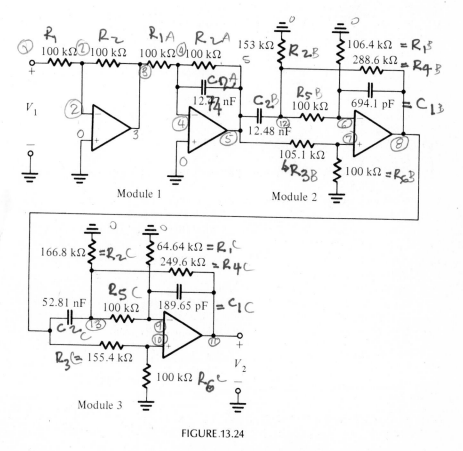

FIGURE.13.24

PROBLEMS

13.1 For this problem, we wish to study an inverse Chebyshev lowpass filter characteristic that meets the following specifications: $\alpha_{max} = 0.25$ dB, $\alpha_{min} = 18$ dB, $\omega_p = 1000$ rad/s, $\omega_s = 1650$ rad/s. Verify the frequencies identified in Fig. P13.1 for infinite attenuation and satisfying α_{min}.

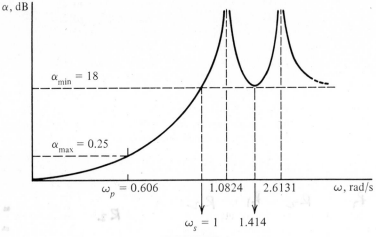

FIGURE P13.1

13.2 Design an inverse Chebyshev lowpass filter that satisfies the specification given in Problem 13.1.

13.3 Design an inverse Chebyshev lowpass filter that satisfies the specifications given in Problem 6.7.

13.4 Design an inverse Chebyshev lowpass filter that satisfies the specifications given in Problem 6.8.

13.5 Design an inverse Chebyshev lowpass filter that satisfies the specifications given in Problem 8.4.

13.6 Design an inverse Chebyshev lowpass filter that satisfies the specifications given in Problem 8.5.

13.7 The circuit shown in Figure P13.7 is the normalized ladder realization of a fifth-order Cauer (elliptic) lowpass filter.
(a) Sketch $R_n(\omega, L)$ for this filter as in Fig. 13.3.
(b) Sketch the attenuation provided by the filter as a function of frequency.
(c) Identify the frequencies of infinite attenuation on the sketch made for part (b).

13.8 Repeat Problem 13.7 for the seventh-order Cauer lowpass filter shown in Fig. P13.8.

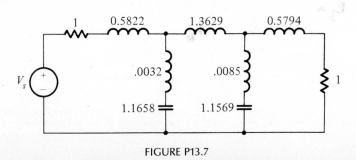

FIGURE P13.7

13.9 A table of Cauer (elliptic) filter parameters gives the following set of values: $\alpha_{max} =$ 0.177 dB, $\alpha_{min} = 40.2$ dB, $\omega_s/\omega_p = 1.37$, which satisfy the specifications given. The table indicates that the pole and zero positions are as follows:

poles: $-0.59676, -0.09225 \pm j\,1.0423, -0.36290 \pm j0.7737$
zeros: $\pm j\,1.4149, \pm j2.0732$

Outline a step-by-step procedure in going from this data to a filter realization.

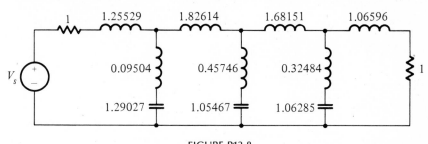

FIGURE P13.8

In desiging Cauer filters using tables, it is necessary to search through various possibilities and make compromises to approximately satisfy the specifications. The following problems are special in that the specifications are designed to match the specifications given in Table 13.5 exactly. Thus the problems are intended to stress finding a circuit that satisfies the specifications. For each problem, let $\omega_p = 1000$ rad/s and $\alpha_{max} = 0.5$ dB.

	Table 13.5 frequency ratio	n
13.10	$\omega_s/\omega_p = 1.5$	3
13.11	$\omega_s/\omega_p = 1.5$	4
13.12	$\omega_s/\omega_p = 1.5$	5
13.13	$\omega_s/\omega_p = 2.0$	3
13.14	$\omega_s/\omega_p = 2.0$	4
13.15	$\omega_s/\omega_p = 2.0$	5
13.16	$\omega_s/\omega_p = 3.0$	3
13.17	$\omega_s/\omega_p = 3.0$	4
13.18	$\omega_s/\omega_p = 3.0$	5

13.19 You are required to design a bandstop filter satisfying the specifications shown in Fig. P13.19. In this case, the prototype is specified as being a Cauer filter. Use magnitude scaling to give appropriate element values.

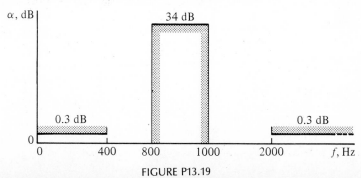

FIGURE P13.19

CHAPTER 14

Prototype and Frequency- Transformed Ladders

The most ancient of wave filters, built late in the last century, were made with inductors and capacitors arranged in ladder form, as shown in Fig. 14.1. Even with roots in antiquity, the lossless ladder remains useful to this day. It is still used in high-frequency applications and as a model for simulation. In this chapter we introduce some of the key concepts in the design of ladder filters. We also show the usefulness of tables, once we have some understanding of how such tables are generated.

14.1 SOME PROPERTIES OF LOSSLESS LADDERS

There are various ways to describe the circuit of Fig. 14.1. It is made up of inductors and capacitors, and so can be described as lossless or as reactance. It is simple in the sense that each branch contains only one element. It is arranged such that all series elements are inductors and all shunt elements are capacitors. We will show that it is a lowpass filter and so may be regarded as the prototype circuit to which the frequency transformations of Chapter 11 may be applied to obtain nonsimple ladder structures.

To begin, we see that since at zero frequency the capacitors are open circuits and the inductors are short circuits, the zero-frequency equivalent of the circuit of Fig. 14.1 is two wires, such that

$$\frac{V_2}{V_1}(0) = 1 \tag{14.1}$$

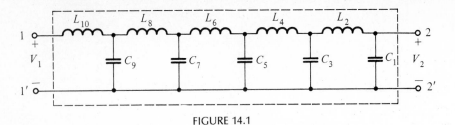

FIGURE 14.1

The general relationship is found by starting at side 2 and working toward side 1. The current in C_1 is

$$I_{C_1} = Y_{C_1}V_2 = C_1sV_2 \qquad (14.2)$$

The current in inductor L_2 is the same as I_{C_1} so that the voltage across L_2 is

$$V_{L_2} = Z_{L_2}I_{L_2} = L_2s(C_1sV_2) \qquad (14.3)$$

By Kirchhoff's voltage law the voltage across capacitor C_3 is

$$V_{C_3} = V_{L_2} + V_2 = (C_1L_2s^2 + 1)V_2 \qquad (14.4)$$

Then the current in C_3 is

$$I_{C_3} = Y_{C_3}V_{C_3} = C_3s(C_1L_2s^2 + 1)V_2 \qquad (14.5)$$

If this process of analysis is continued until side 1 is reached, it is found that

$$V_1(s) = Q(s)\,V_2(s) \qquad (14.6)$$

or

$$\frac{V_2(s)}{V_1(s)} = \frac{1}{Q(s)} \qquad (14.7)$$

where $Q(s)$ is an even polynomial of degree n, n being the total number of inductors plus capacitors. Such a function is known as an *all-pole function*, since all zeros are at infinity. At high frequencies the last equation will have the asymptotic form

$$\frac{V_2(s)}{V_1(s)} \approx \frac{K}{s^n} \qquad (14.8)$$

meaning that the transfer function will roll off at the rate of $-6n$ dB per octave. Since the zero-frequency value is 1 and the roll off rate is negative, the transfer function represents a lowpass filter.

Two specific circuits of the general form of the ladder of Fig. 14.1 are shown in Fig. 14.2. For the circuit in Fig. 14.2a the transfer function is

$$T(s) = \frac{V_2}{V_1} = \frac{1}{s^2 + 1} \qquad (14.9)$$

Similarly, for Fig. 14.2b it is

$$T(s) = \frac{V_2}{V_1} = \frac{1}{s^4 + 3s^2 + 1} \qquad (14.10)$$

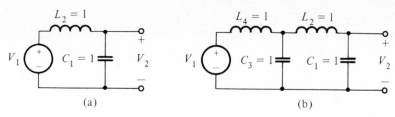

FIGURE 14.2

Both are all-pole functions, and for both the degree of the denominator polynomial is equal to the number of elements in the circuit.

The ladder circuit of Fig. 14.1 is also known as a lossless *coupling circuit*, that usage implying a load resistor at the output side. If there is a resistor at the input side, representing the resistance of the source, for example, then the structure as shown in Fig. 14.3 is known as a *doubly terminated* structure. The input resistor is designated as R_1 and the load resistor as R_2. To distinguish the voltage at the input to the lossless circuit from that of the source, one is designated as V_1 and the other as V_s.

A specific example of a lossless coupling circuit with double terminations is shown in Fig. 14.4. It is made from the circuit of Fig. 14.2a with $R_1 = R_2 = 1\,\Omega$. For that circuit the current in the inductor is simply related to the output voltage:

$$I_L = (s + 1)V_2 \tag{14.11}$$

The source voltage is, by Kirchhoff's voltage law,

$$V_s = (s + 1)I_L + V_2 = (s + 1)^2 V_2 + V_2 \tag{14.12}$$

Then

$$\frac{V_2}{V_s} = \mathbf{T} = \frac{1}{s^2 + 2s + 2} \tag{14.13}$$

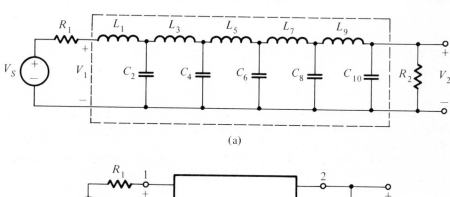

(a)

(b)

FIGURE 14.3

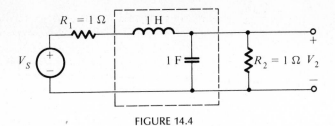

FIGURE 14.4

where $\mathbf{T}$ is used to distinguish the new transfer function from $T = V_2/V_1$. Comparing this equation with Eq. (14.8), we see that the addition of two resistors to the circuit has not increased the degree of the denominator polynomial. However, the polynomial is no longer even, but has both even and odd parts. Further, $\mathbf{T}(0) = \frac{1}{2}$, indicating the voltage-divider action of the circuit at zero frequency.

The transfer function for a general doubly terminated circuit such as that shown in Fig. 14.3a has the form

$$\mathbf{T} = \frac{V_2}{V_s} = \frac{K}{b_n s^n + b_{n-1} s^{n-1} + \cdots + b_1 s + b_0} = \frac{K}{Q(s)} \qquad (14.14)$$

For the doubly terminated circuit of Fig. 14.3a it is necessary that

$$\frac{V_2}{V_s}(0) = \frac{R_2}{R_1 + R_2} \qquad (14.15)$$

For the special case where $R_1 = R_2$, the ratio of Eq. (14.15) has the value of $\frac{1}{2}$, as was the case for Eq. (14.13). For the last two equations to be compatible, it is necessary that

$$\frac{K}{b_0} = \frac{R_2}{R_1 + R_2} \qquad (14.16)$$

This amounts to a constraint on the value of K, and tells us that the gain K of the transfer function of Eq. (14.14) has a specific limit. It should be noted that the condition of Eq. (14.16) assumes that $\mathbf{T}(0) = \mathbf{T}_{max}$ as is the case for the Butterworth response. If this maximum does not occur at $\omega = 0$, then we require that

$$|\mathbf{T}(j\omega)| \leq \mathbf{T}_{max} \qquad \text{all} \quad \omega \qquad (14.17)$$

and that the K in Eq. (14.14) be adjusted such that this is the case.

Comparing the denominator of Eq. (14.14) with the $Q(s)$ in Eq. (14.6), we see that both the lossless circuit and the doubly terminated lossless circuit are represented by transfer functions having the same denominator degree, which is n. Further, for both circuits n is the total number of capacitors and inductors. Thus both transfer functions have the same asymptotic rolloff rate of $-6n$ dB/octave.

In arriving at Eq. (14.13) we analyzed the circuit given in Fig. 14.4 and obtained the transfer function $\mathbf{T}(s)$. The synthesis problem is the inverse: Given the values of R_1 and R_2 and a prescribed form for the transfer function $\mathbf{T}(s)$, we are required to find the lossless ladder circuit element values.

14.2 A SYNTHESIS STRATEGY

The doubly terminated circuit already considered is shown again in Fig. 14.5 with two additional features identified. The current from the source is designated I_1, and its reference direction is shown in the figure. In addition, the input impedance Z_{11} is identified as the impedance of the RLC circuit made up of the lossless ladder and the terminating resistor R_2. We assume that the circuit is operating in the sinusoidal steady state. The input impedance has both a real and an imaginary component

$$Z_{11} = R_{11} + jX_{11} \tag{14.18}$$

The current at the input is

$$I_1 = \frac{V_s}{R_1 + Z_{11}} \tag{14.19}$$

Now since the LC circuit is lossless, we may equate the average power into the circuit to that in the load. Thus,

$$R_{11}|I_1(j\omega)|^2 = \frac{|V_2(j\omega)|^2}{R_2} \tag{14.20}$$

Substituting Eq. (14.19) for I_1 into this equation, gives us

$$\frac{R_{11}|V_s(j\omega)|^2}{|R_1 + Z_{11}|^2} = \frac{|V_2(j\omega)|^2}{R_2} \tag{14.21}$$

From this equation we determine the magnitude squared of the desired transfer function

$$\left|\frac{V_2(j\omega)}{V_s(j\omega)}\right|^2 = |T(j\omega)|^2 = \frac{R_2 R_{11}}{|R_1 + Z_{11}|^2} \tag{14.22}$$

Our end objective is to determine Z_{11} as a function of the $|T(j\omega)|^2$ specification. To accomplish this, we seemingly go a roundabout way by introducing an auxiliary function:

$$|A(j\omega)|^2 = 1 - 4\frac{R_1}{R_2}|T(j\omega)|^2 \tag{14.23}$$

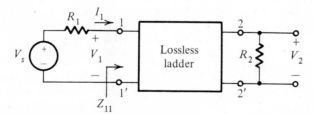

FIGURE 14.5

Into this equation we substitute Eq. (14.22) to give, after some algebraic manipulation,

$$|A(j\omega)|^2 = A(s)A(-s)|_{s=j\omega} = \frac{|R_1 - Z_{11}|^2}{|R_1 + Z_{11}|^2} \tag{14.24}$$

From this equation we see that $A(s)$ may be separated from $A(-s)$ to give

$$A(s) = \pm \frac{R_1 - Z_{11}}{R_1 + Z_{11}} \tag{14.25}$$

Clearly, this equation may be solved for Z_{11} in terms of $A(s)$ and R_1. The result is

$$Z_{11} = R_1 \frac{1 - A(s)}{1 + A(s)} \tag{14.26}$$

or

$$Z_{11} = R_1 \frac{1 + A(s)}{1 - A(s)} \tag{14.27}$$

Here we have reduced the problem to determining a lossless circuit terminated in a resistor R_2 from a specified Z_{11}. Sidney Darlington showed that this was always possible in his classic 1939 paper,* and so the circuit may be found.

To illustrate the application of the strategy, consider the problem of determining circuits having a Butterworth response for which there are equal terminations with $R_1 = R_2 = 1$, a normalized value that may later be magnitude scaled. The Butterworth response is

$$|T_n(j\omega)|^2 = \frac{1}{1 + \omega^{2n}} \tag{14.28}$$

However, this magnitude-squared function must be reduced to satisfy the condition of Eq. (14.16). Since the magnitude-squared function is

$$|T(j\omega)|^2 = T(s)T(-s)|_{s=j\omega} \tag{14.29}$$

then from Eq. (14.14) we have

$$|T(j\omega)|^2 = \frac{K^2}{|Q(j\omega)|^2} = \frac{K^2}{Q(s)Q(-s)}\bigg|_{s=j\omega} \tag{14.30}$$

These quantities are known from our study of the Butterworth functions in Chapter 6, and are shown in the following table:

Order	$|Q(j\omega)|^2$	$Q(s)$
1	$1 + \omega^2$	$s + 1$
2	$1 + \omega^4$	$s^2 + \sqrt{2}s + 1$
3	$1 + \omega^6$	$s^3 + 2s^2 + 2s + 1$

* S. Darlington, "Synthesis of Reactance 4-Poles which Produce Prescribed Insertion Loss Characteristics, *J. Math. Phys.*, pp. 257–353, Sept. 1939.

So we see that b_0 in Eq. (14.14) is 1 for all n, and with equal terminations it is then necessary that $K = \frac{1}{2}$. Then $|T(j\omega)|^2$ in Eq. (14.28) must be multiplied by $K^2 = \frac{1}{4}$. With that accomplished, we may determine the auxiliary function of Eq. (14.23):

$$|A(j\omega)|^2 = 1 - \frac{1}{1 + \omega^{2n}} = \frac{\omega^{2n}}{1 + \omega^{2n}} \qquad (14.31)$$

Since

$$\omega^{2n}\big|_{\omega^2 = -s^2} = s^n(-s)^n \qquad (14.32)$$

we see that

$$A(s) = \frac{s^n}{Q(s)} \qquad (14.33)$$

Substituting this value of $A(s)$ into Eqs. (14.26) and (14.27), we obtain

$$Z_{11} = R_1 \left(\frac{1-A}{1+A}\right)^{\pm 1} = R_1 \left(\frac{Q - s^n}{Q + s^n}\right)^{\pm 1} \qquad (14.34)$$

It is conventional to normalize the impedance level of the circuit by letting $R_1 = 1$. If we do that in the last equation, and then let n have several values, we will see the pattern that is followed in obtaining the circuits.

$n = 1$

For $n = 1$, $Q = s + 1$, and $A(s) = s/(s + 1)$. Hence the impedance Z_{11} is

$$Z_{11} = \frac{1 - s/(s + 1)}{1 + s/(s + 1)} = \frac{1}{2s + 1} \qquad (14.35)$$

or

$$Z_{11} = 2s + 1 \qquad (14.36)$$

The circuit realizations of these two Z_{11} functions are shown in Fig. 14.6 with the section enclosed by the dashed lines being the lossless coupling circuit. In this case the element values are evident from the direct inspection of the expressions for Z_{11}.

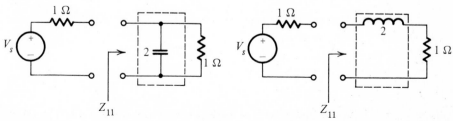

FIGURE 14.6

$n = 2$

As already tabulated, $Q = s^2 + \sqrt{2}s + 1$ for the case $n = 2$, and hence

$$A(s) = \frac{s^2}{s^2 + \sqrt{2}s + 1} \tag{14.37}$$

Then we find that Z_{11} is

$$Z_{11} = \frac{\sqrt{2}s + 1}{2s^2 + \sqrt{2}s + 1} \tag{14.38}$$

or the reciprocal of this quantity. This time we cannot determine the circuit represented by this Z_{11} directly by inspection. Instead, we expand this function, or its reciprocal, as a continued fraction. This is always done by dividing the polynomial of higher degree by the polynomial of lower degree. The continued fraction is carried out by synthetic division in these steps:

$$
\begin{array}{r}
\sqrt{2}s + 1\ \overline{)2s^2 + \sqrt{2}s + 1}\ (\ \sqrt{2}s \leftarrow C_1 s \\
\underline{2s^2 + \sqrt{2}s} \\
1\ \overline{)\sqrt{2}s + 1}\ (\ \sqrt{2}s \leftarrow L_2 s \\
\underline{\sqrt{2}s} \\
1\ \overline{)1}\ (\ 1 \leftarrow R_2 \\
\underline{1} \\
0
\end{array} \tag{14.39}
$$

Thus the continued fraction is

$$Z_{11} = \cfrac{1}{\sqrt{2}s + \cfrac{1}{\sqrt{2}s + \cfrac{1}{1}}} \tag{14.40}$$

or $1/Z_{11}$ is the same continued fraction. The two circuits that have been found by this process are shown in Fig. 14.7, Fig 14.7a being the direct representation of Eq. (14.40) and Fig. 14.7b being the representation of the reciprocal of this function.

$n = 3$

For $n = 3$ we have found that $Q(s) = s^3 + 2s^2 + 2s + 1$, so $A(s)$ is found directly

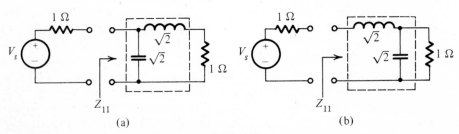

(a)   (b)

FIGURE 14.7

from Eq. (14.33) as

$$A(s) = \frac{s^3}{s^3 + 2s^2 + 2s + 1} \tag{14.41}$$

So we determine Z_{11} and from that its reciprocal:

$$Z_{11} = \frac{2s^2 + 2s + 1}{2s^3 + 2s^2 + 2s + 1} \tag{14.42}$$

We next expand Z_{11} in a continued fraction by dividing the denominator polynomial by the numerator polynomial in the pattern of dividing one step, inverting, dividing one step, and so on. Thus

$$
\begin{array}{r}
2s^2 + 2s + 1 \overline{)\, 2s^3 + 2s^2 + 2s + 1} \quad (\,s \leftarrow L_1 s \\
\underline{2s^3 + 2s^2 + s} \\
s + 1 \,\overline{)\, 2s^2 + 2s + 1} \quad (\, 2s \leftarrow C_2 s \\
\underline{2s^2 + 2s} \\
1\,)\ \ s + 1 \,(\, s \leftarrow L_3 s \\
\underline{s} \\
1\,)\, 1 \,(\, 1 \leftarrow R_2 \\
\underline{1} \\
0
\end{array}
\tag{14.43}
$$

or

$$Z_{11} = \cfrac{1}{s + \cfrac{1}{2s + \cfrac{1}{s + \cfrac{1}{1}}}} \tag{14.44}$$

From this continued fraction expansion, and from its reciprocal, we recognize the two circuits shown in Fig. 14.8, each consisting of a lossless coupling circuit terminated in a 1-Ω resistor. The steps are summarized in Fig. 14.9.

The circuits we have found and their element values are summarized in Table 14.1 for the cases $n = 2$ and $n = 3$ that we have just studied. If this process is continued until $n = 10$, then all of the values shown in the table are obtained. These element values are for the Butterworth case, and once found they may be stored in the table and regarded as completed forever. If necessary, we could derive any one of the circuits with little trouble, but it will never be necessary as long as you have the table.

Comparing the two coupling circuits of Fig. 14.8, we first observe that the impedances of the two circuits have reciprocal relationships and are said to be *duals* of each other. Observe that one has two inductors and one capacitor, while the other has two capacitors and one inductor. Since the two realizations are completely equivalent, there may be some basis for choice depending on whether inductors or capacitors are most easily obtained. One might be called a minimum-capacitance realization, and the other the minimum-inductance realization. These properties hold in general as illustrated by Table 14.2. The first and last element structures are different, depending upon whether n is even or odd, and of

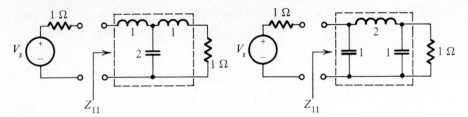

FIGURE 14.8

Procedure to obtain doubly terminated lowpass prototype	Example: Butterworth $n = 3$ $R_1 = R_2 = 1\ \Omega$

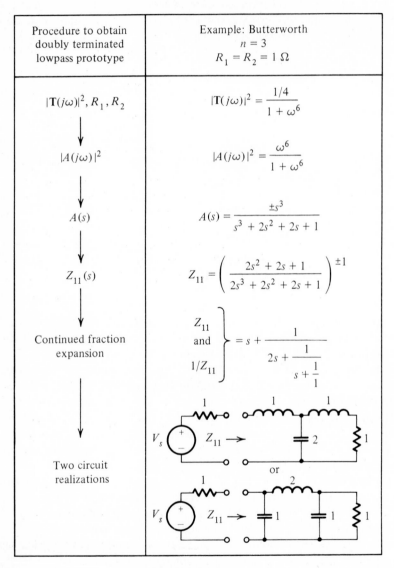

FIGURE 14.9

the two realizations one will always be minimum inductance and the other minimum capacitance.

There are many other interesting properties of ladder circuits which are treated by advanced treatises on the subject.* The general problem of finding a ladder realization is not as simple as might be suggested by the special case of Butterworth filters. In some cases the lossless terminated circuit is found only by including ideal transformers in the realization. Often it is possible to realize different values of R_1 and R_2, and sometimes it is not possible to find a circuit with equal terminations without the ideal transformer.

One additional example will illustrate the case of unequal terminations. Consider the realization of a filter with a Butterworth response for which $n = 3$, $R_1 = 1\,\Omega$, and $R_2 = 2\,\Omega$. From Eq. (14.16) we see that

$$\frac{K}{b_0} = \frac{R_2}{R_1 + R_2} = \frac{2}{3} \quad \text{or} \quad K = \frac{2}{3} \, b_0 = \frac{2}{3} \tag{14.45}$$

Then

$$|A(j\omega)|^2 = 1 - 2 \, \frac{(2/3)^2}{1 + \omega^6} = 1 - \frac{8/9}{1 + \omega^6} \tag{14.46}$$

or

$$A|(j\omega)^2| = \frac{\omega^6 + 1/9}{\omega^6 + 1} \tag{14.47}$$

$$A(s)A(-s) = \frac{(s^3 + 1/3)(-s^3 + 1/3)}{(s^3 + 2s^2 + 2s + 1)(-s^3 + 2s^2 - 2s + 1)} \tag{14.48}$$

Then

$$A(s) = \frac{s^3 + 1/3}{s^3 + 2s^2 + 2s + 1} \tag{14.49}$$

From this we find

$$Z_{11} = \frac{2s^2 + 2s + 2/3}{2s^3 + 2s^2 + 2s + 4/3} \tag{14.50}$$

The continued fraction expansion of this function gives

$$Z_{11} = \cfrac{1}{s + \cfrac{1}{1.5s + \cfrac{1}{2s + \cfrac{1}{0.5}}}} \tag{14.51}$$

The circuit realizations of this Z_{11} and for its reciprocal are shown in Fig. 14.10.

* D. S. Humphreys, *The Analysis, Design and Synthesis of Electrical Filters,* Prentice-Hall, Englewood Cliffs, N.J., 1970.

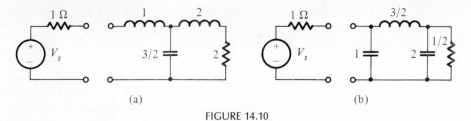

(a)   (b)

FIGURE 14.10

14.3 TABLES FOR OTHER RESPONSES

The last section introduced the philosophy of the generation of tables by a study of the Butterworth response case. In Table 14.1 we have all of the Butterworth circuits from $n = 2$ to $n = 10$.* This table was derived for the half-power frequency of $\omega_0 = 1$ rad/s, normalized in magnitude such that $|T(j0)| = 1$ or $\alpha(0) = 0$ dB. We recognize that the derivations which were illustrated in the last section need not be repeated—we have done it once and for all. If we should need a Butterworth filter of order n, we simply consult Table 14.1 which gives the required circuit. We may then frequency scale and magnitude scale as required.

Since the form of the transfer function $|T(j\omega)|^2$ is known for many other kinds of responses, the general procedure given in the last section can be applied to any responses and thus determine ladder circuits for any specified form of magnitude response. For example, we know from our studies in Chapter 8 that the Chebyshev response has the form

$$|T(j\omega)|^2 = \frac{1}{1 + \varepsilon^2 C_n^{\,2}(\omega)} \tag{14.52}$$

This equation reminds us that the Chebyshev response must be tabulated for different values of n, but also some factor that is equivalent to ε, usually α_{max}. This makes the tabulation of information for the Chebyshev response more complex than it was for the Butterworth response.

Table 14.3 is typical of Chebyshev response tables. In Table 14.3 a passband ripple of $\alpha_{max} = 0.1$ dB is used with n ranging from 2 to 8.

If you object to letting someone else do your work, you can follow the procedure outlined for the Butterworth response and derive your own circuits. Another alternative is to make use of explicit formulas that have been derived for the Butterworth and Chebyshev cases.† The derivations leading to these formulas are very clever and are recommended for the serious reader. The formulas may be used to derive your own tables, or to solve problems for which no convenient tables can be found.

In Chapter 13 we learned of the Cauer (or elliptic) form of response, and found that it was a very efficient magnitude response. Cauer circuits realize a given set of specifications with a lower value of n, in general, and this is accom-

---

* For larger n consult Weinberg, listed in Appendix B.

† A. S. Sedra and P. O. Brackett, *Filter Theory and Design: Active and Passive,* Matrix Publishers, Portland, Ore., pp. 208–212.

TABLE 14.1 Table of element values for doubly terminated Butterworth filters for $n = 2$ to $n = 10$ normalized to half-power frequency of 1 rad/s

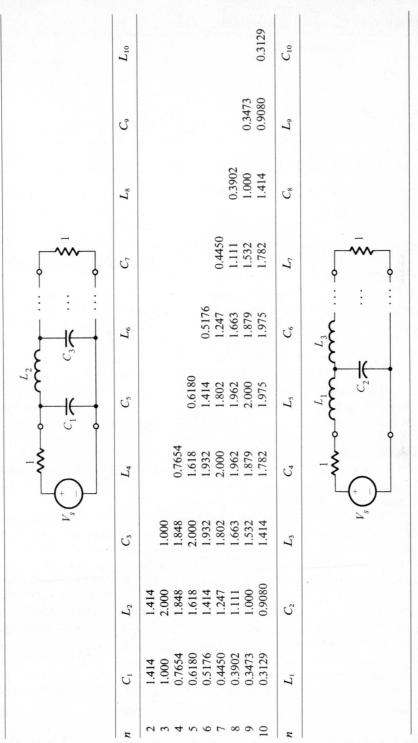

n	C_1	L_2	C_3	L_4	C_5	L_6	C_7	L_8	C_9	L_{10}
2	1.414	1.414								
3	1.000	2.000	1.000							
4	0.7654	1.848	1.848	0.7654						
5	0.6180	1.618	2.000	1.618	0.6180					
6	0.5176	1.414	1.932	1.932	1.414	0.5176				
7	0.4450	1.247	1.802	2.000	1.802	1.247	0.4450			
8	0.3902	1.111	1.663	1.962	1.962	1.663	1.111	0.3902		
9	0.3473	1.000	1.532	1.879	2.000	1.879	1.532	1.000	0.3473	
10	0.3129	0.9080	1.414	1.782	1.975	1.975	1.782	1.532	0.9080	0.3129
n	L_1	C_2	L_3	C_4	L_5	C_6	L_7	C_8	L_9	C_{10}

TABLE 14.2   Doubly terminated lossless ladders

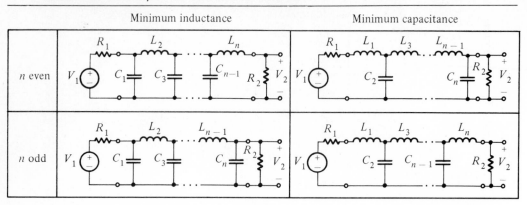

plished by placing zeros of $|T(j\omega)|$ or poles of attenuation in the stop band. Since the Cauer circuits are very efficient, it would be highly desirable to be able to use them as obtained from a table.

We have not studied the strategy by which Cauer circuits are found. The procedure through which such circuits are generated is known as the *Cauer ladder development.* Significant features of the Cauer circuits are illustrated in Fig. 14.11. The circuit is developed in such a way that parallel resonant LC circuits in a series position, as shown in Fig. 14.11a, or series resonant LC circuits in a shunt position, as shown in Fig. 14.11b, produce the required zero of transmission at prescribed frequencies, marked ω_{z1} and ω_{z2} in the figure. The generation of Cauer circuits through the use of computer programs to produce tables of circuits was first accomplished by Saal and Ulbrich in Germany in 1958. Their results were first published as a journal article and later in handbook form. A revised handbook was published by R. Saal and W. Entenmann in 1979. Table 14.4 is a specimen page from this work which illustrates the amount of information available to the designer. The notation employed differs from that we have used, but it is easily translated by designers needing the information.

The Butterworth, Chebyshev, and Cauer circuits that have been given thus far are doubly terminated lossless filters. Another *model* which is applicable to other filter design problems is the singly terminated lossless filter shown in Fig. 14.12. Now singly terminated filters appear in two basic forms, depending on whether the termination is on side 1-1' or on side 2-2'. We ordinarily think of a termination as being on side 2-2', as shown in Fig. 14.12b. However, it is also possible for the termination to be on side 1-1' as shown in Fig. 14.12a, especially if the driver is a current source in shunt with a 1-Ω resistor, obtained from the circuit shown in Fig. 14.12a as the Norton equivalent. We group these two circuits together because all information for the design of either circuit may be placed in a single table.

The theory of the design of singly terminated lossless filters may be found in other references.* Although we will not pursue this theory, it is important to rec-

* M. E. Van Valkenburg, *Introduction to Modern Network Synthesis,* Wiley, New York, 1960, chap. 14.

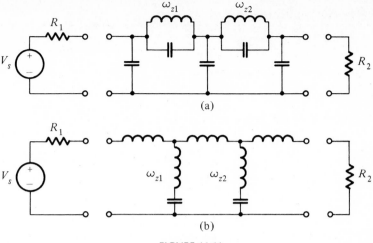

(a)

(b)

FIGURE 14.11

ognize some properties in order to be able to use the tables. The formulas for the transfer function $T(s)$ are shown in Fig. 14.12. We note that Z_{11} plays the same role in the formula in Fig. 14.12a as does Y_{22} for that in Fig. 14.12b. The meaning of these functions may be explained with the aid of Fig. 14.13. As shown in Fig. 14.13a, Z_{11} is the input impedance of the lossless circuit with side 2-2' open circuited. Similarly, Fig. 14.13b explains that Y_{22} is the input admittance as measured at side 2-2', but with side 1-1' short circuited. Since the circuits are derived from either Z_{11} or Y_{22}, it is seen that the order in which the elements appear will be opposite for the two circuits. This statement is illustrated by Fig. 14.14, which is a single table giving the element values for the two kinds of singly terminated circuits.

Now that we have seen the manner in which such tables are constructed, we can see how we can use the tables for singly terminated lossless circuits. Table 14.5 gives the element values for all such networks from $n = 2$ to $n = 10$. Similarly, Table 14.6 gives element values for the Chebyshev form of response for ripple widths of $\alpha_{max} = 0.1$ dB, 0.5 dB, and 1 dB, normalized such that the end of the ripple band is at $\omega = 1$ rad/s.

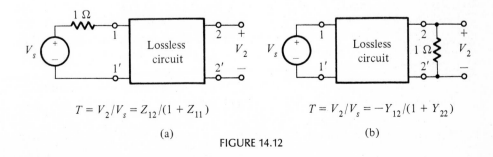

$$T = V_2/V_s = Z_{12}/(1 + Z_{11})$$

(a)

$$T = V_2/V_s = -Y_{12}/(1 + Y_{22})$$

(b)

FIGURE 14.12

TABLE 14.3 Chebyshev lowpass element values (1-rad/s bandwidth)

n	C_1	L_2	C_3	L_4	C_5	L_6	C_7	L_8	R_2
				(A) Ripple width = 0.1 dB					
2	0.84304	0.62201							0.73781
3	1.03156	1.14740	1.03156						1.00000
4	1.10879	1.30618	1.77035	0.81807					0.73781
5	1.14681	1.37121	1.97500	1.37121	1.14681				1.00000
6	1.16811	1.40397	2.05621	1.51709	1.90280	0.86184			0.73781
7	1.18118	1.42281	2.09667	1.57340	2.09667	1.42281	1.18118		1.00000
8	1.18975	1.43465	2.11990	1.60101	2.16995	1.58408	1.94447	0.87781	0.73781
				(B) Ripple width = 0.5 dB					
3	1.5963	1.0967	1.5963						1.0000
5	1.7058	1.2296	2.5408	1.2296	1.7058				1.0000
7	1.7373	1.2582	2.6383	1.3443	2.6383	1.2582	1.7373		1.0000
				(C) Ripple width = 1.0 dB					
3	2.0236	0.9941	2.0236						1.0000
5	2.1349	1.0911	3.0009	1.0911	2.1349				1.0000
7	2.1666	1.1115	3.0936	1.1735	3.0936	1.1115	2.1666		1.0000
n	L_1'	C_2'	L_3'	C_4'	L_5'	C_6'	L_7'		R_2

TABLE 14.4

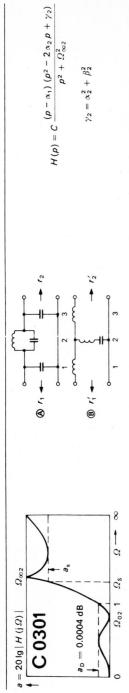

C 0301

$a = 20 \lg |H(j\Omega)|$

$a_D = 0.0004$ dB

$$H(p) = C \frac{(p - \alpha_1)(p^2 - 2\alpha_2 p + \gamma_2)}{p^2 + \Omega_{\infty 2}^2}$$

$$\gamma_2 = \alpha_2^2 + \beta_2^2$$

Ⓐ r_1 , r_2

Ⓑ r_1' , r_2'

Θ	Ω_s	a_s dB	Ⓐ ν	$r_1=1$ $c_{2\nu-1}$	$l_{2\nu}$	$r_2=1$ $c_{2\nu}$	$r_1=\infty$ $c_{2\nu-1}$	$l_{2\nu}$	$r_2=1$ $c_{2\nu}$	$\Omega_{\infty 2\nu}$	$\Omega_{0\nu}$	$-\alpha_\nu$	$\pm\beta_\nu$	C
P			1	0.215447	0.430894		0.323171	0.287763			0.000000000	4.6415114706	0.0000000000	0.010000500
			2	0.215447			0.107724				0.000000000	2.3207557353	4.0196666455	
T			1	0.352300	0.644596		0.498448	0.455596			0.000000000	2.8384938606	0.0000000000	0.040002000
			2	0.352300			0.176150				0.8660254038	1.4192469303	2.6062972889	
6	9.566772234	42.9	1	0.343555	0.627272	0.013082	0.497356	0.433296	0.018938	11.039187451	0.000000000	2.9107397976	0.0000000000	4.854768535
			2	0.343555			0.159835				0.8660192395	1.3523783769	2.6067114834	
7	8.205509048	38.8	1	0.340400	0.621022	0.017970	0.497103	0.425256	0.026243	9.466069419	0.000000000	2.9377169608	0.0000000000	3.564408978
			2	0.340400			0.153808				0.8668338130	1.3285827705	2.6061981810	
8	7.185296535	35.3	1	0.336763	0.613815	0.023724	0.496910	0.415991	0.035006	8.286760417	0.000000000	2.9694482196	0.0000000000	2.726915806
			2	0.336763			0.146760				0.8670814857	1.3013667668	2.6051789380	
9	6.392453222	32.3	1	0.332645	0.605657	0.030398	0.496824	0.405513	0.045401	7.369992055	0.000000000	3.0062088188	0.0000000000	2.152732500
			2	0.332645			0.138650				0.8673622948	1.2708414599	2.6034801598	
10	5.758770483	29.5	1	0.328049	0.598552	0.038055	0.496902	0.393838	0.057642	6.637003450	0.000000000	3.0483216879	0.0000000000	1.742022288
			2	0.328049			0.129424				0.8676762829	1.2371381099	2.6009116270	
11	5.240843064	27.0	1	0.322981	0.586511	0.046772	0.497213	0.380987	0.072003	6.037674189	0.000000000	3.0961614710	0.0000000000	1.438142756
			2	0.322981			0.119023				0.8680234971	1.2004101042	2.5972692200	
12	4.809734345	24.7	1	0.317444	0.575543	0.056640	0.497839	0.366992	0.088827	5.538590797	0.000000000	3.1501588544	0.0000000000	1.207017019
			2	0.317444			0.107376				0.8684403896	1.1608350611	2.5923381613	
13	4.445411483	22.6	1	0.311448	0.563665	0.067766	0.498879	0.351894	0.108548	5.116620990	0.000000000	3.2108050350	0.0000000000	1.027146411
			2	0.311448			0.094402				0.8688178176	1.1186169973	2.5858986634	
14	4.133565494	20.7	1	0.305003	0.550898	0.080276	0.500447	0.336750	0.131716	4.755241832	0.000000000	3.2786561101	0.0000000000	0.884424144
			2	0.305003			0.080004				0.8692650428	1.0739884456	2.5777214924	
15	3.863703305	18.9	1	0.298122	0.537267	0.094316	0.502681	0.318633	0.159033	4.442336609	0.000000000	3.3543370059	0.0000000000	0.769282649
			2	0.298122			0.064074				0.8697457319	1.0272123385	2.5675913167	
16	3.627955279	17.2	1	0.290821	0.522806	0.110061	0.505741	0.300634	0.191398	4.168817895	0.000000000	3.4385445105	0.0000000000	0.675047314
			2	0.290821			0.046482				0.8702599561	0.9785534329	2.5552949416	
17	3.420303620	15.7	1	0.283122	0.507559	0.127711	0.509816	0.281868	0.229969	3.927736584	0.000000000	3.5320487745	0.0000000000	0.596946908
			2	0.283122			0.027086				0.8708077913	0.9284289602	2.5406374431	
18	3.236067978	14.2	1	0.275051	0.491576	0.147502	0.515123	0.262478	0.276246	3.713687911	0.000000000	3.6356925244	0.0000000000	0.531497535
			2	0.275051			0.005716				0.8713893184	0.8771081264	2.5234483540	

Θ	Ω_s	a_s dB	Ⓑ ν	$r_1'=1$ $l_{2\nu-1}$	$c_{2\nu}$	$l_2'=1$ $l_{2\nu}$	$r_1'=0$ $l_{2\nu-1}$	$c_{2\nu}$	$l_2'=1$ $l_{2\nu}$	$\Omega_{\infty 2\nu}$	$\Omega_{0\nu}$	$-\alpha_\nu$	$\pm\beta_\nu$	C

* Reprinted by permission from Rudolf Saal, *Handbook of Filter Design*, AEG-Telefunken Aktiengesellschaft Publishing Department, Frankfurt am Main, West Germany, 1979.

14.4 FREQUENCY TRANSFORMATIONS

The concept of the frequency transformation was studied in detail in Chapter 11. There we outlined the two uses of such transformations. The first is when transforming pole and zero locations; the second is for converting one LC circuit into another. In particular we are interested in converting each capacitor and each inductor in the prototype low-pass circuit into another circuit using the transformations

$$Z_L = L\,A(s) \quad \text{and} \quad Z_C = \frac{1}{C\,A(s)} \tag{14.53}$$

These operations were covered in Chapter 11 and so will be reviewed only briefly in this section.

The most often used frequency transformations were summarized in Table 11.1 which is repeated here as Table 14.7. To illustrate the use of the table, suppose that specifications require that we use an $n = 4$ Chebyshev response with 0.1 dB ripple width as a prototype. Such a circuit is found in Table 14.3 and is shown in Fig. 14.15a. If the prototype is to be transformed to bandpass filter, then the appropriate transformation is found from Table 14.11 and is shown in Fig. 14.15b.

To obtain numerical values suppose that it is specified that the normalized values describing the bandpass filter are $\omega_0 = 1$ and bw = 0.20. The use of the formulas listed in Table 14.11 gives the element values shown in the filter of Fig. 14.15b. Suitable frequency and magnitude scaling remains to be done before the design of the bandpass filter is complete.

The step-by-step procedure for the design of a bandpass filter, for example, is given in Fig. 14.16. We begin with the filter specifications for the bandpass filter; these are first reduced to the equivalent specifications of the lowpass prototype filter, from which we obtain the required value of n as well as α_{max} for a Chebyshev response. From this information we consult a table and look up a lowpass prototype that meets our specifications. To the prototype the appropriate

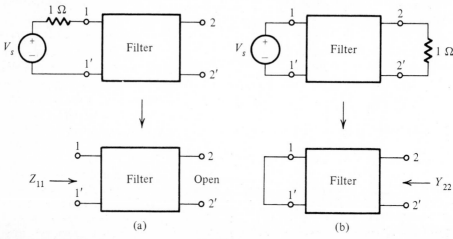

(a)   (b)

FIGURE 14.13

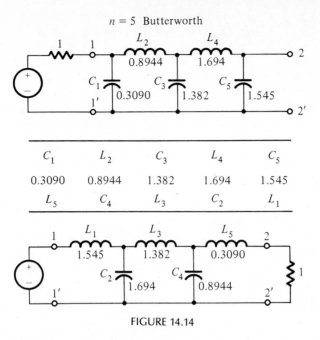

FIGURE 14.14

frequency transformation is applied, giving the final design in an element normalized form. To this normalized circuit we apply frequency scaling, and magnitude scaling if required, and the design is complete.

14.5 PASSIVE FILTER DESIGN

Throughout our studies we have stressed filter design without inductors, especially using op amps in combination with R and C. There remain applications for which the op amp is not suited. In one class of such applications, the required power is beyond the capability of the op amp. An extreme example is the inductor used for system stabilization at the Mojave power station in California; this massive structure covers more than an acre of desert land. The other applications for which the op amp is often not suited are those operating at high frequency. At frequencies approaching the megahertz range it often becomes necessary to use inductors, either as coils, as coaxial line or waveguides, or as realized on a chip. Figure 14.17 depicts a radio-frequency inductor designed to operate at 400 MHz, with Q's in the range of 40–50, and with inductance as great as 200 nH. At high frequencies there is a different technology than that stressed so far, but the basic ideas remain the same. We will illustrate this by examples.

Example 14.1 For our first example we will assume that a local amateur radio station operating at a frequency around 30 MHz is interfering with our reception on TV channel 2 (around 60 MHz). We need a filter to minimize interference. We decide that his transmissions should be attenuated by 20 dB, and that our own TV set input can stand attenuation of 2 dB. The model of the filter we require is shown in Fig. 14.18 with a termination of 50 Ω. The attenuation requirements are shown in Fig. 14.19a, and when transformed to the equivalent lowpass prototype, requirement is as shown in Fig. 14.19b. For simplicity we

TABLE 14.5 n-pole Butterworth networks

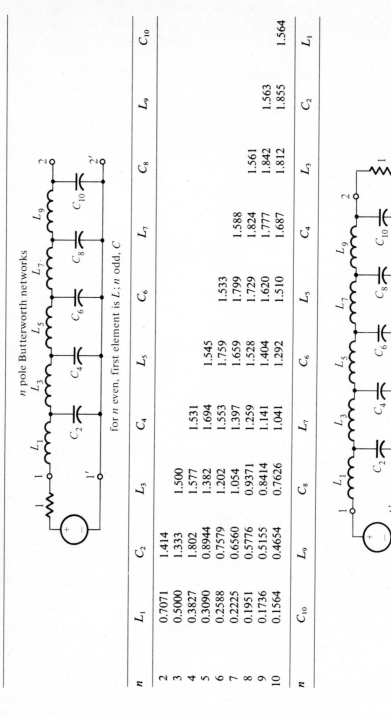

n pole Butterworth networks

for n even, first element is L; n odd, C

n	L_1	C_2	L_3	C_4	L_5	C_6	L_7	C_8	L_9	C_{10}
2	0.7071	1.414								
3	0.5000	1.333	1.500							
4	0.3827	1.802	1.577	1.531						
5	0.3090	0.8944	1.382	1.694	1.545					
6	0.2588	0.7579	1.202	1.553	1.759	1.533				
7	0.2225	0.6560	1.054	1.397	1.659	1.799	1.588			
8	0.1951	0.5776	0.9371	1.259	1.528	1.729	1.824	1.561		
9	0.1736	0.5155	0.8414	1.141	1.404	1.620	1.777	1.842	1.563	
10	0.1564	0.4654	0.7626	1.041	1.292	1.510	1.687	1.812	1.855	1.564

n	C_{10}	L_9	C_8	L_7	C_6	L_5	C_4	L_3	C_2	L_1

TABLE 14.6   Element values for singly-terminated lossless circuits in ladder form with a Chebyshev response; α_{max} = ripple width in dB, ω_0 = 1 rad/s (end of ripple band)

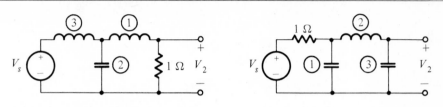

α_{max}	Element 1	Element 2	Element 3
0.1 dB	0.51577	1.08640	1.08947
0.5 dB	0.79814	1.30013	1.34650
1.0 dB	1.0118	1.33325	1.50882

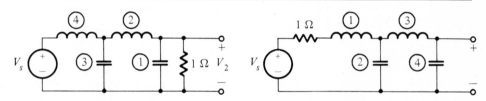

α_{max}	Element 1	Element 2	Element 3	Element 4
0.1 dB	0.5543944	1.1994155	1.4575641	1.2453347
0.5 dB	0.83515	1.39157	1.72789	1.31376
1.0 dB	1.04953	1.41257	1.90936	1.28168

TABLE 14.7

Prototype (lowpass) elements	Highpass elements	Bandpass elements	Bandstop elements
L_p	$1/\omega_0 L_p$	L_p/bw $\quad$ $\text{bw}/\omega_0^2 L_p$	$\text{bw}L_p/\omega_0^2$ $\quad$ $1/\text{bw}L_p$
C_p	$1/\omega_0 C_p$	$\text{bw}/C_p\omega_0^2$ $\quad$ C_p/bw	$1/\text{bw}C_p$ $\quad$ $\text{bw}/C_p\omega_0^2$

will use a Butterworth response. To determine n, we make this calculation:

$$n = \frac{\log\left[(10^{20/10} - 1)/(10^{2/10} - 1)\right]}{2 \log 2} = 3.7 \text{ (round up to 4)} \qquad (14.54)$$

To find a suitable lowpass prototype circuit, we consult Table 14.10, which for $n = 4$ gives the circuit shown in Fig. 14.20. As given, the circuit elements are in henry (H), farads (F)

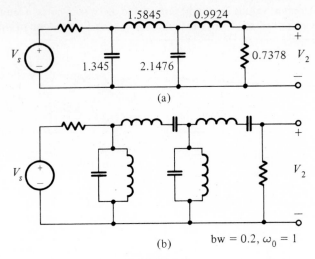

(a)

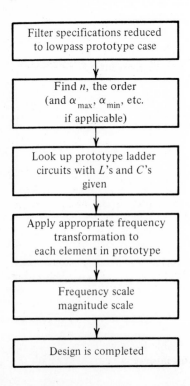

(b) $bw = 0.2, \omega_0 = 1$

FIGURE 14.15

and ohms (Ω). The magnitude response is Butterworth, and the half-power or 3-dB frequency is $\omega_0 = 1$ for the circuit. Since $\alpha_{max} = 2$ is the specification shown in Fig. 14.19b, we must first find Ω_0, and then use frequency scaling to adjust the circuit elements to give the highpass response of Fig. 14.19b. Thus we have

$$\Omega_0 = \frac{1}{(10^{2/10} - 1)^{1/8}} = 1.069 \qquad (14.55)$$

```
┌─────────────────────────────┐
│ Filter specifications reduced │
│   to lowpass prototype case   │
└─────────────────────────────┘
               │
               ▼
┌─────────────────────────────┐
│      Find n, the order        │
│  (and α max, α min, etc.      │
│       if applicable)          │
└─────────────────────────────┘
               │
               ▼
┌─────────────────────────────┐
│  Look up prototype ladder     │
│   circuits with L's and C's   │
│            given              │
└─────────────────────────────┘
               │
               ▼
┌─────────────────────────────┐
│ Apply appropriate frequency   │
│       transformation to       │
│  each element in prototype    │
└─────────────────────────────┘
               │
               ▼
┌─────────────────────────────┐
│     Frequency scale           │
│     magnitude scale           │
└─────────────────────────────┘
               │
               ▼
┌─────────────────────────────┐
│     Design is completed       │
└─────────────────────────────┘
```

FIGURE 14.16

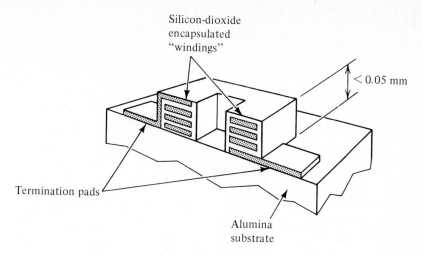

Silicon-dioxide
encapsulated
"windings"

< 0.05 mm

Termination pads

Alumina
substrate

FIGURE 14.17

Applying this number to Fig. 14.19a, we see that the 3-dB frequency for the high-pass filter is

$$f_0 = 56.11 \text{ MHz} \quad \text{or} \quad 353.55 \times 10^6 \text{ rad/s} \tag{14.56}$$

We can achieve this 3-dB frequency if we frequency scale by letting

$$k_f = 3.5355 \times 10^8 \tag{14.57}$$

At the same time it is required that the lossless circuit be terminated in a 50-Ω resistor, which we accomplish by magnitude scaling with

$$k_m = 50 \tag{14.58}$$

Combining the scaling equations with those of the lowpass to highpass transformation, we see that

$$C_n = \frac{1}{k_f k_m} \frac{1}{L_n} = 5.673 \times 10^{-11} \frac{1}{L_n} \tag{14.59}$$

and similarly,

$$L_n = \frac{k_m}{k_f} \frac{1}{C_n} = 1.4182 \times 10^{-7} \frac{1}{C_n} \tag{14.60}$$

Using these equations, we find the values of the elements shown in Fig. 14.21. The filter design is complete, assuming only that we have access to capacitors in the appropriate picofarad range and inductors in the nanohenry range.

Example 14.2 As our second example consider the design of a so-called crossover circuit for a loudspeaker system, depicted in Fig. 14.22. There are to be two loudspeakers

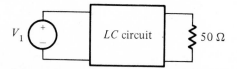

V_1 LC circuit 50 Ω

FIGURE 14.18

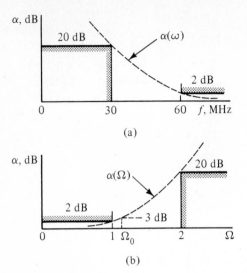

(a)

(b)   FIGURE 14.19

(marked *woofer* and *tweeter*) connected in parallel to the output of an audio amplifier, marked V_s. The loudspeakers are represented by the 1-Ω resistors. The system uses a low-pass filter and a highpass filter, as shown. We will let this crossover frequency be 1 rad/s for the time being, and later frequency scale as well as magnitude scale. In addition it is desired that the entire system appear as a 1-Ω resistor as far as the voltage source V_s is concerned. In other words, we desire that the load connected to V_s be frequency independent. Without going through the specifications, suppose that we find that an $n = 3$ Butterworth response is sufficient. Then the prototype is found from Table 14.9 and is shown in Fig. 14.23, along with the corresponding highpass filter obtained with $\omega_0 = 1$. These are to be connected in parallel, as shown in Fig. 14.23c.

Let us first study $Z_{in}(s)$ as identified in Fig. 14.23c. The impedance of the lowpass section is

$$Z_L = \frac{3s}{2} + \frac{1}{(4/3)\,s + 1/(s/2 + 1)} = \frac{12s^3 + 24s^2 + 24s + 12}{8s^2 + 16s + 12} \tag{14.61}$$

Similarly, the impedance of the highpass section is

$$Z_H = \frac{3}{2s} + \frac{1}{(4/3)\,s + 1/(1/2s + 1)} = \frac{12s^3 + 24s^2 + 24s + 12}{12s^3 + 16s^2 + 8s} \tag{14.62}$$

Because of the parallel connection

$$Y_{in} = Y_L + Y_H \tag{14.63}$$

substituting Eqs. (14.61) and (14.62) into Eq. (14.63), we obtain

$$Y_{in} = \frac{12s^3 + 24s^2 + 24s + 12}{12s^3 + 24s^2 + 24s + 12} = 1 \tag{14.64}$$

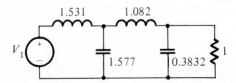

FIGURE 14.20

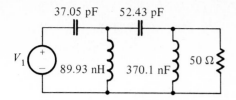

FIGURE 14.21

so that

$$Z_{\text{in}} = 1 \qquad (14.65)$$

which is independent of frequency, as desired.

To complete the design, we need only frequency and magnitude scale. Suppose that we wish the cutoff frequency to be 2500 Hz; the choice of this frequency is a personal matter and will vary from one audio buff to another. Suppose that we also have in mind using loudspeakers represented by 8-Ω resistors. This means that our scaling factors will be

$$k_f = 2\pi \times 2500 = 5000\,\pi \qquad \text{and} \qquad k_m = 8 \qquad (14.66)$$

Using these values to scale the element sizes shown in Fig. 14.23c will give the final design, which is shown in Fig. 14.24. With this the design is complete.

14.6   SENSITIVITY OF DOUBLY TERMINATED LADDERS

The first example in Chapter 9 considered the case of a simple RLC series circuit for which ω_0 and Q had been defined. It was shown that the sensitivity functions

$$S_x^{\omega_0} \qquad \text{and} \qquad S_x^{Q} \qquad (14.67)$$

where x is R, L, or C, were typically of value of $\frac{1}{2}$, as given in Eqs. (9.34)–(9.36). A similar analysis of RC op-amp circuits in Chapter 9 showed that the value of $\frac{1}{2}$ was small compared to many values for the circuit with an active element, the op amp.

Our discussion in this section is limited to the doubly terminated lossless filter, and we are interested in the sensitivity of $|T(j\omega)|$ or $\alpha(\omega)$ with respect to the values of L_k or C_k in the lossless circuit. The heuristic analysis that we will follow

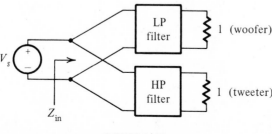

FIGURE 14.22

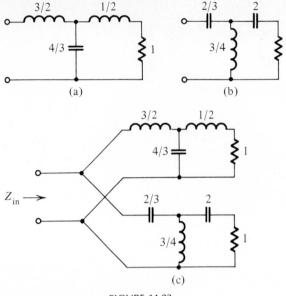

FIGURE 14.23

was given by Orchard* in 1966 who pointed out the inherent advantage of passive circuits over the corresponding active circuits.

Figure 14.25 shows a typical pass-band attenuation characteristic for a Chebyshev bandpass filter. For such filters the pass-band attenuation α_{max} is often small, perhaps 0.1 dB, and at several frequencies $\alpha = 0$, so that there is no loss due to the filter. These frequencies are identified as ω_1, ω_2, and ω_3 in the figure. The corresponding transfer function magnitude response $|T(j\omega)|$ shown in Fig. 14.25b will have the value $|T| = 1$ at the frequencies for which $\alpha = 0$.

At any frequency at which the attenuation vanishes, we consider the change in attenuation caused by any element, capacitor or inductor, which changes from its nominal or design value. Whether the element value increases or decreases, the attenuation must increase. This is illustrated in Fig. 14.26 with an exaggerated increase. At the design value of the element it is clear that

$$\frac{\partial \alpha}{\partial L_k} = 0 \qquad \text{or} \qquad \frac{\partial \alpha}{\partial C_k} = 0 \qquad (14.68)$$

* H. J. Orchard, "Inductorless Filters," *Electron. Lett.*, vol. 2, pp. 224–225, Sept. 1966.

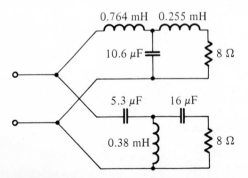

FIGURE 14.24

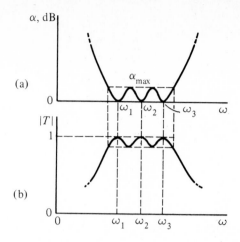

FIGURE 14.25

The same arguments apply to the corresponding values of $|T(j\omega)|$ at these same frequencies; that is,

$$\frac{\partial |T|}{\partial L_k} = 0 \qquad \text{or} \qquad \frac{\partial |T|}{\partial C_k} = 0 \qquad (14.69)$$

as shown in Fig. 14.27. Since the sensitivity functions of interest are

$$S_{L_k}^{|T|} = \frac{L_k}{|T|} \frac{\partial |T|}{\partial L_k} \qquad (14.70)$$

$$S_{C_k}^{|T|} = \frac{C_k}{|T|} \frac{\partial |T|}{\partial C_k} \qquad (14.71)$$

We see from Eq. (14.69) that the sensitivity functions are indeed zero at frequencies at which $\alpha = 0$ and $|T| = 1$. Since the values of $\alpha_{\max}$ remain small throughout the pass band, sensitivities are seen to remain small over the frequency range that is identified as the pass band. It is true that the sensitivities may become large in the stop band, but this is of less concern to the designer since requirements on loss are less stringent in the stop band in most filter applications.

It was pointed out by Orchard that the conclusions reached for the doubly

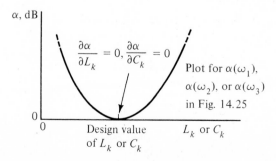

FIGURE 14.26

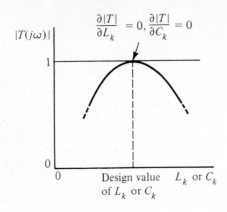

FIGURE 14.27

terminated lossless filter do not apply to the singly terminated circuits, and thus these circuits are inherently more sensitive.

The low sensitivity that characterizes the doubly terminated ladder circuits has implications with respect to their use in design, namely, that the doubly terminated circuits are those most commonly used in applications for filtering, phase correction, and delay equalizers. As we will learn in the next chapter, this also means that the circuit is used as a model for circuits that use active elements to simulate inductors, resistors, and other elements.

PROBLEMS

14.1 In this problem, you are given $A(s)$ derived from Eq. (14.22) and asked to find the two doubly terminated lossless realizations with $R_1 = 1$ ohm. The function is

$$A(s) = \frac{s^2 + 1.35s + 0.35}{s^2 + 3s + 1}$$

In working the problem, ignore small numerical differences that might have resulted from improper roundoff.

14.2 Repeat Problem 14.1 for the function

$$A(s) = \frac{(s^2 + 0.6934s + 0.4808)(s + 0.6934)}{s^3 + 2s^2\,2s + 1}$$

14.3 For the doubly terminated lossless realization shown in Fig. P14.3:
(a) Find another doubly terminated circuit that meets the same specifications.
(b) Find $A(s)$ that describes the circuit.

14.4 Repeat Problem 14.3 for the circuit given in Fig. P14.4.

14.5 The specification function for a doubly terminated lossless network is

$$|\mathbf{T}(j\omega)|^2 = \frac{K}{1 + \omega^8}$$

which indicates a fourth-order Butterworth filter. Find two circuit realizations. Carry through the steps of finding $A(s)$ and Z_{11} to find the circuits. You may wish to check your result in an appropriate table.

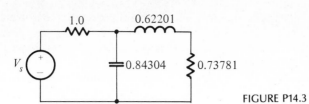

FIGURE P14.3

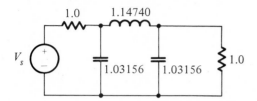

FIGURE P14.4

14.6 Repeat Problem 14.5 for a fifth-order Butterworth filter where

$$|\mathbf{T}(j\omega)|^2 = \frac{K}{1 + \omega^{10}}$$

14.7 The specification function for a doubly terminated lossless filter has the form

$$|\mathbf{T}(j\omega)|^2 = \frac{K}{1 + \omega^6}$$

indicating a third-order Butterworth response.
  (a)   Find two realizations if it is required that $R_1 = 1$ and $R_2 = 2$.
  (b)   Repeat (a) if it is required that $R_1 = 1$ and $R_2 = 3$.
14.8 This problem is concerned with the application of frequency-transformation techniques to obtain a given frequency response from a lowpass prototype. For this
problem, assume that the prototype is a third-order Butterworth doubly terminated
lossless circuit. Determine the element values in the frequency-transformed circuit
and sketch $|T(j\omega)|$ as a function of frequency, if the transformation is that specified
by:
  (a)   Fig. 11.16.
  (b)   Fig. 11.18.
  (c)   Fig. 11.23.
  (d)   Fig. 11.26.
14.9 Repeat Problem 14.8 if the lowpass prototype is that given in Problem 14.3.
14.10 This problem relates to the speaker crossover system shown in Fig. 14.22. For this
system, find the average power in both the lowpass load and the highpass load, and
show that the sum of these two powers is a constant.
14.11 This problem has the objective of eliminating interference from a television set.
From a spectrum analyzer, it is found that reception will be improved if the following attenuations can be obtained by a filter:

f, MHz	α, dB
48	20
60	2
66	20

Design a passive filter to put on the front end of a television receiver to match a 50-ohm transmission line.

14.12 The model for a given design problem is a singly terminated lossless circuit given in tables with $R_1 = 1$. Assume that the prototype is a third-order Butterworth. A bandstop characteristic is desired for which $f_0 = 100$ kHz and the 3-dB bandwidth is BW = 10 kHz. If it is required that $R_1 = 50$ ohms, find a filter realization.

CHAPTER 15

Ladder
Design
with
Simulated
Elements

In the last chapter we extolled the virtues of the doubly terminated lossless ladder: low sensitivity, the circuit for which frequency transformations are directly applicable. As design engineers we see that these advantages are obtained at a high price: we use inductors which are hard to build and heavy in weight, and which are difficult to adapt to integrated-circuit realizations. In the next three chapters we consider methods which have been discovered to get rid of inductors. We will see that it is possible, by some power akin to that of King Midas, that all inductors we touch will turn not to gold but to inductorless circuits.

15.1 THE IDEAL GYRATOR AND RIORDAN'S CIRCUIT

Some electrical devices operate by causing the roles of the electric and magnetic fields to be interchanged. Two well-known examples are the Hall-effect devices and some waveguide configurations operating at microwave frequencies. In 1948 Tellegen proposed a model for such devices which is known as the *ideal gyrator*. We define his model in terms of the currents and voltages shown in Fig. 15.1, where

$$v = Ki_2 \tag{15.1}$$

$$v_2 = -Ki_1 \tag{15.2}$$

where K is a real constant. The symbol that is used to indicate a Tellegen gyrator is shown in Fig. 15.2.

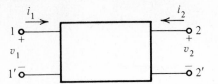

FIGURE 15.1

Now suppose that we terminate the gyrator in a capacitor, as shown in Fig. 15.3. This places a constraint between voltage v_2 and current i_2 which is that

$$i_2 = -\frac{d}{dt} C v_2 \qquad (15.3)$$

If we substitute this relationship into Eq. (15.1), we have

$$v_1 = K i_2 = K\left(-\frac{d}{dt} C v_2\right) \qquad (15.4)$$

We next substitute Eq. (15.2) into this equation, giving

$$v_1 = K\frac{d}{dt}(CKi_1) = \frac{d}{dt}(K^2 C i_1) = \frac{d}{dt}(L_{eq} i_1) \qquad (15.5)$$

Here $L_{eq} = K^2 C$ is the value of the equivalent inductance. Thus we see that by means of a gyrator, a capacitor becomes the equivalent of an inductor. This is an important conclusion.

Unfortunately Hall-effect devices do not operate at the frequency range of our interest, and so we seek a gyrator based on an op-amp circuit. Of several that have been proposed, the most successful was given by Riordan* in 1967. This circuit is shown in Fig. 15.4 and is seen to consist of two op amps and five impedances. To analyze this circuit, we observe that

$$V_2 = V_1\left(1 + \frac{Z_4}{Z_5}\right) \qquad (15.6)$$

and that

$$V_3 = V_1\left(1 + \frac{Z_2}{Z_3}\right) - V_2\left(\frac{Z_2}{Z_3}\right) \qquad (15.7)$$

or

$$V_3 = V_1\left(1 - \frac{Z_2 Z_4}{Z_3 Z_5}\right) \qquad (15.8)$$

* R. H. S. Riordan, "Simulated Inductors Using Differential Amplifiers," *Electron. Lett.*, vol. 3, pp. 50–51, 1967.

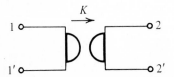

FIGURE 15.2

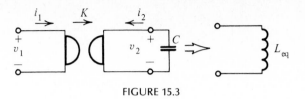

FIGURE 15.3

Now the input current is

$$I_1 = \frac{V_1 - V_3}{Z_1} = V_1 \frac{Z_2 Z_4}{Z_1 Z_3 Z_5} \tag{15.9}$$

Thus the input impedance is seen to be

$$Z_{in} = \frac{V_1}{I_1} = \frac{Z_1 Z_3 Z_5}{Z_2 Z_4} \tag{15.10}$$

From this equation we see that if either Z_2 or Z_4 are capacitors, such that $Z = 1/Cs$ and all remaining elements are resistors of value R, then the last equation becomes

$$Z_{in} = (CR^2)s = L_{eq}s \tag{15.11}$$

and the circuit behaves as if it were an inductor of value

$$L_{eq} = CR^2 \tag{15.12}$$

For example, if $C = 0.01 \ \mu F$ and the four resistors have the value $R = 1 \ k\Omega$, then at the inputs the circuit appears to be a 10-mH inductor.

The circuit of Fig. 15.5 is identical to the Riordan circuit of Fig. 15.4, as can be verified by simply tracing out the connections. Another circuit that is very similar to the Riordan circuit is given in Fig. 15.6a. The two remaining circuits of Fig. 15.6 are due to Antoniou* and these are considered extensively in the next

---

* A. Antoniou, "Realization of Gyrators Using Operational Amplifiers, and Their Use in RC-Active-Network Synthesis," *Proc. IEE*, vol. 116, pp. 1838–1850, 1969.

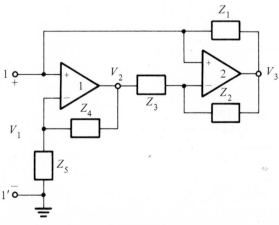

FIGURE 15.4

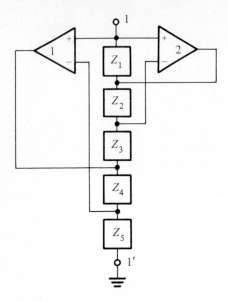

FIGURE 15.5

section. The circuits are similar in that they are all described by Eq. (15.10). Thus any of them can be used to simulate an inductor. But that is not all.

15.2 ANTONIOU'S GIC AND BRUTON'S FDNR

Of the two Antoniou circuits of Fig. 15.6 we will consider that of Fig. 15.6b, which has been shown to be the "best" circuit.[†] This circuit is redrawn in Fig. 15.7, with the circuit enclosed in dashed lines excluding Z_5. This part is known as a *generalized impedance converter* (GIC). We have identified part of the circuit as a GIC, but this does not change the circuit which has an input impedance as given by Eq. (15.10):

$$Z_{11'} = \frac{Z_1 Z_3}{Z_2 Z_4} Z_5 \qquad (15.13)$$

However, if we remove Z_5 and identify the input terminals 2-2' and at the same time terminate terminals 1-1' in Z_0, then we find that

$$Z_{22'} = \frac{Z_4 Z_2}{Z_3 Z_1} Z_0 \qquad (15.14)$$

these quantities being identified in Fig. 15.8.

If we now make the choice that $Z_4 = 1/C_4 s$ and that all other elements in Fig. 15.7 are resistors, then

$$Z_{11'} = \frac{R_1 R_3 R_5}{R_2} C_4 s = L_{eq} s \qquad (15.15)$$

[†] A. S. Sedra and P. O. Brackett, *Filter Theory and Design: Active and Passive*, Matrix Publishers, Portland, Ore., 1978, sec. 8.4.

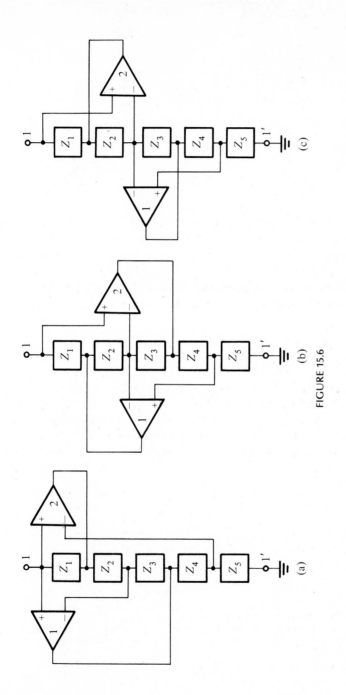

FIGURE 15.6

433

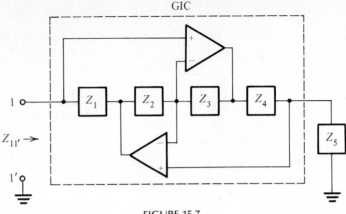

FIGURE 15.7

This is the impedance of the circuit shown in Fig. 15.9a. The result is the same as that found in the last section. Conceptually we may regard the GIC as a circuit that converts the resistor R_5 into an inductor of value $L_{eq} = R_1 R_3 R_5 C_4 / R_2$, as given by Eq. (15.15). The surprise comes when we turn the GIC end for end, remove R_5, and terminate the opposite side in a capacitor of value C_0, as shown in Fig. 15.9c. Then Eq. (15.14) gives us

$$Z_{22'} = \frac{R_2}{R_1 R_3 C_4 C_0} \frac{1}{s^2} = \frac{1}{Ds^2} \qquad (15.16)$$

This is something new. We note that when $s = j\omega$, then

$$Z_{22'}(j\omega) = \frac{-1}{D\omega^2} \qquad (15.17)$$

We see that this function is negative and varies inversely with ω^2. The circuit of Fig. 15.9c is known as a *frequency-dependent negative resistor* (FDNR). The FDNR concept was introduced by Bruton[*] along with a method for the simula-

[*] L. T. Bruton, "Network Transfer Functions Using the Concept of Frequency-Dependent Negative Resistance," *IEEE Trans. Circuit Theory*, vol. CT-16, pp. 406–408, 1969.

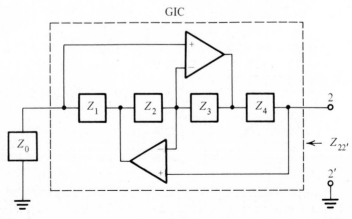

FIGURE 15.8

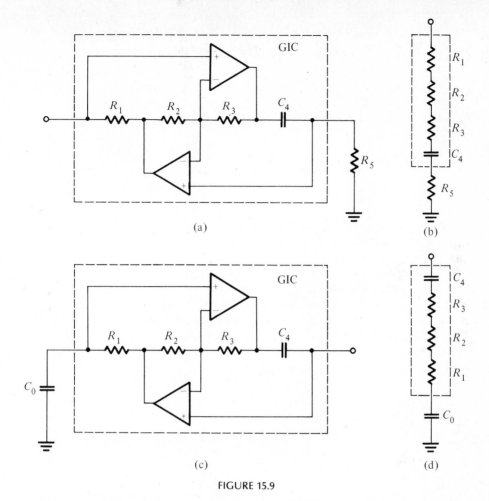

FIGURE 15.9

tion of ladder filters. Since the circuits of Fig. 15.9a and c are complicated, they may be replaced by the shorthand notation of Fig. 15.9b and d in which the op-amp connections are removed.

The other new concept introduced by Bruton is that of magnitude scaling all elements by the factor $1/s$. Past experience has shown us that circuits may be magnitude scaled without changing the transfer function $T(s)$. In the past all such scaling has been by a constant k_m. Since the elements are given by the equations

$$Z_R = R, \qquad Z_L = Ls, \qquad Z_C = \frac{1}{Cs} \tag{15.18}$$

scaling each of these by $1/s$ gives us

$$Z_R' = \frac{R}{s}, \qquad Z_L' = L, \qquad Z_C' = \frac{1}{Cs^2} \tag{15.19}$$

Thus we see that such scaling actually results in a transformation of elements: a resistor becomes a capacitor, an inductor becomes a resistor, and a capacitor becomes an FDNR. Note carefully that none of the impedances in Eq. (15.19) de-

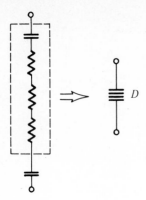

FIGURE 15.10

scribes an inductor. We can eliminate inductors, and the only price we must pay is to use FDNRs instead. These can be realized by the circuit of Fig. 15.9c. We require a new symbol for the new element just introduced. It is four parallel lines, as shown in Fig. 15.10.

In summary the elements resulting from the Bruton transformation are shown in Fig. 15.11. We now have R and L transformed into familiar elements, and a new element introduced in transforming C. The new element is, of course, the FDNR described by Eq. (15.16) which is realized by the circuit shown in Fig. 15.9c. The manner in which this element and also the simulated inductor may be realized using integrated-circuit technology is suggested by Fig. 15.12, which shows the pin connections of the AF120, a commercial implementation produced by National Semiconductor. Fig. 15.12a shows a standard gyrator module; two external capacitors makes it an FDNR, as shown in Fig. 15.12b; one external capacitor makes it into a grounded inductor, as shown in Fig. 15.12c.

Example 15.1 Figure 15.13a shows a simple RLC series circuit arranged with the output voltage across the capacitor. This circuit has been studied extensively and is well known to be a lowpass filter. The voltage-ratio transfer function is

$$\frac{V_2}{V_1} = \frac{1/Cs}{R + Ls + 1/Cs} \qquad (15.20)$$

The circuit in Fig. 15.13b is obtained from that in Fig. 15.13a by applying the Bruton

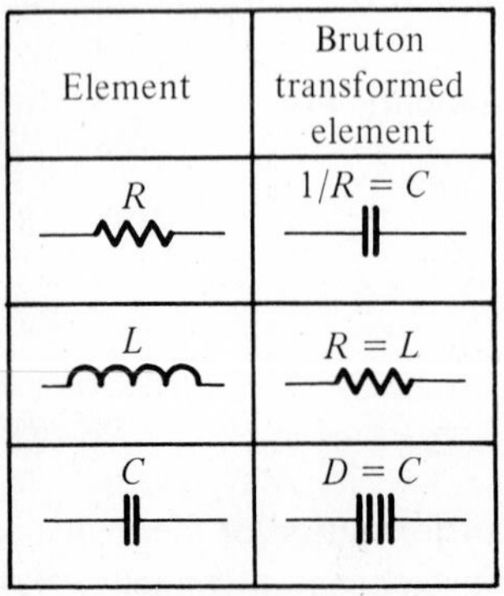

Element	Bruton transformed element
R ⌇	$1/R = C$ ⊣⊢
L ⌇	$R = L$ ⌇
C ⊣⊢	$D = C$ ⫴

FIGURE 15.11

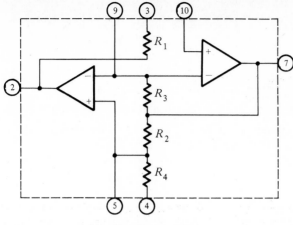

(a)   Gyrator module

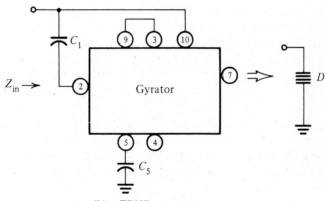

(b)   FDNR connection

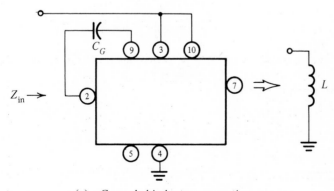

(c)   Grounded inductor connection

FIGURE 15.12

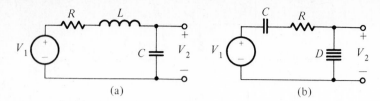

FIGURE 15.13

transformation, as shown in Fig. 15.11. The voltage-ratio transfer function for the circuit in Fig. 15.13b is

$$\frac{V_2}{V_1} = \frac{1/Cs^2}{R/s + L + 1/Cs^2} \tag{15.21}$$

which is clearly the same equation as Eq. (15.20). Thus we see that the two circuits of Fig. 15.13 have identical transfer functions. This illustrates the objective in applying the Bruton transformation.

Example 15.2 The problem to be considered is the design of a fourth-order Butterworth lowpass filter making use of FDNRs. The prototype for such a filter is found from Table 14.1, and is shown in Fig. 15.14a. Using the Bruton transformation of elements as in Fig. 15.11, we obtain the equivalent circuit of Fig. 15.14b complete with element values for R, C, and D. To realize the FDNR, we make use of the circuit of Fig. 15.9c, which is described by Eq. (15.16). We make the choices

$$C_0 = C_4 = 1 \text{ F}, \qquad R_2 = R_3 = 1 \, \Omega, \qquad R_1 = D \tag{15.22}$$

and the realization of the normalized circuit of Fig. 15.14b becomes that shown in Fig. 15.15a.

Suppose that we next specify that the half-power frequency be 1000 Hz, and that the final design make use of 0.01-μF capacitors only. This means that scaling is accomplished by the choices

$$k_f = 2\pi \times 1000 \quad \text{and} \quad k_m = \frac{1}{(2\pi \times 1000)(0.01 \times 10^{-6})} = 1.59 \times 10^4 \tag{15.23}$$

With this scaling the circuit becomes that shown in Fig. 15.15b. It is seen that all of the

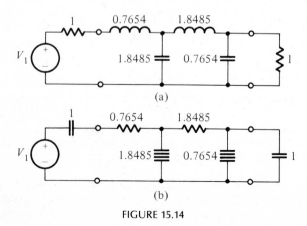

(a)

(b)

FIGURE 15.14

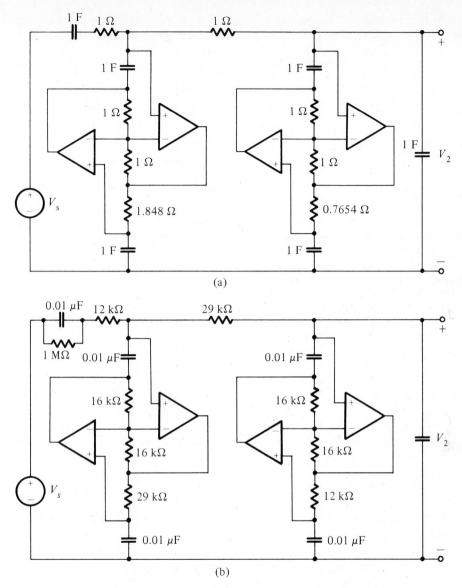

FIGURE 15.15

elements in the circuit are in a practical range and can be realized with ordinary op amps. Experiments on laboratory circuits such as that of Fig. 15.15b show excellent agreement between the specified Butterworth fourth-order response and that obtained. Since this circuit is completely equivalent, except for scaling, to that shown in Fig. 15.14a, this circuit will have the same low sensitivity values as those found for a passive ladder circuit. In other words, this is a very practical realization technique.

Example 15.3 We require a highpass filter with a half-power frequency of 10 Hz. If constructed using an inductor, this would be a very heavy filter, and so we wish to consider the possibility of using simulated inductors. Suppose that it is further specified that a third-order Butterworth filter be used as the prototype in this design. This prototype with element values determined from Table 14.1 is shown in Fig. 15.16a. Making use of the lowpass to highpass frequency transformation, we obtain the highpass circuit shown in Fig. 15.16b for $\omega_0 = 1$. Using the circuit of Fig. 15.9, we have the equation for impedance,

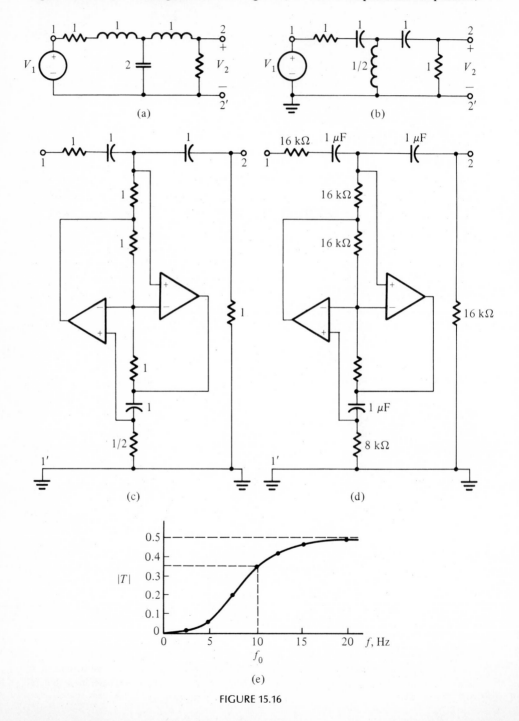

FIGURE 15.16

Eq. (15.15). In that equation let all element values be unity except for R_s, which is then equal to L_{eq}. Since we require $L_{eq} = \frac{1}{2}$ H, we obtain the circuit shown in Fig. 15.16c.

Next we scale, using the required scaling from $\omega_0 = 1$ to $f_0 = 10$, so that $k_f = 62.83$. Magnitude scaling is set by the requirement that the capacitors have the value of 1.0 μF, so that

$$k_m = \frac{1}{10^{-6} \times 20\pi} = 15{,}915 \qquad (15.24)$$

The circuit resulting from this scaling is given in Fig. 15.16d, and this circuit has the response shown in Fig. 15.16e. Again, all element values are in a practical range, and this realization would be light in weight compared to one with a physical inductor.

15.3 CREATING NEGATIVE ELEMENTS

In some of our filter designs we will require negative elements. How are such elements created? To answer our question, we begin with the circuit given in Fig. 15.17, to which we will apply nodal analysis. At node 1 the Kirchhoff current law gives us

$$I_1 + \frac{V_2 - V_1}{R_1} = 0 \qquad (15.25)$$

which, at node 2, is

$$\frac{V_2 - V_1}{R_2} + (0 - V_1)\frac{1}{Z_L} = 0 \qquad (15.26)$$

Eliminating V_2 from these two equations, gives us

$$-I_1 R_1 - V_1 R_2 \frac{1}{Z_L} = 0 \qquad (15.27)$$

Solving for the ratio I_1/V_1, we obtain

$$\frac{V_1}{I_1} = Z_{in} = -\frac{R_1}{R_2} Z_L \qquad (15.28)$$

or

$$Y_{in} = -\frac{R_2}{R_1} Y_L \qquad (15.29)$$

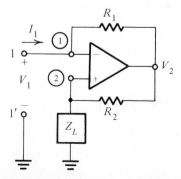

FIGURE 15.17

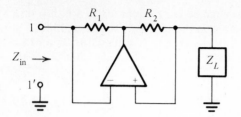

FIGURE 15.18

From these equations we see that the circuit, shown in alternative form in Fig. 15.18, creates the negative of Z_L or Y_L and also scales the value by the ratio of resistances. Thus if $Z_L = R$, then R_{in} is negative and can be thought of as a negative resistance at terminals 1-1', as shown in Fig. 15.19. Similarly, the termination of the circuit in a capacitor makes Z_{in} appear as a negative capacitor. The circuit we have studied is one of a general class of circuits known as *negative impedance convertors* (NIC), as represented in a general form in Fig. 15.20.

To illustrate potential applications of negative elements, consider the section of a ladder circuit in Fig. 15.21a. This may be shown to be equivalent to the circuit of Fig. 15.21b.* The circuit of Fig. 15.21a may be difficult to realize because of the floating inductor L_2. The equivalent circuit of Fig. 15.21b has no floating inductor, but it does have a negative capacitor and also an ideal transformer. In general the circuit of Fig. 15.21b is easier to realize than that of Fig. 15.21a and the circuit we have studied would be used to realize the negative capacitor.

As a second example, the filter circuit shown in Fig. 15.22a, was obtained[†] using a sequence of circuit transformations. As given, it has problems in realization. First, it contains floating inductors, and it also contains a negative inductor. Both problems are easily solved by making use of the Bruton transformation which gives the circuit shown in Fig. 15.22b. In this circuit both FDNRs are grounded, and the negative resistor is easily realized using the circuit under study, that of Fig. 15.18.

15.4 CREATING FLOATING ELEMENTS

The floating element of our present interest is the floating inductor for which the two terminals may be at different voltages, neither equal to zero (or grounded). A representation of such an inductor is shown in Fig. 15.23. Here terminals 1 and 2 are floating, while terminals 1' and 2' are grounded. Any circuit we propose to replace the inductor of this figure must satisfy the following tests: (1) if terminals 2-

* H. J. Orchard and D. F. Sheahan, "Inductorless Bandpass Filters," *IEEE J. Solid-State Circuits*, vol. SC-5, pp. 108–118, 1970.
† Due to P. Geffe.

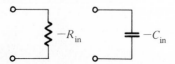

FIGURE 15.19

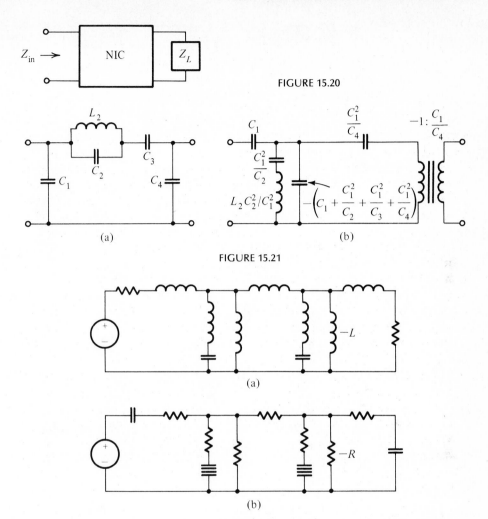

FIGURE 15.20

FIGURE 15.21

FIGURE 15.22

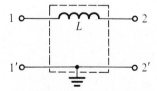

FIGURE 15.23

$2'$ are shorted, then $Z_{11'} = Ls$, or (2) if terminals 1-1' are shorted, then $Z_{22'} = Ls$. These conditions are illustrated in Fig. 15.24.

One possible solution to our problem is illustrated in Fig. 15.25, which shows two gyrators, both terminated in the same capacitor C. From the basic relationship for the gyrator defined in terms of Fig. 15.26,

$$Z_{in} = K \frac{1}{Z_L} \qquad (15.30)$$

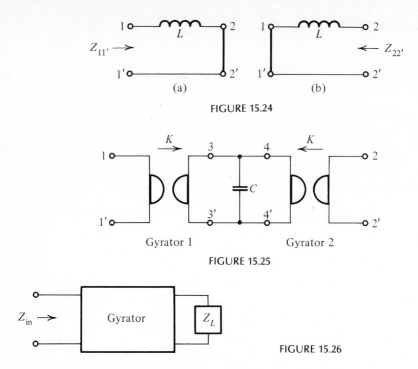

FIGURE 15.24

FIGURE 15.25

FIGURE 15.26

we see that when $Z_L = 0$, corresponding to a short circuit, then $Z_{in} = \infty$, or an open circuit. Thus if we short terminals 2-2′, then at terminal pair 4-4′ there will be an open circuit, and the input impedance at terminals 1-1′ will be as found in Eq. (15.5), that of an inductor. The same argument works in reverse, of course, so that the criterion illustrated by Fig. 15.24 is satisfied.

Even if the circuit of Fig. 15.25 does work, it turns out that there is a better way of accomplishing the same objective. The new solution was suggested by Riordan in his original paper, and can now be understood in terms of the properties of GICs as first pointed out by Gorski-Popiel.* Consider the circuit shown in Fig. 15.27, where Fig. 15.27a reminds us of the concept, and Fig. 15.27b shows us the circuit that will actually be used. First let us consider the consequences of shorting the output terminals 2-2′, as was done for the two-gyrator circuit of Fig. 15.25. We study the input impedance identified in Fig. 15.28, noting that since the voltage across the input terminals to the op amp is zero, then terminals 1 and 2 are actually connected together. Thus $Z_{in} = 0$ as shown in Fig. 15.28b. Returning to Fig. 15.27b, we see that shorting output terminals 2-2′ causes the resistor R to be shorted to ground. Then the circuit becomes the usual one for simulating an inductor, and $Z_{11'}$ is the impedance of an inductor, as required by the criterion of Fig. 15.24. Of course the same arguments work in reverse, with terminals 1-1′ shorted. Then in the circuit of Fig. 15.27 we have a floating inductor. It does take more elements than the realization of a grounded inductor, but it is no more difficult conceptually.

* J. Gorski-Popiel, "RC-active synthesis using positive-immittance converters," *Electron. Lett.,* vol. 2, pp. 381–382, Aug. 1967.

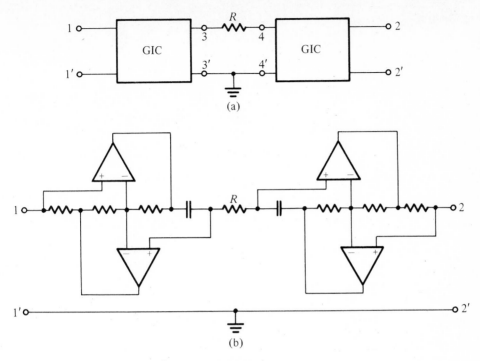

FIGURE 15.27

We can now generalize the discussion just given by noting that any circuit that is connected in place of the resistor R in Fig. 15.27b will be transformed by the NIC. The NIC we are using in the circuit for simulating floating inductors is the one that changes a resistor into an inductor. If a resistive circuit replaces R, then every element in that circuit is transformed from a resistor to an inductor. The manner in which this applies to a T-circuit and a π-circuit is illustrated by Fig. 15.29. We will soon show that the NIC can be used for other kinds of such transformations.

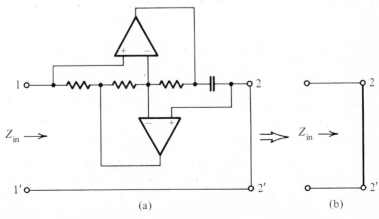

FIGURE 15.28

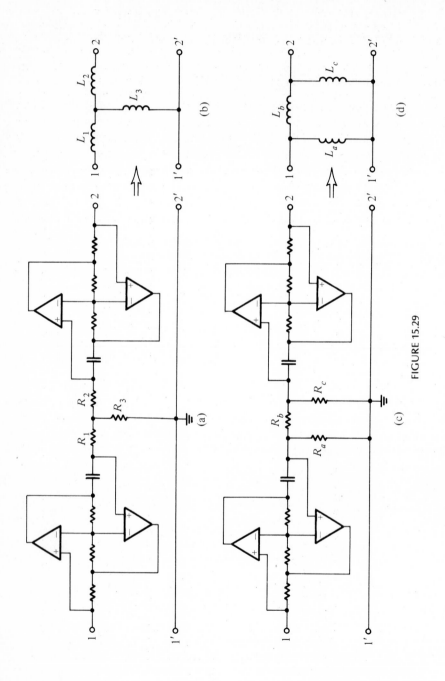

FIGURE 15.29

446

Example 15.4   The circuit shown in Fig. 15.30a is known as a double-tuned bandpass filter in which the three inductors L_p, L_s, and L_M represent the π equivalent of a transformer. We may replace the inductors by resistors, using the concepts just developed. As shown in Fig. 15.30b, the π-connection of resistors together with two GICs are equivalent to the π-connection of inductors, and so a realization may be found without inductors.

15.5   GIC BANDPASS FILTERS

The lowpass to bandpass frequency transformation studied in Chapter 14 causes shunt capacitors to become parallel LC circuits and series inductors to become series LC circuits. The bandpass filter shown in Fig. 15.31a is then the standard form for a ladder bandpass filter, and so it is the starting point in a search for a simulation of this circuit making use of GICs. We first apply the Bruton transformation to obtain the circuit shown in Fig. 15.31b. In this circuit we can identify two canonical sections, which are labeled as section A and section B. Bandpass filters of higher order will be made up of an alternation of these two kinds of sec-

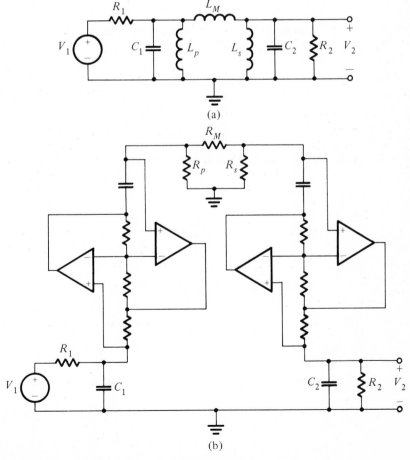

(a)

(b)

FIGURE 15.30

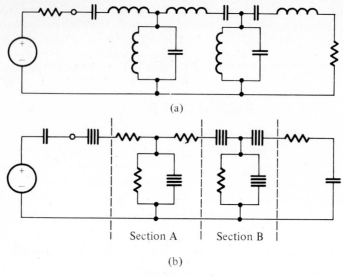

FIGURE 15.31

tions, plus the terminations shown on either end. If we understand the manner in which these two sections may be simulated, then we can simulate a bandpass filter of any order.

Figure 15.32 shows the standard section A drawn in somewhat different form than in Fig. 15.31b. Section B, which is shown in Fig. 15.33, requires further discussion. In section B we first remove the resistor from the FDNRs and then note that we have a T-section of FDNRs. We then ask the question as to what GIC would transform a T-section of resistors into a T-section of FDNRs. The answer to this question may be found by returning to Fig. 15.7, which is described by Eq. (15.13) and repeated here:

$$Z_{11'} = \frac{Z_1 Z_3}{Z_2 Z_4} Z_5 \qquad (15.31)$$

and also to Fig. 15.8, which is described by Eq. (15.14):

$$Z_{22'} = \frac{Z_2 Z_4}{Z_1 Z_3} Z_0 \qquad (15.32)$$

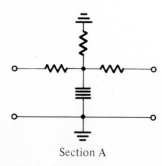

Section A

FIGURE 15.32

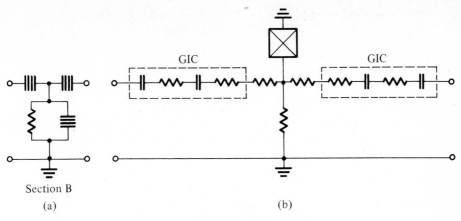

GIC   GIC

Section B

(a)   (b)

FIGURE 15.33

Thinking in terms of the resistive T-section in Fig. 15.33b, we select $Z_0 = R_0$ and also $Z_2 = Z_4 = 1/Cs$. If $Z_1 = R_1$ and $Z_3 = R_3$, then

$$Z_{22'} = \frac{R_0}{R_1 R_3 C^2} \frac{1}{s^2} \tag{15.33}$$

which is the equation for an FDNR. However, this particular FDNR converts resistors, such as R_0 in Fig. 15.34, into an FDNR, and thus it is directly applicable to the circuit of Fig. 15.33b. It does transform the T-section of resistors into the T-section of FDNRs by multiplying each element by $1/Ds^2$.

The new symbol in Fig. 15.33b represents a new element which, when multiplied by $1/Ds^2$, becomes a constant. Then it is clear that the impedance of this new element must be

$$Z_E = Es^2 \tag{15.34}$$

such that the product is

$$Es^2 \times \frac{1}{Ds^2} = \frac{E}{D} = R \tag{15.35}$$

GIC

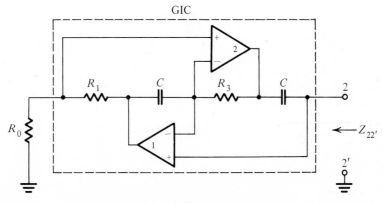

FIGURE 15.34

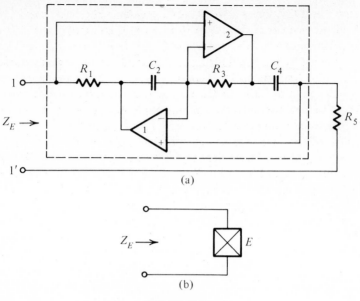

(a)

(b)

FIGURE 15.35

The circuit satisfying this specification is shown in Fig. 15.35 for which Eq. (15.31) applies. Here we let $Z_5 = R_5$, $Z_1 = R_1$, $Z_3 = R_3$, $Z_2 = 1/C_2 s$, and $Z_4 = 1/C_4 s$, such that Eq. (15.31) becomes

$$Z_{11'} = R_1 R_3 R_5 C_2 C_4 s^2 = Es^2 \qquad (15.36)$$

When $s = j\omega$, then

$$Z_{11'} = -E\omega^2 \qquad (15.37)$$

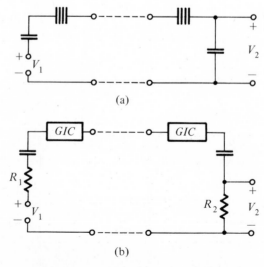

(a)

(b)

FIGURE 15.36

which shows that this too is a frequency-dependent negative resistor, but in this case the impedance varies directly with frequency squared, rather than inversely with frequency squared. Whether the × in the box element is new or not, it is routinely realized by the circuit of Fig. 15.35a and can be as easily constructed as any of the other elements in this chapter.

From the circuit of Fig. 15.31b extended to the right for higher-order cases, we see that the bandpass filter will consist of alternating A and B sections and, in addition, the termination sections marked T at both ends of the network. These termination sections are represented in Fig. 15.36a with each consisting of an FDNR and a capacitor. These elements came about through applying the Bruton transformation to R_1 and R_2 and capacitors. This is equivalent to the operation of a GIC on these circuit elements as indicated in Fig. 15.36b. One of the appropriate GIC's to perform this operation is that shown in Fig. 15.17a. The input impedance is described by Eq. (15.31). If we let the elements of the GIC have unit

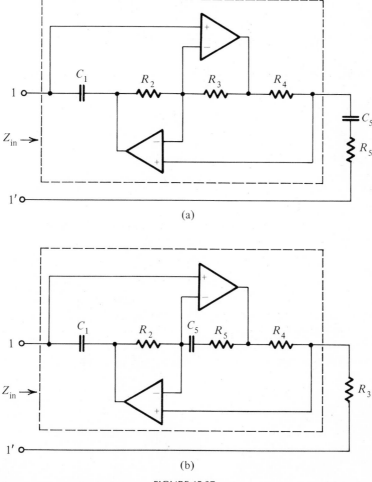

(a)

(b)

FIGURE 15.37

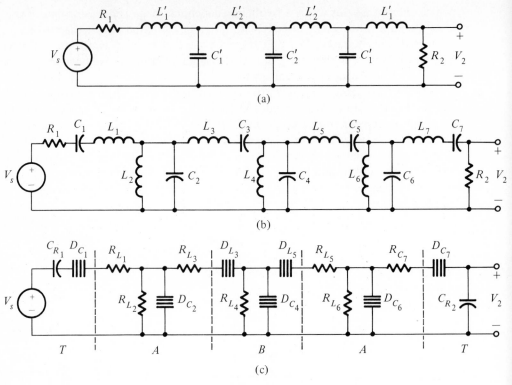

FIGURE 15.38

values, then Z_{in} represents a series FDNR and capacitor since

$$Z_{in} = \frac{Z_1 Z_3 Z_5}{Z_2 Z_4} = \frac{1}{s}\left(\frac{1}{C_5 s} + R_s\right) = \frac{1}{Ds^2} + \frac{1}{Cs} \qquad (15.38)$$

where $D = C_5$ and $C = 1/R_s$. It is interesting to note that Z_5 and Z_3 (or Z_1) can be interchanged without changing the input impedance Z_{in}. Thus the two circuits of Fig. 15.37 are equivalent. We make use of this equivalence in the next example of filter design.

Example 15.5* From the specifications given for a bandpass filter, it is found that for a Butterworth response, the prototype must be of seventh order such that the bandpass filter will be of fourteenth order. This high-order requirement suggests that there will be an advantage in the methods of this chapter when compared to using inductors. For this filter, the center frequency is to be 150 Hz, the bandwidth is 150 Hz, and the terminations should be 1000 Ω.

We will first outline the method to be used before becoming involved in actual numbers. Figure 15.38 shows (a) the seventh-order prototype, (b) the fourteenth-order bandpass filter, and (c) the Bruton-transformed filter. In terms of the parameters of the bandpass filter, the filter consisting of two A sections, one B section, and the two T sections is that shown in Fig. 15.39. The section nearest V_s has been transformed using the circuit equivalance of Fig. 15.37.

*As noted earlier, the methods of this section are due to Antoniou. This particular example follows one provided by L. T. Bruton and David Treleaven, "Active Filter Design Using Generalized Impedance Converters," *EDN*, pp. 68–75, 1973.

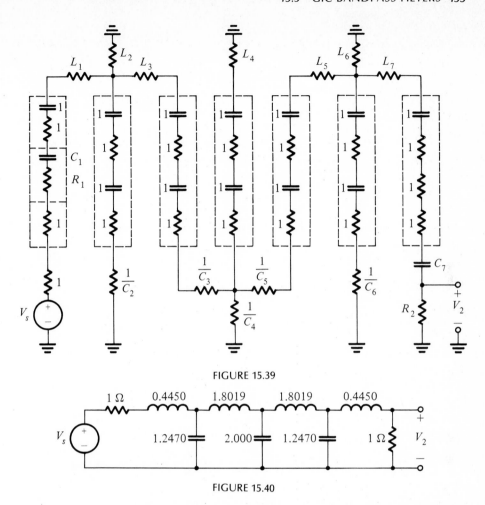

FIGURE 15.39

FIGURE 15.40

The prototype Butterworth filter is given in Table 14.2 and is shown in Fig. 15.40 for a bandwidth of 1 rad/s. If we frequency scale 150 Hz to 1 rad/s, then we note that $\omega_0 = 1$ and BW = 1. This particular choice greatly simplifies the frequency transformation equations, so that in transforming from Fig. 15.38a to 15.38b each inductor L_p transforms into a series-connected inductor L_p and capacitor $1/L_p$, while a capacitor transforms into a parallel-connected inductor $1/C_p$ and capacitor C_p. Thus the elements for Fig. 15.38b are routinely expressed in terms of the values from the prototype of Fig. 15.40 as follows:

$$C_1 = C_7 = 1/L_1' = 2.2472$$

$$C_3 = C_5 = 1/L_2' = 0.5549$$

$$L_1 = L_7 = L_1' \quad = 0.4450$$

$$C_2 = C_6 = C_1' \quad = 1.2470$$

$$L_3 = L_5 = L_2' \quad = 1.8019$$

$$C_4 = C_2' \quad = 2.0000$$ (15.39)

$$L_2 = L_6 = 1/C_1' = 0.8019$$

$$L_4 = 1/C_2' \quad = 0.5000$$

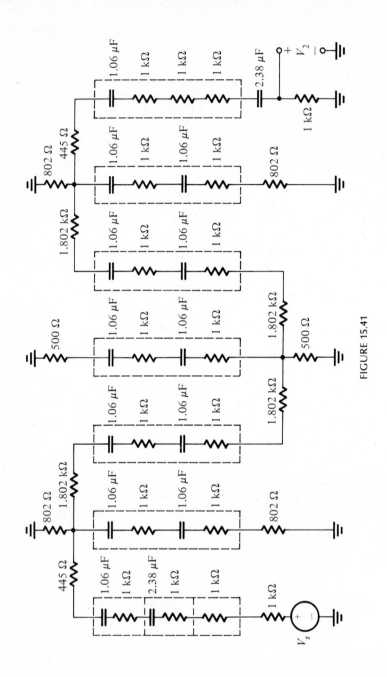

FIGURE 15.41

These values along with $R_1 = R_2 = 1$ complete element values for the circuit of Fig. 15.39.

To complete this example, we scale frequency so that 1 rad/s becomes 150 Hz and 1 Ω becomes 1000 Ω. Thus

$$k_f = 2\pi \times 150 \quad \text{and} \quad k_m = 1000 \quad (15.40)$$

giving the final design shown in Fig. 15.41.

PROBLEMS

15.1 For a particular filtering problem, the prototype shown in Fig. P15.1 is chosen. This is a singly terminated lowpass Chebyshev filter having a ripple width of 0.9 dB and a bandwidth of 1 rad/s. For the final design, the filter is to be frequency-scaled by a factor of 1000 and magnitude-scaled to give practical element sizes. Find the filter using the methods of this chapter.

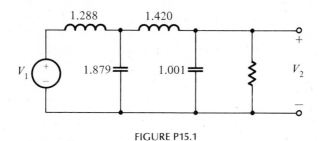

FIGURE P15.1

15.2 A filtering problem is found to require the choice of a fourth-order lowpass Butterworth filter as the prototype. The filter is to have a half-power frequency of 1000 rad/s. Magnitude scale to give practical element sizes. Design the filter making use of the methods of this chapter.

15.3 Using simulated inductors, design a doubly terminated highpass filter based on a fourth-order Butterworth prototype. The half-power frequency of the filter is to be 5 kHz and the terminating resistors are each 100 ohms.

15.4 The specifications for a highpass filter are shown in Fig. P15.4 and it is further specified that the response be Butterworth. Using the methods of this chapter, find a filter realization and magnitude scale to give element values in a practical range.

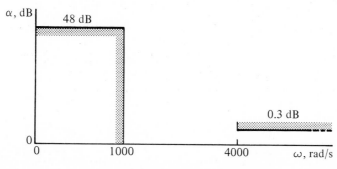

FIGURE P15.4

15.5 A Chebyshev highpass filter is required to satisfy the specifications given in Fig. P15.5. Using the methods of this chapter, find a filter realization. Magnitude scale to give element values in a practical range.

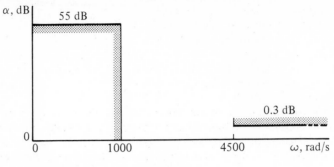

FIGURE P15.5

15.6 A filter found in old equipment has the circuit shown in Fig. P15.6. The design calculations have been lost, but the nameplate indicates $f_0 = 1218$ Hz, $f_L = 752.8$ Hz, and $f_H = 1970.8$ Hz (half-power frequencies). As an engineer, you are assigned to replace this filter with a new one based on the design methods of this chapter. Describe the attenuation characteristics you achieve.

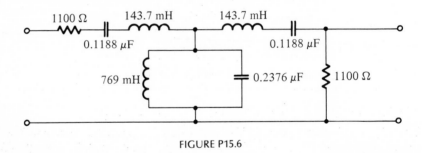

FIGURE P15.6

15.7 The following specifications are made on a bandpass filter which is to have a Butterworth response. The prototype is to be third order, $f_0 = 1$ kHz, and the two half-power frequencies are 1250 and 800 Hz. The terminating resistors are $R_1 = R_2 = 1000 \ \Omega$. Design the filter using magnitude scaling to achieve elements in a practical range.

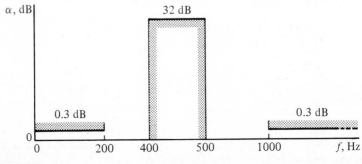

FIGURE P15.11

15.8 The prototype for a filter is an $n = 2$ Butterworth. A bandpass characteristic is required in which $\omega_0 = 1000$ rad/s and the half-power bandwidth is 1000 rad/s. Using the methods of this chapter, design the bandpass filter.

15.9 This problem requires the design of a bandpass filter to meet the specifications set for Problem 7.9. In contrast to the design of Chapter 7, this design is to be made using element simulation methods based on a doubly terminated ladder filter. Arrange your design in the general form as shown in Fig. 15.41.

15.10 The starting point for this problem is the circuit given in Problem 13.8 which is a lowpass Cauer (elliptic) filter. Find a realization which is equivalent to the passive filter using the methods of this chapter. Magnitude scale to obtain practical element sizes in your design.

15.11 Design a bandstop filter which satisfies the attenuation specifications of Fig. P15.11. Use a Butterworth prototype in your design, and make use of the methods of this chapter in completing your design. The use of 1-kΩ terminating resistors is suggested.

15.12 The design required by this problem is similar to that of Example 15.5 except that the prototype is that shown in Fig. P15.12 which is a seventh-order Chebyshev lowpass filter with a ripple width of 0.5 dB normalized so that the 3-dB bandwidth is 1.0 rad/s. The bandpass filter is required to have $\omega_0 = 1000$ rad/s and a 3-dB bandwidth of 1000 rad/s. The filter is to be driven by a voltage source with an internal resistance of 1000 Ω. Arrange your design in the form of the filter given in Fig. 15.41.

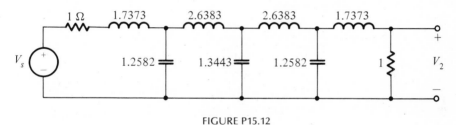

FIGURE P15.12

15.13 Repeat Problem 15.12 using the fifth-order prototype shown in Fig. P15.13. This prototype has a Chebyshev lowpass response with a ripple width of 0.5 dB normalized such that the 3-dB bandwidth is 1.0 rad/s.

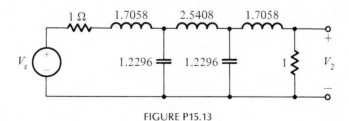

FIGURE P15.13

15.14 Repeat Problem 15.12 using as the prototype the entry in Table 14.3 for $n = 6$.

15.15 Repeat Problem 15.12 using as the prototype the entry in Table 14.3 for $n = 7$.

15.16 What passive circuit has the same input impedance as the circuit given in Fig. P15.16?

15.17 The circuit shown in Fig. P15.17 is a model for a tunnel diode. We wish to simulate this circuit for an experiment using the concepts of this chapter. It is specified that the simulation circuit must not contain inductors or negative resistors. Find an equivalent circuit meeting these specifications and having the same Z_{in}.

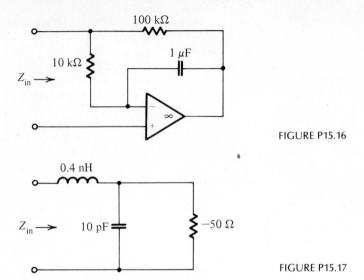

FIGURE P15.16

FIGURE P15.17

15.18 The circuit shown in Fig. P15.18 is one of low sensitivity and is due to Fliege.* By applying Kirchhoff's current law to nodes a, b, and c, show that

$$\frac{V_2}{V_1} = \frac{Y_A(Y_1Y_3 - Y_0Y_2) + Y_BY_2(Y_4 + Y_5)}{Y_1Y_3(Y_A + Y_5) + Y_2Y_4(Y_B + Y_0)}$$

Further, show that a lowpass filter results when the choice is made $Y_0 = 0$, $Y_1 = G_1 + C_1s$, $Y_2 = G_2$, $Y_3 = C_3s$, $Y_4 = G_4$, $Y_5 = G_5$, $Y_A = 0$, $Y_B = G_B$.

* Gabor C. Temes and Jack W. LaPatra, *Introduction to Circuit Synthesis and Design,* McGraw-Hill, New York, 1977, pp. 303–309.

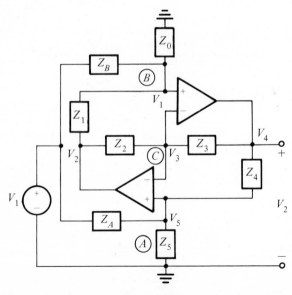

FIGURE P15.18

CHAPTER 16

Leapfrog
Simulation
of
Ladders

In the last chapter we introduced surprising innovations in the design of filters: we showed that it was possible to produce synthetic or fake elements such as an inductor or negative capacitor, and we showed that new circuit elements such as the FDNR were of great utility. The methods to be introduced in this chapter date back to an era in which analog computers were widely used. To use an analog computer to study a system—an electric circuit, a mechanical system such as the automobile, a chemical processing plant—that system was modeled by writing equations based on conservation laws. Then the equations were represented by block diagrams (or signal flow graphs). Each block represented some analog operation such as summing, differencing, or integration. When properly patched, the analog computer is the analog of the system under study. The innovation we introduce in this chapter is the use of the analog computer itself as *the* filter. Although this has a ring of the impractical, such is not the case at all; the filters that result are competitive with those studied in previous chapters.

16.1 LADDER SIMULATION

The circuit we wish to simulate is the ladder represented in Fig. 16.1. Our motivation in using this circuit is that we wish to exploit its low sensitivity to parameter changes, as found in Chapter 14. In Fig. 16.1 all series elements or combinations of elements are represented by their admittances, all shunt elements or combinations of elements are represented by impedances. Using the reference directions indicated, the branch relationships are

$$I_1 = Y_1(V_1 - V_2) \tag{16.1}$$

$$V_2 = Z_2(I_1 - I_3) \tag{16.2}$$

$$I_3 = Y_3(V_2 - V_4) \tag{16.3}$$

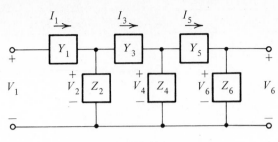

FIGURE 16.1

$$V_4 = Z_4(I_3 - I_5) \tag{16.4}$$

$$I_5 = Y_5(V_4 - V_6) \tag{16.5}$$

$$V_6 = Z_6 I_5 \tag{16.6}$$

We recognize that these equations are not a unique description of the ladder network, but the pattern of writing successive voltage and current relationships will prove to be useful.

The basic operations used to represent algebraic equations as block diagrams are shown in Fig. 16.2a. Here G operates on X to produce Y_a; X_1 and X_2 are added to produce Y_b. To route a quantity Y to several other blocks, a pick-off point is employed.* These elementary operations will be applied to Eqs. (16.1) and (16.2). If Eq. (16.1) is written in the form

$$I_1 = Y_1[V_1 + (-V_2)] \tag{16.7}$$

then these operations are as depicted in Fig. 16.3a. In a similar manner, Eq. (16.2) is represented by Fig. 16.3b. To produce $-V_2$ and $-I_3$ as required by these block diagrams, V_2 and I_3 are inverted by the operation -1. If we continue this procedure until all six equations are drawn as in Fig. 16.3, then the block diagram shown in Fig. 16.4 is the result. Notice that in this figure the repeated use of the two operations of Fig. 16.3 makes up the complete circuit. For larger ladders than that of Fig. 16.1, the same pattern exists with a larger block diagram needed for a complete representation.

The repeated use of negative feedback in Fig. 16.4 is shown in a slightly different form in Fig. 16.5. To Girling and Good,[†] who originated the methods described in this chapter, this pattern suggested the children's game called *leapfrog*, and they so named the circuits that result.

Our next objective is to convert the block diagram representation of the ladder network into a form that can be implemented. In an active circuit analog of the block diagram we cannot realize currents, and so we must simulate the cur-

---

* Some writers prefer to use signal flow graphs; the corresponding operations are shown in Fig. 16.2b. We use block diagrams since the form relates directly to the active circuits used to realize the blocks.
† F. E. J. Girling and E. F. Good, "The Leapfrog or Active-Ladder Synthesis," *Wireless World,* vol 76, pp. 341–345, July 1970. This presentation is continued in two related papers which also appeared in *Wireless World:* "Applications of the Active-Ladder Synthesis," pp. 445–450, Sept. 1970, and "Bandpass Types," pp. 505–510, Oct. 1970. Both Girling and Good were engineers with the Royal Radar Establishment, Great Malvern, England.

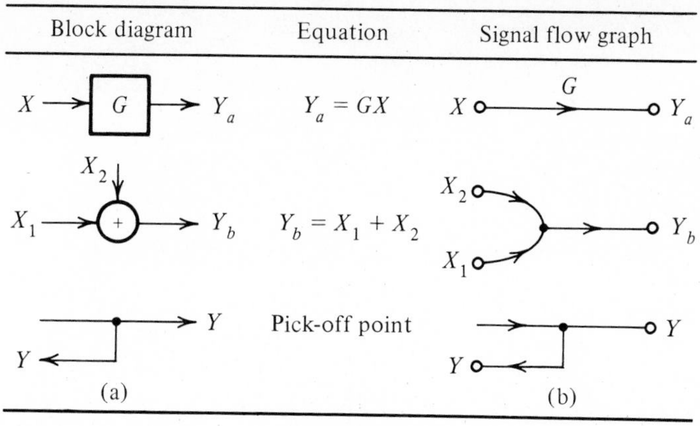

Block diagram
Equation
Signal flow graph
X
G
Y_a
Y_a = GX
X
G
Y_a
X_2
X_1
+
Y_b
Y_b = X_1 + X_2
X_2
X_1
Y_b
Y
Y
Pick-off point
Y
Y
(a)
(b)

FIGURE 16.2

rents as voltages. In terms of symbols, we will let I_1 be replaced by V_{I_1}. Similarly, the blocks will be implemented with active amplifiers characterized by transfer functions. Hence we will substitute T_{Z_2} for Z_2. With these changes the branch relationships of Eqs. (16.1)–(16.6) have the analog forms:

$$V_{I_1} = T_{Y_1}(V_1 - V_2) \tag{16.8}$$

$$V_2 = T_{Z_2}(V_{I_1} - V_{I_3}) \tag{16.9}$$

$$V_{I_3} = T_{Y_3}(V_2 - V_4) \tag{16.10}$$

$$V_4 = T_{Z_4}(V_{I_3} - V_{I_5}) \tag{16.11}$$

$$V_{I_5} = T_{Y_5}(V_4 - V_6) \tag{16.12}$$

$$V_6 = T_{Z_6}V_{I_5} \tag{16.13}$$

In this form the transfer functions can be realized as *RC*-op-amp circuits with all variables being voltages. The resulting simulation of the ladder network is now shown in Fig. 16.6.

There remains one additional change to be made before we turn to actual implementation. The network of Fig. 16.6 contains five unity-gain inverting amplifiers in each of the five feedback paths. It is usually preferred that these amplifiers be in the feed-forward parts of the network. This can be accomplished by

−V_2
V_1
+
Y_1
I_1
V_1 −V_2
I_1
+
I_1 −I_3
Z_2
V_2
−I_3
(a)
(b)

FIGURE 16.3

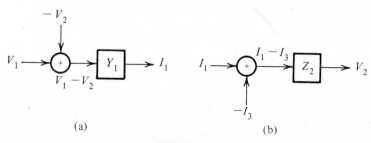

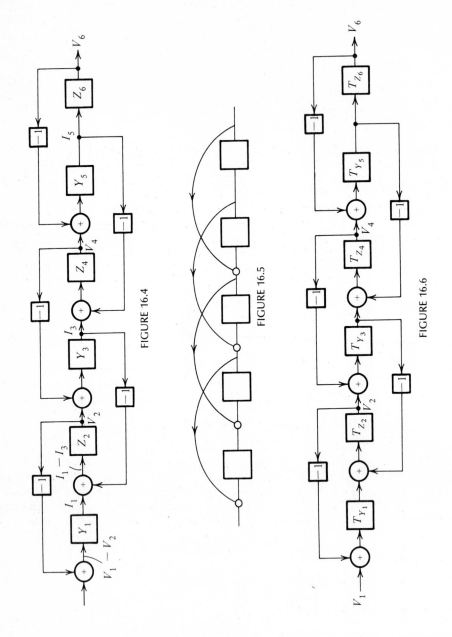

FIGURE 16.4

FIGURE 16.5

FIGURE 16.6

462

modifying Eqs. (16.8)–(16.13) and then observing the corresponding circuit interpretation. Thus,

$$V_{I_1} = T_{Y_1}(V_1 - V_2) \qquad (16.14)$$

$$-V_2 = -T_{Z_2}(V_{I_1} - V_{I_3}) \qquad (16.15)$$

$$-V_{I_3} = T_{Y_3}(-V_2 + V_4) \qquad (16.16)$$

$$V_4 = -T_{Z_4}(-V_{I_3} + V_{I_5}) \qquad (16.17)$$

$$V_{I_5} = T_{Y_5}(V_4 - V_6) \qquad (16.18)$$

$$-V_6 = -T_{Z_6}V_{I_5} \qquad (16.19)$$

These equations are interpreted in Fig. 16.7a, and the only change from the circuit of Fig. 16.6 is that the overall leapfrog realization is now an inverting one with $-V_6$ replacing V_6. Similarly, Eqs. (16.8)–(16.13) may be modified in a somewhat different form:

$$-V_{I_1} = -T_{Y_1}(V_1 - V_2) \qquad (16.20)$$

$$-V_2 = T_{Z_2}(-V_{I_1} + V_{I_3}) \qquad (16.21)$$

$$V_{I_3} = -T_{Y_3}(-V_2 + V_4) \qquad (16.22)$$

$$V_4 = T_{Z_4}(V_{I_3} - V_{I_5}) \qquad (16.23)$$

$$-V_{I_5} = -T_{Y_5}(V_4 - V_6) \qquad (16.24)$$

$$-V_6 = T_{Z_6}(-V_{I_5}) \qquad (16.25)$$

which may be interpreted as the network of Fig. 16.7b. It is interesting to note that the signs of the transfer functions alternate as we progress from V_1 to V_6, and that around each of the loops there is one negative transfer function resulting in the required negative feedback.

To illustrate the procedure for the design of a filter by the leapfrog method, suppose that we are required to realize a fourth-order Butterworth filter with a half-power frequency of 10,000 rad/s, and a doubly terminated realization is required. From the tables of Chapter 14, we obtain the network shown in Fig. 16.8a. Comparing this network with that of Fig. 16.1, we identify Y_1, Z_2, Y_3, and Z_4 as shown in Fig. 16.8b. From these identifications, we see that

$$Y_1 = \frac{1/L_1}{s + R_1/L_1} \qquad (16.26)$$

$$Z_2 = \frac{1}{C_2 s} \qquad (16.27)$$

$$Y_3 = \frac{1}{L_3 s} \qquad (16.28)$$

$$Z_4 = \frac{1/C_4}{s + 1/C_4 R_2} \qquad (16.29)$$

where the numerical values of all circuit parameters are known.

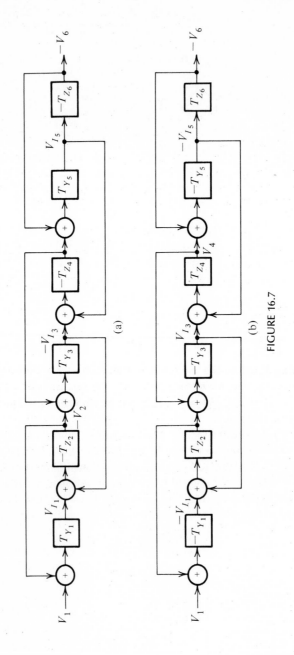

FIGURE 16.7

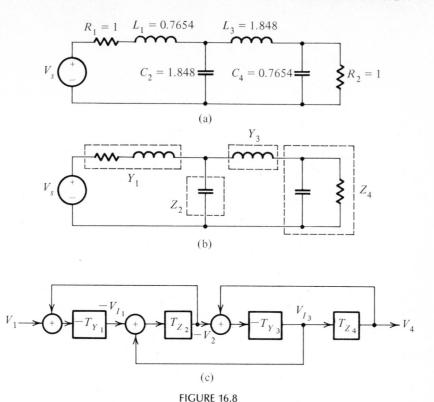

FIGURE 16.8

Our next objective is to determine the transfer functions for use in the leap-frog realization of Fig. 16.8, and also to determine circuits that can be used in each of the blocks. From Eqs. (16.26)–(16.29) it is evident that we will require integrators and lossy integrators, in addition to circuits that will accomplish the summing and sign reversal. A catalog of the required circuits is given in Fig. 16.9. Since all of these circuits are inverting, we will rearrange the equations for Y_1, Z_2, Y_3, and Z_4 with the objective of matching the transfer functions required in the leapfrog circuit of Fig. 16.8c. Identifying the impedances and admittances with transfer functions, we have

$$T_{Y_1} = \frac{-1/L_1}{s + R_1/L_1} \tag{16.30}$$

$$T_{Z_2} = \frac{-1}{C_2 s} \times (-1) \tag{16.31}$$

$$T_{Y_3} = \frac{-1}{L_3 s} \tag{16.32}$$

$$T_{Z_4} = \frac{-1/C_4}{s + 1/R_2 C_4} \times (-1) \tag{16.33}$$

We let all resistors have the value of 1 Ω. We do so because $R_1 = R_2 = 1$ in the ladder network, and we arbitrarily select 1 Ω for the circuits given in Fig. 16.9,

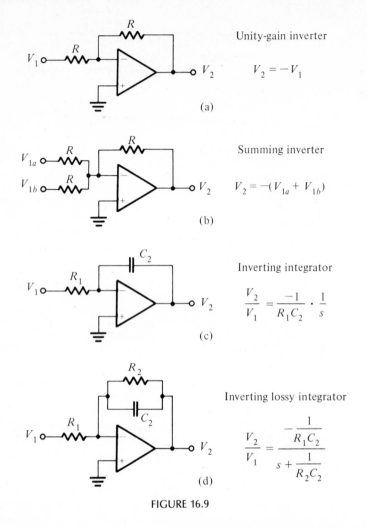

FIGURE 16.9

knowing that we may use magnitude scaling in the final design. The resulting design is shown in Fig. 16.10 with a designation of the purpose of each of the circuits used. The final step in the design is scaling. Since it is specified that the cutoff frequency be 10,000 rad/s, it is required that $k_f = 10^4$. If we select $k_m = 10^4$, then the final design becomes that of Fig. 16.11. We see that all resistors have one value, and that only two different capacitor sizes are required. Since the design was based on a simulation of a doubly terminated ladder network, the sensitivity of this realization is the same as that of the ladder which has low sensitivity.

In this particular design the realization given in Fig. 16.7b was used. A different design results by using the leapfrog realization given in Fig. 16.7a. It is also possible to use noninverting circuits in place of those given in Fig. 16.9. One such circuit is shown in Fig. 16.12, which is a noninverting integrator for which

$$\frac{V_2}{V_1} = T = \frac{+2}{RC}\frac{1}{s} \qquad (16.34)$$

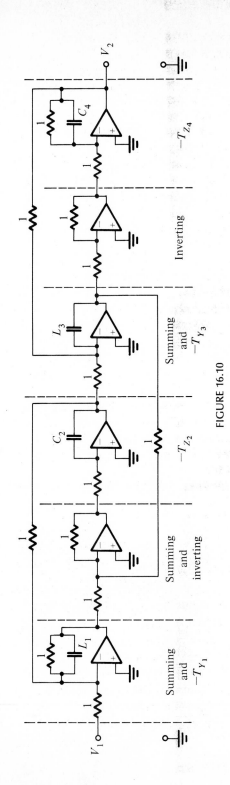

FIGURE 16.10

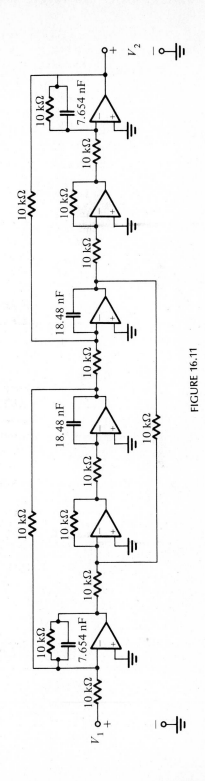

FIGURE 16.11

467

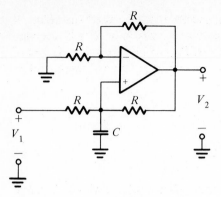

FIGURE 16.12

Clearly such a circuit can be used to replace an inverting integrator in cascade with a unity-gain inverting circuit.

In summary, leapfrog design of lowpass filters is accomplished in the following steps:

1. From specifications for the lowpass filter, determine a suitable lowpass prototype. It is usual to employ a doubly terminated prototype, but other circuits such as a singly terminated prototype can be used.
2. Identify the various Y_i and Z_i in the form of Fig. 16.1.
3. Select a leapfrog block diagram to simulate the circuit. These are given in Fig. 16.7.
4. Find an active circuit that realizes each of the blocks. These may be taken from a catalog of realizations. Adjust the gain if necessary.
5. Arrange the circuit with necessary summers and inverters.
6. Frequency scale to meet specifications; magnitude scale to give convenient element values.

16.2 BANDPASS LEAPFROG FILTERS

The leapfrog method is especially well suited to the design of bandpass filters. Starting with the prototype filter as in the last section in the design of lowpass filters, we apply the lowpass to bandpass frequency transformation as developed in Chapter 11 and shown in Fig. 16.13. In this figure BW is the bandwidth and ω_0

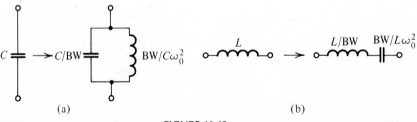

(a)   (b)

FIGURE 16.13

the resonant frequency, which are the usual specifications for the bandwidth filter. It is often possible to combine the terminating resistors R_1 or R_2 with the resonant circuits of Fig. 16.13, giving the circuits of Fig. 16.14. We will make use of the admittance of the circuit of Fig. 16.14a and the impedance of the circuit of Fig. 16.14b. Routine analysis gives these as

$$Y = \frac{(1/L)s}{s^2 + (R/L)s + 1/LC} \tag{16.35}$$

$$Z = \frac{(1/C)s}{s^2 + (1/RC)s + 1/LC} \tag{16.36}$$

Letting $R = 0$ in Eq. (16.35) gives the admittance of the lossless series circuit, and $R = \infty$ in Eq. (16.36) gives the impedance of the lossless parallel circuit.

In Chapter 14 we learned that doubly terminated lossless ladder networks always come in pairs, two alternative realizations. In selecting the prototype to be used for bandpass filter design we have our choice as to which of the two to use. We can also alter the form of the realization by a source transformation from a voltage source to a current source. There is also some flexibility in deciding how to incorporate the terminating resistors into subnetworks.

A fourth-order doubly terminated prototype filter is shown in Fig. 16.15a together with the network that results from the lowpass to bandpass frequency transformation. We see that R_1 is incorporated in the admittance Y_1 and also that R_2 is part of Z_4. Similarly the third-order structure shown in Fig. 16.15b shows R_1 as part of Z_1 and R_2 as part of Z_3.

Examining Fig. 16.15 we see that there are two kinds of RLC circuits which must be simulated to find a leapfrog realization having the forms shown in Fig. 16.14 and described by Eqs. (16.35) and (16.36). Those located at the ends of the network incorporate resistors, and the internal networks do not. Before we can proceed, we will review active networks that can be used to simulate the RLC passive networks of Fig. 16.15.

16.3   ACTIVE RESONATORS

The form of the transfer functions which corresponds to Eqs. (16.35) and (16.36) is that of a second-order bandpass filter. Of the bandpass filters we have studied in the past, we single out two for this application: the biquad circuit of Chapter 5

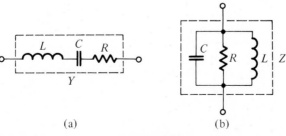

(a)   (b)

FIGURE 16.14

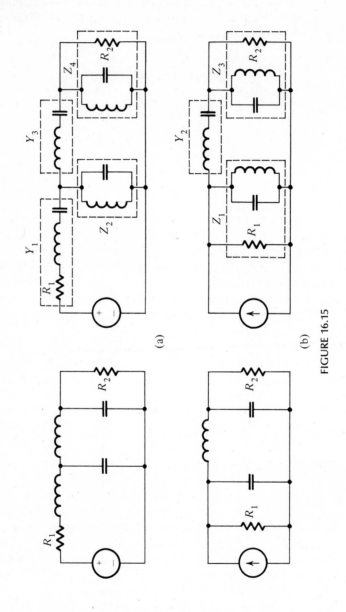

FIGURE 16.15

(a)

(b)

470

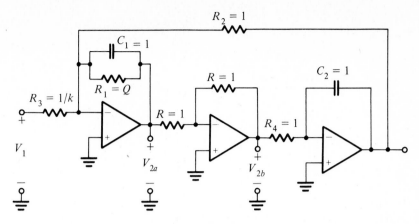

FIGURE 16.16

and the Friend circuit of Chapter 7 together with methods for obtaining enhanced Q.

We first rearrange the stages of the biquad filter shown in Fig. 5.9 by interchanging the second and third stages as illustrated in Fig. 16.16. This does not change the transfer function of the filter, but it does provide us with an inverting output shown as V_{2a} and a noninverting output V_{2b}. The general form of the transfer function given by Eq. (5.31) is repeated here:

$$T = \frac{V_2}{V_1} = \frac{(-1/R_3C_1)s}{s^2 + (1/R_1C_1)s + 1/R_2R_4C_1C_2} \qquad (16.37)$$

With the selection of elements indicated in Fig. 16.16, this transfer function simplifies to a normalized form with $\omega_0 = 1$, which is

$$T = \frac{ks}{s^2 + (1/Q)s + 1} \qquad (16.38)$$

where k may be adjusted by R_3 to meet gain requirements since

$$k = \frac{1}{R_3} \qquad (16.39)$$

The Q specification is met by adjusting R_1. Frequency scaling is employed to meet the ω_0 specification. Finally the output connection determines whether the filter is inverting or noninverting. We observe that if R_1 is made infinite (by removal), this corresponds to the $Q = \infty$ case representing LC circuits without resistors.

The second circuit to be considered is the Friend circuit shown in Fig. 16.17 for which the transfer function is

$$T = \frac{-2Qs}{s^2 + (1/Q)s + 1} \qquad (16.40)$$

From this equation we see two problems compared to the biquad circuit: there is no simple way to adjust the gain without additional elements, and there is no way

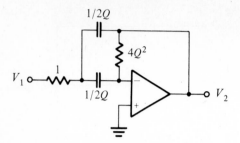

FIGURE 16.17

to make this circuit into a noninverting circuit without using an extra unity-gain inverting stage in cascade.

To meet the gain adjustment requirement, we make use of the technique introduced in Chapter 7. The 1-Ω resistor of the Friend circuit is replaced by a voltage-divider circuit which reduces the gain by a factor of k and at the same time appears as a 1-Ω resistor in the Thévenin sense. The modified circuit is shown in Fig. 16.18, for which

$$T = \frac{-2kQs}{s^2 + (1/Q)s + 1} \qquad (16.41)$$

If we are required to sum voltages and reduce the gain at the same time, then the circuit given in Fig. 16.19 may be used. If $T(s)$ is as given in Eq. (16.40), then

$$V_2(s) = T(s)(k_1 V_{1a} + k_2 V_{1b}) \qquad (16.42)$$

These quantities are defined in the figure.

Finally we consider a technique for realizing infinite Q in order that we may simulate the lossless LC circuits. This is accomplished using the Q enhancement techniques of Chapter 7. We let Q_0 be the design value of Q in the Friend circuit, which we wish to enhance to the value Q_{new}. The circuit that accomplishes this is shown in Fig. 16.20. If we let $Q_{new} = \infty$, then the transfer function becomes

$$T(s) = \frac{[(1 + 2Q_0^2)/Q_0])s}{s^2 + 1} \qquad (16.43)$$

Here we may select Q_0 to obtain the required gain, or we may combine this circuit with the input resistive circuits of Fig. 16.18 or 16.19 and choose k_1 and k_2 to reduce the gain appropriately.

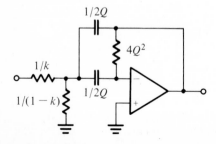

FIGURE 16.18

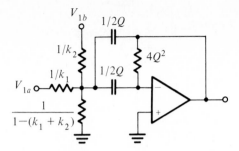

FIGURE 16.19

16.4 BANDPASS LEAPFROG DESIGN

Suppose that we are required to design a filter using the leapfrog method. From specifications it is found that a fourth-order Butterworth prototype will suffice. The bandpass filter is to have half-power frequencies at 800 and 1250 Hz. From this information we see that

$$\omega_0 = (\omega_1 \omega_2)^{1/2} \times 2\pi = 6283.2 \text{ rad/s} \tag{16.44}$$

and

$$BW = (1250 - 800) \times 2\pi = 2827 \text{ rad/s}$$

For our design we make use of the normalized frequencies of $\omega_0 = 1$ and BW = 0.45 and later frequency scale with $k_f = 6283.2$.

The fourth-order Butterworth prototype circuit is given in Fig. 16.8b. Making use of the frequency transformations given in Fig. 16.13 together with the prescribed values of ω_0 and BW, we obtain the circuit given in Fig. 16.21, which we wish to simulate using the leapfrog arrangement given in Fig. 16.8c. Using the numerical values of Fig. 16.21, we find that

$$Y_1 = \frac{0.5879s}{s^2 + 0.5879s + 1} \tag{16.45}$$

$$Z_2 = \frac{0.2435s}{s^2 + 1} \tag{16.46}$$

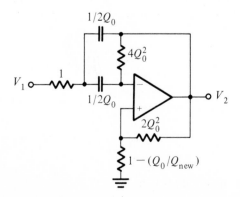

FIGURE 16.20

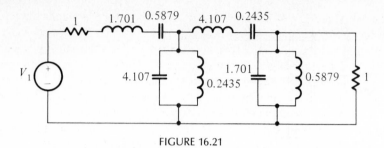

FIGURE 16.21

Further,

$$Y_3 = Z_2 \quad \text{and} \quad Z_4 = Y_1 \tag{16.47}$$

Thus we need circuits to give transfer functions corresponding only to the Y_1 and Z_2 given by these equations.

We first consider the design of T_{Y_1} and T_{Z_4}. Comparing Eq. (16.45) with that given in Fig. 16.22a, we see that $1/Q = 0.5879$, so that $Q = 1.701$. Equating numerators,

$$2k_1 Q = 0.5879 \quad \text{so that} \quad k_1 = 0.1728 \tag{16.48}$$

Turning to the design of T_{Z_2} and T_{Y_3} we compare Eq. (16.47) with the equation of Fig. 16.22b. We must assume a value for Q_0 and we let it be 1.0. This means that $k_2 = 0.2435/3 = 0.0812$. With this we have determined all of the parameters of the modified Friend circuits of Fig. 16.22.

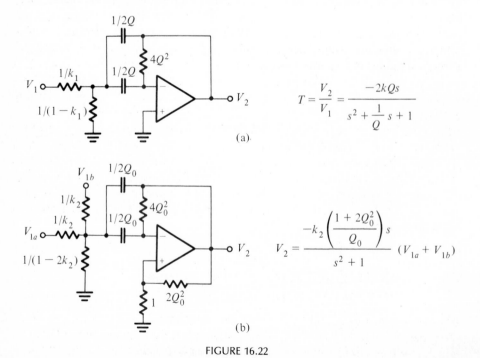

$$T = \frac{V_2}{V_1} = \frac{-2kQs}{s^2 + \dfrac{1}{Q}s + 1}$$

(a)

$$V_2 = \frac{-k_2 \left(\dfrac{1 + 2Q_0^2}{Q_0} \right) s}{s^2 + 1} (V_{1a} + V_{1b})$$

(b)

FIGURE 16.22

The next question is how to fit these together to form a leapfrog circuit. The particular circuit we select is shown in Fig. 16.23b, which is a special case of that originally given in Fig. 16.7b. From this figure we see that we must pay particular attention to inverting or noninverting circuits, and we must make provision for summing at the three specified points in the circuit. The manner in which this is accomplished is shown in Fig. 16.23a. To satisfy all of the summing requirements, one circuit had to be used for both summing and inverting, as well as two circuits of the form shown in Fig. 16.22b. One circuit provides the needed inverting. In general this step is carried out by fitting the circuits together to realize the leapfrog form; it will require following a cut-and-try procedure.

Once the choice of configuration as in Fig. 15.23a is made, we use the parameters determined from the transfer functions as related to the circuits of Fig. 16.22. The resulting normalized parameters are shown in Fig. 16.24a. For the ordinary inverting stages we have chosen a unit value for the resistors, but clearly this choice is arbitrary.

The final step in the design is scaling. We are constrained to make $k_f = 6283.2$, as discussed in connection with Eq. (16.44). However, we may magnitude scale to provide convenient element values. After a few trials, the choice is made to make half of the capacitors have the value of 0.01 μF. Using the scaling equation

$$C_{\text{new}} = \frac{1}{k_f k_m} C_{\text{old}} \tag{16.49}$$

we find that $k_m = 7960$. This choice results in the design circuit given in Fig. 16.24b. This example illustrates the ease with which realizations are found using the leapfrog method.

Doubly terminated lossless networks always appear in two alternative forms, which we have previously identified as minimum inductance and minimum capacitance forms. The form that is an alternative to that used in the last example is shown in Fig. 16.25a, a third-order Butterworth prototype filter. Once the lowpass to bandpass transformation is applied to this circuit, we obtain the one shown in Fig. 16.25b, in which Z_1, Y_2, and Z_3 are identified. We have also used source transformation such that it is driven by a current source and the input resistor appears in a parallel position. Since our objective is to find the analog of the passive bandpass circuit, it does not matter whether it is driven by a current source or by a voltage source. Thus the leapfrog realization is shown in Fig. 16.25c, and the procedure for completing the filter design remains exactly as illustrated by the last example. In Fig. 16.25b element values were obtained with $\omega_0 = 1$ and BW = 0.45.

16.5 GIRLING-GOOD FORM OF LEAPFROG

The method of this section was completely described in the Girling and Good 1970 papers cited in Sec. 16.1. There they introduced an alternative form in which leapfrog circuits can be drawn. We introduce this alternative form now

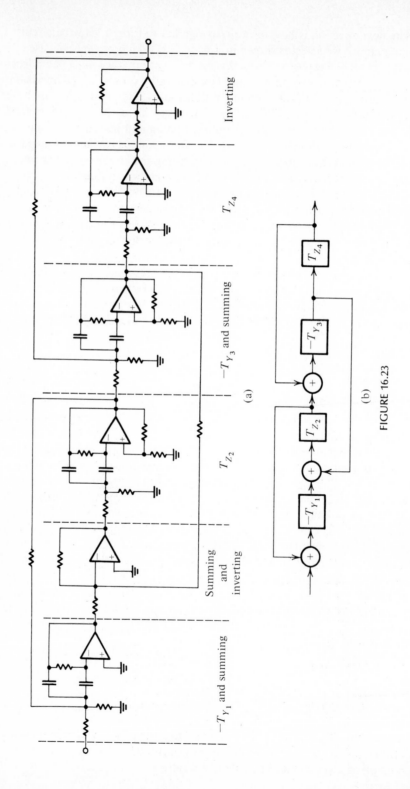

FIGURE 16.23

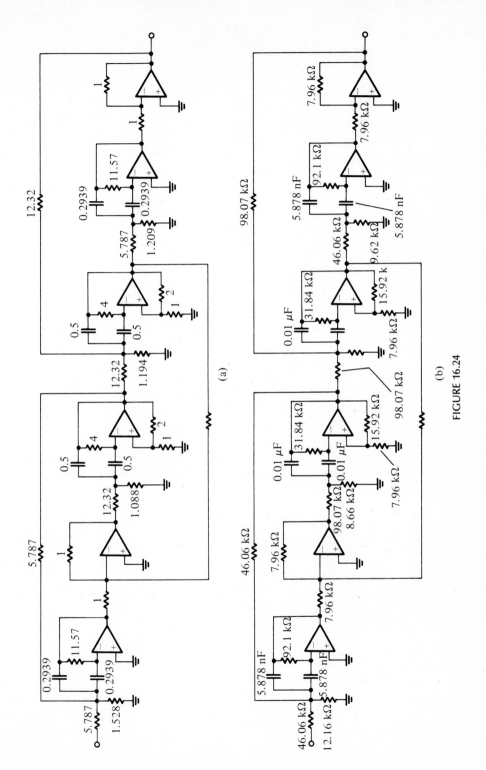

FIGURE 16.24

477

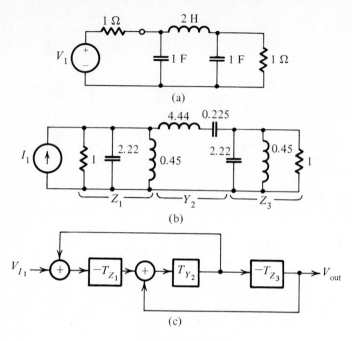

(a)

(b)

(c)

FIGURE 16.25

in preparation for the next chapter, which makes extensive use of it. In doing so, we make use of the lowpass case, although its extension to the bandpass case considered earlier in the chapter is routine.

A simple leapfrog structure is shown in Fig. 16.26a. In Fig. 16.26b we show the Girling–Good form which is completely equivalent and might be called a *topological transformation*. For comparison, specific points on the two block diagrams are represented by the letters $A, B, \ldots, J$. We see that the Girling–Good

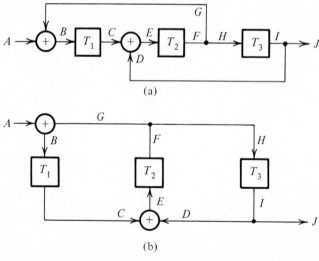

(a)

(b)

FIGURE 16.26

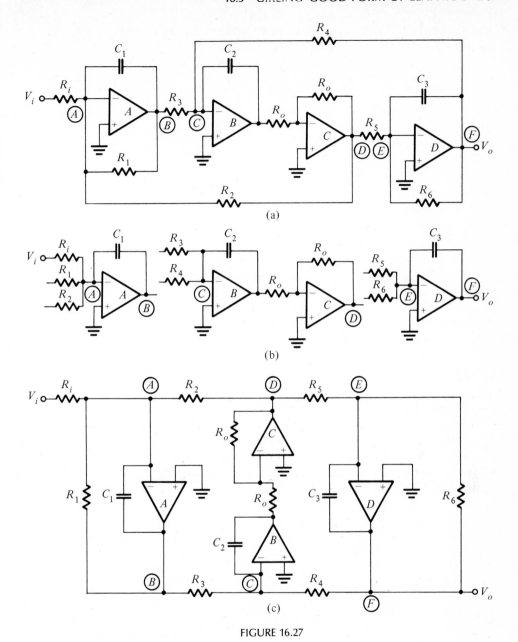

FIGURE 16.27

representation is more compact, but it is less evident, at least to the beginner, what is being accomplished in each stage.

A specific case of a leapfrog realization of a lowpass filter is shown in Fig. 16.27a. In making this drawing, the conventional lossy integrator has been modified by separating R_1 from C_1 and R_6 from C_3. The three stages of the circuit are shown in Fig. 16.27b in a form that emphasizes the various inputs to each stage. Finally in Fig. 16.27c we show the Girling–Good circuit. The reader should check all points in the three figures to understand the equivalence of the connections to the three integrator circuits.

Before continuing, we will review some of the general properties of the doubly terminated lowpass filter as related to the leapfrog realization. A fourth-order filter together with the block diagram representation of its leapfrog realization is shown in Fig. 16.28. Some of these properties are different for odd-ordered filters, but these relate primarily to the form of the resistive terminations and the sign of the output voltage, and will be evident by comparing results with those shown in Fig. 16.27.

1. Excluding provisions for the terminating resistors, all block diagrams represent integrators, some inverting and some noninverting. The realization is sometimes known as an all-integrator circuit.
2. As we progress from input toward output, the signs of the integrators alternate. Noninverting integrators are used to simulate series inductors, and inverting integrators are used to simulate shunt capacitors.
3. All summing is accomplished before integration takes place. By the sign changes described under 2, only summing is required (not differencing), and this is done by the common summing circuit introduced in Chapter 2.
4. The sign of the output voltage compared to the input may be positive or negative. This is ordinarily not important in design, but can be corrected by an additional stage of inversion if necessary.

The block diagram of Fig. 16.28b is general with one block for each circuit element in Fig. 16.28a, but it is usual practice to incorporate the terminating resistors in other blocks. The manner in which this is accomplished is shown by starting with the groupings in Fig. 16.29 for which the leapfrog flow graph is as shown in Fig. 16.30. Our objective is to modify the leapfrog flow graph of Fig. 16.28 to match that of Fig. 16.30. One of the ways in which this can be accomplished is shown in Fig. 16.31a with the Girling–Good form of Fig. 16.31b. One of the possible circuit realizations of the flow graph of Fig. 16.31 is shown in Fig. 16.32. In this case all summing resistors were chosen to have the value of R_i. Although this is not necessary, it is convenient since $R_i = 1$ in all tables and all resistors with the exception of R_o can be scaled to the same practical value. The same choice is made for the value of R in the inverting stages. Finally the modification of the circuit to that shown in Fig. 16.29 changed the output voltage to that voltage across R_o and L_4, identified simply as V_x. The voltage across R_o, which is the desired V_o, is easily identified in Fig. 16.32. With the choice we have made of $R_i = R = 1$, there are only five variables remaining in determining a realization; these are C_1, L_2, C_3, L_4, and R_o, the values of four capacitors and one resistor.

As an example of leapfrog design, consider the problem of designing a fourth-order Chebyshev lowpass filter for which the ripple width is 0.1 dB and the ripple bandwidth is 10,000 rad/s. Further it is required that the attenuation be adjusted such that the minimum value in the pass band is 0 dB.

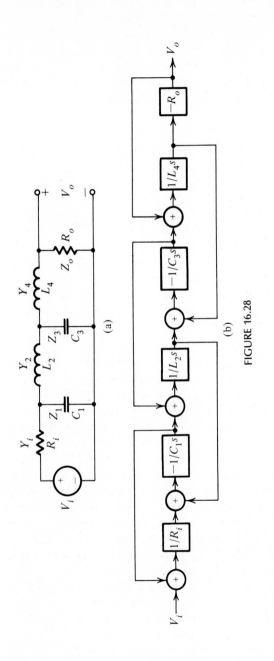

(a)

(b)

FIGURE 16.28

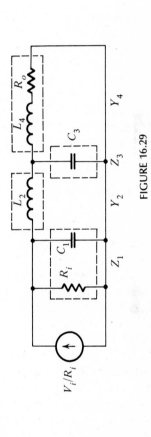

FIGURE 16.29

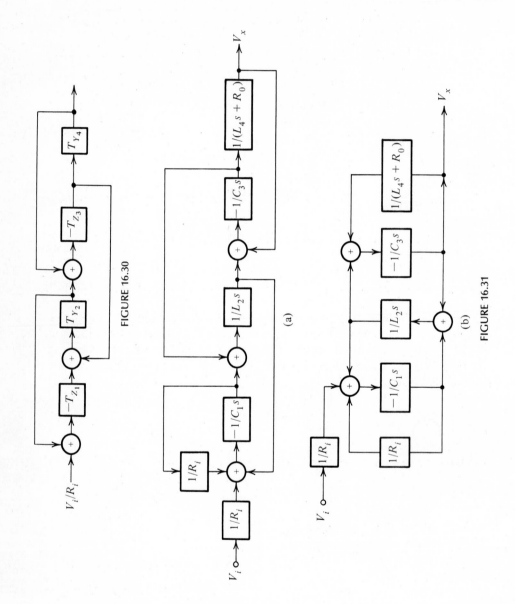

FIGURE 16.30

(a)

(b)

FIGURE 16.31

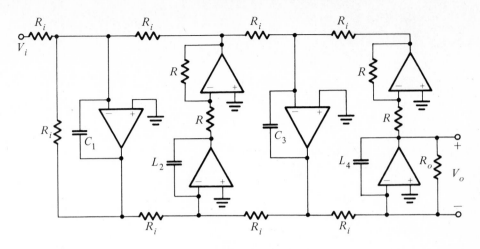

FIGURE 16.32

From Table 14.3 we find that the following element values are specified for $n = 4$ and $\alpha_{max} = 0.1$ dB:

$$R_i = 1, \quad C_1 = 1.108787, \quad L_2 = 1.306184, \quad C_3 = 1.770351,$$
$$L_4 = 0.818075, \quad R_o = 0.737811 \tag{16.50}$$

Inserting these values into Fig. 16.32 gives a normalized realization as shown in Fig. 16.33.

Next we consider the problem of adjusting the overall circuit gain such that the minimum attenuation is 0 dB in the pass band. In Chapter 14 we discussed the fact that a lowpass ladder at zero frequency (dc) has the form shown in Fig. 16.34, since all inductors become short circuits and all capacitors open circuits. For this circuit

$$\frac{V_o}{V_i}(0) = T(0) = \frac{R_o}{R_i + R_o} \tag{16.51}$$

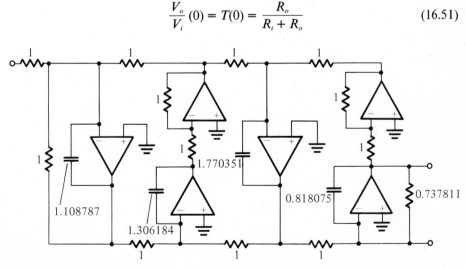

FIGURE 16.33

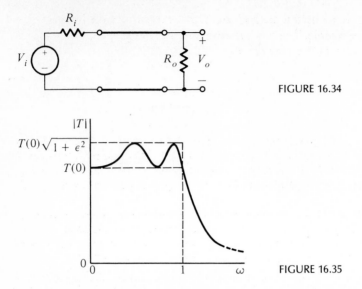

FIGURE 16.34

FIGURE 16.35

For the particular problem under study, $T(0) = 0.42456$. In terms of the Chebyshev response shown in Fig. 16.35, the maximum value of $|T|$ is greater than $T(0)$ by the factor $(1 + \epsilon^2)^{1/2}$, as found in Chapter 7, where $\epsilon^2 = 10^{0.1\alpha_{max}} - 1$. For the values of this problem $|T|_{max} = 0.42952$. But we require that this value be 1.0 corresponding to the 0-dB requirement, which means that the overall gain must be increased by $1/0.42952 = 2.3282$. This is accomplished by changing R_i to $R_{in} = 0.42952R_i$, while at the same time leaving all of the other other resistors in Fig. 16.32 as $R_i = 1$.

The specifications require that in denormalizing we scale such that $k_f = 10^4$. We make the arbitrary choice that $k_m = 10^4$ (to make most of the resistors in the circuit have the value of 10 k). Using Eq. (16.49) to obtain capacitor values gives us the final design shown in Fig. 16.36.

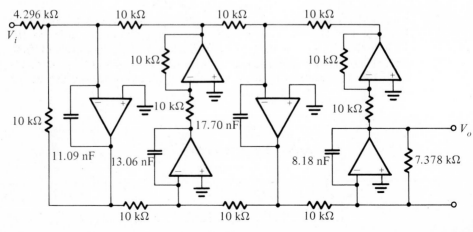

FIGURE 16.36

The simplicity of the design method for leapfrog circuits in the lowpass case is evident from this example. The method with a new element substituting for resistors will be extended in the next chapter.

PROBLEMS

16.1   Design a lowpass filter using the leapfrog method. The prototype for this problem is a fifth-order Butterworth as described in Table 14.1. The half-power frequency is to be 1000 rad/s. Your design should have the appearance of the circuit of Fig. 16.10 and should be magnitude-scaled to give elements in a practical range of values.

16.2   Repeat Problem 16.1 if the prototype is changed to a sixth-order Butterworth filter as described in Table 14.1, but other conditions remain unchanged.

16.3   Repeat Problem 16.1 if the prototype is changed to a fifth-order Chebyshev lowpass filter having a ripple width of 0.1 dB as described in Table 14.3

16.4   This problem requires the design of a bandpass filter making use of the leapfrog method. The prototype for this problem is a third-order Butterworth filter. The two half-power frequencies of the bandpass filter are 1960 and 2040 Hz. Magnitude scale to obtain practical element values.

16.5   Design a bandpass filter using the leapfrog method for which the prototype is a fourth-order Butterworth filter. For the filter, $f_0 = 1000$ Hz and one of the half-power frequencies is 1344 Hz. Magnitude scale your filter to give reasonable element values.

16.6   Make use of the leapfrog method to design a bandpass filter for which the prototype is a fourth-order Butterworth filter. For this problem, $f_0 = 1$ kHz and the 3-dB bandwidth is 100 Hz. Magnitude scale your filter to give reasonable element values. Your filter should have the general appearance of Fig. 16.24.

16.7   This problem requires that the lowpass filter be designed to have the appearance of the all-integrator or Girling–Good filter shown in Fig. 16.33. For this problem, the prototype is the fifth-order Chebyshev filter with 0.1-dB ripple width that is described in Table 14.3. Magnitude scale to give reasonable element values.

16.8   Repeat Problem 16.7 if the prototype is changed to the seventh-order Chebyshev filter with 0.1-dB ripple width described in Table 14.3.

16.9   Design a filter that has a bandpass characteristic by the leapfrog method using an all-integrator or Girling–Good realization in a form shown in Fig. 16.48. For this problem, the prototype is the third-order Chebyshev filter with a 0.5-dB ripple width as shown in Fig. P16.8. Design for the center frequency of $\Omega_0 = 1000$ rad/s, with a 3-dB bandwidth of 1000 rad/s. Magnitude scale to give reasonable element values.

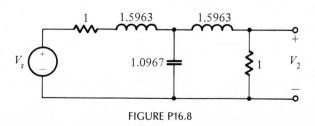

FIGURE P16.8

16.10 Repeat Problem 16.8 with the prototype changed to a third-order Chebyshev filter with a 1-dB ripple width as shown in Fig. P16.9.

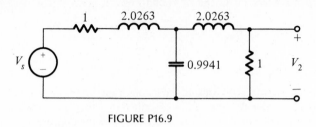

FIGURE P16.9

16.11 Repeat Problem 16.8 with the prototype changed to a fifth-order Chebyshev filter with a 1-dB ripple width as shown in Fig. P16.10.

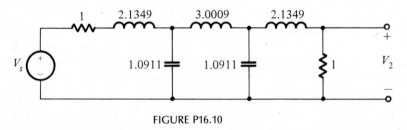

FIGURE P16.10

16.12 Repeat Problem 16.8 with the prototype changed to a fifth-order Chebyshev filter with a 0.1-dB ripple width as shown in Fig. P16.11 as obtained from Table 14.3.

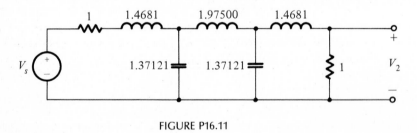

FIGURE P16.11

CHAPTER 17

Switched-
Capacitor
Filters

Innovations often result from urgent needs. The need that gave rise to the topic of this chapter was that of implementing voice-frequency filters using MOS (metal-oxide-semiconductor) technology. In this technology it is relatively easy to implement capacitors and switches as well as op amps, but difficult to construct resistors with the required accuracy. The recognition that a resistor could be approximated with two MOS switches and one capacitor was key to solving this problem, and this has given rise to a new field of filter design, especially suited to integrated-circuit implementation.

17.1 THE MOS SWITCH

Figure 17.1a shows the cross section of the integrated-circuit construction of a MOS transistor. In one specialized use of this transistor, the voltage between the source and the gate is either zero or a value larger than a threshold voltage V_{cr}, typically 1 or 2 V. In this mode of operation, the device is known as the *MOS switch*.

A circuit representation of the transistor is shown in Fig. 17.1b. The voltage that controls the switching action is shown as v_{GS}, and the path of interest is that between S and D, having resistance R_{GS}. When the transistor is in the so-called *off mode*, then R_{GS} is large, perhaps 100–1000 MΩ. When the transistor is in the *on mode*, then R_{GS} is much smaller, perhaps 10 kΩ depending on the size of the transistor on the chip. Thus the ratio of these two resistor values is of the order of 10^5 (or 100 dB). A representation of these facts is given in Fig. 17.2, along with a simplified model for the MOS switch: either a short circuit or an open circuit. This is shown in Fig. 17.3 with the switch open or closed, depending on the value of v_{GS}. Such a switch is also known as a single-pole single-throw (SPST) switch.

The voltage waveform that is used to activate the MOS switch is shown in Fig. 17.4. Here we have changed variables by letting v_{GS} be φ to conform to common usage. This waveform is generated by a clock which is an important part of digital systems. It is a pulse train which is periodic and of period T_c, as shown in

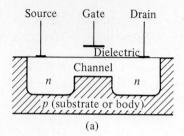

Source · Gate · Drain · Dielectric · Channel · n · n · p (substrate or body)

(a)

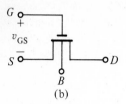

(b)

FIGURE 17.1

Condition	State	Equivalent Resistor	Model
$v_{GS} > V_{cr}$	On	10 kΩ	Short
$v_{GS} < V_{cr}$	Off	100 MΩ	Open

FIGURE 17.2

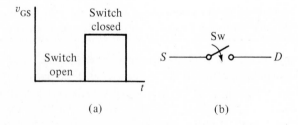

v_{GS} — Switch closed — Switch open — t

S — Sw — D

(a)   (b)

FIGURE 17.3

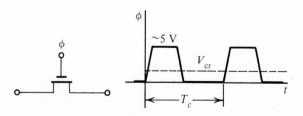

ϕ

ϕ — ~5 V — V_{cr} — T_c — t

FIGURE 17.4

the figure. The quantity $f_c = 1/T_c$ is known as the *clock frequency* of the pulse train. In digital systems the clock provides a timing standard to start or stop operations. Here we use it to turn on and then off the MOS switch.

Figure 17.5 shows the waveforms of a *two-phase clock*. It is required that ϕ_1 and ϕ_2 have the same frequency and also that they be nonoverlapping, so that when ϕ_1 is on, ϕ_2 will be off, and vice versa. If two MOS switches are connected in series as shown in Fig. 17.6 and then driven by the two waveforms of Fig. 17.5, then there will never be a direct connection from 1 to 2, since one of the two switches will always be open. As shown in Fig. 17.7a, when S_1 is open, S_2 will be closed, and when S_1 is closed, S_2 will be open. This switching action is equivalent to that shown in Fig. 17.7b, and that switch is also identified as a single-pole double-throw (SPDT) switch. This kind of switch is thus implemented by means of two SPST switches operated by a two-phase clock.

Such a scheme for switching is easily extended to four MOS switches, as shown in Fig. 17.8. With the two clock voltages ϕ_1 and ϕ_2 connected as shown, we now have a double-pole double-throw (DPDT) switch. Such a switch is useful; two applications are illustrated in Fig. 17.9. In Fig. 17.9a the switches are closed in position *a*, and the capacitor charges to the value $v_C = v_{1a} - v_{1b}$. When the switch is moved to positions *b*, then

$$v_2 = v_{1b} - v_{1a} \tag{17.1}$$

or the *difference* of two voltages is formed. If $v_{1b} = 0$ and $v_{1a} = v_1$, as in Fig. 17.9b, then

$$v_2 = -v_1 \tag{17.2}$$

or an *inverter* has been formed such that the output is the negative of the input.

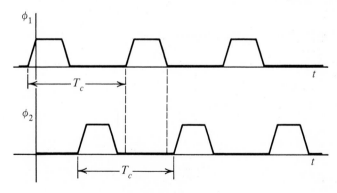

FIGURE 17.5

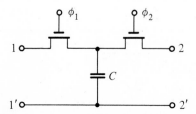

FIGURE 17.6

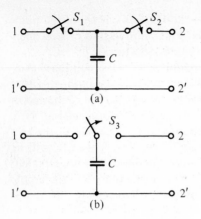

(a)

(b)

FIGURE 17.7

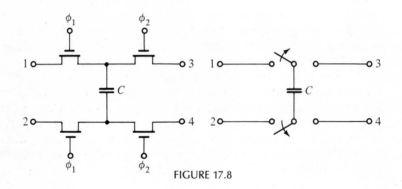

FIGURE 17.8

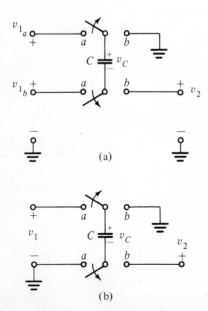

(a)

(b)

FIGURE 17.9

In the implementation of switched-capacitor circuits, consideration must be given to stray or parasitic capacitance associated with the integrated-circuit form of the capacitor. Two circuits that provide stray insensitivity are shown in Fig. 17.10. The manner in which these switching arrangements make it possible to construct a stray-insensitive inverting integrator is shown in Fig. 17.11. In that circuit, C_B represents the stray capacitance from the bottom plate to ground (the substrate or body), and C_p represents the capacitance from the top plate to ground. Typically C_B has a value of 10 percent of C_1, while C_p is smaller, perhaps 1 percent of C_1.* Consider the operation of the switches with respect to C_p. With ϕ_b closed, C_p is connected to the voltage source and will charge to V_1. However, when ϕ_a is closed, C_p will discharge and there will be no coupling to the op amp. In the case of C_B, it will always be connected to ground, either directly through ϕ_a or through the virtual ground of the op amp. Hence its presence will not influence the operation of the integrating circuit.

17.2 THE SWITCHED CAPACITOR

Figure 17.7b is repeated as Fig. 17.12 and shows a capacitor C_R with a periodic SPDT switch. Assume that the input voltage $v_1(t)$ is time varying, and that at the initial instant the switch is connected to a, thus forming the circuit shown in Fig. 17.13a. If v_1 is constant, then the voltage of the capacitor will increase, as shown in Fig. 17.13b with the time constant $\tau = R_1 C_R$. We have stated that R_1 is about $10^4 \ \Omega$, and a typical value for C_R is 1 pF; then v_C will reach 63 percent of its final value in

$$\tau = R_1 C_R = 10^4 \times 10^{-12} = 10 \text{ ns} \qquad (17.3)$$

We will assume that this is small compared to variations in $v_1(t)$. If the switch is now changed to position b and discharged at voltage v_2, then the charge transferred will be

$$q_C = C_R(v_1 - v_2) \qquad (17.4)$$

and this will be accomplished in time T_c. The current will be on the average

$$i(t) = \frac{\Delta q}{\Delta t} \simeq \frac{C_R(v_1 - v_2)}{T_c} \qquad (17.5)$$

The size of an equivalent resistor to give the same value of current is then

$$R_C = \frac{v_1 - v_2}{i} = \frac{T_c}{C_R} = \frac{1}{f_c C_R} \qquad (17.6)$$

From the equation

$$i(t) = \frac{1}{R_C}(v_1 - v_2) \qquad (17.7)$$

* For a discussion of MOS integrated-circuit technology, see Robert W. Brodersen, Paul R. Gray, and David A. Hodges, "MOS Switched Capacitor Filters," *Proc. IEEE*, vol. 67, pp. 61–75, Jan. 1979.

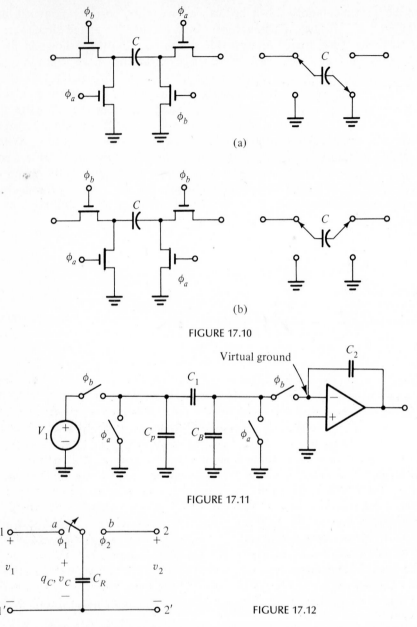

FIGURE 17.10

Virtual ground

FIGURE 17.11

FIGURE 17.12

we see that the position of R_C in the equivalent circuit is that shown in Fig. 17.14. In other words, a switched capacitor as shown in Fig. 17.12 is approximately equivalent to a series resistor as shown in Fig. 17.14.

Before we consider the significance of the word *approximately*, let us use the typical values given previously to determine the range of values of R_C that might be realized. Using $C_R = 1$ pF, and a typical value of $f_c = 100$ kHz, then R_C by Eq. 17.6 is 10 MΩ. This will be found to be in a useful range. To realize this value of C_R requires a silicon area of about 0.01 mm².

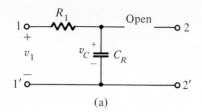

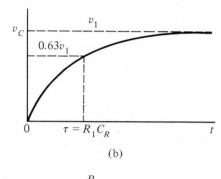

FIGURE 17.13

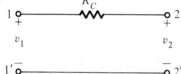

FIGURE 17.14

For the approximation of Eq. (17.5) to be valid, it is necessary for the switching frequency f_c to be much larger than the signal frequencies of interest of $v_1(t)$ and $v_2(t)$. This turns out to be the case for voice-frequency filters, and the switched capacitor may be regarded as a direct replacement for the resistor.

In the studies of this chapter, we will assume that Eq. (17.5) is valid. This implies that a high clock frequency is used, and that there is a sample and hold at the input to the switched-capacitor circuits. As a further assumption, we will neglect stray capacitances. We will use simple switching schemes assuming that stray-insensitive switching can be substituted before actual implementation. We will stress the similarity of switched-capacitor circuits to analog circuits and work with the transfer function of the analog circuit rather than introduce the z-transform concepts of sampled-data systems. We believe that these assumptions will make it possible to present basic ideas while avoiding as many complications as possible.

17.3   ANALOG OPERATIONS

In this section we are concerned with the four analog operations on voltages: addition, subtraction, multiplication, and integration. We are concerned with accomplishing these operations with switched-capacitor circuits, but in each case we will show the corresponding analog circuits using resistors, capacitors, and op amps as derived in earlier chapters.

We begin with the integrator circuit of Fig. 17.15 for which the transfer function is

$$\frac{V_2}{V_1} = \frac{-1}{R_1 C_2 s} \tag{17.8}$$

The corresponding switched-capacitor circuit has a transfer function which is obtained by substituting $R_1 = R_C$ from Eq. (17.6) such that

$$\frac{V_2}{V_1} = -f_c \frac{C_1}{C_2} \frac{1}{s} \tag{17.9}$$

Both circuits are inverting, and both represent integrators. It is significant that the last equation involves the ratio C_1/C_2. In MOS technology, the shapes of capacitors can be controlled with accuracy. Hence the ratio of capacitances can be realized with accuracy even though the value of capacitance can not. If addition of two voltages is required in addition to integration, then the circuit of Fig. 17.16 may be used, paying special attention to maintain the same switching sequence for the two inputs. For this circuit

$$V_2 = \frac{-1}{R_0 C_2 s} V_0 + \frac{-1}{R_1 C_2 s} V_1 \tag{17.10}$$

To find the corresponding equation for the switched-capacitor circuit of Fig. 17.16b, we substitute for the resistor from Eq. (17.6):

$$V_2 = -f_c \frac{C_0}{C_2} \frac{1}{s} V_0 - f_c \frac{C_1}{C_2} \frac{1}{s} V_1 \tag{17.11}$$

For the special case $C_0 = C_1 = C$

$$V_2 = -f_c \frac{C}{C_2} \frac{1}{s} (V_0 + V_1) \tag{17.12}$$

so that the circuit inverts the sum of two voltages and multiplies by a constant. It is interesting to observe that for $f_c = 100$ kHz and $C_2 = 10\ C$ the gain is 10^6!

The circuit shown in Fig. 17.17a is the familiar lossy integrator for which

$$\frac{V_2}{V_1} = \frac{-1/R_1 C_2}{s + 1/R_3 C_2} \tag{17.13}$$

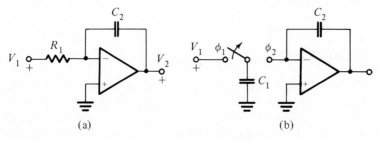

(a) (b)

FIGURE 17.15

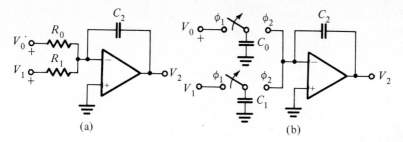

FIGURE 17.16

The corresponding switched-capacitor circuit is shown in Fig. 17.17b, and the gain is

$$\frac{V_2}{V_1} = \frac{-f_c\, C_1/C_2}{s + f_c\, C_3/C_2}$$

(17.14)

for which the dc gain ($s = 0$) is C_1/C_3.

Suppose that V_0 is the voltage being fed back from some other part in the circuit, as shown in Fig. 17.18. Then we may use the concept of superposition to write

$$V_2 = -f_c\, \frac{C_1}{C_2} \frac{1}{s} V_1 - \frac{C_3}{C_2} V_0$$

(17.15)

The circuits considered thus far in this section have all been inverting circuits. Consider next a number of noninverting circuits. The first of these is shown in Fig. 17.19. Figure 17.19a shows the noninverting integrator circuit for which

$$\frac{V_2}{V_1} = \frac{1}{R_1 C_2} \frac{1}{s}.$$

(17.16)

The corresponding switched-capacitor circuit shown in Fig. 17.19b is seen to be a combination of the inverting integrator of Fig. 17.15b and the inverting switching arrangement of Fig. 17.9, described by Eq. (17.2). Combining Eq. (17.2) with Eq.

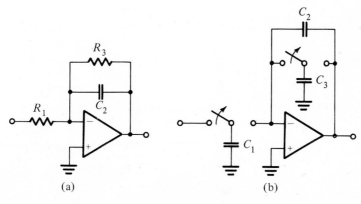

(a)   (b)

FIGURE 17.17

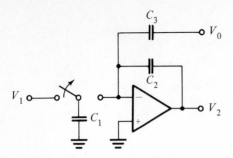

FIGURE 17.18

(17.9) we have

$$\frac{V_2}{V_1} = f_c \frac{C_1}{C_2} \frac{1}{s}$$ (17.17)

If we use the other switching arrangement shown in Fig. 17.9 in combination with the standard integrator circuit of Fig. 17.15b, as shown in Fig. 17.20b, then the transfer function becomes

$$V_2 = f_c \frac{C_1}{C_2} \frac{1}{s} (V_1 - V_0)$$ (17.18)

The equation indicates that this circuit forms a difference of voltages, multiplies by $f_c C_1 / C_2$, and integrates. This circuit will find frequent use in a later section. The corresponding RC–op-amp circuit is shown in Fig. 17.20a, and for this circuit

$$V_2 = \frac{1}{R_1 C_2} \frac{1}{s} (V_1 - V_0)$$ (17.19)

As in Fig. 17.18, a circuit with several inputs, one of which is fed back from another part of the circuit, as shown in Fig. 17.21, may be analyzed by superposition. Thus we make use of Eq. (17.18) and write

$$V_2 = f_c \frac{C_1}{C_2} \frac{1}{s} (V_1 - V_0) - \frac{C_3}{C_2} V_3$$ (17.20)

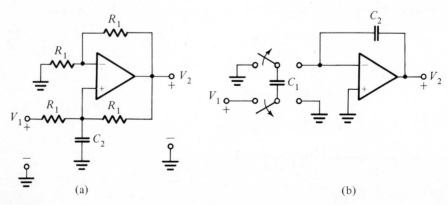

(a)　　　　　　　　　　　　　　　　(b)

FIGURE 17.19

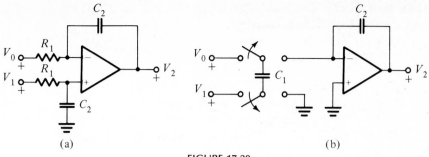

FIGURE 17.20

This concept may be extended to cases where there are more than three inputs to a circuit under analysis.

17.4   RANGE OF CIRCUIT ELEMENT SIZES

The original motivation in introducing switched capacitor circuits was to achieve integrated-circuit realizations at voice-band frequencies. These two factors place constraints on the element sizes to be used, as we shall see. Consider the integrating circuit shown in Fig. 17.15 for which

$$\frac{V_2}{V_1} = \frac{1}{R_1 C_2}\frac{1}{s} \qquad (17.21)$$

As in many analog circuits, this result involves the product $R_1 C_2$. In earlier chapters typical values are $C = 0.01 \ \mu F$ and $R = 10 \ k\Omega$. For these values,

$$R_1 C_2 = 10^{-8} \times 10^4 = 10^{-4} \qquad (17.22)$$

For an integrated-circuit realization suppose that a typical value of capacitor available is 10 pF. Then to give the same RC product, the value of R must be

$$R_1 = \frac{10^{-4}}{C_2} = \frac{10^{-4}}{10^{-11}} = 10 \ M\Omega \qquad (17.23)$$

To achieve this value of resistance, the frequency of the clock switch must be

$$f_c = \frac{1}{RC_R} = \frac{1}{10^7 \times 10^{-11}} = 10 \ kHz \qquad (17.24)$$

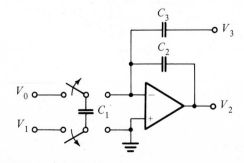

FIGURE 17.21

So here we deal with a different range of element values: small capacitors, large resistors, and a modest frequency range of switching.

First let us consider the generation of the two-phase signals shown in Fig. 17.5, which operate at the clock frequency. One method that is commonly used is shown in Fig. 17.22, which makes use of the chroma crystal oscillator used in television receivers and operating at the standard frequency of 3.579 MHz. Count-down circuits are then employed to reduce this frequency to the actual clock frequency f_c. The signal ϕ_2 is then generated from ϕ_1 by a circuit having the logic shown in Fig. 17.22b, which is actually realized by a circuit of MOS transistors. We will assume that the count-down circuit is tunable and that ϕ_1 and ϕ_2 can be generated with an f_c range up to 2 MHz. Let f_0 be some characterizing frequency of the filter, such as the cutoff frequency of the lowpass filter or the resonant frequency of the bandpass filter. Then we are interested in the frequency ratio f_c/f_0. The lowest frequency which we may use is known as the *Nyquist frequency*. We must sample at a frequency that is twice the highest component of signal frequency in order to avoid aliasing, a fact familiar in communications and first given by Harry Nyquist in 1928. On the other hand, the clock frequency should be as high as possible to avoid distortion due to discrete-time sampling. We will assume a frequency range for the clocks

$$2 < \frac{f_c}{f_0} < 500 \qquad (17.25)$$

A typical value for f_c will be taken as 100 kHz.

We have seen that the ratio of capacitances is important in the design of switched-capacitor circuits. The fact that a ratio such as C_2/C_1 can be formed with great accuracy is one of the advantages of a switched-capacitor realization. This is accomplished by the two capacitors being formed on the same silicon chip, in near proximity, so that the dielectrics separating the two plates of the capacitor are uniform and the capacitor ratios are proportional to areas. Capacitors

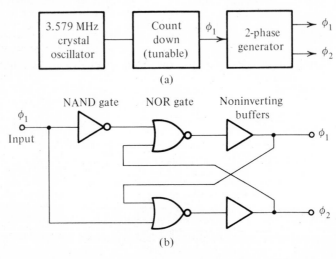

(a)

(b)

FIGURE 17.22

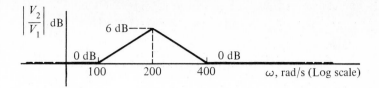

FIGURE 17.29

while for the second section

$$C_3 = 0.4 \text{ pF} \quad \text{and} \quad C_4 = 0.2 \text{ pF} \tag{17.41}$$

The final filter design which realizes $T(s)$ given by Eq. (17.38) is shown in Fig. 17.31.

The circuit shown in Fig. 17.32 is a bandpass filter of the general type classified as biquad filters in Chapter 5. For this circuit we see that

$$V_3 = -V_1 - \left(1 + \frac{1}{\alpha^2 Cs}\right)V_2 \tag{17.42}$$

The second and third units together result in noninverting integration described by the equation

$$V_2 = \frac{1}{Cs} V_3 \tag{17.43}$$

Combining these two equations, we find that

$$\frac{V_2}{V_1} = \frac{-(1/c)s}{s^2 + (1/C)s + (1/\alpha^2 C^2)} \tag{17.44}$$

From this result, we see that the bandpass filter has parameters

$$\omega_0 = \frac{1}{\alpha C} \quad \text{and} \quad Q = \frac{1}{\alpha} \tag{17.45}$$

The switched-capacitor equivalent is shown in Fig. 17.33. We see that the non-inverting integrator of Fig. 17.20b is used to substitute for two modules in Fig. 17.32, and that the feedback resistor is realized using an SPDT switched capacitor.

In completing a design based on the circuit of Fig. 17.33, the capacitance may be magnitude-scaled to give a value in the range desired. In integrated-circuit technology, such values are in the picofarad range. Let the magnitude-scaled capacitor have the value C_m which is obtained from Eq. (17.45).

$$C_m = \frac{1}{\alpha \omega_0 k_m} \tag{17.46}$$

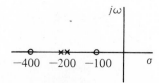

FIGURE 17.30

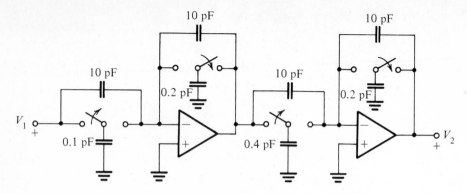

FIGURE 17.31

The equivalent resistance of the switched capacitor is given by Eq. (17.6), and this equation may be solved for the required value of capacitance.

$$C_R = \frac{1}{f_c R_c} \tag{17.47}$$

When this is magnitude-scaled, the result is

$$C_{Rm} = \frac{1}{f_c R_c k_m} \tag{17.48}$$

Consider the problem of designing a filter such that $\omega_0 = 2{,}000$ rad/s with a Q of 2. If we select C to have the value of 1 pF and the clock frequency to be $f_c = 10^4$ Hz, then it is necessary that $k_m = 10^9$. From Eqs. (17.46) and (17.48), we obtain the values

$$C_1 = 0.1 \text{ pF} \qquad \text{and} \qquad C_2 = 0.4 \text{ pF} \tag{17.49}$$

The design filter is then as shown in Fig. 17.34.

There are many different circuits that realize the properties of the biquad circuit as studied in Chapter 5. An example of such a circuit is that due to El-Masry* shown in Fig. 17.35. Observe that it makes use of the strays-insensitive switching arrangement of Fig. 17.10.

In selecting analog circuits for conversion into equivalent switched-capacitor circuits, we have avoided circuits with floating nodes, such as those with two passive elements in series. This has excluded Sallen–Key circuits and the first-order noninverting circuits.

The first- and second-order filters that have been presented can now be used in cascade connections to realize a desired form of response as usual. Such a design procedure is very useful when specifications are not tight and great accuracy is not required. We next turn to other procedures which are useful for high-accuracy low-sensitivity requirements.

---

* E. I. El-Masry, "Strays-Insensitive Active Switched-Capacitor Biquad," *Electron. Lett.*, vol. 16, pp. 480–481, June 5, 1980.

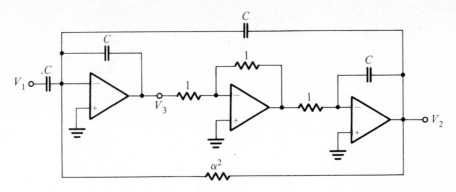

FIGURE 17.32

17.6 LEAPFROG SWITCHED-CAPACITOR FILTERS

The filters studied in the last section were well suited to low-precision applications, much as the universal filters studied in Chapter 5. When precision is required as well as low sensitivity, then a filter that simulates the passive doubly terminated ladder has the advantage. Among the first to realize this were D. A. Hodges, R. W. Brodersen, and P. R. Gray at the University of California at Berkeley. They made use of the leapfrog realization of Girling and Good to find a structural simulation of the passive ladder, and then employed switched-capacitor elements to find a realization. These steps are illustrated in Fig. 17.36.

We begin with the lowpass doubly terminated ladder network, an example of which is shown in Fig. 17.37a. As we know from Chapter 13, the response of this network is determined by the size of the elements. By the appropriate choice, the response may be made Butterworth, Chebyshev, or any other. The leapfrog simulation of this circuit is shown in Fig. 17.37b. The particular form of this simulation was studied in Chapter 16. Some of the blocks have a negative sign, indicating the choice of an inverting stage; this was done so that all signal combina-

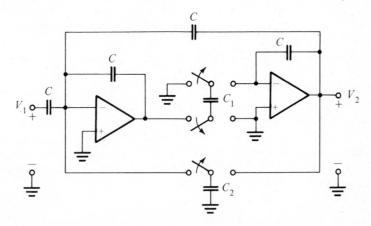

FIGURE 17.33

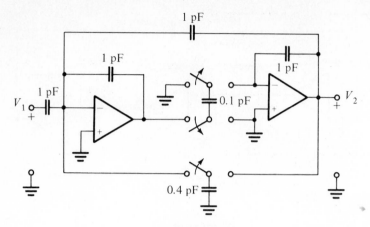

FIGURE 17.34

tions would be summations instead of differences. The Girling–Good form of this leapfrog circuit is given in Fig. 17.38a.

Now we have switched-capacitor circuits to add signals as well as those to form differences. The one shown in Fig. 17.16b is used to add V_0 and V_1 while that of Fig. 17.20b forms the difference of V_1 and V_0. It turns out that the differencing circuit is best suited to our purposes in contrast to the situation in Chapter 16. If we return to the discussion of Chapter 16, we recall that the price paid for the use of summing circuits was that every other block was required to be inverting. Then if we are to use differencing operations to subtract the feedback signal from that being fed forward, then the sign of all blocks returns to being positive. The block diagram that results is shown in Fig. 17.38b. This block diagram consists only of integrators with difference inputs, except for the two terminating resistors R_i and R_o.

At this point we assume that the terminations are equal and that both have unit value, $R_i = R_o = 1$ (not always the case, of course). We can then simplify the

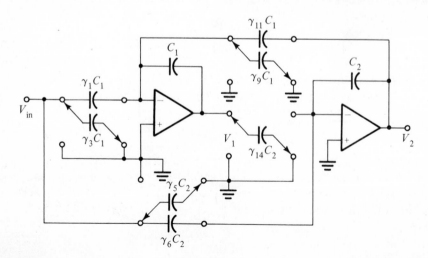

FIGURE 17.35

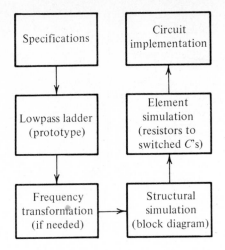

FIGURE 17.36

block diagram as illustrated by Fig. 16.31. The result is the block diagram of Fig. 17.39. This is the block diagram representation of the ladder that we wish to implement using switched-capacitor units.

Let us return to the differencing integrator circuit last given in Fig. 17.20, repeated as Fig. 17.40 in which we have changed the notation so that capital C's are replaced by lowercase c's for reasons that will soon become apparent. Then Eq. (17.18) becomes

$$V_2 = f_c \frac{c_1}{c_2} \frac{1}{s} (V_1 - V_0) \qquad (17.50)$$

In the form of a transfer function this may be written

$$\frac{V_2}{V_1 - V_0} = \frac{1}{(c_2/c_1)(1/f_c)s} \qquad (17.51)$$

This must be equated to each of the integrator-block transfer functions shown in Fig. 17.39, which are

$$\frac{1}{C_1 s}, \qquad \frac{1}{L_2 s}, \qquad \frac{1}{C_3 s}, \qquad \frac{1}{L_4 s} \qquad (17.52)$$

Designating the first stage as 1 and adding a second subscript to c_1 and c_2 to so indicate, we arrive at a typical result:

$$C_1 = \frac{c_{21}}{c_{11}} \frac{1}{f_c} \qquad (17.53)$$

Now C_1, L_2, C_3, and L_4 are parameters obtained from tables tabulated on the basis that $\omega_0 = 1$, a half-power frequency or a frequency indicating the end of the ripple band. Frequency scaling to some specified ω_0 is accomplished using the scaling equations. Assuming that the magnitude is not scaled so that $k_m = 1$, we have

$$L_{\text{new}} = \frac{1}{k_f} L_{\text{old}} \qquad (17.54)$$

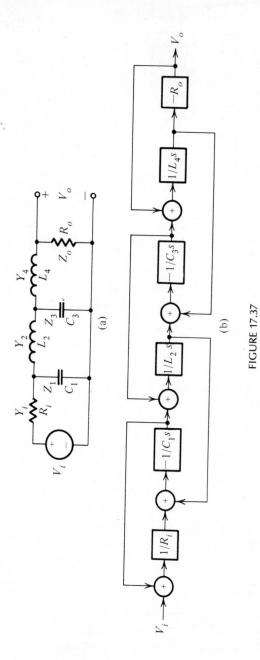

FIGURE 17.37

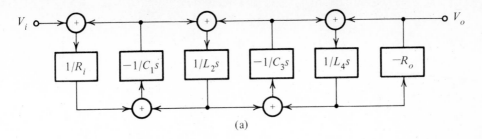

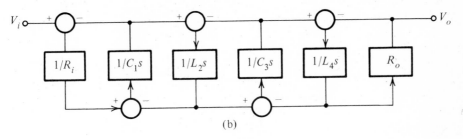

FIGURE 17.38

$$C_{new} = \frac{1}{k_f} C_{old} \tag{17.55}$$

where $\omega_0 = k_f \times 1$. Then the scaled capacitance is

$$C_1' = \frac{C_1}{\omega_0} = \frac{C_1}{2\pi f_0} \tag{17.56}$$

Equating this to Eq. (17.53) and solving for the ratio c_{21}/c_{11} gives

$$\frac{c_{21}}{c_{11}} = \frac{1}{2\pi} \frac{f_c}{f_0} C_1 \tag{17.57}$$

In general,

$$\frac{c_{2n}}{c_{1n}} = \frac{1}{2\pi} \frac{f_c}{f_0} X_n \tag{17.58}$$

where X is either L or C, the information obtained from tables.

The terminating resistors will be simulated using the circuit of Fig. 17.12, for

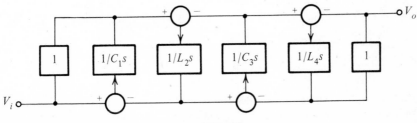

FIGURE 17.39

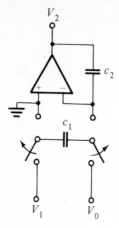

FIGURE 17.40

which

$$R_i = \frac{1}{f_c C_i} \qquad \text{or} \qquad R_o = \frac{1}{f_c C_o} \tag{17.59}$$

Since R_i and R_o are given in tables, C_i may be chosen for a given f_c to meet the required value.

The circuit that simulates the one given in Fig. 17.37a is now represented in Fig. 17.41. Before we use the equations just derived to design a filter, we consider some aspects of the switched-capacitor filter illustrated by the special case of Fig. 17.41.

Consider two adjacent integrator circuits of Fig. 17.41 shown in Fig. 17.42. The switching of the DPDT switches is indicated by the directions of the arrows. Since the directions are upward in both cases, the switches are in opposition in a loop sense and are sometimes described as having "alternate phase." This is required to compensate for a time delay between the input and output of the switched-capacitor integrator and was first reported by Jacobs, Allstot, Brodersen, and Gray.* The manner in which the resistive components of the input and output are implemented as required in the block diagram of Fig. 17.29 is illustrated in Fig. 17.43. These are shown at the input and output of the switched-capacitor filter of Fig. 17.41.

As an example of the design of a switched-capacitor filter we consider the following specifications: The response of the lowpass filter is to be Butterworth, with a half-power frequency of 1 kHz. The clock frequency is 40 kHz. The low-pass prototype doubly terminated ladder has the following values for the elements of Fig. 17.37a: $R_i = 1$, $C_1 = 0.7654$, $L_2 = 1.848$, $C_3 = 1.848$, $L_4 = 0.7654$, and $R_o = 1$, these values being in ohms, farads and henrys. Using Eq. (17.52), we obtain the following capacitive ratios:

$$\frac{c_{21}}{c_{11}} = 4.87, \qquad \frac{c_{22}}{c_{12}} = 11.76, \qquad \frac{c_{23}}{c_{13}} = 11.76, \qquad \frac{c_{24}}{c_{24}} = 4.87 \tag{17.60}$$

* G. M. Jacobs, D. J. Allstot, R. W. Brodersen, and P. R. Gray, "Design Techniques for MOS Switched-Capacitor Ladder Filters," *IEEE Trans. Circuits Syst.*, vol CAS-25, pp. 1014–1021, Dec. 1978.

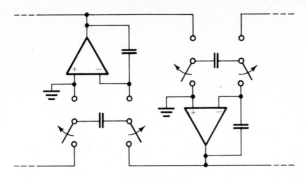

FIGURE 17.41

The range of capacitors commonly used in MOS technology is 1 pF to 100 pF, so that any values in this range might be used so long as the ratios are realized. Normally a small value of c_{11} will be selected in order to keep the area small.

Return now to Eq. (17.59) and the realization of a switched-capacitor filter to satisfy the specifications for R_i and R_o. We must frequency scale the same as was done for c_1 and c_2, which requires that

$$C_i = \frac{C_i'}{\omega_0} \qquad (17.61)$$

resulting in

$$C_i = \frac{1}{R_i} 2\pi \frac{f_0}{f_c} \qquad (17.62)$$

Here $R_i = 1$ is always the case, but R_o may not be of unit value. If we let $R_i = 1$, then

$$C_i = 2\pi \frac{f_0}{f_c} \qquad (17.63)$$

But by Eq. (17.58) (dropping the second subscript)

$$\frac{c_2}{c_1} = \frac{1}{2\pi} \frac{f_c}{f_0} C_1 \qquad (17.64)$$

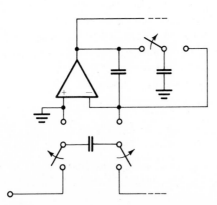

FIGURE 17.42

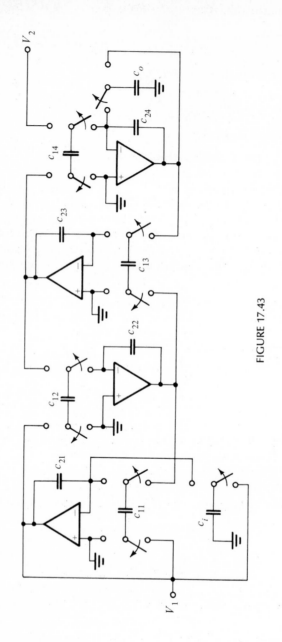

FIGURE 17.43

512

For simplicity in integrated-circuit implementation, let $C_i = c_1$ such that $c_2 = C_1$, which is the specified value from tables appropriately scaled. If we let $c_1/c_2 = \alpha$, $\alpha < 1$, then

$$c_1 = C_i = \alpha c_2, \qquad \alpha < 1 \tag{17.65}$$

and the two designs are thus coordinated.

17.7 BANDPASS SWITCHED-CAPACITOR FILTERS

The procedure for the design of a switched-capacitor filter, which was given in the last section for the lowpass case, is equally applicable for other kinds of responses. The only additional step is the use of a frequency transformation from the lowpass to the desired response. To illustrate procedure, we will consider the design of a bandpass switched-capacitor filter.

For this example we assume that this application requires a second-order prototype, as shown in Fig. 17.44. This prototype contains both a shunt capacitor and a series inductor, and so a prototype of higher order will be analyzed by the repetition of some of the steps to be illustrated. The lowpass to bandpass frequency transformation was last illustrated by Fig. 16.13. From this figure we see that the element values in Fig. 17.45 are given in terms of those in Fig. 17.44 by the equations

$$C_A = \frac{C_1}{BW}, \qquad L_A = \frac{BW}{\omega_0^2 C_1} \tag{17.66}$$

$$C_B = \frac{BW}{\omega_0^2 L_2}, \qquad L_B = \frac{L_2}{BW} \tag{17.67}$$

Given these values, we next represent the circuit of Fig. 17.45 by an equivalent block diagram. However, we place a restriction on the block diagram to be found, namely, that it contain only integrators, along with suitable summing or differencing operations. One way this may be accomplished is to start at the "far" end of the circuit, the output, and use the two Kirchhoff laws in working back toward the input. Thus,

$$V_4 = V_{\text{out}} \tag{17.68}$$

$$V_4 = R_o I_4 \tag{17.69}$$

$$I_4 = I_3 \tag{17.70}$$

$$V_3 = \frac{1}{C_B s} I_3 \tag{17.71}$$

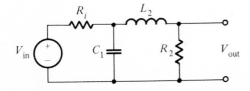

FIGURE 17.44

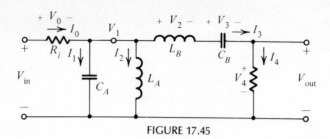

FIGURE 17.45

$$I_3 = \frac{1}{L_B s} V_2 \tag{17.72}$$

$$V_2 = V_1 - V_3 - V_4 \tag{17.73}$$

$$I_2 = \frac{1}{L_A s} V_1 \tag{17.74}$$

$$V_1 = \frac{1}{C_A s} I_1 \tag{17.75}$$

$$I_1 = I_0 - I_2 - I_3 \tag{17.76}$$

$$I_0 = \frac{1}{R_i} V_0 \tag{17.77}$$

$$V_0 = V_{in} - V_1 \tag{17.78}$$

Observe the manner in which these equations are written, with voltage and current alternating, each equation defining some quantity used in previous equations.

For purposes of simulation, every quantity in this set of equations is a voltage. If we accept this fact, then we are spared writing V_{I_x} in place of I_x. (Some authors scale the current equations by multiplying each current equation by a constant R.) A block diagram representation of these equations is given in Fig. 17.46. A possible switched-capacitor realization is shown in Fig. 17.47. While the

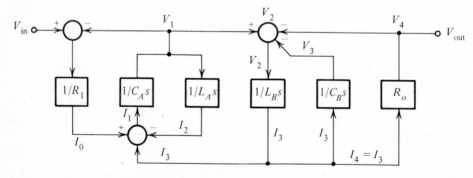

FIGURE 17.46

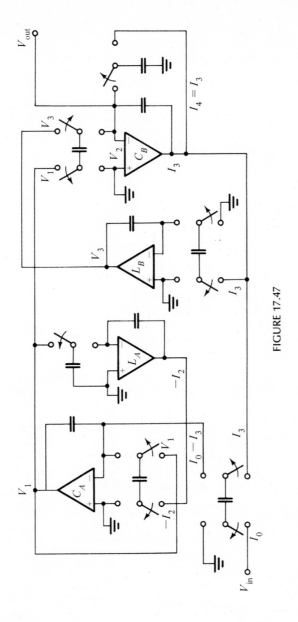

FIGURE 17.47

515

FIGURE 17.48

number of operations and the kinds of operation differ from those illustrated in the design of a lowpass filter, the procedure is generally the same. Once the design is complete, it should be tested on a computer simulation program designed for sampled-data systems.

The design procedure which has been illustrated for a simple fourth-order bandpass filter may be generalized to apply to filters of higher order. The same steps are involved, simply repeated more times. For example, suppose that the specifications for a filter are such that a fifth-order prototype such as that of Fig. 17.48 may be used. The element values for this filter come from Table 14.3 and provide a Chebyshev response with 0.5-dB ripple width and a 1-rad/s ripple bandwidth. The tenth-order bandpass filter that is required has a center frequency of $f_0 = 1000$ Hz and a ripple bandwidth of 600 Hz. The available clock frequency for the switching of capacitors is 100 kHz.

The lowpass-to-bandpass frequency transformation of the filter of Fig. 17.48 gives the element values shown in Fig. 17.49a using the values $\omega_0 = 1$ rad/s and bw = 0.6. Figure 17.49b shows the same bandpass filter but with each voltage and current identified. From this circuit, we write the Kirchhoff law equations starting

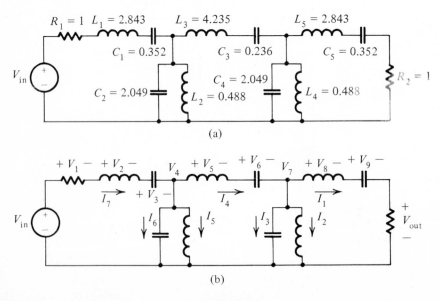

(a)

(b)

FIGURE 17.49

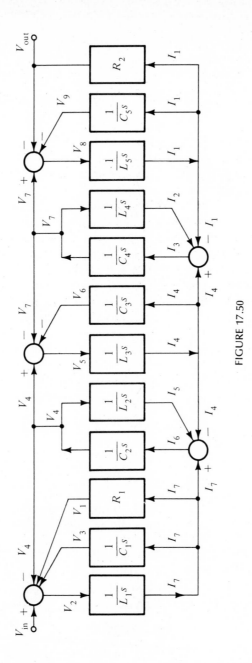

FIGURE 17.50

at V_{out} and working toward V_{in}. The result is

$$V_{\text{out}} = R_2 I_1 \tag{17.79}$$

$$V_9 = \frac{1}{C_5 s} I_1 \tag{17.80}$$

$$I_1 = \frac{1}{L_5 s} V_8; \qquad V_8 = V_7 - V_9 - V_{\text{out}} \tag{17.81}$$

$$V_7 = \frac{1}{C_4 s} I_3; \qquad I_3 = I_4 - I_2 - I_1 \tag{17.82}$$

$$I_2 = \frac{1}{L_4 s} V_7 \tag{17.83}$$

$$V_6 = \frac{1}{C_3 s} I_4 \tag{17.84}$$

$$I_4 = \frac{1}{L_3 s} V_5; \qquad V_5 = V_4 - V_6 - V_7 \tag{17.85}$$

$$V_4 = \frac{1}{C_2 s} I_6; \qquad I_6 = I_7 - I_4 - I_5 \tag{17.86}$$

$$I_5 = \frac{1}{L_2 s} V_4 \tag{17.87}$$

$$V_3 = \frac{1}{C_1 s} I_7 \tag{17.88}$$

$$I_1 = \frac{1}{L_1 s} V_2; \qquad V_2 = V_{\text{in}} - V_1 - V_3 - V_4 \tag{17.89}$$

and

$$V_1 = R_1 I_1 \tag{17.90}$$

These equations are in proper form for an all-integrator realization with summing equations written to the right on the page. The block diagram realization is given in Fig. 17.50. From the block diagram, we construct the switched-capacitor realization as shown in Fig. 17.51.

There remains the problem of selecting the values of capacitors to be used in the switched-capacitor filter. The design equation is Eq. (17.57). With $f_0 = 1000$ Hz and $f_c = 100$ kHz, this equation becomes

$$\frac{c_{2n}}{c_{1n}} = 15.92 \, X_n \tag{17.91}$$

where X_n is either C_n or L_n to be taken from Fig. 17.49a for this particular problem. If we select $c_{1n} = 1$ pF, then we obtain the values in Table 17.1. In addition to the values of Table 17.1, we will require capacitors which realize the equiva-

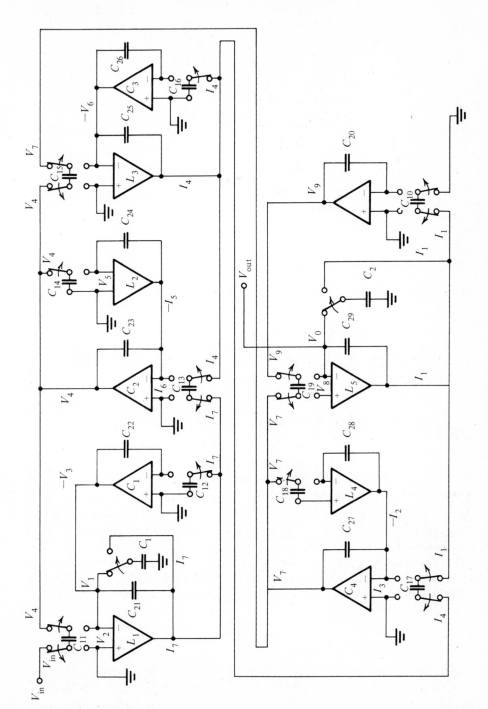

FIGURE 17.51

Table 17.1

Capacitor	Value, pF
C_{21}	45.361
C_{22}	5.604
C_{23}	32.620
C_{24}	7.769
C_{25}	67.421
C_{26}	3.757
C_{27}	32.620
C_{28}	7.769
C_{29}	45.261
C_{20}	5.604

lent of R_1 and R_2. Both have the value

$$C_1 = C_2 = \frac{1}{2\pi f_0 f_c R} = 1.59 \text{ nF} \qquad (17.92)$$

These values complete the design.

PROBLEMS

17.1 Find the switched-capacitor filter which is equivalent to the filter shown in Problem 4.2.

17.2 Find the switched-capacitor filter which is equivalent to that shown in Problem 4.3.

In the following problems, the statement of the filter design problem is given in Chapter 4. For these problems, find the solution to the design problem in the form of a switched-capacitor filter, paying special attention to the ratios of capacitors and the clock frequency.

17.3 Rework Problem 4.5 as indicated above.

17.4 Rework Problem 4.6 as indicated above.

17.5 Rework Problem 4.7 as indicated above.

17.6 Rework Problem 4.8 as indicated above.

17.7 Rework Problem 4.9 as indicated above.

17.8 Rework Problem 4.10 as indicated above.

17.9 Rework Problem 4.11 as indicated above.

17.10 Rework Problem 4.12 as indicated above.

17.11 Rework Problem 4.13 as indicated above.

17.12 Rework Problem 4.16 as indicated above.

17.13 Design a second-order or biquad circuit realization in switched-capacitor form for a lowpass filter with $\omega_0 = 10{,}000$ rad/s and $Q = 5$. Pay special attention to the ratio of capacitance in the filter and the clock frequency chosen.

17.14 Repeat Problem 17.13 for $f_0 = 1000$ Hz and $Q = 8$.

17.15 Repeat Problem 17.13 for $\omega_0 = 7500$ rad/s and $Q = 10$.

17.16 Find a switched-capacitor filter to satisfy the requirements given in Problem 16.1. The clock frequency may be chosen, and the capacitance ratios should be a design consideration.

17.17 Repeat Problem 17.16 to satisfy the requirements given in Problem 16.2.

17.18 Repeat Problem 17.16 to satisfy the requirements given in Problem 16.3.

17.19 Find a switched-capacitor realization of a lowpass filter which has a Butterworth $n = 5$ response in the form of a cascade of lower-order switched-capacitor filters. The clock frequency may be selected as a design consideration.

17.20 Design a switched-capacitor bandpass filter to satisfy the requirements given in Problem 16.8. The clock frequency may be selected in your design, and the capacitance ratios should be determined to be as small as practical.

17.21 Repeat Problem 17.20 for the requirements given in Problem 16.9.

17.22 Repeat Problem 17.20 for the requirements given in Problem 16.10.

17.23 Repeat Problem 17.10 for the requirements given in Problem 16.11.

CHAPTER 18

Delay

Equalization

We now continue our studies of Chapter 10 concerning filters designed specifically to achieve delay specifications. In Chapter 10 we sought maximally flat
delay, accepting any magnitude response that resulted. The studies of this chapter
are different. Here we specify the desired magnitude response, and then "patch
up" the delay through a process known as equalization. The techniques we will
study had their origins in the telephone industry. The first summary of these techniques was given by Bode* in his classic textbook published in 1945.

18.1 EQUALIZATION PROCEDURES

The limitations of the Bessel–Thomson filters studied in Chapter 10 are illustrated by Fig. 18.1. The delay achieved decreases with increasing frequency, and
the magnitude of the transfer function also decreases with frequency. The depar-
ture from the ideal response is a function of the order of the filter n, and better as
n increases. In contrast, delay equalization is accomplished in two steps:

1. A filter is designed to provide the desired magnitude response.
2. The phase of the resulting filter is then supplemented over a frequency range
   of importance by a cascade of first- or second-order all-pass filters such that
   the attenuation is not changed, but the summation of delays is approximately
   constant over a prescribed frequency range. This delay equalization is illustrated in Fig. 18.2 for two different forms of system delay.

   To further illustrate the procedure, consider the waveforms shown in Fig.
18.3. The input is a triangular waveform which is distorted by high-frequency
noise. Filtering is required to remove the high-frequency noise, but we wish to
maintain the integrity of the signal insofar as possible. We cannot do this per-
fectly, but we wish the distortion to be as small as is practical. Suppose that the

---

* H. W. Bode, *Network Analysis and Feedback Amplifier Design,* Van Nostrand, Princeton, N.J., 1945,
chap. 12.

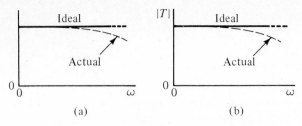

FIGURE 18.1

noise is removed by a fifth-order Butterworth filter having the delay response shown in Fig. 18.4. If additional delay is introduced by a cascade-connected circuit, then equalization is accomplished and the total delay is approximately constant in the range of frequencies from $\omega = 0$ to $\omega = 0.9$ (normalized frequencies). The result is the response marked "output," which is not triangular but easily recognizable.

It is clear that equalization will always result in additional delay, and this fact is illustrated by Fig. 18.5. The figure shows several signal waveforms. There is the step input to the system, the response without equalization, and the response with equalization. In many applications such as telephone communications, additional delay is not important, and so equalization is frequently employed.

The discussion thus far has been in terms of a Butterworth magnitude response. Figure 18.6 reminds us that the Butterworth response is relatively smooth compared to other responses we have studied, such as the inverse Chebyshev response and the Chebyshev response, shown in Fig. 18.6b and c. Further, the Butterworth response has a narrow peak for larger values of n. This was shown earlier in Fig. 10.7, repeated as Fig. 18.7 but with each curve normalized such that $D_n(0) = 0$ for each n.

There is another general class of applications in which delay of a given communication system is determined experimentally, and from the experimentally determined values of delay a smooth curve may be drawn. For this system, delay equalization is to be provided within a specified error. This information is the beginning of an interesting design problem which will be illustrated later in the chapter.

Equalization will be accomplished by a cascade-connected circuit which provides delay as a function of frequency, but does not introduce attenuation as a function of frequency. As we have seen many times in earlier chapters, such cir-

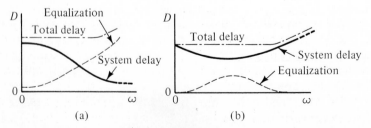

FIGURE 18.2

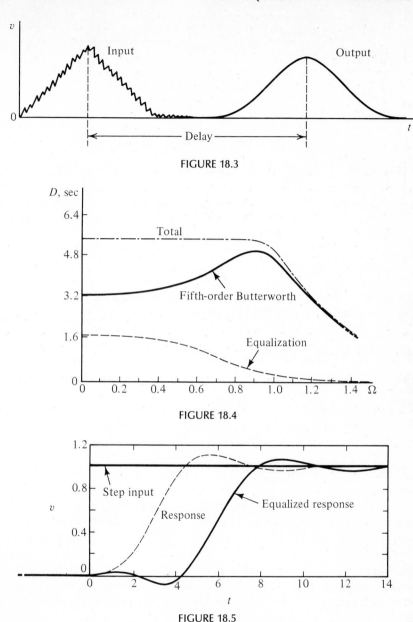

FIGURE 18.3

FIGURE 18.4

FIGURE 18.5

cuits are called allpass circuits and are distinguished by pole–zero patterns as shown in Fig. 18.8. The pattern of Fig. 18.8a has one pole and one zero at equal distances from the origin. That shown in Fig. 18.8b is said to have quadrantal symmetry, where the poles and zeros have equal but opposite real and imaginary parts. The patterns shown in Fig. 18.8c and d are merely the superposition of the patterns of Fig. 18.8a and b. If a filter having the pole locations shown in Fig. 18.9a, say a Butterworth third-order filter, has added to it second-order poles and zeros, as shown in Fig. 18.9b, then the two pole–zero patterns will represent the

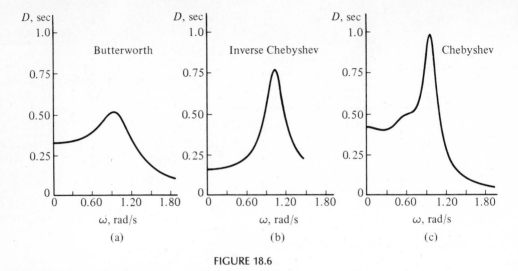

FIGURE 18.6

same magnitude response, but the phase and so the delay characteristics will be very different. The studies of this chapter are based on this observation.

18.2 EQUALIZATION WITH FIRST-ORDER CIRCUITS

Consider the first-order allpass function

$$T_1(s) = K \frac{s - \sigma_1}{s + \sigma_1} \tag{18.1}$$

corresponding to the pole–zero pattern of Fig. 18.8a. For this function,

$$T_1(j\omega) = K \quad \text{and} \quad \theta = -2 \tan^{-1} \left| \frac{\omega}{\sigma_1} \right| \tag{18.2}$$

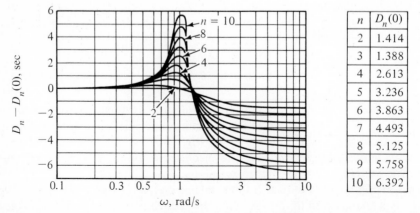

n	$D_n(0)$
2	1.414
3	1.388
4	2.613
5	3.236
6	3.863
7	4.493
8	5.125
9	5.758
10	6.392

FIGURE 18.7 Reprinted by permission from Jiri Vlach, *Computerized Approximation and Synthesis of Linear Networks,* John Wiley & Sons, Inc., New York, 1976.

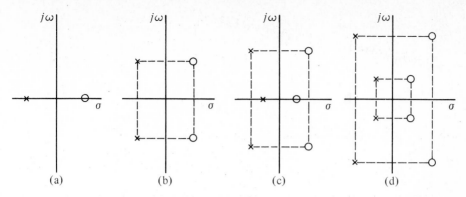

FIGURE 18.8

the delay is found from θ in this equation as

$$D_1(\omega) = -\frac{d}{d\omega}\,\theta(\omega) = \frac{2/\sigma_1}{1 + (\omega/\sigma_1)^2} \tag{18.3}$$

This delay is plotted as a function of frequency in Fig. 18.10. Since D_1 decreases with increasing frequency, we see that

$$D_{\max} = D_1(0) = \frac{2}{\sigma_1} \tag{18.4}$$

Next we review circuit realizations of the first-order transfer function of Eq. (18.1). A passive realization of this equation is given in Fig. 3.9 for which the parameters of Eq. (18.1) are $K = -1$ and $\sigma_1 = 1/RC$. Circuits employing op amps were given in Chapter 4 and are repeated in Fig. 18.11. That shown in Fig. 18.11a was given as Fig. 4.28 and described by Eq. (4.63), from which we see that the parameters are $K = 1$ and $\sigma_1 = 1/RC$. For Fig. 18.11b a comparison with Eq. (4.65) gives $K = -\frac{1}{2}$ with $\sigma_1 = 1/RC$. Thus we have in the two circuits of Fig. 18.11 a noninverting and an inverting realization, and in each case the design equation is $\sigma_1 = 1/RC$.

Now that we have an expression for D_1 in Eq. (18.3), we will make use of it to equalize the delay of a filter. Assume that only one section of a first-order all-

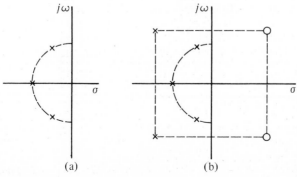

FIGURE 18.9

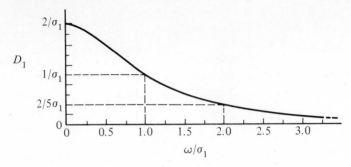

FIGURE 18.10

pass filter may be used, and we select the Butterworth lowpass filter to equalize. Then the overall or total delay will be the sum of the two delays

$$D_t(\omega) = D_B(\omega) + D_1(\omega) \tag{18.5}$$

Our objective is to select D_1 such that the lowest nonzero derivative of D_t vanishes at $\omega = 0$. Accomplishing this objective is complicated,* but the constraint on D_1 is very simple; it is

$$\sigma_1 = \left(2 \sin \frac{3\pi}{2n}\right)^{1/3} \tag{18.6}$$

Another useful relationship involves adding $D_B(0)$ to Eq. (18.4), so that Eq. (18.5) reduces to

$$D_t(0) = \frac{1}{\sin (\pi/2n)} + \frac{2}{\sigma_1} \tag{18.7}$$

From this initial value of D_t the overall delay will be maximally flat at $\omega = 0$, but will eventually increase, as shown in Fig. 18.12.

Example 18.1   For this example we wish a fourth-order Butterworth delay response to be maximally flat at $\omega = 0$, employing one section of allpass filter. Substituting $n = 4$ into Eq. (18.6), we find that $\sigma_1 = 1.2271$. The delay at $\omega = 0$ is determined from Eq. (18.7) to be

$$D_t(0) = 2.613 + 1.630 = 4.243 \text{ seconds} \tag{18.8}$$

The poles for the fourth-order Butterworth response are shown in Fig. 18.13 together with the allpass pole and zero corresponding to $\sigma_1 = 1.2271$. The plots of three delays are given in Fig. 18.14, with D_1 and D_B summed to given D_t. The quantity

$$\Delta D = D_{\max} - D(0) \tag{18.9}$$

is an interesting measure of the effectiveness of the equalization. From the curve we see that the change in D_B is 1.3 for the Butterworth delay response, but the addition of the allpass section results in $\Delta D_t = 0.75$, a significant improvement. We use the circuit of Fig. 18.11a which is designed from the equation

$$\sigma_1 = 1.2271 = \frac{1}{RC} \tag{18.10}$$

* H. J. Blinchikoff and A. I. Zverev, *Filtering in the Time and Frequency Domains*, Wiley, New York, 1976, p. 217 ff.

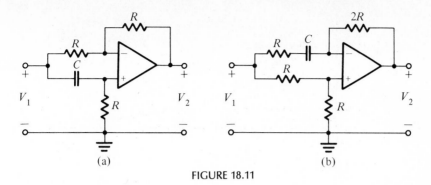

FIGURE 18.11

Values satisfying this requirement are $C = 1$ F and $R = 0.8149$ Ω. These values can be scaled at the same time that the Butterworth filter is scaled.

Example 18.2   Consider a fifth-order Butterworth filter which is to be made maximally flat in delay at $\omega = 0$ by the addition of a first-order allpass circuit. For this example we will determine equations from which the various delay curves can be plotted. From Eq. (18.6) we find that $\sigma_1 = 1.174$. Substituting this value into Eq. (18.3) gives us

$$D_1 = 1.704 \, \frac{1}{1 + (\omega/1.174)^2} \tag{18.11}$$

To determine the delay of the fifth-order Butterworth transfer function

$$T_{B_5}(s) = \frac{1}{(s + 1)(s^2 + 1.625s + 1)(s^2 + 0.625s + 1)} \tag{18.12}$$

we first determine the angle functions corresponding to each of the three factors in the de-

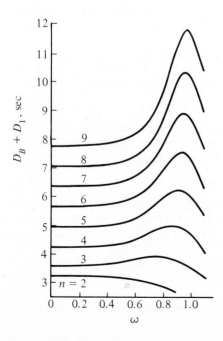

FIGURE 18.12   Reprinted by permission from H. J. Blinchikoff and A. I. Zverev, *Filtering in the Time and Frequency Domains*, John Wiley & Sons, Inc., New York, 1976.

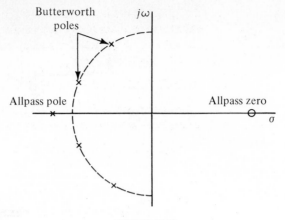

FIGURE 18.13

nominator. These are

$$\theta_t = \tan^{-1}\omega + \tan^{-1}\left(\frac{0.625}{1 - \omega^2}\right) + \tan^{-1}\left(\frac{1.625}{1 - \omega^2}\right).$$ (18.13)

Differentiating this equation with respect to ω gives

$$D_{B_5}(\omega) = \frac{1}{1 + \omega^2} + \frac{0.625(1 + \omega^2)}{(1 - \omega^2)^2 + 0.384\omega^2} + \frac{1.625(1 + \omega^2)}{(1 - \omega^2)^2 + 2.62\omega^2}$$ (18.14)

From these equations the two curves shown in Fig. 18.15 marked D_1 and D_{B_5} are plotted, and these are added to give D_t. In this example the addition of the allpass section has reduced ΔD from 1.8 to 1.2.

18.3 DELAY AND SECOND-ORDER ALLPASS FUNCTIONS

The second-order allpass function was introduced in Chapter 5, given by Eq. (5.74), which is repeated here:

$$T_2(s) = \frac{s^2 - (\omega_0/Q)s + \omega_0^2}{s^2 + (\omega_0/Q)s + \omega_0^2}$$ (18.15)

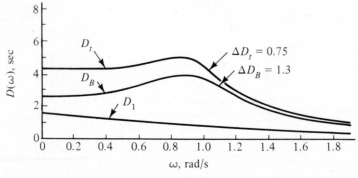

FIGURE 18.14

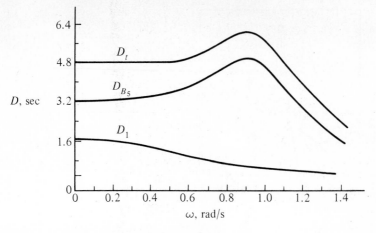

FIGURE 18.15

From this we see that T_2 is expressed in terms of only two parameters, ω_0 and Q, and so we know that delay will also be expressed in terms of these parameters. The phase function is

$$\arg T_2(j\omega) = \theta_2(\omega) = -2 \tan^{-1}\left|\frac{\omega_0/Q}{\omega_0{}^2 - \omega^2}\right| \qquad (18.16)$$

This equation may be expressed in simpler form by letting $\Omega = \omega/\omega_0$ so that

$$\theta_2(\Omega) = -2 \tan^{-1}\left|\frac{\Omega/Q}{1 - \Omega^2}\right| \qquad (18.17)$$

The delay function is found by differentiating the phase function. Hence

$$D_2(\Omega) = \frac{d\theta_2}{d\Omega}\frac{d\Omega}{d\omega} = \frac{2}{\omega_0 Q}\frac{1 + \Omega^2}{(1 - \Omega^2)^2 + \Omega^2/Q^2} \qquad (18.18)$$

Let us now determine the form of the plot of $\omega_0 D_2$ as a function of Ω, with Q as the parameter. From the last equation it is clear that

$$\omega_0 D_2(0) = \frac{2}{Q} \qquad (18.19)$$

The maximum value of $\omega_0 D_2$ is found by setting the derivative to zero. If we differentiate Eq. (18.18), then setting the expression to zero gives the condition of the maximum value as

$$\Omega^4 + 2\Omega^2 + \left(\frac{1}{Q^2} - 3\right) = 0 \qquad (18.20)$$

The solution of this equation is

$$\Omega_{max} = \sqrt{-1 + \sqrt{4 - \frac{1}{Q^2}}} \qquad (18.21)$$

From this we see that when $Q > 1$, then $\Omega_{max} \approx 1$, or $\omega_{max} = \omega_0$. Finally as Ω be-

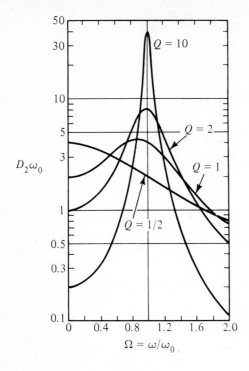

FIGURE 18.16 Reprinted by permission from H. J. Blinchikoff and A. I. Zverev, *Filtering in the Time and Frequency Domains,* John Wiley & Sons, Inc., New York, 1976.

comes large, we see that

$$\omega_0 D_2(\infty) \to 0 \qquad (\text{slope} = -12 \text{ dB per octave}) \qquad (18.22)$$

The plot of Eq. (18.18) for four values of Q illustrates our conclusions, as shown in Fig. 18.16. In addition we see that when the solution of Eq. (18.21) is approximately 1, then

$$(D_2\omega_0)_{\text{max}} = 4Q \qquad (18.23)$$

which is also evident in the curves of Fig. 18.16.

Another interesting relationship comes from the last equation. Let the poles and zeros of the quad shown in Fig. 18.17 have a real part which is $\pm \alpha$. From Chapter 5 we know that this real part is $\alpha = \omega_0/2Q$, so that

$$D_{2\text{max}} = \frac{2}{\alpha} \qquad (18.24)$$

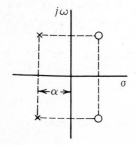

FIGURE 18.17

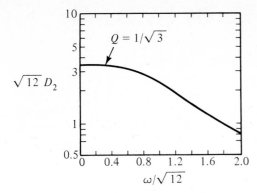

FIGURE 18.18

This relationship between maximum delay and s-plane location of the poles and zeros applies only for Q larger than 1.

Example 18.3   We wish to study the characteristics of the second-order transfer function

$$T_2(s) = \frac{s^2 - 6s + 12}{s^2 + 6s + 12} \tag{18.25}$$

Comparing this equation with Eq. (18.15) we see that $\omega_0 = \sqrt{12} = 2\sqrt{3}$, and $Q = 1/\sqrt{3}$. Substituting this value of Q into Eq. (18.21), we see that $\Omega_{max} = 0$, which shows us that this value of Q is the smallest for which there is a maximum with zero derivative. From Eq. (18.19) we find that $\omega_0 D_2(0) = 2/Q = 2\sqrt{3} = 3.4641$, and so $D_{2max} = 1$ second. This response is shown in Fig. 18.18.

As in the case of first-order allpass sections, the objective is to use one or more sections connected in cascade, as represented in Fig. 18.19a. Each of the cascaded sections will be characterized by its own ω_0 and Q, illustrated by Fig. 18.19b, and the total delay of the cascaded sections is found by simply adding the delays of the individual sections. For simple equalization problems it will be possible to use cut-and-try methods (to "eyeball" the design), but for complicated equalization problems, or those with rigid specifications, it will be found necessary to use computer optimization programs. This will be illustrated by Example 18.5.

It is interesting to compare the equalization of the Butterworth delay response with a first-order allpass section, as shown in Fig. 18.12, with that attainable with a second-order allpass section. We will use the same requirement in designing the second-order section, namely, that the derivative of the combined delays with respect to frequency has zero value at $\omega = 0$. The result* is shown in Fig. 18.20 in which the equalization for first-order sections is compared to that for second-order sections. From the figure it is seen that there is greater delay with a second-order section, but that the "flatness" of the response extends over a greater range of frequencies.

* H. J. Blinchikoff and A. I. Zverev, *Filtering in the Time and Frequency Domains,* Wiley, New York, 1976, pp. 225–230.

18.4 SECOND-ORDER ALLPASS CIRCUITS

The circuit that we will use to realize second-order allpass transfer functions is due to Delyiannis;[†] we should recognize that it is not unique and that many other such circuits, both active and passive, are available. The Delyiannis circuit is shown in Fig. 18.21, and we see that it is the circuit we have previously identified as a bandpass Friend circuit with the addition of the voltage-dividing resistors R_3 and R_4. The voltage at the + terminal of the op amp is

$$V_C = \frac{R_4}{R_3 + R_4} V_1 = k V_1 \qquad (18.26)$$

Also, $V_C = V_B$ since there is no voltage across the input terminals of the op amp. We will analyze the circuit using node analysis, applied at nodes A and B. Then at node A, Kirchhoff's current law gives us

$$\left(2Cs + \frac{1}{R_1}\right) V_A + (-Cs) V_2 - \left(\frac{1}{R_1} + kCs\right) V_1 = 0 \qquad (18.27)$$

Similarly, at node B we have

$$(-Cs) V_A + \left(\frac{-1}{R_2}\right) V_2 + \left[k\left(Cs + \frac{1}{R_2}\right) \right] V_1 = 0 \qquad (18.28)$$

If we eliminate V_A and then solve for V_2/V_1, we obtain

$$\frac{V_2}{V_1} = k \frac{s^2 + [2/R_2C + (1 - 1/k)/R_1C]s + 1/R_1R_2C^2}{s^2 + (2/R_2C)s + 1/R_1R_2C^2} \qquad (18.29)$$

For the circuit to be an allpass filter having a transfer function of the form of Eq. (18.15) it is necessary that the coefficient of s in the numerator be the negative of the coefficient of s in the denominator of Eq. (18.29). This requires that

$$\frac{2}{R_2C} + \frac{1 - 1/k}{R_1C} = -\frac{2}{R_2C} \qquad (18.30)$$

[†] T. Delyiannis, "High-Q Factor Circuit with Reduced Sensitivity," *Electron. Lett.*, vol. 4, p. 577, 1968; T. Delyiannis, "RC Active All-Pass Sections," *Electron. Lett.*, vol. 5, pp. 59–60, 1969.

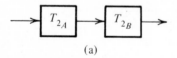

(a)

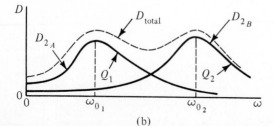

(b)

FIGURE 18.19

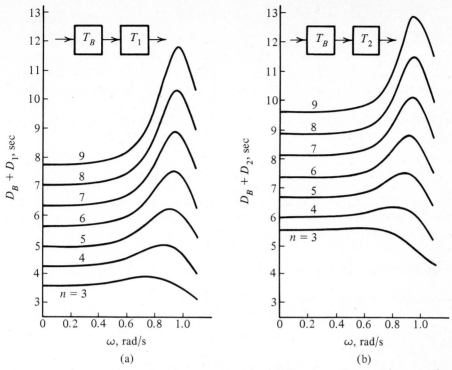

FIGURE 18.20   Reprinted by permission from H. J. Blinchikoff and A. I. Zverev, *Filtering in the Time and Frequency Domains*, John Wiley & Sons, Inc., New York, 1976.

This equation simplifies to

$$\frac{R_2}{R_1} = 4\,\frac{R_4}{R_3} \qquad \text{or} \qquad \frac{R_4}{R_3} = \frac{1}{4}\frac{R_2}{R_1} \qquad (18.31)$$

This is the condition that must be satisfied for Fig. 18.21 to be an allpass filter.

In our earlier study of the Friend circuit in Chapter 7 we scaled the frequency such that $\omega_0 = 1$, and then selected the values

$$R_1 = 1 \qquad \text{and} \qquad R_2 = 4Q^2 \qquad (18.32)$$

Substituting these values into Eq. (18.31), we see that it is necessary that

$$\frac{R_4}{R_3} = Q^2 \qquad (18.33)$$

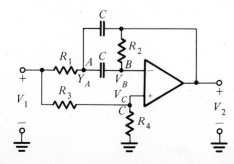

FIGURE 18.21

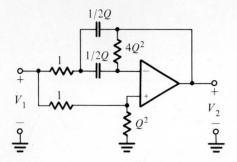

FIGURE 18.22

To satisfy this requirement, we let $R_4 = Q^2$ and $R_3 = 1$. Recalling that for the Friend circuit the capacitors had the value $C = 1/2Q$, we obtain the frequency normalized circuit shown in Fig. 18.22. From Eq. (18.29) we see that

$$T(0) = k = \frac{R_4}{R_3 + R_4} = \frac{Q^2}{1 + Q^2} \tag{18.34}$$

Hence the circuit of Fig. 18.22 realizes the scaled transfer function

$$T(s) = \frac{Q^2}{1 + Q^2} \frac{s^2 - (1/Q)s + 1}{s^2 + (1/Q)s + 1} \tag{18.35}$$

It is interesting to observe that if we set Eq. (18.30) to 0 rather than $-2/R_2C$, then we obtained the values

$$R_3 = 1 \quad \text{and} \quad R_4 = 2Q^2 \tag{18.36}$$

Since several arbitrary choices of element values were made, it should be evident that other possibilities exist. For example, if we satisfy Eq. (18.31) with the choices

$$R_1 = \frac{1}{4Q^2} \quad \text{and} \quad R_2 = 1 \tag{18.37}$$

then we find that it is necessary that

$$R_4 = \frac{1}{4} \quad \text{and} \quad R_3 = \frac{1}{4Q^2} \tag{18.38}$$

The capacitor values are then found from the equation, for $\omega_0 = 1$, such that

$$\frac{1}{R_1R_2C^2} = 1 \quad \text{or} \quad C^2 = \frac{1}{R_1R_2} = 4Q^2 \quad \text{or} \quad C = 2Q \tag{18.39}$$

This information is shown in Fig. 18.23. This design may be made using either of the circuits shown in Fig. 18.22 and Fig. 18.23. We do note, however, that the circuit of Fig. 18.23 may be obtained from that of Fig. 18.22 by magnitude scaling with $k_m = 1/4Q^2$.

18.5 CIRCUIT ADJUSTMENT TECHNIQUES

In the design of cascaded sections of delay equalizers it is often necessary to make fine adjustments experimentally. In this case designers resort to a cut-and-try pro-

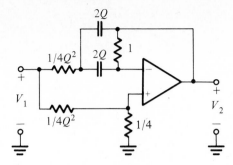

FIGURE 18.23

cedure, called *functional tuning*. The manner in which this may be accomplished may be shown with the aid of Fig. 18.24. Here an additional resistor has been added to the circuit of Fig. 18.22 having the value Q^2, connected between node A and ground by the closing of a switch. If the switch is closed, then a bridge circuit is formed as an approximation and is balanced such that the voltages at nodes A and C are equal, and

$$V_A = V_C = kV_1 \qquad (18.40)$$

by Eq. (18.26). If we substitute this equation for V_A into Eq. (18.28), we obtain

$$\left(\frac{-1}{R_2}\right) V_2 + kV_1Cs - kV_1Cs + \left(\frac{1}{R_2}\right) V_1 = 0 \qquad (18.41)$$

or

$$\frac{V_2}{V_1} = 1 \qquad (18.42)$$

Hence with the switch closed, the circuit provides neither attenuation nor phase shift and so provides a reference by which the time delay may be measured.*

The same concept applies to the first-order all-phase circuit shown in Fig. 18.11a. If this circuit is modified by the addition of a switch, as shown in Fig. 18.25, then with the switch connected to R, $V_2/V_1 = 1$, and there is no time delay. Then by switching from one position to the other and observing the time delay, a final adjustment may be made.

* C. R. Hoffman, "Develop Your Own Effective Delay-Distortion Equalizer," *EDN*, pp. 74–77, Apr. 5, 1977.

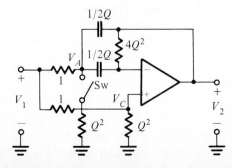

FIGURE 18.24

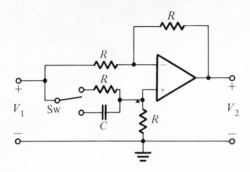

FIGURE 18.25

18.6 ESTIMATING THE NUMBER OF SECTIONS NEEDED

We have developed the strategy that an arbitrary delay characteristic required for equalization may be realized by the cascade connection of a number of first- and second-order allpass sections. In designing the equalizer, it is necessary to first decide on the number of sections that will be required, and then determine characterizing parameters for each of the sections. We next present a method by which the number of sections may be estimated.

We begin with the second-order delay characteristic shown in Fig. 18.26, in which the delay D_2 is plotted as a function of ω. The area under the curve will be

$$A_2 = \int_0^\infty D_2(\omega)\, d\omega \tag{18.43}$$

This area may also be expressed in terms of the phase function θ_2. Then since

$$D_2 = -\frac{d\theta_2}{d\omega} \tag{18.44}$$

we have

$$A_2 = \int_0^\infty \frac{-d}{d\omega} \theta_2(\omega)\, d\omega = \theta_2(\omega)\Big|_\infty^0 \tag{18.45}$$

or

$$A_2 = \theta_2(0) - \theta_2(\infty) \tag{18.46}$$

Now the phase function associated with $T_2(s)$ given in Eq. (18.15) may be written

$$\theta_2(\omega) = \tan^{-1}\left(\frac{-\omega_0/Q}{\omega_0^2 - \omega^2}\right) - \tan^{-1}\left(\frac{\omega_0/Q}{\omega_0^2 - \omega^2}\right) \tag{18.47}$$

Hence we obtain the value for A_2 from Eq. (18.46):

$$A_2 = (0 + \pi) - (-\pi + 0) = 2\pi \tag{18.48}$$

We see that for any value of ω_0 and Q

$$\int_0^\infty D_2(\omega)\, d\omega = 2\pi \tag{18.49}$$

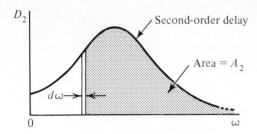

FIGURE 18.26

By the same reasoning we also find that

$$\int_0^\infty D_1(\omega)\, d\omega = \pi \tag{18.50}$$

for first-order delay sections.

On the basis of these two equations we see that the number of sections required may be estimated by determining the delay function that produces a constant delay when added to the filter or equipment delay. The area under this equalizer response is then determined, perhaps by graphical integration or simply by estimating, and is

$$A_{eq} = \int_0^\infty D_{eq}(\omega)\, d\omega \tag{18.51}$$

Then the number of sections required is, approximately,

$$n_2 = \frac{A_{eq}}{2\pi} \tag{18.52}$$

for second-order sections, or twice this value for first-order sections. This is an approximate result because some of the delay equalizing area will be outside of the frequency range of interest. However, it is very useful in making an initial determination of the value of n_2.

Example 18.4 It is determined experimentally that the equalization delay required for a given communication system is that shown in Fig. 18.27 by the solid curve. We wish to determine the number of sections of allpass equalization that will be required. The area is first determined graphically by approximating the actual curve by the straight-line segments shown on the figure. From these we estimate the area to be

$$A_{eq} = 8.4 \times 0.76 + \tfrac{1}{2}\,[8.4(1.6 - 0.76)] = 9.91 \tag{18.53}$$

Dividing this number by 2π gives $n_2 = 1.6$, which we must round up to 2. The two sections which do actually give the required delay equalization are shown in Fig. 18.28 together with required values of ω_0 and Q to completely specify the equalizer sections. The manner by which these numbers may be determined is considered in the next section.

18.7 STRATEGIES FOR EQUALIZER DESIGN

For a specified form of response and the requirement that equalization be provided such that the total delay is maximally flat at $\omega = 0$, a closed form of solu-

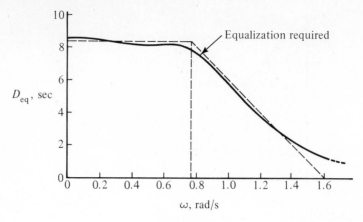

FIGURE 18.27

tion is possible, as we have seen in Section 18.2. Such an analytical approach becomes very complicated when more than one allpass equalization is used. If the equalization problem is relatively simple, then the experimental approach of Section 18.5 may be used, or a cut-and-try procedure may be used if only a few iterations are required. For complex problems it is usual to use computer optimization methods. Programs are available commercially to perform this optimization.

In developing a strategy, the first decision required is the sense in which the design is to be optimal. Optimization criteria are usually expressed in terms of a number which may be the difference between the result and the desired response expressed as an average error, a mean-square error, or a peak error. If the criterion selected is that the mean-square error be minimized, then standard procedures* may be followed. Let D_d be the desired delay over a range of frequencies from ω_1 to ω_2. We then select the number of first-order or second-order sections to be used, each characterized by σ_1 for first-order stations and ω_0 and Q for second-

* Blinchikoff and Zverev, *op. cit.*, pp. 103–106, 233–235.

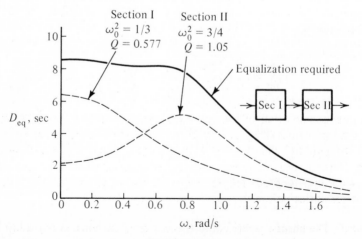

FIGURE 18.28

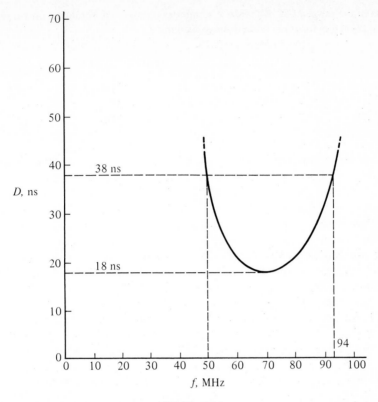

FIGURE 18.29

order sections. Let D_a be the delay available from the assumed configuration of sections. The least-squares error is defined as

$$\varepsilon = \int_{\omega_1}^{\omega_2} [D_d(\omega) - D_a(\omega)]^2 \, d\omega \qquad (18.54)$$

We then require that

$$\frac{\partial \varepsilon}{\partial \sigma_i} = 0, \qquad \frac{\partial \varepsilon}{\partial \omega_i} = 0, \qquad \frac{\partial \varepsilon}{\partial Q_i} = 0 \qquad (18.55)$$

The simultaneous solution of the equations that result from this differentiation gives the optimal values of σ_i, ω_i, and Q_i. From these values the response is calculated and compared to that desired. If this is not adequate, then a different configuration of sections is assumed, and the calculations are repeated. The efficiency of computer methods makes it possible for a designer to try a number of different possibilities before deciding which design will be used.

Example 18.5   The channel delay of certain equipment is shown in Fig. 18.29 as measured experimentally. Equalization is required such that the total delay approximates a constant in the range of 50–94 MHz. A computer optimization program provides the following parameters for three second-order sections:

Section	ω_0	f_0	Q
I	3.652×10^8	58.1 MHz	2.815
II	4.516×10^8	71.8 MHz	3.272
III	5.392×10^8	85.8 MHz	4.200

The three responses corresponding to these values are plotted from Eq. (18.18) and shown in Fig. 18.30 together with the delay due to the equipment. Adding these four delays gives a total delay which is approximately constant over the specified band of frequencies, having an average value of 68.6 ns. A circuit realization based on the normalized circuit given in Fig. 18.22 is shown in Fig. 18.31. The final element values were determined from Fig. 18.22 with appropriate frequency and magnitude scaling. The last stage is provided to compensate for the gain reduction due to the terms of the form $Q^2/(1 + Q^2)$ in Eq. (18.35). We have assumed that it was desirable that all capacitors have the same value, and have chosen the values of k_m accordingly.

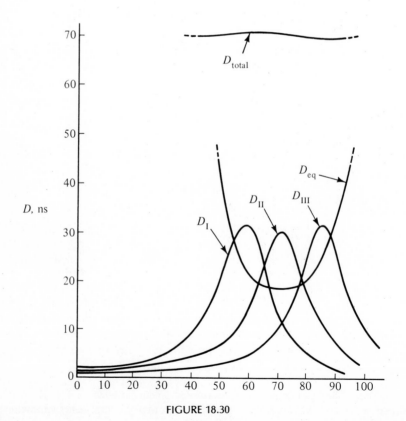

FIGURE 18.30

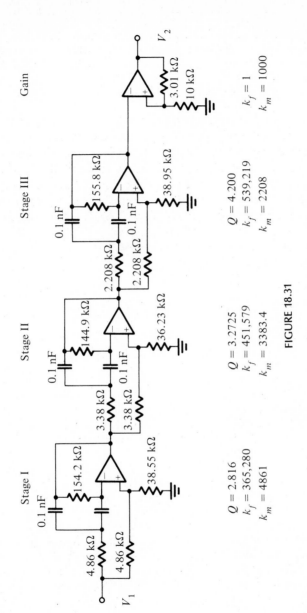

FIGURE 18.31

PROBLEMS

18.1 Figure P18.1 shows the circuit originally considered in Chapter 7 and shown in Fig. 7.18 modified by the addition of a second op-amp circuit. Determine the values for k_1 and k_2 such that the transfer function $T = V_2/V_1$ is an allpass function.

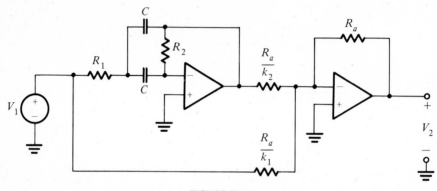

FIGURE P18.1

18.2 In Fig. 14.22, let the output voltage of the woofer and tweeter be V_{2LP} and V_{2HP} respectively, and let the filters be third-order Butterworth lowpass and highpass. Consider the function

$$\frac{V_{2LP} \pm F_{2HP}}{V_s}$$

and show that the plus sign results in a second-order allpass function, while the minus sign results in a first-order all pass function.

18.3 Twenty-five identical first-order allpass sections are connected in cascade. What is the resulting value of σ_0 if the overall delay at $\omega = 10^4$ rad/s is 2 ms, and the delay at 1.5×10^4 rad/s is 1 ms.

18.4 The measurement of the delay introduced by communications equipment is shown in Fig. P18.4 with normalized delay. We wish to design a circuit to provide delay equalization so that the total delay is approximately constant from $\omega = 0.1$ to $\omega = 1.0$. Design a circuit that will provide the required delay equalization. Scale the circuit so that 1 sec becomes 1 ms.

18.5 The measurement of the delay in communication equipment provides the curve shown in Fig. P18.5. We wish to introduce delay equalization so that the total delay is approximately constant from $\omega = 0$ to $\omega = 1$ (normalized frequency). Design the circuit to provide the desired equalization. Scale the circuit so that 1 sec becomes 10 μs.

18.6 The measurement of the delay introduced by communication equipment is shown in Fig. P18.6. We wish to provide delay equalization so that the total delay is approximately constant over the range of frequencies shown in the figure.
(a) Design a circuit that will provide the required equalization.
(b) Scale the circuit so that 1 sec becomes 1 μs.

18.7 It is required to design an allpass filter satisfying the given linear delay

$$D(\omega) = -0.48\omega + 4.6 \qquad 1 \le \omega \le 6$$

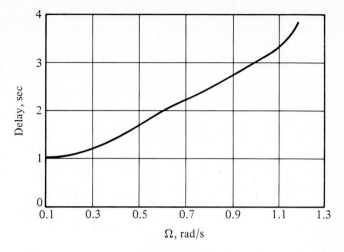

FIGURE P18.4

A computer optimization program provides the needed parameters as follows: one first-order section with $\sigma = 3.5$, four second-order sections each having a Q of 0.5 and frequencies 3.2, 3.8, 4.5 and 4.8.

(a) Design the circuit to satisfy the computed parameters.

(b) Frequency scale so that $\omega = 1$ becomes $\omega = 1000$ rad/s, and determine the circuit element values for the circuit designed in part (a).

18.8 For high-frequency equalization, it is necessary to use passive elements. For the

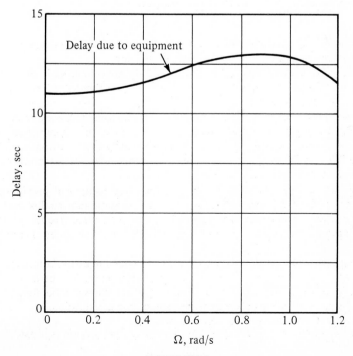

FIGURE P18.5

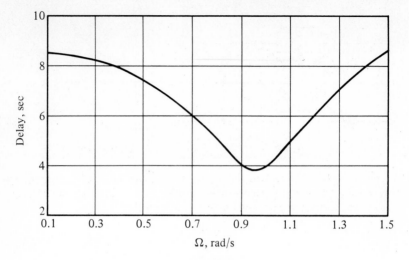

FIGURE P18.6

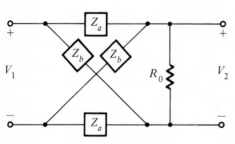

FIGURE P18.8

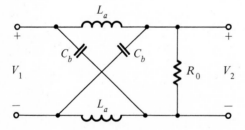

FIGURE P18.9

balanced lattice shown in Fig. P18.8, show that if $R_0 = 1$ and $Z_a Z_b = R_0^2 = 1$, then

$$T(s) = \frac{V_2}{V_1} = \frac{1 - Z_a}{1 + Z_a}$$

From this, it follows that if Z_a is a reactance function, then

$$|T(j\omega)| = 1$$

which is the requirement of an allpass circuit.

18.9 Apply the results of Problem 18.8 to the circuit shown in Fig. P18.9, and show that the transfer function represents a first-order allpass circuit with a pole at $-\sigma_0$ and a zero at $+\sigma_0$. Obtain design equations for L and C.

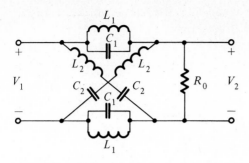

FIGURE P18.10

18.10 Apply the results of Problem 18.8 to the circuit given in Fig. P18.10 and show that the transfer function is a second-order allpass function. Determine design equations for the inductors and capacitors in terms of ω_0 and Q.

CHAPTER 19

Op-Amp
Oscillators

Those with laboratory experience in adjusting active filters will understand the connection between filters and oscillators. If the filter parameters are adjusted so that the poles of the transfer function $T(s)$ are placed on the imaginary axis, corresponding to $Q \to \infty$, then the filter oscillates. In this chapter we study some of the design considerations in designing a filterlike circuit to oscillate in a stable manner with low distortion of the output waveform.

19.1 A SIMPLE OSCILLATOR

In the circuit shown in Fig. 19.1 the input is the voltage V_1 and the output is V_2. Writing the equations for each of the cascade-connected modules, we have

$$V_a = - \frac{R_2}{R_1} (V_1 + V_2) \tag{19.1}$$

$$V_b = - \frac{1/C_1 s}{R_3} V_a \tag{19.2}$$

$$V_2 = - \frac{1/C_2 s}{R_4} V_b \tag{19.3}$$

We may eliminate V_a and V_b from these equations to obtain

$$- \frac{R_2}{R_1} (V_1 + V_2) = R_3 R_4 C_1 C_2 s^2 \, V_2 \tag{19.4}$$

Finally, solving for $T(s) = V_2/V_1$, we have

$$T(s) = \frac{-R_2/R_1 R_3 R_4 C_1 C_2}{s^2 + R_2/R_1 R_3 R_4 C_1 C_2} \tag{19.5}$$

Comparing this equation with the standard form of a second-order transfer function, we see that $Q = \infty$ and

$$\omega_0{}^2 = \frac{R_2}{R_1 R_3 R_4 C_1 C_2} \tag{19.6}$$

Thus the circuit will oscillate at the frequency given by Eq. (19.6) for any combination of element values. In studying other oscillators, we will find that there is always a condition for oscillation, but there is none for this circuit.

The circuit in Fig. 19.1 is drawn in the form of a filter with an input and an output identified. In oscillators there is no input identified, and so to modify this circuit to look like an oscillator, we set $V_1 = 0$ and remove one of the resistors R_1, retaining that used for feedback. We recognize that even a small input to the feedback circuit will initiate the feedback process and that sustained oscillations will result. This "small input" is provided by the noise that is always present in electric circuits, from pickup, power-supply transients, thermal noise, and so on. So in practical oscillators no input will be shown, although we may wish to restore an input to aid in analysis.

In many realizations a resistor network manufactured such that all resistors are equal with precision will be used. Then we may set all resistors equal:

$$R_1 = R_2 = R_3 = R_4 = R \tag{19.7}$$

The frequency of oscillation is now simplified to

$$\omega_0{}^2 = \frac{1}{R^2 C_1 C_2} \tag{19.8}$$

If we also match the capacitors such that $C_1 = C_2 = C$, then

$$\omega_0 = \frac{1}{RC} \quad \text{or} \quad f_0 = \frac{1}{2\pi RC} \quad \text{Hz} \tag{19.9}$$

The circuit that corresponds to these simplifications is shown in Fig. 19.2. The circuit is designed by simply selecting values of R and C to match a specified frequency.

We should next consider the problem of *tuning*, should tuning be necessary in some application. For Eq. (19.9) to apply, it is necessary that $C_1 = C_2$ so that if one capacitor is to be adjusted, then the other must be adjusted in exactly the same way. This may be accomplished by using *ganged* capacitors for which an adjustment of one capacitor automatically results in the same adjustment for the other. If we prefer to use only one capacitor for final tuning of the oscillator, then we must use Eq. (19.8) from which

$$f_0 = \frac{1}{2\pi R \sqrt{C_1 C_2}} \quad \text{Hz} \tag{19.10}$$

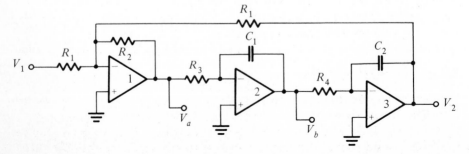

FIGURE 19.1

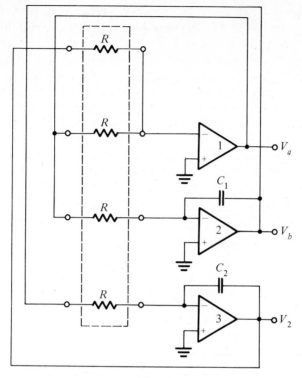

FIGURE 19.2

To illustrate oscillator design, suppose that we wish an oscillator to operate at a frequency of 10 kHz. If we employ a resistor network with $R = 10$ kΩ, then from Eq. (19.9) the required value of C is

$$C = \frac{1}{2\pi R f_0} = 1.591 \times 10^{-9} \text{ F} \qquad (19.11)$$

$$= 1.591 \text{ nF}$$

Our problem now reduces to finding identical capacitors of this value.

Returning now to the circuit of Fig. 19.2, we observe that three output terminals are shown. The three voltages oscillate at the same frequency, of course, but have an interesting phase relationship. If we select $v_2(t)$ as a reference, then we see from Eq. (19.3) that $v_b(t)$ lags v_2 by 90° and that, similarly, from Eq. (19.2) $v_a(t)$ lags v_b by 90°. This is an important feature of the circuit in applications requiring oscillating voltages with phase quadrature.

19.2 LOOP GAIN

We next generalize the results obtained for the simple oscillator of the last section. The circuit shown in Fig. 19.1 may be drawn in block diagram form as shown in Fig. 19.3a with $A(s)$ representing the voltage-ratio transfer function for the feed-forward part of the circuit, and $B(s)$ representing the transfer function for the feedback circuit. The block diagram representation is equivalent to the equations

$$V_e = V_1 + B(s)V_2 \qquad (19.12)$$

$$V_2 = A(s)V_e \qquad (19.13)$$

Eliminating V_e from the two equations and solving for V_2/V_1, we have

$$\frac{V_2}{V_1} = T(s) = \frac{1}{1 - A(s)B(s)} \qquad (19.14)$$

In this equation $A(s)$ and $B(s)$ must carry their own sign. Now the denominator of this equation when set equal to 0 is the *characteristic equation* for the system under study; that is,

$$1 - A(s)B(s) = 0 \qquad (19.15)$$

is the characteristic equation, and the roots of this equation determine the time-domain behavior of the system. We are interested in this chapter in the conditions under which a pair of these roots are on the imaginary axis.

In the last section we set $V_1 = 0$ as depicted in Fig. 19.3b. Now we observe that if we break the loop and insert a test voltage V_T as shown, then from Fig. 19.3 we have

$$\frac{V_2}{V_T} = B(s)A(s) \qquad (19.16)$$

This quantity is known as the open-loop gain, to be called LG. This has been a very simple mathematical manipulation, but an important concept has been introduced. To analyze an oscillator circuit, we break the feedback loop as shown in Fig. 19.3b and then determine the open-loop gain. This quantity is shown to be

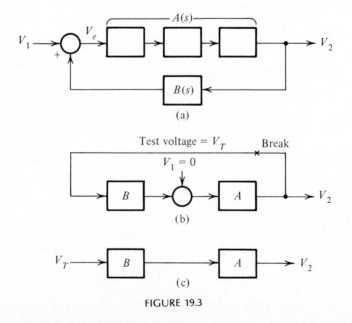

FIGURE 19.3

equal to $A(s)B(s)$, which may be used to modify Eq. (19.15) to the form

$$1 - LG = 0 \tag{19.17}$$

from which the condition for oscillation and the frequency of oscillation may be determined.

We will illustrate the use of this concept by reexamining the circuit of Fig. 19.1, shown in Fig. 19.4 with the feedback loop broken. Under this condition the loop gain is

$$LG = \left(-\frac{R_2}{R_1}\right)\left(-\frac{1}{R_3 C_1 s}\right)\left(-\frac{1}{R_4 C_2 s}\right) \tag{19.18}$$

Substituting this into Eq. (19.17), we have

$$1 - LG = 1 + \frac{R_2}{R_1 R_3 R_4 C_1 C_2}\frac{1}{s^2} = 0 \tag{19.19}$$

or

$$s^2 + \frac{R_2}{R_1 R_3 R_4 C_1 C_2} = 0 \tag{19.20}$$

which is the characteristic equation. It tells us that the system is oscillatory and that the frequency of oscillation is

$$\omega_0 = \sqrt{\frac{R_2}{R_1 R_3 R_4 C_1 C_2}} \quad \text{rad/s} \tag{19.21}$$

as found previously.

To extend our understanding of the use of loop gain, let us inquire as to the conditions that a system made up from a bandpass filter transfer function can represent an oscillator. To do so, assume that

$$LG = \frac{Ks}{s^2 + (\omega_0/Q)s + \omega_0^2} \tag{19.22}$$

Then the characteristic equation is

$$1 - LG = 1 - \frac{Ks}{D} = \frac{s^2 + (\omega_0/Q - K)s + \omega_0^2}{D} \tag{19.23}$$

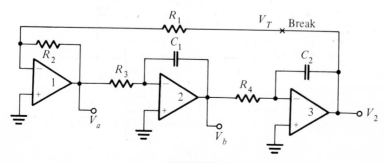

FIGURE 19.4

where we have used D to represent the denominator of Eq. (19.22). From this equation we see that the condition for oscillation is

$$K = \frac{\omega_0}{Q} \tag{19.24}$$

and that when this condition is fulfilled, the system will oscillate at the frequency ω_0.

The Friend circuit studied in Chapter 7 has a transfer function of the form of Eq. (19.22) provided that K is negative. This sign inversion is accomplished as shown in Fig. 19.5 by the addition of an inverting amplifier of gain $-R_B/R_A$. The resistors R_c and R_d provide voltage reduction and are to be chosen such that

$$R_1 = \frac{R_c R_d}{R_c + R_d} \tag{19.25}$$

Then we may determine the loop gain as follows:

$$LG = \left(\frac{-R_B}{R_A}\right)\left(\frac{-R_d}{R_c + R_d}\right)\left(\frac{-2Q\omega_0 s}{s^2 + (\omega_0/Q)s + \omega_0{}^2}\right) \tag{19.26}$$

Substituting this equation into the characteristic equation $1 - LG = 0$, gives the following condition for oscillation:

$$\frac{\omega_0}{Q} = \frac{R_B}{R_A}\frac{R_d}{R_c + R_d}2Q\omega_0 \tag{19.27}$$

From our study of the Friend circuit in Chapter 7 we know that

$$Q = \frac{1}{2}\sqrt{\frac{R_2}{R_1}} \tag{19.28}$$

If we substitute this equation into Eq. (19.27), we obtain

$$\frac{R_B}{R_A} = 2\frac{R_c}{R_2} \tag{19.29}$$

as the condition for oscillation. Also from Chapter 7 we know that the frequency of oscillation is

$$\omega_0 = \frac{1}{C\sqrt{R_1 R_2}} = \frac{1}{C\sqrt{R_2\, R_c R_d/(R_c + R_d)}} \tag{19.30}$$

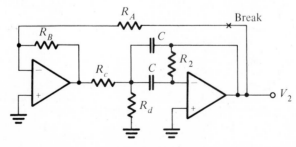

FIGURE 19.5

From the last two equations we see that R_c may be set to obtain the condition for oscillation, and R_d then used to set the frequency of oscillation, provided only that Eq. (19.25) is satisfied.

19.3 WEIN BRIDGE OSCILLATOR

Figure 19.6a shows a passive circuit known as a Wein bridge. This circuit has long been used for making measurements of unknown resistors or capacitors. Assuming that V_1 represents a sinusoidal source, it may be shown that when $V_2 = 0$ and so a null has been obtained, then the relationship among the elements is

$$\frac{R_1}{R_2} + \frac{C_2}{C_1} = \frac{R_3}{R_4} \qquad (19.31)$$

Thus if all component values are known except one, then the value of that one may be determined from this equation. In making measurements, either R_3 or R_4 is a calibrated resistor which is varied until the null voltage is found. When an op amp is inserted into the bridge, as shown in Fig. 19.6b, then the circuit is known as a *Wein bridge oscillator* provided that the elements are adjusted so that oscillation is possible.

In the Wein bridge oscillator we let $R_1 = R_2 = R$ and $C_1 = C_2 = C$, and show the circuit in the form illustrated in Fig. 19.7. To determine the loop gain, we break the circuit as shown, and then determine LG. First observe that the op amp with resistors R_3 and R_4 constitutes a noninverting amplifier of gain

$$K = 1 + \frac{R_3}{R_4} \qquad (19.32)$$

From the circuit of Fig. 19.7, with the feedback path broken, we see that the loop gain may be found considering the series RC circuit together with the shunt RC

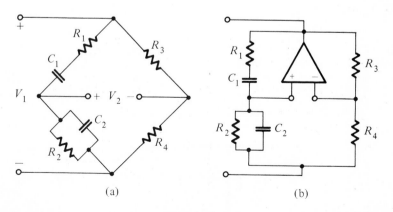

(a)   (b)

FIGURE 19.6

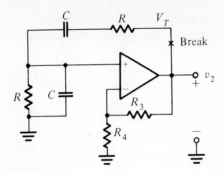

FIGURE 19.7

circuit as constituting a voltage divider. Thus

$$LG = \frac{K[R(1/Cs)/(R+1/Cs)]}{R(1/Cs)/(R+1/Cs)+R+1/Cs} = \frac{(K/RC)s}{s^2+(3/RC)s+1/R^2C^2} \quad (19.33)$$

Then the characteristic equation is

$$1 - LG = 1 - \frac{(K/RC)s}{s^2+(3/RC)s+1/R^2C^2} \quad (19.34)$$

The numerator of this equation thus becomes

$$s^2 + (3-K)\frac{1}{RC}s + \frac{1}{R^2C^2} = 0 \quad (19.35)$$

The roots of this equation are the poles of the transfer function $T(s)$. The locus of these poles is shown in Fig. 19.8 for all positive values of K. Clearly the value of K of interest in making the circuit into an oscillator is $K = 3$. Then the condition for oscillation is

$$K = 3 = 1 + \frac{R_3}{R_4} \quad \text{or} \quad R_3 = 2R_4 \quad (19.36)$$

If we make $R_4 = R$, then $R_3 = 2R$. Then the Wein bridge oscillator may be constructed with five equal resistors. If we use an R network, then the realization is

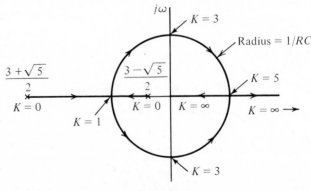

FIGURE 19.8

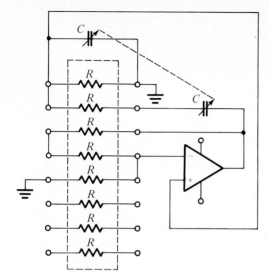

FIGURE 19.9

as shown in Fig. 19.9. For this circuit the frequency of oscillation is

$$\omega_0 = \frac{1}{RC} \quad \text{or} \quad f_0 = \frac{1}{2\pi RC} \quad \text{Hz} \tag{19.37}$$

The condition $C_1 = C_2 = C$ requires comment if C is used to adjust the frequency of oscillation. The dashed line connecting the two capacitors in Fig. 19.9 implies that the two capacitors are *ganged* or connected mechanically such that they always have the same value.

19.4   CONDITION FOR THIRD-ORDER CIRCUIT OSCILLATION

The oscillator circuits studied in previous sections have been second-order oscillators for which the condition for oscillation was apparent. Such is not the case for higher order circuits, as we shall see. We begin with Eq. (19.17) which is

$$1 - LG = \frac{N(s)}{D(s)} \tag{19.38}$$

The characteristic equation is $N(s) = 0$, and the roots of this equation are the poles of $T(s)$. We are interested in the case $N(j\omega_0) = 0$, for this implies poles on the imaginary axis at $\pm j\omega_0$, indicating that the circuit oscillates at the frequency ω_0.

To review the conditions for a second-order $N(s)$, let that function be

$$N_2(s) = a_0 s^2 + a_1 s + a_2$$
$$= a_0 \left(s^2 + \frac{a_1}{a_0} s + \frac{a_2}{a_0} \right) \tag{19.39}$$

The condition for oscillation is apparently $a_1 = 0$, for then

$$N_2(s) = a_0\left(s^2 + \frac{a_2}{a_0}\right) \tag{19.40}$$

from which we see that

$$N_2\left(j\sqrt{\frac{a_2}{a_0}}\right) = 0 \tag{19.41}$$

The general situation and the special condition of Eq. (19.41) are represented in Fig. 19.10. The general case of Eq. (19.39) is shown in Fig. 19.41a and the special case of Eq. (19.40) is shown in Fig. 19.41b.

The third-order $N(s)$ of Eq. (19.38) is

$$N_3(s) = a_0s^3 + a_1s^2 + a_2s + a_3 \tag{19.42}$$

Let $s = j\omega_0$ be the frequency at which $N_3(j\omega_0) = 0$. We arrange this last equation into its real and imaginary parts:

$$N(j\omega_0) = (-a_1\omega_0^2 + a_3) + j\omega_0(-a_0\omega_0^2 + a_2) = 0 \tag{19.43}$$

This equation can be satisfied only if both the real and the imaginary parts vanish. In other words, we require that

$$-\omega_0^2a_1 + a_3 = 0 \quad \text{and} \quad -\omega_0^2a_0 + a_2 = 0 \tag{19.44}$$

These conditions are satisfied if

$$\omega_0^2 = \frac{a_3}{a_1} = \frac{a_2}{a_0} \tag{19.45}$$

This requires that

$$a_1a_2 = a_0a_3 \quad \text{or} \quad a_1a_2 - a_0a_3 = 0 \tag{19.46}$$

We see then that Eq. (19.46) is the condition for oscillation, and Eq. (19.45) is the frequency of oscillation. The result is illustrated in Fig. 19.11. The pole locations shown in Fig. 19.11a represent the general case for a third-order polynomial in which a pair of poles and complex conjugates and the third pole are real. When $a_1a_2 - a_0a_3 = 0$, then the conjugate poles move to the imaginary axis, as shown in

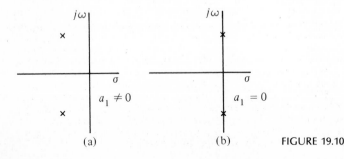

(a) (b) FIGURE 19.10

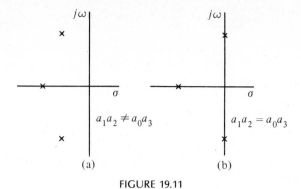

FIGURE 19.11

Fig. 19.11b, and Eq. (19.42) becomes

$$a_0(s + \alpha_1)(s^2 + \omega_0^2) = 0 \qquad (19.47)$$

The circuit having a $T(s)$ with poles described by this equation represents one oscillating at the frequency ω_0.

Most practical oscillators are either second- or third-order circuits. For circuits of higher order, which may well be oscillating circuits, analysis should be carried out making use of the Routh criterion* which is simple to apply and is, of course, identical with Eq. (19.46) for the case $n = 3$.

To illustrate the use of the criterion of Eq. (19.46), consider the polynomial

$$(s + 1)(s + 2)^2 = s^3 + 5s^2 + 8s + 4 \qquad (19.48)$$

We see that Eq. (19.46) is not satisfied since $5 \times 8 - 1 \times 4 \neq 0$. However, the polynomial

$$(s + 1)(s^2 + 4) = s^3 + s^2 + 4s + 4 \qquad (19.49)$$

does satisfy Eq. (19.46) since $4 \times 1 - 4 \times 1 = 0$.

19.5 RC PHASE-SHIFT OSCILLATORS

We consider a phase-shift oscillator first proposed in 1941 by Ginzton and Hollingsworth, which makes use of the *RC* ladder circuit shown in Fig. 19.12. This circuit may be analyzed by assuming the output V_2 and then applying the two Kirchhoff laws to find V_1. The result is

---

* See M. Van Valkenburg, *Network Analysis,* 3rd ed., Prentice-Hall, Englewood Cliffs, N.J., 1974, pp. 309–317, or J. G. Truxal, *Introductory System Engineering,* McGraw-Hill, New York, 1972, pp. 516–536. Truxal relates a delightful story on the origins of the Routh criterion. "A story about the origins of the Routh test is interesting, whether true or not. Routh and Maxwell originally met as classmates at Cambridge University, where they competed vigorously (Routh graduated first in his group, just ahead of Maxwell, in 1854). Even though they went their separate ways, the bitter professional competition continued. In the late 1870s Maxwell remarked (in a published paper) that determining stability from the coefficients of the characteristic polynomial (without factoring) was, unfortunately, an insoluble problem. After reading this, Routh worked diligently for three years to develop the Routh test. He presented it with the opening remarks, 'It has recently come to my attention that my good friend James Clerk Maxwell has had difficulty with a rather trivial problem. . . .' "

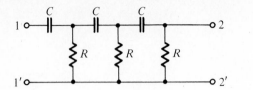

FIGURE 19.12

two Kirchhoff laws to find V_1. The result is

$$\frac{V_2}{V_1} = \frac{(RCs)^3}{(RCs)^3 + 6\,(RCs)^2 + 5\,(RCs) + 1} \tag{19.50}$$

If this circuit is connected in cascade with an inverting op amp of gain $-R_2/R$, as shown in Fig. 19.13, then the loop gain is

$$\text{LG} = \frac{V_2}{V_1}\,\frac{-R_2}{R} \tag{19.51}$$

We note in the circuit of Fig. 19.13 that the input resistor of the inverting amplifier, $R_1 = R$, replaces the resistor at end 2-2' in Fig. 19.12, thereby saving one resistor in the circuit. An alternative form of the same circuit is shown in Fig. 19.14. If we now form the characteristic equation $1 - \text{LG} = 0$, the numerator polynomial is, with $K = -R_2/R$,

$$(1 + K)(RCs)^3 + 6\,(RCs)^2 + 5\,(RCs) + 1 = 0 \tag{19.52}$$

Applying the criterion of Eq. (19.46), we see that the circuit will oscillate when $a_1a_2 = a_0a_3$ or

$$5 \times 6 - (1 + K) \times 1 = 0 \quad \text{or} \quad K = 29 \tag{19.53}$$

The frequency of oscillation is then given by Eq. (19.45) as

$$\omega_0{}^2 = \frac{1}{6R^2C^2} \quad \text{or} \quad \omega_0 = \frac{1}{\sqrt{6}\,RC} \tag{19.54}$$

Why call it a phase-shift oscillator? The condition for oscillation requires that $R_2 = 29R$. If we find LG by multiplying Eq. (19.50) by -29, then at the frequency $s = j\omega_0$ we have $1 - LG = 0$ since

$$\text{LG}\left(\frac{j}{\sqrt{6}RC}\right) = \frac{-29\,(j/\sqrt{6})^3}{(j/\sqrt{6})^3 + 6(j/\sqrt{6})^2 + 5(j/\sqrt{6}) + 1} = 1 \tag{19.55}$$

In other words, the circuit of Fig. 19.12 provides 180° of phase shift at the frequency of oscillation $\omega_0 = 1/\sqrt{6}RC$, which just compensates for the 180° phase

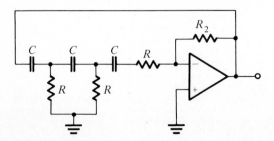

FIGURE 19.13

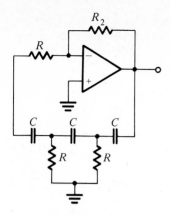

FIGURE 19.14

shift provided by the inverting op amp. Since the passive RC circuit provides exactly 180° of phase shift at ω_0, the circuit used as an oscillator is characterized by this phase shift.

The second phase-shift oscillator we will consider is that shown in Fig. 19.15. We observe that it consists of three "lossy integrator" circuits connected in cascade with simple feedback. The transfer function of each of the circuits is

$$T_1(s) = T_2(s) = T_3(s) = \frac{-1}{R_1 C \left(s + 1/R_2 C \right)} \tag{19.56}$$

If we break the feedback path, then the loop gain is this transfer function cubed. Carrying out the algebra associated with cubing Eq. (19.56) gives

$$LG = \frac{-(R_2/R_1)^3}{(R_2 Cs)^3 + 3(R_2 Cs) + 3\,(R_2 Cs)^2 + 1} \tag{19.57}$$

Then the characteristic equation, which is the numerator of $1 - LG = 0$, is

$$(R_2 Cs)^3 + 3(R_2 Cs)^2 + 3(R_2 Cs) + [1 + (R_2/R_1)^3] = 0$$

The criterion of Eq. (19.46) requires that

$$3 \times 3 - 1 - \left(\frac{R_2}{R_1}\right)^3 = 0 \quad \text{or} \quad \left(\frac{R_2}{R_1}\right)^3 = 8 \tag{19.58}$$

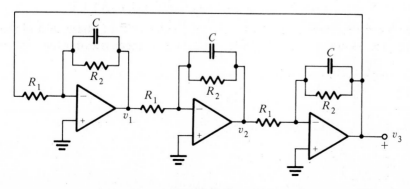

FIGURE 19.15

This is realized when $R_2 = 2R_1$, and this is the condition for oscillation. Then Eq. (19.45) gives the frequency of oscillation. If we let $R_1 = R$ in the circuit of Fig. 19.15, then $R_2 = 2R$, and this frequency is

$$\omega_0^2 = \frac{\sqrt{3}}{R_2^2 C^2} \quad \text{or} \quad \omega_0 = \frac{\sqrt{3}}{2RC} \tag{19.59}$$

Next we return to Eq. (19.56) and determine the value of the transfer function when the circuit is oscillating at the frequency just determined. For each circuit we find that

$$T(j\omega_0) = 1 \angle 120° \tag{19.60}$$

If we let the voltage v_1 be selected as the reference so that

$$v_1(t) = V_m \sin \omega_0 t \tag{19.61}$$

then this phase shift we have just determined will mean that

$$v_2(t) = V_m \sin(\omega_0 t + 120°) \tag{19.62}$$

$$v_3(t) = V_m \sin(\omega_0 t - 120°) \tag{19.63}$$

These three voltages are represented in phasor form in Fig. 19.16. The three voltages are thus equal in amplitude, but displaced by 120° from each other. They have the same form as the voltages in a three-phase power system.

19.6 AMPLITUDE STABILIZATION

In our analysis of the Wein bridge oscillator we found that the condition for oscillation was $K = 3$. This caused the roots of the characteristic equation, which are the poles of the transfer function describing the circuit, to move to the imaginary axis, as shown in Fig. 19.17. Setting the condition to cause $K = 3$ by adjusting a resistor may be a precarious operation, for we see from this figure that causing K to exceed 3.00 will cause the poles to migrate into the right half of the s plane. Theoretically this would cause oscillations to continue with ever-increasing amplitude and without limit. As a practical matter, the nonlinear nature of

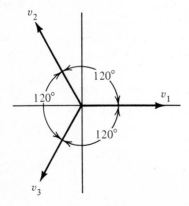

FIGURE 19.16

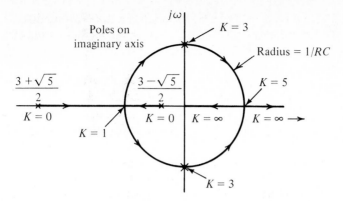

FIGURE 19.17

the op amp limits the amplitude of oscillation, and in doing so introduces considerable distortion, as depicted by the waveform of a clipped sinusoid shown in Fig. 19.18. If we modify the circuit such that this distortion is greatly limited, then it is said that we have stabilized the amplitude.

Frederick E. Terman relates the interesting story that the first invention of William Hewlett related to amplitude stabilization of an audio oscillator.* Hewlett introduced a small incandescent lamp operating below a luminous level into the circuit. His circuit employed vacuum tubes, but the corresponding op-amp circuit is shown in Fig. 19.19, where R_a is now made up of a resistor in series with the lamp. The resistance characteristic of the lamp is illustrated in Fig. 19.20, which shows that when the filament of the lamp is cold, the resistance is small, but when it becomes hot, the resistance becomes larger. Since the gain of the op-amp circuit is

$$K = 1 + \frac{R_b}{R_a} \qquad (19.64)$$

we see that making R_a larger will make K smaller. Making K smaller will cause the poles of the oscillator to migrate back to the imaginary axis and thus stabilize the system. This automatic gain adjustment causes the distortion of the amplifier to be relatively low. Hewlett's invention resulted in the first product of the then tiny Hewlett-Packard Company and was the beginning of an exponential growth to the present-day huge corporation.

In practice an oscillator is designed with a value of K that is about 5 percent greater than the value that would place the poles on the imaginary axis. When some transient in the circuit causes oscillation to begin, the output voltage in-

* F. E. Terman, "Educator Remembers Hewlett-Packard in Early Days," *Electron. J.,* Aug. 1977.

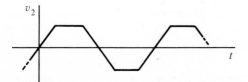

FIGURE 19.18

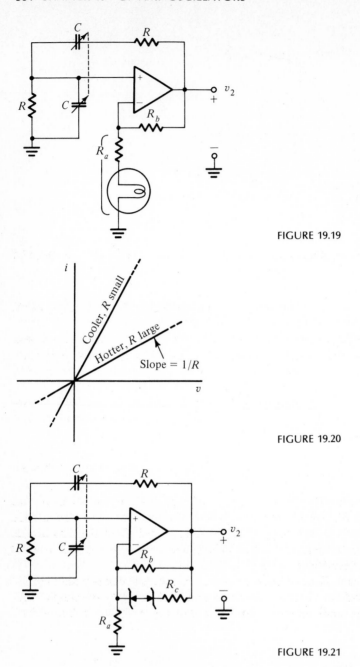

FIGURE 19.19

FIGURE 19.20

FIGURE 19.21

creases exponentially until limiting is encountered, and then the output voltage stabilizes to the required sinusoidal output.

The nonlinear characteristic provided by the incandescent lamp is now replaced in most designs by two back-to-back Zener diodes, as shown in the circuit of Fig. 19.21. As long as the voltage across R_b has a magnitude that is less than

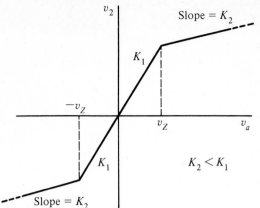

FIGURE 19.22

the Zener breakdown voltage v_Z, the Zener diodes appear as an open circuit, and the gain is

$$K_1 = 1 + \frac{R_b}{R_a} \qquad (19.65)$$

As soon as the voltage across the Zener diodes exceeds breakdown, then the diodes conduct, and the resistor R_c is suddenly in parallel with R_b. Let their combined value be

$$R_d = \frac{R_b R_c}{R_b + R_c} < R_b \qquad (19.66)$$

so that the gain becomes

$$K_2 = 1 + \frac{R_d}{R_a} \qquad (19.67)$$

The two conditions are shown in the circuit of Fig. 19.22, which is part of that given in Fig. 19.21. Since we have shown that $K_2 < K_1$ and these K values are the slopes relating the voltage v_a to v_2, the overall characteristic of the circuit is as shown in Fig. 19.23. Amplitude limiting takes place in the same manner as it did for the lamp in series with R_a, as in Fig. 19.21.

The back-to-back Zener diodes are placed in different places within oscillator circuits to provide this amplitude stabilization. For example, the simple oscillator of Fig. 19.2 is modified to the form shown in Fig. 19.24 for stabilization.

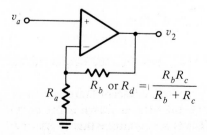

FIGURE 19.23

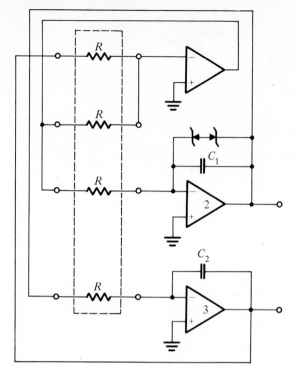

FIGURE 19.24

PROBLEMS

19.1 Given that the following R networks as used in the circuit of Fig. 19.9 are available: 3.3 kΩ, 4.7 kΩ, 6.8 kΩ, 10 kΩ, 15 kΩ, 22 kΩ, 47 kΩ, and 100 kΩ.
(a) Design a Wein-bridge oscillator for a frequency of 1000 Hz.
(b) Repeat part (a) for a frequency of 10,000 Hz.

19.2 The circuit shown in Fig. P19.2 consists of a cascade connection of three sections. In the circuit, all capacitors have the value C and all resistors have the value R.
(a) Let $V_0 = 0$ and find the transfer function $T(s) = V_2/V_1$. Sketch the asymptotic Bode plot for this transfer function.
(b) Connect the node marked V_2 to that marked V_0 such that $V_0 = V_2$, and then determine the pole and zero locations for $T(s)$.
(c) Determine the frequency of oscillation of the circuit.

19.3 Consider the oscillator circuit shown in Fig. P19.3.
(a) Show that the condition for oscillator is $R_2/R_1 = 2C_1/C_2$.
(b) When this condition is met, show that the frequency of oscillation is

$$\omega_0{}^2 = \frac{1}{R^2 C_1 (C_2 + C_3)}$$

(c) Show that the capacitor C_3 may be used to tune the circuit since it will change ω_0 but not move the poles from the imaginary axis of the s plane.

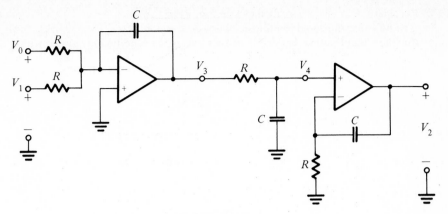

FIGURE P19.2

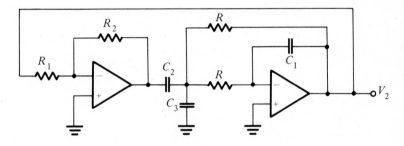

FIGURE P19.3

19.4 Consider the circuit shown in Fig. P19.4.
(a) Show that the condition for oscillation is $K = 2C_1/C_2$.
(b) Show that the frequency of oscillation is

$$\omega_0{}^2 = \frac{1}{R^2 C_1(C_2 + C_3)}$$

(c) Show that C_3 may be used to tune the circuit.

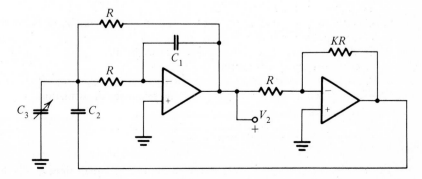

FIGURE P19.4

19.5 Consider the oscillator circuit shown in Fig. P19.5.

(a) Show that the condition for oscillation is $K = 3 + C_1/C_2$.

(b) Show that the frequency of oscillation under the condition of (a) and with $C_1 = C_2$ is

$$\omega_0 = \frac{\sqrt{2}}{RC}$$

(c) What advantage is gained by ganging C_1 and C_2?

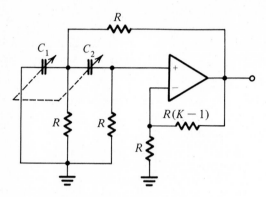

FIGURE P19.5

19.6 Consider the oscillator circuit shown in Fig. P19.6.

(a) Show that the condition for oscillation is $C_1/C_2 = 2 + R_1/R$.

(b) Show that the frequency of oscillation is

$$\omega_0{}^2 = \frac{1}{R_1{}^2 2 C_0 C_2}$$

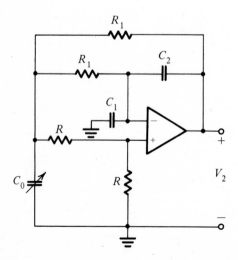

FIGURE P19.6

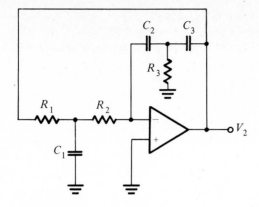

FIGURE P19.7

19.7 Consider the oscillator circuit shown in Fig. P19.7.

(a) Show that the condition for oscillation is given as two equalities: $C_1 = C_2 + C_3$ and $R_3 = R_1 R_2 / (R_1 + R_2)$.

(b) Show that the frequency of oscillation is

$$\omega_0^2 = \frac{1}{R_1 R_2 C_2 C_3}$$

CHAPTER 20

Better
Op-Amp
Models

*There must be an ideal world, a sort of
mathematician's paradise where everything
happens as it does in textbooks.*
—Bertrand Russell (1872–1970)

Indeed there is such a world, and it is the laboratory that ordinarily accompanies a course based on such a textbook as this. It is surprising how closely the measured performance of a filter will agree with specifications on which the design is based. Agreement, yes, but only so long as the frequency range is limited to the voiceband, perhaps 10 kHz or more. At higher frequencies the designer must take another look at the model being used for the op amp. It is this subject that we explore in this chapter. As we will discover, this often leads to unusually complicated algebra. It leads to the design technique of using the op amp of infinite gain as the "first cut," followed by simulation using a computer program such as SPICE,* which incorporates a very accurate representation of the op amp. At this point the filter can sometimes be adjusted to meet specifications.

20.1 ANOTHER LOOK AT THE OP AMP

In Chapter 2 we discussed the op amp and its operation, but concluded that we would use the model in which the gain is infinite at all frequencies. Suppose that we return to the model for the op amp shown in Fig. 20.1 and measure the characteristics of the op amp operating open loop such that

$$V_2 = A(s) \, V_{\text{in}} \tag{20.1}$$

where V_{in} is the voltage applied between the input terminals to the op amp. If the op amp is a dominant-pole compensated type, such as the 741, then the gain and phase characteristics will be those shown in Fig. 20.2. Because the op amp contains a small integrated capacitor for reasons of stability, there will appear a low-

---

* SPICE was developed by D. O. Pederson and his colleagues at the Electronics Research Laboratories of the University of California at Berkeley.

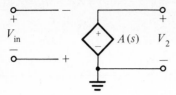

FIGURE 20.1

frequency dominant pole[†] in the open-loop transfer function. This frequency is designated as f_a (or ω_a) in Fig. 20.2, and for this particular characteristic it is seen to be at about 5 Hz. Thus the transfer function appears to be

$$\frac{V_2}{V_{\text{in}}} = A(s) = \frac{K}{s + \omega_a} \tag{20.2}$$

where ω_a is the -3-dB or half-power frequency. The low-frequency gain of the op amp is

$$A(0) = A_0 \tag{20.3}$$

where A_0 is typically 10^5 or 100 dB. Combining the last two equations, we see that

$$K = A_0\omega_a = \text{GB} \tag{20.4}$$

where the *gain bandwidth* (GB) is defined as the product of the zero-frequency gain and the half-power frequency of the open-loop transfer function of the op amp. Then Eq. (20.2) becomes

$$A(s) = \frac{\text{GB}}{s + \omega_a} \tag{20.5}$$

This is sometimes known as the *one-pole rolloff model.*

For frequencies much larger than ω_a, the last equation has a simple form:

$$A(s) = \frac{\text{GB}}{s} \tag{20.6}$$

This model is valid over the large frequency range for which the gain characteristic has a slope of -20 dB per octave and a phase of $-90°$, as shown in Fig. 20.2. This model is seen to adequately describe the op amp over a large fraction of the frequencies of interest. From this equation we see that when $|A(j\omega)| = 1$, then

$$\omega_u = \text{GB (gain} = 0 \text{ dB}) \tag{20.7}$$

This is known as the unity-gain frequency, and the relationship is a useful one to characterize an op amp. For 741-type op amps a typical value is seen to be 1 MHz.

At frequencies ordinarily in excess of ω_u, the slope of the gain curve again changes, indicating a high-frequency pole for which a typical value is 2.75 MHz,

[†]For an excellent discussion of the need for such a pole, see A. Budak, *Passive and Active Network Analysis and Synthesis,* Houghton Mifflin, Boston, 1974, pp. 222–240.

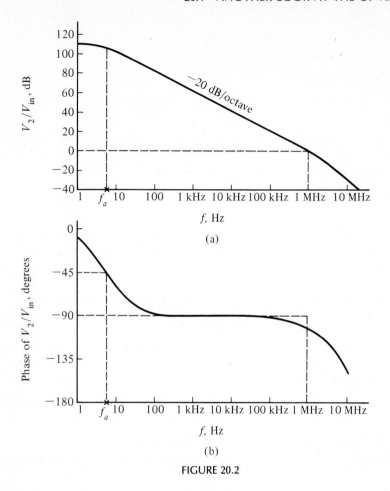

FIGURE 20.2

which we designate as f_c (or ω_c). The two-pole rolloff model then becomes

$$A(s) = \frac{GB \, \omega_p}{(s + \omega_a)(s + \omega_p)} \tag{20.8}$$

With the typical values that have been cited, this is

$$A(s) = \frac{17.28 \times 10^{12}}{(s + 31.4)(s + 17.28 \times 10^6)} \tag{20.9}$$

The model parameters in this transfer function and other values given apply only to the inexpensive 741-type op amp. For the more expensive instrument-type op amps the characteristics are likely to be as shown in Fig. 20.3 with an f_a of about 70 kHz and a unity-gain frequency of 0.5 GHz. These op amps are ordinarily too expensive to be used in filter applications.

At the other extreme in terms of quality, switch-capacitor filters as discussed in Chapter 17 make use of MOS op amps. MOS technology is useful for in-

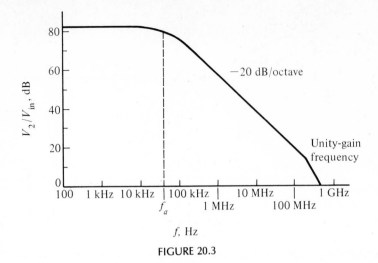

FIGURE 20.3

tegrated-circuit realizations because high density is possible and the cost is relatively low. However, the gain of MOS op amps is low, 50–60 dB (400–1000), and the unity-gain bandwidth is typically less than 1 MHz.

20.2   INVERTING AMPLIFIERS

We next return to the op-amp connections considered in Section 2.2 and replace A as a constant by $A(s)$ indicating that gain varies with frequency. This is shown on the op amp in Fig. 20.4a for the familiar inverting stage. For analysis we replace the op amp by a controlled-voltage source, as shown in Fig. 20.4b for which

$$V_2(s) = -A(s)V_a(s) \qquad (20.10)$$

where V_a is the input voltage to the op amp. We see that this voltage is

$$V_a = \left(\frac{V_2 - V_1}{R_1 + R_2}\right) R_1 + V_1 \qquad (20.11)$$

Combining these last two equations and letting $R_2/R_1 = K$, we have

$$\frac{V_2 - V_1}{1 + K} + V_1 + \frac{V_2}{A} = 0 \qquad (20.12)$$

From this equation we may solve for the ratio of V_2 to V_1, which is

$$\frac{V_2}{V_1} = -K \frac{A/(1 + K)}{1 + A/(1 + K)} \qquad (20.13)$$

Observe that if $A \to \infty$, then

$$\frac{V_2}{V_1} = -K = \frac{-R_2}{R_1} \qquad (20.14)$$

as in Chapter 2. As a first-order approximation, let $A(s) = GB/s$ in Eq. (20.13),

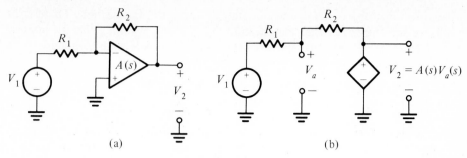

(a)

(b)

FIGURE 20.4

which reduces to

$$\frac{V_2}{V_1} = -K \frac{GB/(1 + K)}{s + GB/(1 + K)} \tag{20.15}$$

Since a typical value of GB is 10^6 and a typical value of K is 9 (for simplicity), we observe that the transfer function now has a pole at -10^5 or 15.92 kHz, and that this is the half-power frequency for the inverting amplifier. Now we see why this amplifier works so well, even with an open-loop half-power frequency of 5 Hz. The next transfer function for the op amp in order of complexity is

$$A(s) = \frac{GB}{s + \omega_a} \tag{20.16}$$

If this is substituted into Eq. (20.13), then we find that

$$\frac{V_2}{V_1} = -K \frac{GB/(1 + K)}{s + \omega_a + GB/(1 + K)} \tag{20.17}$$

Since ω_a is typically 30 compared to a $GB/(1 + K)$ of 10^5, the ω_a term can be neglected with little error. Hence for either form of $A(s)$, the transfer function for the inverting amplifier is given by Eq. (20.15). Plotted as a function of frequency, this response is as shown in Fig. 20.5, which is a typical one-pole rolloff characteristic.

The circuit of Fig. 20.4a has only two external resistors shown, and yet be-

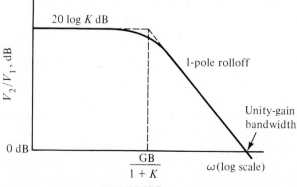

FIGURE 20.5

haves as a first-order circuit, one containing a capacitor, for example. To describe this, we sometimes say that the op amp "has a pole," meaning that an internal capacitor does indeed cause the circuit to have a one-pole rolloff response.

20.3 NONINVERTING AMPLIFIER

The noninverting amplifier may be analyzed following the procedure just completed for the inverting amplifier. The circuit under consideration is shown in Fig. 20.6a with the controlled-source representation shown in Fig. 20.6b. Analysis makes use of Kirchhoff's voltage law and the voltage-divider action at the input to the op amp. Thus,

$$V_a + V_1 = \frac{V_2}{R_1 + R_2} R_1 \qquad (20.18)$$

where

$$V_2 = A(s)V_a \qquad (20.19)$$

We anticipate the final result by letting

$$K = 1 + \frac{R_2}{R_1} \qquad (20.20)$$

as in Chapter 2. Combining these three equations and simplifying, we have

$$\frac{V_2}{V_1} = K \frac{-A}{-A + K} \qquad (20.21)$$

The simplest model for the op amp is

$$A_1 = \frac{GB}{s} \qquad (20.22)$$

Substituting this equation into Eq. (20.21), we have

$$\frac{V_2}{V_1} = K \frac{GB/K}{s + GB/K} \qquad (20.23)$$

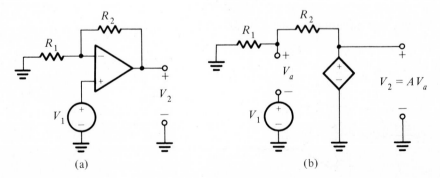

(a)   (b)

FIGURE 20.6

Observe again that for an infinite GB, this equation reduces to the familiar

$$\frac{V_2}{V_1} = K = 1 + \frac{R_2}{R_1} \tag{20.24}$$

If the model for the op amp is taken to be

$$A(s) = \frac{GB}{s + \omega_a} \tag{20.25}$$

then the transfer function is

$$\frac{V_2}{V_1} = K \frac{GB/K}{s + \omega_a + GB/K} \tag{20.26}$$

If we compare this result with that obtained for the inverting amplifier, we first recognize that K for the inverting amplifier is R_2/R_1, while K for the noninverting case is $1 + R_2/R_1$. With this recognition we see that the half-power frequencies are the same, and that ω_a can usually be neglected. Hence the response shown in Fig. 20.5 applies to both cases.

The pattern begins to emerge that using a first-order $A(s)$ rather than a constant value (usually infinity) causes the transfer function to increase its order by 1. We will also see that the complexity of the algebraic equations increases.

20.4   TRANSFER FUNCTIONS WITH $A = GB/s$

In this section we study the transfer functions for two circuits that have been used extensively in earlier studies, the integrator and the bandpass Friend circuit. Rather than assume that $A = \infty$ (which is equivalent to saying that GB $= \infty$) we will assume the op-amp model in which $A(s) = GB/s$, and examine the consequences of this assumption.

First consider the integrator circuit of Fig. 20.7. To analyze this circuit, we return to the controlled-source representation of Fig. 20.4b and replace R_2 by C_2. Using the same method of analysis, we write

$$V_a = \frac{V_2 - V_1}{R_1 + 1/C_2 s} R_1 + V_1 = \frac{V_2}{A} \tag{20.27}$$

Solving this equation for the transfer function V_2/V_1, we obtain

$$\frac{V_2}{V_1} = -\frac{1}{R_1 C_2} \frac{1}{s - (1/A)(s + 1/R_1 C_2)} \tag{20.28}$$

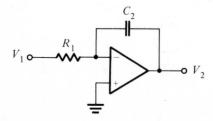

FIGURE 20.7

Next we let $A = -GB/s$, such that

$$\frac{V_2}{V_1} = -\frac{1}{R_1 C_2} \frac{GB}{s(s + GB + 1/R_1 C_2 GB)} \qquad (20.29)$$

Since GB is nominally much larger than $1/R_1 C_2 GB$ (typical values being GB = 10^6, $R_1 = 10^4$, $C_2 = 10^{-6}$), we may use the approximation

$$\frac{V_2}{V_1} = -\frac{1}{R_1 C_2} \frac{GB}{s(s + GB)} \qquad (20.30)$$

This function is represented as a pole at the origin and another at $s = -GB$, as shown in Fig. 20.8a. We see that letting GB $\to \infty$ causes the second pole to move to infinity, and we are left with the usual

$$\frac{V_2}{V_1} = -\frac{1}{R_1 C_2 s} \qquad (20.31)$$

which we associate with an integrator.

Let us now use the more accurate transfer function

$$A(s) = \frac{GB}{s + \omega_a} \qquad (20.32)$$

in Eq. (20.28). The result is

$$\frac{V_2}{V_1} = -\frac{1}{R_1 C_2} \frac{GB}{s^2 + (GB + \omega_a + 1/R_1 C_2)s + \omega_a/R_1 C_2} \qquad (20.33)$$

Rather than use the quadratic formula to factor this second-order equation, we assume that one of the poles is at $-GB$, as in Eq. (20.30), and then approximate the other pole. A good approximation is

$$s = -\frac{1}{R_1 C_2} \frac{\omega_a}{GB} \qquad (20.34)$$

If we use the value $\omega_a = 30$ and the values just given for the other factors, then the poles of Eq. (20.33) are as shown in Fig. 20.8b and at

$$s = -10^6 \quad \text{and} \quad s = -3 \times 10^{-3} \qquad (20.35)$$

In other words, one pole is far removed from the origin, the other is very near to

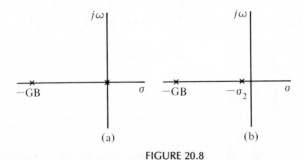

(a)   (b)

FIGURE 20.8

the origin. As GB → ∞, one pole moves to infinity, the other moves to the origin, and Eq. (20.31) represents the integrator circuit.

As our second example, we will study the Friend circuit shown in Fig. 20.9 using $A(s) = -\text{GB}/s$ rather than ∞, as used in Chapter 7. From our analysis of Chapter 7 we have, before normalization,

$$\omega_0 = \frac{1}{C\sqrt{R_1 R_2}} \quad \text{and} \quad Q = \frac{1}{2}\sqrt{\frac{R_2}{R_1}} \tag{20.36}$$

Using these quantities, the transfer function becomes

$$\frac{V_2}{V_1} = \frac{-2Q\omega_0 s}{s^2 + (\omega_0/Q)s + \omega_0^2 + (s/\text{GB})[s^2 + \omega_0(2Q + 1/Q)s + \omega_0^2]} \tag{20.37}$$

Now this is a third-order denominator, and overall the concepts of ω_0 and Q do not apply, these being second-order concepts. To obtain some estimate of pole location, we expand the denominator of Eq. (20.37) and normalize to give

$$s^3 + \left[\text{GB} + \left(2Q + \frac{1}{Q}\right)\omega_0\right]s^2 + \cdots = 0 \tag{20.38}$$

The general form of the denominator of Eq. (20.37) is

$$(s + s_1)(s + s_1^*)(s + \sigma_2) = 0 \tag{20.39}$$

where s_1 and s_1^* are the complex conjugate poles and σ_2 is the pole on the negative real axis. If we multiply the factors of Eq. (20.39), we obtain

$$s^3 + (s_1 + s_1^* + \sigma_2)s^2 + \cdots = 0 \tag{20.40}$$

which we may compare with Eq. (20.38). Let us assume that the displacement of the complex conjugate poles is small so that

$$s_1 + s_1^* = \frac{\omega_0}{Q} \tag{20.41}$$

(remembering s-plane geometry). So

$$\sigma_2 + \frac{\omega_0}{Q} = \text{GB} + 2Q\omega_0 + \frac{\omega_0}{Q} \tag{20.42}$$

Then

$$\sigma_2 = \text{GB} + 2Q\omega_0 \tag{20.43}$$

In stating that the displacement of the poles representing the second-order factors

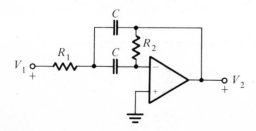

FIGURE 20.9

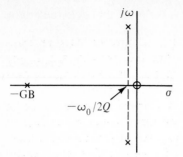

FIGURE 20.10

was small, we assumed that a factor in the denominator of Eq. (20.37) was the usual second-order factor. Then the denominator is

$$\left(s^2 + \frac{\omega_0}{Q} s + \omega_0^2\right)(s + GB), \qquad 2Q\omega_0 \ll GB \tag{20.44}$$

In other words, a more accurate representation of the transfer function for the Friend circuit is the usual pair of complex conjugate poles described by the quantities ω_0 and Q, a zero at the origin, and the new pole on the negative real axis at $\sigma_2 = -GB$ [more accurately, at $-(GB + 2Q\omega_0)$], as shown in Fig. 20.10. Again, as GB becomes infinite, the real pole vanishes to infinity.

The analyses of the two circuits have given approximate results, intended to suggest the consequences of a more accurate model for the op amp in terms of the poles that represent it. The results we have obtained have been approximations to Eqs. (20.33) or (20.37). But given these equations, an exact plot of the magnitude or the phase can be made, or the circuit can be simulated using some suitable computer program such as SPICE. For Eq. (20.37), for the Friend circuit, the response of the magnitude ratio plotted for the assumption of infinite GB is shown in Fig. 20.11. However, for a finite GB, the typical value for 741-type op amps being 10^6, there may be a significant change in the magnitude response, as shown by the dashed line response of the figure. This plot would suggest that the actual resonant frequency is less than ω_0 and that the peak amplitude is less than anticipated from an approximate model. This is, of course, important information for the designer.

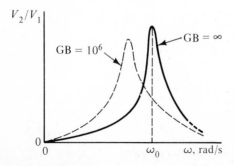

FIGURE 20.11

20.5 OP-AMP SLEW RATE

Almost all of our earlier discussion of specification quantities has been in terms of frequency-domain concepts, such as Q, ω_0, BW, or GB. We now turn to the time domain and introduce quantities that are important in specifying the performance of op amps. The most important of these is the *slew rate*.

It may seem strange to use the term *slew* to describe the performance of a tiny device such as the op amp. The dictionary meaning of slew is "to turn or twist," or "to swing around." In the navy this term is used to describe the turning motion of artillery. The term was also used in radar systems developed in the 1940s where a radar system's antenna turning in space was said to be slewing.* Since early op-amp circuits were often developed for radar applications, it is not surprising that a radar term should be adopted for op-amp characterization. An electronics system does not turn or twist, but the analog concept applies.

Figure 20.12 shows a simplified representation of the op amp, giving only quantities that are important in discussing slew rate. The most important of these is a coupling capacitor C_c through which passes the current i_1 which is generated by an internal current source. The output voltage v_2 is related to these quantities by the equation.

$$i_1 = C_c \frac{dv_2}{dt} \tag{20.45}$$

We define the slew rate (SR) as

$$SR = \frac{dv_2}{dt}\bigg|_{max} \tag{20.46}$$

so that

$$SR = \frac{i_1}{C_c} \tag{20.47}$$

For the 741-type op amp, typical values of the quantities involved are $i_1 = 20\ \mu A$ and $C_c = 30$ pF, so that

$$SR = \frac{2 \times 10^{-5}}{30 \times 10^{-12}} = 0.67\ V/\mu s \tag{20.48}$$

* H. M. James, N. B. Nichols, and R. S. Phillips, *Theory of Servomechanisms*, McGraw-Hill, New York, 1947, p. 18, for example.

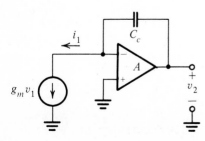

FIGURE 20.12

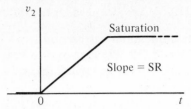

FIGURE 20.13

If we integrate Eq. (20.46), then

$$v_2(t) = \text{SR } t \qquad (20.49)$$

assuming that $v_2(0) = 0$. Hence a step applied to the input of the op amp produces a ramp of output voltage increasing with time, with a slope that is equal to SR. This will increase until the output voltage reaches saturation, so that the output voltage will appear as a truncated ramp, as shown in Fig. 20.13. The time required for the output voltage to reach saturation is called the *slew time*, as shown in the figure.

Clearly SR is a measure of excellence of an op amp since we desire the output to follow the input closely. If fact, we considered SR to be infinite in discussions prior to this chapter. To relate SR to the characteristics of the op amp, we return to the frequency domain and observe that i_1 of Eq. (20.45) has the transform

$$I_1 = g_m V_1 \qquad (20.50)$$

where g_m is the first-stage transconductance in the op amp and V_1 is the input voltage. The transform of Eq. (20.45) may be equated to the I_1 just written, so that

$$g_m V_1 = C_c s V_2 \qquad (20.51)$$

or

$$T = \frac{V_2}{V_1} = \frac{g_m}{\omega C_c} \qquad (20.52)$$

The frequency at which $|T| = 1$ defines the unity-gain frequency as discussed earlier in this chapter. Let this frequency be ω_u. Then we may solve the last equation to obtain

$$C_c = \frac{g_m}{\omega_u} \qquad (20.53)$$

Substituting this value in Eq. (20.47), we obtain

$$\text{SR} = \frac{\omega_u i_1}{g_m} \qquad (20.54)$$

We see from this equation that SR is determined by the designer of the op amp and that there is little that a circuit designer can do about it, other than try to select a better op amp. The range of SR values available are indicated by the following chart:

Type of op amp	Slew rate
Instrument type	250–1000 V/μs
741 type	1 V/μs
MOS type	0.4 V/μs

Another important time-domain specification used to characterize the op amp is the *settling time*. This quantity may be defined in terms of the quantities shown on Fig. 20.14, which is the response of the output of the op amp when subjected to a step input. The error band is defined as a given percentage of the final value. For 741-type op amps this may be 1 percent or more, while for instrument op amps this may be as small as 0.01 percent. These times range from 70 ns for instrument-type op amps (to 0.01 percent of final value) to 300 μs for MOS-type op amps (to 1 percent of final value).

20.6   DESIGN OPTIMIZATION

We began this chapter with a quotation from Bertrand Russell relating to the ideal world in which everything happens as it does in textbooks. In earlier sections of this chapter we have confessed that the op amp is not as ideal as we may have implied in other chapters, nor as ideal as it appears in a laboratory in which frequencies are carefully selected on the low side of the spectrum. In this final section we confess the possibility that we may have given other impressions of the idealness of analog filter design.

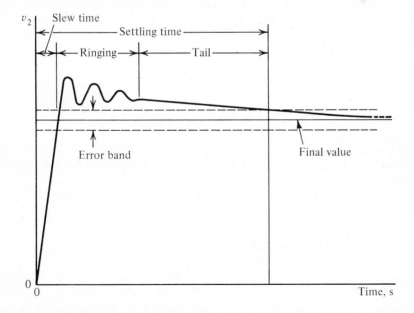

FIGURE 20.14

The typical design situation may be described as having more parameters to adjust than we have specifications to satisfy. We encountered this situation early where Q and ω_0 were selected as proper specifications, but the circuits used to realize the specifications often had 5 or more parameters to adjust. The procedure then was to set all but two of these parameters to unity, recognizing that the elements finally used would be in a proper range through frequency and magnitude scaling. It should be explicitly stated at this point that there was no guarantee that just any choice would give the best solution to the problem. To find the best requires that all parameters be considered, as well as alternative circuits that might be used to satisfy the specifications. To be philosophical, such knowledge is best learned from experience with real-world constraints. Some of the factors that will be important in this case are listed here.

1. *Sensitivity.* The sensitivity to element change may be different for different choices of elements. To optimize all sensitivities will require that all possibilities be examined.
2. *Element size.* The range of element sizes is important, especially if an integrated-circuit realization is being considered. What is the ratio of the largest to the smallest capacitor? Further, can the elements be found in the stockroom, or must a special order be placed? Other factors relating to the elements to be selected: cost, size, availability, quality, ...
3. *Linear range.* Will the elements selected result in the circuit operating in its linear range? Or will some op amp saturate and produce strange results?
4. *How many are to be produced.* If only one of a kind is need, then almost any solution will do. If a production line is being set up to produce thousands or perhaps millions, then a careful optimal design is a must.
5. *Signal-to-noise ratio.* We have never mentioned noise or signal-to-noise ratio. In many designs this is important.

In a typical design you will always forget something important. What have we forgotten in making this list?

PROBLEMS

20.1 The model used to describe the op amp shown in the circuit of Fig. P20.1 is $A(s) = GB/s$. Find an expression for the output voltage, V_3, as a function of V_1 and V_2.

20.2 In the circuit shown in Fig. P20.2a, the output is fed back to the positive input terminal through a resistive circuit. Let the op amp be characterized by the $A(s)$ having the characteristics shown in Fig. P20.2b.
(a) For this circuit, find an expression for $T(s)$ V_2/V_1, and
(b) For this circuit, show that when

$$A_0 > 20 \log (1 + R_2/R_1)$$

the circuit is unstable.

20.3 The model used to describe the op amps is $A(s) = GB/s$. (see Fig. P20.3). Assume

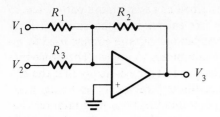

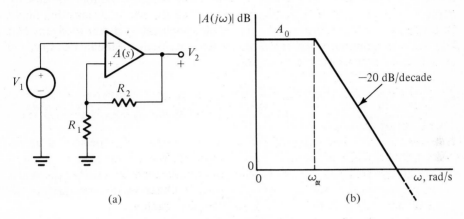

FIGURE P20.1

FIGURE P20.2

(a)   (b)

that the two op amps are identical and that

$$1 + R_2/R_1 = 1 + R_4/R_3$$

Determine the expression for the transfer function $T = V_2/V_1$.

20.4 The model used to describe the two identical op amps in the circuit of Fig. P20.4 is

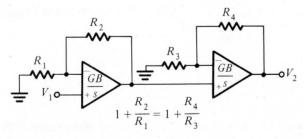

$$1 + \frac{R_2}{R_1} = 1 + \frac{R_4}{R_3}$$

FIGURE P20.3

$A(s) = GB/s$. Determine the expression for the transfer function $T = V_2/V_1$.

20.5 The amplifier circuit shown in Fig. P20.5 has op amps with frequency-dependent gains described by the transfer function $A_i(s) = GB_i/s$, $i = 1, 2, 3$. Obtain the expression for the voltage-ratio transfer function in terms of GB_1, GB_2, GB_3, R_1, and R_2.

20.6 A lowpass filter used in an anti-aliasing application makes use of the Sallen–Key circuit with $K = 1$ and has the following element values: $R_1 = R_2 = 225$ kΩ, $C_1 = 50$ pF, and $C_2 = 25$ pF.

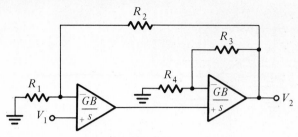

FIGURE P20.4

(a) Assuming an ideal op amp, what will be the attenuation at the frequency $f =$ 253 kHz?

(b) Assuming an op amp described by the characteristic of Fig. 20.2a, what will be the attenuation at this frequency?

20.7 The circuit given in Fig. P20.7 is a second-order filter containing no capacitors external to the op amps. Let the op amps be described by the models $A_1 = GB_1/s$ and $A_2 = GB_2/s$.

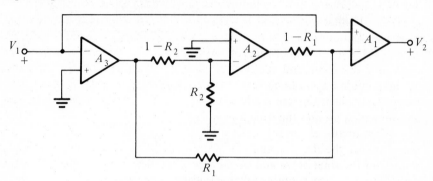

FIGURE P20.5

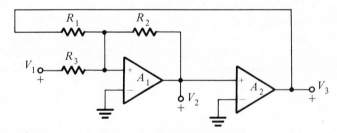

FIGURE P20.7

(a) Show that the bandpass transfer function is

$$\frac{V_2}{V_1} = \frac{\dfrac{-GB_1}{R_3}\left[\dfrac{1}{R_1} + \dfrac{1}{R_2} + \dfrac{1}{R_3}\right]^{-1} s}{s^2 + \dfrac{GB_1}{R_2}\left[\dfrac{1}{R_1} + \dfrac{1}{R_2} + \dfrac{1}{R_3}\right]^{-1} s + \dfrac{GB_1 GB_2}{R_1}\left[\dfrac{1}{R_1} + \dfrac{1}{R_2} + \dfrac{1}{R_3}\right]^{-1}}$$

(b) Obtain design equations for R_1, R_2, and R_3 in terms of the ω_0 and Q of the specifications and GB_1 and GB_2.

20.8 Repeat Problem 20.7 for the lowpass transfer function V_3/V_1.

APPENDIX A
Scaling

A.1 MAGNITUDE AND FREQUENCY SCALING

To begin, we remind ourselves that the use of scaling is commonly encountered in everyday experience. We have all seen scale models of airplanes or automobiles, in this case reduced from their usual size to a much smaller size. And we have all viewed statues of famous persons ordinarily larger than life size. We may speak of the first example as scaling down, the second as scaling up. In some cases we are interested in both. Civil engineers make scale models of construction sites to aid in the design of roads, bridges, dams. They ordinarily use one scale for horizontal distances, and another one for vertical distances, for otherwise the relief map might appear to be very flat. This relief map is made by scaling distances down. Once decisions are made with the aid of the scale model, then distances will be scaled up and the design completed.

What quantities do we scale in filter design? Ordinarily we scale the size of the elements, the R's, L's, and C's of the circuit, and as a result we scale the frequency or the time delay and magnitude. The way in which these objectives are accomplished is the subject of this appendix. Since scaling will be a part of every design, you will likely refer to this appendix frequently.

Figure A.1 is the representation of a 2-port circuit. Port 1-1' is the input. If the circuit is operating in the sinusoidal steady state, then the input voltage and input current can be represented by the phasors V_1 and I_1. Similarly, 2-2' is the output port, characterized by the phasors V_2 and I_2 having reference directions as shown. The voltage ratio transfer function for the 2-port is defined as

$$T(j\omega) = \frac{V_2(j\omega)}{V_1(j\omega)} \qquad (A.1)$$

for some prescribed condition at port 2-2' such as (1) with open terminals, as shown in the figure, (2) with a load such as a resistor connected from 2 to 2', or (3) with another 2-port connected in *cascade* (or tandem). The phasor $T(j\omega)$ may be represented in polar form:

$$T(j\omega) = |T(j\omega)| \angle\theta(j\omega) \qquad (A.2)$$

The specifications for a filter are ordinarily given in terms of the quantities in this equation, the magnitude or phase.

A common form of structure for the 2-port is shown in Fig. A.1 and is known as a ladder circuit. Each of the blocks of this ladder represents one or more elements. If single elements, then each block is a 1-port circuit, as represented in Fig. A.2. The 1-port is characterized by the driving-point impedance

$$Z(j\omega) = \frac{V(j\omega)}{I(j\omega)} \tag{A.3}$$

and its reciprocal is the driving-point admittance

$$Y(j\omega) = \frac{1}{Z(j\omega)} \tag{A.4}$$

The impedances of the passive elements are

$$Z_R = R, \qquad Z_L = j\omega L, \qquad Z_C = \frac{1}{j\omega C} \tag{A.5}$$

and their reciprocals are

$$Y_R = \frac{1}{R} = G, \qquad Y_L = \frac{1}{j\omega L}, \qquad Y_C = j\omega C \tag{A.6}$$

Clearly T will be determined by the Z's and the Y's of these last two equations and a prescribed structure for the 2-port. Scaling the Z's and the Y's will change T. How?

Consider the following definitions relating to the expansion or compression of frequency, magnitude, or time. When

$$v(t) \text{ is replaced by } v(10t) \tag{A.7}$$

or

$$V(j\omega) \text{ is replaced by } V(j10\omega) \tag{A.8}$$

then we say that v or V has been *compressed* by a factor of 10 in time or frequency. Similarly, when

$$v(t) \text{ is replaced by } v(t/10) \tag{A.9}$$

or

$$V(j\omega) \text{ is replaced by } V(j\omega/10) \tag{A.10}$$

then we say that v or V has been *expanded* by a factor of 10 in time or frequency. Such changes of scale for the frequency case are illustrated in Fig. A.3. The magnitude of impedance $|Z(j\omega)|$ may be changed. When

$$|Z(j\omega)| \text{ is replaced by } 10\,|Z(j\omega)| \tag{A.11}$$

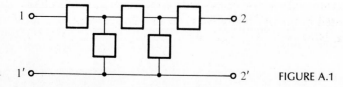

FIGURE A.1

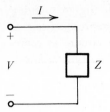

FIGURE A.2

then we say that the magnitude of the impedance is *scaled up* by a factor of 10, while if

$$|Z(j\omega)| \text{ is replaced by } \frac{1}{10} |Z(j\omega)| \qquad (A.12)$$

then we say that the magnitude of the impedance is *scaled down* by a factor of 1/10.

We say that $|Z(j\omega)|$ has been magnitude scaled when it is multiplied by a real constant k_m; it is *scaled up* if $k_m > 1$, *down* if $k_m < 1$. This is illustrated by Fig. A.4a, which shows a Bode plot* for some impedance magnitude. Magnitude scaling this $|Z(j\omega)|$ simply moves this characteristic up or down on the figure. To change the impedance magnitude by k_m, we change the magnitude of the impedance for every element in the circuit. The impedance magnitudes for the passive elements are, from Eq. (A.5),

$$Z_R = R, \qquad |Z_L| = \omega L, \qquad |Z_C| = \frac{1}{\omega C} \qquad (A.13)$$

If each of these is multiplied by k_m, then we have

$$k_m Z_R = k_m R, \qquad k_m |Z_L| = k_m \omega L, \qquad k_m |Z_C| = \frac{1}{\omega \, C/k_m} \qquad (A.14)$$

If we designate the elements after scaling as "new" and those before as "old," then we have the equations that we obtain by comparing Eqs. (A.4) and (A.14) term by term:

$$R_{\text{new}} = k_m \, R_{\text{old}} \qquad (A.15)$$

$$L_{\text{new}} = k_m \, L_{\text{old}} \qquad (A.16)$$

$$C_{\text{new}} = \frac{1}{k_m} \, C_{\text{old}} \qquad (A.17)$$

If the elements in the old circuit are changed according to these equations, then all impedances in that circuit will be scaled in magnitude.

Our next objective is to scale the frequency without affecting the magnitude. This objective is illustrated in Fig. A.4b. We have changed the magnitude of the impedance in Fig. A.4a. Now we wish to scale the frequency in such a way that

---

* Bode plots are reviewed in Chapter 3.

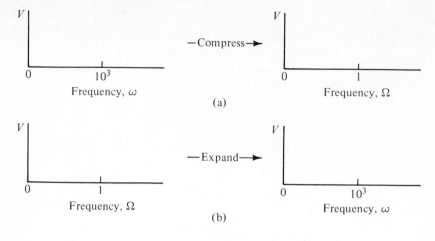

FIGURE A.3

the magnitude is not further scaled. For the inductor we must keep

$$|Z_L| = \omega L \tag{A.18}$$

constant. To do so, any change in ω must be compensated by a corresponding change in L. This we do in terms of a real constant k_f as follows:

$$|Z_L| = \omega L = (k_f \omega)\frac{1}{k_f} L = (k_f \omega)L_{new} \tag{A.19}$$

In other words, if we increase the frequency by the amount k_f, then we must reduce the inductance by that amount to keep the magnitude of the impedance constant. Similarly, for the capacitor:

$$|Z_C| = \frac{1}{\omega C} = \frac{1}{(k_f \omega)(1/k_f)C} = \frac{1}{(k_f \omega)\ C_{new}} \tag{A.20}$$

Here we must decrease the capacitance by the amount $1/k_f$ while increasing the frequency by the amount k_f if the magnitude of the impedance is to remain constant. From the last two equations we see that new element values may be expressed in terms of old values as follows:

$$L_{new} = \frac{1}{k_f} L_{old} \tag{A.21}$$

$$C_{new} = \frac{1}{k_f} C_{old} \tag{A.22}$$

Since resistance is unaffected by frequency scaling,

$$R_{new} = R_{old} \tag{A.23}$$

Referring once more to Fig. A.4b, we see that it is not necessary that we scale in magnitude and scale in frequency separately. We can do both at once. Com-

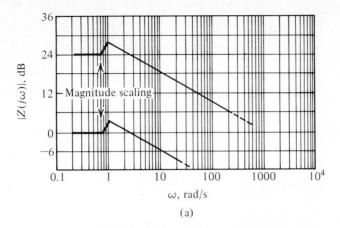

(a)

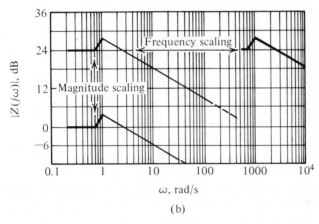

(b)

FIGURE A.4

bining the results of Eqs. (A.15)–(A.17) with Eqs. (A.21)–(A.37),

$$L_{\text{new}} = \frac{k_m}{k_f} L_{\text{old}} \tag{A.24}$$

$$C_{\text{new}} = \frac{1}{k_m k_f} C_{\text{old}} \tag{A.25}$$

$$R_{\text{new}} = k_m R_{\text{old}} \tag{A.26}$$

These three equations are known as the *element scaling equations.*

Example A.1   Using the method developed in Chapter 3, the circuit realization shown in Fig. A.5 is found. It is clear that these are not element values in the practical range, but this will be characteristic of unscaled circuits. This particular realization is special in another way. The circuit consists of two separate circuits in cascade, indicated by labeling one as T_1 and the other as T_2. Further, we will show later that these two circuits do not interact. When the circuits are isolated, then we may select different values of k_m for each of the stages, each to accomplish a desired circuit element size.

For this particular circuit suppose that we are not required to frequency scale, but we

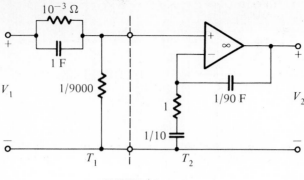

FIGURE A.5

wish to magnitude scale in order to realize more practical element values. We decide to favor 1-μF capacitors as far as possible. This means that C_{old} values are shown in the figure, and that we choose $C_{\text{new}} = 10^{-6}$F. With $k_f = 1$ (no frequency scaling), Eq. (A.25) becomes

$$k_m = \frac{C_{\text{old}}}{C_{\text{new}}} \tag{A.27}$$

The values for C_{old} of 1 F and 0.1 F give the values of k_m:

$$k_{m1} = 10^6 \quad \text{and} \quad k_{m2} = 10^5 \tag{A.28}$$

Now Eqs. (A.24) and (A.26) are used to obtain the element values shown in Fig. A.6. The "new" element sizes are of a more practical range of sizes.

We have scaled impedances, but there remains the question of scaling the voltage-ratio transfer function $T(j\omega)$ of Eq. (A.1) and its magnitude given in Eq. (A.2). Consider, for the present, the first part of the circuit that has been our example, the part designated as T_1 in Fig. A.5. This circuit is known as a voltage-divider circuit, the general form of which is shown in Fig. A.7. For this simple circuit we have

$$V_1 = I(Z_1 + Z_2) \quad \text{and} \quad V_2 = IZ_2 \tag{A.29}$$

Dividing these equations gives

$$T = \frac{V_2}{V_1} = \frac{Z_2}{Z_1 + Z_2} = \frac{1}{1 + Z_2/Z_1} \tag{A.30}$$

From this result we see that if both Z_1 and Z_2 are scaled by k_m, T is not affected. However, frequency scaling Z_1 and Z_2 would frequency scale T.

The result that has been given for the voltage-divider circuit, also called an inverted-L circuit, holds in general for any voltage-ratio or current-ratio transfer function.* That is, any linear circuit is unaffected by magnitude scaling as far as the ratio of output to input voltages is concerned. We get exactly the same T function for any k_m we selected. This is an important result and the basis for design strategies to give practical element sizes.

---

* M. E. Van Valkenburg, *Introduction to Modern Network Synthesis*, Wiley, New York, 1960, p. 53 ff.

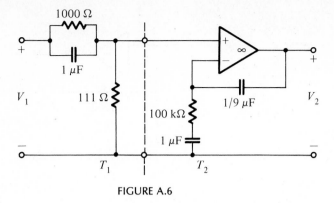

FIGURE A.6

It should be clear that if $T(j\omega)$ is not changed by magnitude scaling, then neither are the magnitude and phase functions

$$T(j\omega) = |T(j\omega)|/\theta(j\omega) \qquad (A.31)$$

Since the rectangular equivalent of this polar form for $T(j\omega)$ is

$$T(j\omega) = \text{Re } T(j\omega) + j \text{ Im } T(j\omega) \qquad (A.32)$$

then the phase function is

$$\theta = \tan^{-1}\left[\frac{\text{Im } T(j\omega)}{\text{Re } T(j\omega)}\right] \qquad (A.33)$$

From this we see that even if $T(j\omega)$ is multiplied by a constant, the phase function is not affected, since that constant cancels in the quotient of imaginary to real parts. Thus magnitude scaling does not affect $T(j\omega)$, $|T(j\omega)|$, or $\theta(j\omega)$. However, any of these functions may be frequency scaled by replacing ω by $k_f\omega$.

A.2   TIME AND DELAY SCALING

Figure A.8 shows two sine waves, one of frequency 4 times the other. Since

$$\omega_0 = 2\pi f_0 = \frac{2\pi}{T_0} \qquad (A.34)$$

we see that the period of the lower frequency sinusoid is the distance A–C. In that same period A–C, the higher frequency sinusoid completes four periods in the same time that the first completes one. If we let $T_0/2\pi = t_0$ be a characterizing

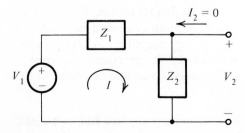

FIGURE A.7

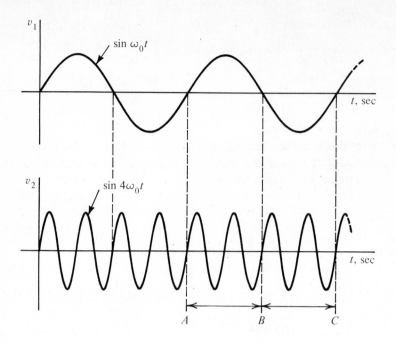

FIGURE A.8

time, then

$$v_1 = A_1 \sin \omega_0 t = A_1 \sin \left(\frac{t}{1/\omega_0} \right) = A_1 \sin \left(\frac{t}{t_0} \right) \qquad (A.35)$$

such that

$$t_0 = \frac{1}{\omega_0} \qquad (A.36)$$

This inverse relationship between time and frequency suggests that when frequency is expanded, time is compressed, and vice versa. Let us define this expansion or compression in terms of the real positive constant

$$k_t = \frac{\text{time response required}}{\text{normalized time}} \qquad (A.37)$$

such that $k_t < 1$ corresponds to time compression and $k_t > 1$ to time expansion. From Eq. (A.36) we see that k_t is inversely related to the k_f of the last section,

$$k_t = \frac{1}{k_f} \qquad (A.38)$$

In comparison to old and new quantities given in Eq. (A.37), we have

$$k_f = \frac{\text{frequency response required}}{\text{normalized frequency}} \qquad (A.39)$$

Because of the relationship of Eq. (A.38) we may modify the scaling equations,

Eqs. (A.24)–(A.26), to

$$L_{\text{new}} = k_m k_t L_{\text{old}} \tag{A.40}$$

$$C_{\text{new}} = \frac{1}{k_m} k_t C_{\text{old}} \tag{A.41}$$

$$R_{\text{new}} = k_m R_{\text{old}} \tag{A.42}$$

We may think of time scaling as a means to speed up the response of a circuit or, for that matter, of any physical system that can be represented by the kinds of equations with which we have been dealing. For example, the circuit shown in Fig. A.9 has a very slow time response due to the large size of the elements. This has been designed on the basis of a characterizing time of $t_0 = 1$ second. If we wish to speed up the time of response by a factor of 10 with $k_t = 1/10$, then Eqs. (A.40)–(A.42) give these values:

$$L_{\text{new}} = 0.2 \text{ H}, \qquad C_{\text{new}} = 0.1 \text{ F}, \qquad R_{\text{new}} = 1 \text{ }\Omega \tag{A.43}$$

To speed up the system by a factor of 10^6, $k_t = 10^{-6}$ and the element values become

$$L_{\text{new}} = 2000 \text{ mH}, \qquad C_{\text{new}} = 1 \text{ }\mu\text{F}, \qquad R_{\text{new}} = 1 \text{ }\Omega \tag{A.44}$$

The time-scaled circuits are shown in Fig. A.9b and c.

The inverse time and frequency relationship just discussed for frequency ω applies for complex frequency $s = \sigma + j\omega$. The exponential

$$v_1(t) = A_1 e^{-\sigma t} = A_1 e^{-t/T} \tag{A.45}$$

where T is the time constant, a characterizing time, and

$$\sigma = \frac{1}{T} \tag{A.46}$$

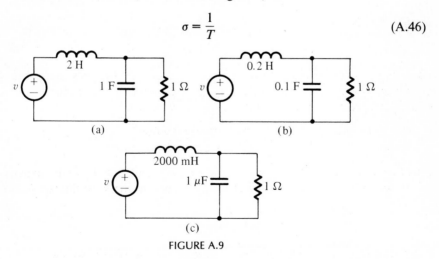

FIGURE A.9

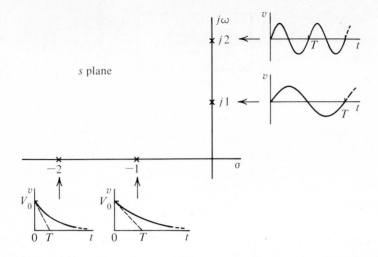

FIGURE A.10

is known as neper frequency. We see that as σ increases, the rate of decrease of the exponential function decreases. This is shown for the imaginary and real axes of the s plane in Fig. A.10, and for these two axes a characterizing time decreases as frequency increases. This holds for the entire complex plane, suggested by Fig. A.11, where a circle of increasing radius represents frequencies of increasing value. Thus we have a general relationship between time and frequency; *frequency expansion corresponds to time compression.*

An application of time scaling is suggested by the system representations of Fig. A.12. In Fig. A.12a, we see that for input v_1 of frequency ω_0, the output v_2 is changed in amplitude and phase angle. If the original system is time-scaled with new elements replacing the old elements, we obtain the system of Fig. A.12b. If the only change in the input to the system with new elements is that the frequency has been increased (or the time scale decreased), then the output of the system in Fig. A.12b is exactly the same as that of Fig. A.12a, except that the frequency is again scaled just as v_1 has been scaled.

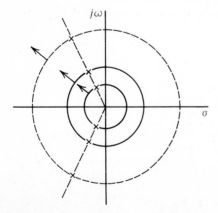

FIGURE A.11

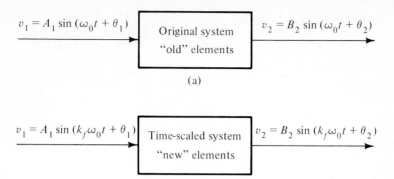

$v_1 = A_1 \sin(\omega_0 t + \theta_1)$ → Original system "old" elements → $v_2 = B_2 \sin(\omega_0 t + \theta_2)$

(a)

$v_1 = A_1 \sin(k_f \omega_0 t + \theta_1)$ → Time-scaled system "new" elements → $v_2 = B_2 \sin(k_f \omega_0 t + \theta_2)$

(b)

FIGURE A.12

An important class of circuits are designed to provide a time delay between the input and output signals. With a step input as shown in Fig. A.13, the output is also a step function, but delayed in time by the amount D_0. Similarly, a pulse shown in Fig. A.14 is also delayed by an amount D_0 by an ideal delay circuit. The time scaling we have just discussed applies in the case of delay scaling. It is usual practice to design the circuit for a delay of 1 second and then to time scale to give the actually desired time delay, perhaps 1 μs. We have shown in Chapter 18 that the ideal delay circuit has a transfer function for which the magnitude is a constant and the phase is given by the equation

$$\theta = -\omega D_0 \tag{A.47}$$

or the delay is

$$D_0 = \frac{-\theta}{\omega} \tag{A.48}$$

This equation shows the inverse relationship between D_0 and ω that was observed in Eq. (A.36). If we define a scaling factor k_D as

$$k_D = \frac{\text{actual delay required}}{\text{normalized delay}} \tag{A.49}$$

then we may write the scaling equations in a form very similar (that is, identical) to those of Eqs. (A.40)–(A.42):

$$L_{\text{new}} = k_m k_D L_{\text{old}} \tag{A.50}$$

$$C_{\text{new}} = \frac{1}{k_m} k_D C_{\text{old}} \tag{A.51}$$

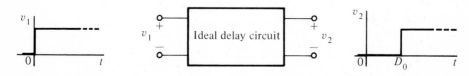

FIGURE A.13

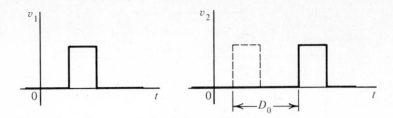

FIGURE A.14

$$R_{\text{new}} = k_m R_{\text{old}} \tag{A.52}$$

To illustrate the use of these scaling equations, consider the filter shown in Fig. A.15a. This filter has a response known as a third-order Bessel, has been designed for $D_0 = 1$ second, and was obtained from a table complete with the element values given. Suppose that we are required to design a filter with 1-μs delay, and with terminating resistors of 1 kΩ. From the problem $D_0 = 10^{-6}$ and $k_m = 10^3$. Substituting these values into Eqs. (A.50)–(A.52), we obtain

$$L_{\text{new}} = 10^{-3} L_{\text{old}} \tag{A.53}$$

$$C_{\text{new}} = 10^{-9} C_{\text{old}} \tag{A.54}$$

$$R_{\text{new}} = 10^3 R_{\text{old}} \tag{A.55}$$

From these equations the circuit elements shown in Fig. A.15 are determined, and the design is complete.

The idealization shown in Figs. A.13 and A.14 of delay without distortion is never attained with circuits made from lumped elements. For example, if the step input shown in Fig. A.13 is applied to a lumped-element circuit, the response will be of the form shown for $v_2(t)$ in Fig. A.16. For such a response the rise time is defined as the time interval between the times that v_2 attains 10

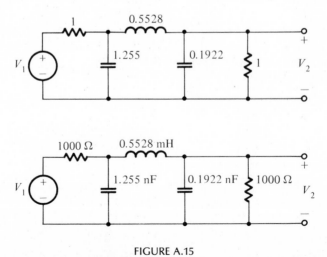

FIGURE A.15

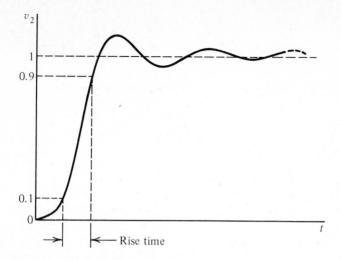

FIGURE A.16

percent and 90 percent of its final value. This rise time is frequently an engineering specification. The concepts of time scaling apply to this problem, and the circuit may be scaled to adjust rise time to its specified value.

From this discussion we see that the scaling constants k_f and k_D are fixed by the frequency range in which the filter is to operate. The role of the magnitude scaling constant k_m is to place the values of the elements in a realizable range. What values of elements can be considered as realizable will change with application and with technological advances.

APPENDIX B

REFERENCES:
WHEN
YOU
NEED
MORE
INFORMATION

B.1. BOOKS CONTAINING FILTER DESIGN TABLES

M. Biey and A. Premoli, *Cauer and MCPER* Functions for Low-Q Filter Design*, Georgi Publishing Co., Switzerland, 1980, 624 pp.

Erich Christian and Egon Eisenmann, *Filter Design Tables and Graphs*, John Wiley & Sons, Inc., New York, 1966; reissued by Transmission Networks International, Inc., Knightdale, N.C., 1977, 310 pp.

J. W. Craig, *Design of Lossy Filters*, The MIT Press, Cambridge, Mass., 1970, 197 pp.

Philip R. Geffe, *Simplified Modern Filter Design*, John F. Rider Publisher, Inc., New York, 1963, 182 pp.

DeVerl S. Humphreys, *The Analysis, Design and Synthesis of Electrical Filters*, Prentice-Hall, Inc., Englewood Cliffs, N.J., 1970, 675 pp.

David E. Johnson and J. L. Hilburn, *Rapid Practical Designs of Active Filters*, John Wiley & Sons, Inc., New York, 1975, 264 pp.

David E. Johnson, J. R. Johnson, and H. P. Moore, *A Handbook of Active Filters*, Prentice-Hall, Inc., Englewood Cliffs, N.J., 1980, 244 pp.

George S. Moschytz and Petr Horn, *Active Filter Design Handbook*, John Wiley & Sons, Inc., New York, 1981, 296 pp.

*MCPER is an abbreviation for multiple-critical-pole equal-ripple-rational.

Rudolf Saal, *Handbook of Filter Design*, AEG-Telefunken, Berlin, West Germany, 1979, 662 pp.

Rudolf Saal and E. Ulbrich, "On the Design of Filters by Synthesis," *IRE Trans. on Circuit Theory*, vol. CT-5, pp. 284–327, December 1958; reprinted in the volume: George Szentirmai, (Ed.), *Computer-Aided Filter Design*, IEEE Press, New York, 1973.

J. K. Skwirzynski, *Design Theory and Data for Electrical Filers*, Van Nostrand Reinhold Co., New York, 1965., 701 pp.

Louis Weinberg, *Network Analysis and Synthesis*, McGraw-Hill Book Co., New York, 1962, reissued by Robert E. Krieger Publishing, Inc., Melbourne, Fla., 1975, 692 pp.

Arthur B. Williams, *Active Filter Design*, Artech House, Inc., Dedham, Mass., 1975, 183 pp.

Arthur B. Williams, *Electronic Filter Design Handbook*, McGraw-Hill Book Co., New York, 1981, 576 pp.

A. I. Zverev, *Handbook of Filter Synthesis*, John Wiley & Sons, Inc., New York, 1967, 576 pp.

B.2. ANALOG FILTER COMPREHENSIVE TEXTBOOKS

Herman J. Blinchikoff and Anatol I. Zverev, *Filtering in the Time and Frequency Domains*, John Wiley & Sons, Inc., New York, 1976, 494 pp.

P. Bowron and F. W. Stephenson, *Active Filters for Communications and Instrumentation*, McGraw-Hill Book Co., New York, 1979, 285 pp.

Leonard T. Bruton, *RC-Active Circuits*, Prentice-Hall, Inc., Englewood Cliffs, N.J., 1980, 523 pp.

Aram Budak, *Passive and Active Network Analysis and Synthesis*, Houghton Mifflin Co., Boston, 1974, 733 pp.

Richard W. Daniels, *Approximation Methods for Electronic Filter Design*, McGraw-Hill Book Co., New York, 1974, 388 pp.

Gobind Daryanani, *Active and Passive Network Synthesis*, John Wiley & Sons, Inc., New York, 1976, 495 pp.

Philip R. Geffe, *Simplified Modern Filter Design*, John F. Rider, Publisher, New York, 1963, 182 pp.

M. S. Ghausi and Kenneth R. Laker, *Modern Filter Design: Active RC and Switched Capacitor*, Prentice-Hall, Inc., 1981, 608 pp.

L. P. Huelsman and P. E. Allen, *Introduction to the Theory and Design of Active Filters*, McGraw-Hill Book Co., New York, 1980, 429 pp.

W. E. Heinlein and W. H. Holmes, *Active Filters for Integrated Circuits,* R. Oldenbourg, Munich, West Germany, 1974, 668 pp.

David E. Johnson, *Introduction to Filter Theory,* Prentice-Hall, Inc., Englewood Cliffs, N.J., 1976, 306 pp.

Harry Y-F Lam, *Analog and Digital Filters: Design and Realization*, Prentice-Hall, Inc., Englewood Cliffs, N.J. 1979, 632 pp.

Claude S. Lindquist, *Active Network Design with Signal Filtering Applications*, Steward & Sons, Long Beach, Calif., 1976, 749 pp.

George S. Moschytz, *Linear Integrated Networks: Fundamentals*, Van Nostrand Reinhold Co., New York, 1974, 583 pp.

George S. Moschytz, *Linear Integrated Networks: Design*, Van Nostrand Reinhold Co., New York, 1975, 694 pp.

Adel S. Sedra and Peter O. Brackett, *Filter Theory and Design: Active and Passive*, Matrix Publishers, Inc., Forest Grove, Ore., 1978, 785 pp.

Gabor C. Temes and Jack W. LaPatra, *Introduction to Circuit Synthesis and Design*, McGraw-Hill Book Co., New York, 1977, 598 pp.

B.3. REFERENCE BOOKS ON ANALOG FILTERS

Andreas Antoniou, *Digital Filters: Analysis and Design*, McGraw-Hill Book Co., New York, 1979, 523 pp. (See chapters on approximation.)

Lawrence P. Huelsman (Ed.), *Active Filters: Lumped, Distributed, Integrated, Digital and Parametric*, McGraw-Hill Book Co., New York, 1974, 372 pp.

Lawrence P. Huelsman (Ed.), *Active RC Filters*, Dowden, Hutchinson and Ross, Inc., Stroudsburg, Pa. 1976, 318 pp. A Benchmark volume.

Sanjit K. Mitra (Ed.), *Active Inductorless Filters*, IEEE Press, New York, 1971, 224 pp.

Rolf Schaumann, M. A. Soderstand, and Kenneth R. Laker (Ed.), *Modern Active Filter Design*, IEEE Press, New York, 1981, 426 pp.

M. E. Van Valkenburg (Ed.), *Circuit Theory: Foundations and Classical Contributions*, Dowden, Hutchinson & Ross, Stroudsburg, Pa., 1974, 450 pp. A Benchmark volume.

Gabor C. Temes and Sanjit K. Mitra (Eds.), *Modern Filter Theory and Design*, John Wiley & Sons, Inc., New York, 1973, 566 pp.

B.4. OPERATIONAL AMPLIFIERS

J. G. Graeme, *Operational Amplifiers: Third-generation Techniques*, McGraw-Hill Book Co., New York, 1973, 233 pp.

Robert G. Irvine, *Operational Amplifier: Characteristics and Applications*, Prentice-Hall, Inc., Englewood Cliffs, N.J., 1981, 462 pp.

E. Moustakas and S-P. Chan, *Introduction to the Applications of the Operational Amplifier*, Academic Cultural Co., Santa Clara, Calif., 1974, 206 pp.

James K. Roberge, *Operational Amplifiers*, John Wiley & Sons, Inc., New York, 1975, 659 pp.

John I. Smith, *Modern Operational Circuit Design*, John Wiley & Sons, Inc., New York, 1971, 256 pp.

G. E. Tobey, J. G. Graeme, and L. P. Huelsman (Eds.), *Operational Amplifiers—Design and Applications*, McGraw-Hill Book Co., New York, 1971, 473 pp.

John V. Wait, Lawrence P. Huelsman, and Granino A. Korn, *Introduction to Operational Amplifier Theory and Applications*, McGraw-Hill Book Co., New York, 1975, 396 pp.

B.5.   ACTIVE CIRCUIT THEORY

Hendrick W. Bode, *Network Analysis and Feedback Amplifier Design*, Van Nostrand Reinhold Co., New York, 1945, 551 pp.

Wai-Kai Chen, *Active Network and Feedback Amplifier Theory*, McGraw-Hill Book Co., New York, 1980, 481 pp.

Lawrence P. Huelsman, *Theory and Design of Active RC Circuits*, McGraw-Hill Book Co., New York, 1968, 297 pp.

Ernst S. Kuh and Ronald A. Rohrer, *Theory of Linear Active Networks*, Holden-Day, San Francisco, Calif., 1967, 650 pp.

Sanjit K. Mitra, *Analysis and Synthesis of Linear Active Networks*, John Wiley & Sons, Inc., New York, 1969, 565 pp.

Robert Spence, *Linear Active Networks*, John Wiley & Sons, Inc., New York, 1970, 359 pp.

B.6.   PASSIVE CIRCUIT THEORY

Norman Balabanian, *Network Synthesis*, Prentice-Hall, Inc., Englewood Cliffs, N.J., 1958. 433 pp.

DeVerl S. Humphreys, *The Analysis, Design and Synthesis of Electrical Filters*, Prentice-Hall, Inc., Englewood Cliffs, N.J., 1970, 675 pp.

K. Geher, *Theory of Network Tolerances*, Akademiai Kiado, Budapest, 1971, 184 pp.

Ernst A. Guillemin, *Synthesis of Passive Networks*, John Wiley & Sons, Inc., New York, 1957, 741 pp.

Shlomo Karni, *Network Theory: Analysis and Synthesis*, Allyn and Bacon, Inc., Boston, 1966, 483 pp.

George Szentermai (Ed.), *Computer-Aided Filter Design*, IEEE Press, New York, 1973, 437 pp.

Kendell L. Su, *Time Domain Synthesis of Linear Networks*, Prentice-Hall, Inc., Englewood Cliffs, N.J., 1971, 382 pp.

M. E. Van Valkenburg, *Introduction to Modern Network Synthesis*, John Wiley & Sons, Inc., New York, 1960, 498 pp.

Jiri Vlach, *Computerized Approximation and Synthesis of Linear Networks*, John Wiley & Sons, Inc., New York, 1969, 477 pp.

Louis Weinberg, *Network Analysis and Synthesis*, McGraw-Hill Book Co., New York; reissued by R. E. Krieger Publishing Co., Melbourne, Fla., 1975, 692 pp.

Anatol I. Zverev, *Handbook of Filter Synthesis*, John Wiley & Sons, Inc., New York, 1967, 576 pp.

Index